Introduction to Air Pollution Science

A Public Health Perspective

Robert F. Phalen, PhD
Professor, Department of Medicine
Center for Occupational and Environmental Health
University of California, Irvine
Irvine, California

Robert N. Phalen, PhD, CIH
Assistant Professor, Health Science and Human Ecology
College of Natural Sciences
California State University, San Bernardino
San Bernardino, California

JONES & BARTLETT
LEARNING

World Headquarters
Jones & Bartlett Learning
5 Wall Street
Burlington, MA 01803
978-443-5000
info@jblearning.com
www.jblearning.com

Jones & Bartlett Learning books and products are available through most bookstores and online booksellers. To contact Jones & Bartlett Learning directly, call 800-832-0034, fax 978-443-8000, or visit our website, www.jblearning.com.

Substantial discounts on bulk quantities of Jones & Bartlett Learning publications are available to corporations, professional associations, and other qualified organizations. For details and specific discount information, contact the special sales department at Jones & Bartlett Learning via the above contact information or send an email to specialsales@jblearning.com.

Production Credits
Publisher: Michael Brown
Managing Editor: Maro Gartside
Editorial Assistant: Chloe Falivene
Production Assistant: Rebekah Linga
Senior Marketing Manager: Sophie Fleck Teague
Associate Marketing Manager: Jody Sullivan
Composition: Circle Graphics, Inc.
Cover Design: Kristin E. Parker
Photo Researcher: Sarah Cebulski
Cover Image: Smog over highway: © elwynn/ShutterStock, Inc.; Smoke: © Daniiel/ShutterStock, Inc.
Printing and Binding: Malloy, Inc.
Cover Printing: Malloy, Inc.

Library of Congress Cataloging-in-Publication Data
Phalen, Robert F. , 1940-
Introduction to air pollution science: a public health perspective / Robert F. Phalen and Robert N. Phalen.
p. cm.
Includes bibliographical references and index.
ISBN-13: 978-0-7637-8044-9 (pbk.)
ISBN-10: 0-7637-8044-8 (ibid.)
1. Air—Pollution—Textbooks. I. Phalen, Robert N. II. Title.
TD883.143.P43 2013
577.27′6—dc23

2011027799

6048
Printed in the United States of America
19 18 17 16 15 10 9 8 7 6 5 4 3

Dedication

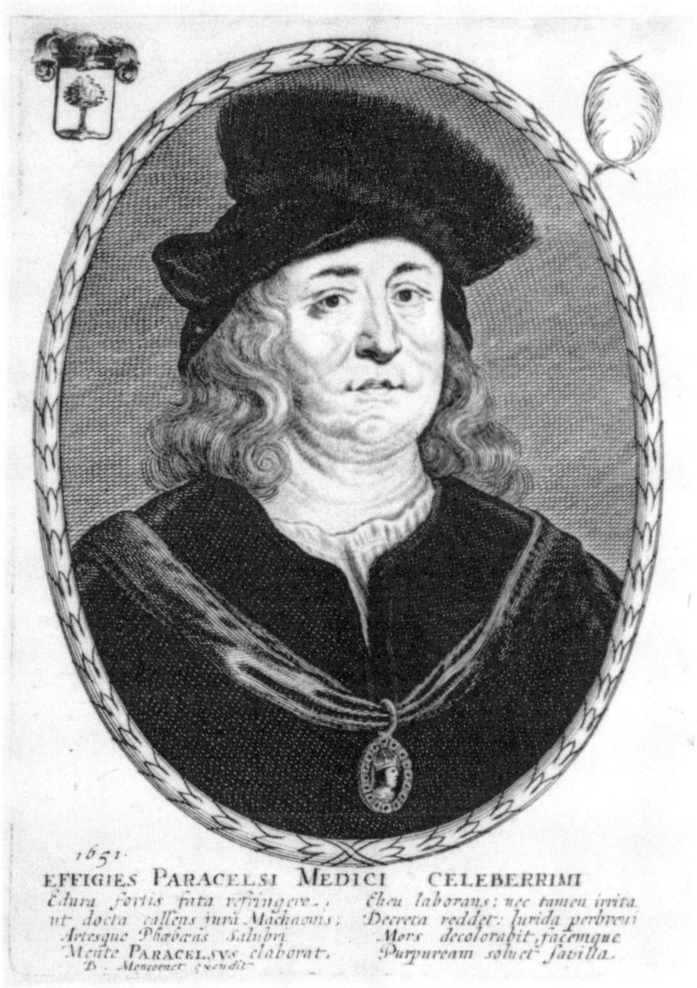

Figure FM–1 Paracelsus
Source: © National Library of Medicine

To Philippus Aurelus Theophrastus Bombastus von Hohenheim—Paracelsus (1493–1541), who probably made the single most important contribution to air pollution science by proclaiming:

> *"All substances are poisons; there is none which is not a poison. The right dose differentiates a poison from a remedy."*

(Quote from Gallo, M. A. (2008) in Casarett and Doull's *Toxicology, The Basic Science of Poisons, 7th Edition,* Klaassen, C. D., Ed., McGraw–Hill Medical, New York.)

Table of Contents

Preface

Air pollution science is both interesting and elegant because it integrates many disciplines. Responsibly managing air pollution requires the expertise and cooperation of a diverse array of specialists. Chemists, physicists, and engineers must have a working knowledge of public health, as well as the basic principles of toxicology, epidemiology, and the regulatory process. Also, public health professionals (including epidemiologists, toxicologists, and regulators) need to acquire a working knowledge of air pollution chemistry, physics, and engineering in order to be relevant and effective. In the interest of public health and welfare, it is no longer acceptable to pursue and promote one's own scientific discipline in isolation. A holistic approach is necessary, with the ultimate goal of making sound decisions that will best protect public health and the environment. To serve this end, this book covers essential traditional topics, as well as some that are new to air pollution textbooks. For example, full chapters are dedicated to risk assessment, toxicology, epidemiology, and ethics.

Traditional topics have been updated to address current issues in air pollution science (e.g., climate change). Individual chapters cover Sources and Emissions; Properties of Air Pollutants (Chemistry and Physics); Sampling and Analysis; Visibility, Climate, and the Ozone Layer; Regulation and Abatement; Human Exposures; Effects on Human Health; Toxicology; Epidemiology; Risk Assessment; and Ethics. The authors believe these are essential basic topics that students and professionals must appreciate in order to understand air pollution science. The chapters are scientifically current, and they introduce important basic concepts, online databases, and even some of the relevant peer-reviewed literature.

The authors have proven records in research and education in the air pollution sciences, and their formal scientific training, professional experience, and viewpoints are complementary. Their combined expertise and interests include air sampling, chemical analyses, aerosol science, industrial hygiene, inhalation toxicology, occupational health, biophysics, dermal toxicology, pollutant control technologies, applied ethics, and undergraduate, graduate, and post-graduate education. They have discussed, reviewed, and edited each other's contributions, and they have had many stimulating discussions and debates regarding the content presented in this text.

This textbook is necessary because it is: (1) motivated by a concern for public health and welfare, but also (2) current from a basic science viewpoint. As the Earth's population expands, air quality will worsen unless cleaner and/or more efficient technologies are developed for generating power, providing food, manufacturing goods, transporting goods, and enjoying life. On the other hand, many people throughout the world are still dealing with serious and very real health problems that are not associated with air pollution. These problems include poor nutrition, infectious disease, and natural disasters. Therefore, air pollution must be placed into a proper perspective within each society or community. Presenting this public health perspective is an important goal of this textbook. After all, the public must deal with all of the potential consequences of a regulatory action, not just the intended benefits.

The authors must thank more people than they can name. First is Mrs. Leslie Kimura, who word-processed every chapter many times and served as our Administrative Assistant. Leslie's young daughter Kayla inspired us all with her patience and healthy scientific curiosity. She was also an invaluable companion to the authors' children and grandchildren, Joseph and Samuel.

Without the advice and help of Dr. Robert H. Friis, our role model as a textbook author, this project could neither have been begun nor completed. We are also appreciative of the guidance from Michael Brown, Maro Gartside, Rebekah Linga, Chloe Falivene, Grace Richards, Sophie Fleck, Teresa Reilly, Catie Heverling, and several other Jones & Bartlett Learning staff. Erica Martinetti, Robert N. Phalen, and Joshua Bracks expertly prepared several figures. The authors' families, Katherine Phalen, Michelle Phalen, and Nancy Phalen, tirelessly performed essential research, editing, and checking. Kathryn E. Terry, attorney at law and member of the California State Bar Committee on Ethics, offered expert suggestions on our ethics chapter. Rowe Yates contributed quiz questions. Dr. Loyda Mendez provided valuable advice.

The authors' children and grandchildren gave up valuable time with their parents and grandparents. They inspired us and helped us to relax during tense times by playing baseball with us. They deserve our most sincere appreciation.

Robert F. Phalen and Robert N. Phalen

Reviewers

April L. Hiscox, PhD
Department of Geography
The University of South Carolina

Robert G. Keesee, PhD
Associate Professor
Department of Atmospheric & Environmental Sciences
State University of New York at Albany

Chris J. Walcek, PhD
Senior Research Scientist
Department of Atmospheric & Environmental Sciences
State University of New York at Albany

Foreword

Air, water, and earth sustain all living things, both plants and animals. They are the source of foodstuffs and energy critical to the well-being of human kind. Constant availability of oxygen within narrow concentration limits is essential for humans and all other mammalian organisms. Likewise, the constant availability of carbon dioxide within critical concentration limits is essential fuel for plants. The evolution of humankind has been strongly influenced by combustion, the interaction of carbonaceous materials and oxygen and the release of thermal energy. Primitive man learned to use fire to enhance the well-being of individuals and small communities of hunters and gatherers. This was soon followed by development of agricultural-based communities. The industrial revolution soon emerged with its strong dependence on the use of energy from available natural resources. That revolution was initially fueled by wood, then coal, and continues today with extensive use of coal, oil, natural gas, and to a lesser extent, uranium fuel for nuclear reactors as the primary energy sources. The availability of refined oil products, gasoline, diesel, and aviation fuel, has been the cornerstone of a transportation sector that has helped create a global economy. Increased agricultural productivity has been key to feeding a growing global population. Enhanced agricultural productivity has benefited from improved germ stocks, the use of petroleum product fueled equipment, and increased use of fertilizers. Nitrogen, extracted from the air, has played a critical role as a fertilizer.

Uses of carbon-based fuel stocks were initially very inefficient and resulted in significant emissions of a variety of gaseous and particulate pollutants primarily to air. Initially, the impacts were local, then observed regionally, and now are recognized as being of global concern. Air pollutants may directly impact the health of individuals and, in some cases, only be identified by studying very large populations. Other impacts on human populations may arise indirectly via contamination of water, soil, and plants. It is clear that the development of modern society has been dependent on the complex inter-relationships between air, water, earth, energy, and food production, and these, in turn, impact the health of the world's population. In both developed and developing countries, people are living longer on average than at any time in the history of human kind.

This text, by a father-son team, Robert F. and Robert N. Phalen, will be useful for undergraduate and graduate students and the lay public who want to better understand the multi-faceted nature of air pollution, its impact on society, and how the impacts can be mitigated. Their decades of experience as researchers studying the health effects of air pollution and as teachers have provided them with a valuable perspective often lacking in textbooks. They understand the scientific information being communicated. Equally as important, they understand the importance of communicating basic principles and using specific examples of the science to illustrate the principles.

Readers of the text will quickly identify a series of conceptual paradigms highly relevant to air quality that are recurring in the book. These include an emphasis on studying air pollution linkages from the sources of pollutants to the ambient air to the breathing zone of people to how inhaled materials are deposited and impact the respiratory tract and remote tissues. The individual chapters on toxicological and epidemiological studies help the reader understand the strengths and weaknesses of each approach and how the resulting knowledge can be integrated. The Phalens wisely provide a chapter on the risk assessment process, which has emerged over the past half century as an approach to synthesizing information from multiple sources to understand human health hazards and risks. Every chapter provides not only coverage of science but, most importantly, places that science in the context of the global society in which we all live.

The senior Dr. Phalen received his undergraduate and early graduate education in Physics and then received his PhD in Radiation Biology and Biophysics. I had the pleasure of working with Robert F. Phalen at the Lovelace Inhalation Toxicology Research Institute (now the Lovelace Respiratory Research Institute), an institute whose successful research program was grounded on issue-resolving multi-disciplinary collaboration. At the University of California, Irvine, he has had an outstanding career as a research scientist and teacher. The junior Dr. Phalen received his undergraduate education in biology, gained experience as an Industrial Hygienist, and then received his PhD in Environmental Health. He has worked at the interface of applying science to resolving environmental and occupational health issues. The rich and varied experiences of the Phalens have taught them the importance of applying the skills of multiple disciplines in the physical, biological, and biomedical sciences, mathematics, information technology, engineering, the societal sciences, and philosophy to increase our knowledge base on air pollution and then use that knowledge to assist in resolving important societal issues. Students with an inquiring mind will identify many potential opportunities for developing a future career in one of the disciplines key to developing and using scientific knowledge of air pollution science.

Unlike many texts in the field, this book is not an encyclopedia of the knowledge of air pollution written from the perspective of multiple super specialists. Neither is this a doom and despair text with finger pointing to establish blame and advocate narrow viewpoints as to how society should move forward. This book exemplifies how science has an important role in helping human kind prosper and live healthy lives with thoughtful attention given to the quality of our air, water, and earth and the wise use of energy resources. Air quality is a crucial interface issue for the future of human kind. As William Shakespeare noted, "The golden age is before us and not behind us," and "What is past is prologue." This textbook will provide readers with an understanding of the past and current science of air pollution so they can be better contributors in the future.

Roger O. McClellan, DVM, MMS, DSc (Honorary)
Inhalation Toxicologist, Aerosol Scientist, and Risk Analyst
Albuquerque, New Mexico

About the Authors

Robert F. Phalen, PhD, is a Fellow of the Academy of Toxicological Sciences and a Professor of Medicine in the Center for Occupational and Environmental Health at the University of California, Irvine. He codirects the Air Pollution Health Effects Laboratory, and teaches undergraduate, graduate, and medical students. Among his numerous duties, he has organized several international conferences and chaired research ethics committees for human and animal subjects.

Robert N. Phalen, PhD, a Certified Industrial Hygienist, is an Assistant Professor of Environmental Health Sciences in the Health Science and Human Ecology Department at California State University, San Bernardino. In addition to his research on air quality, pesticides, sampling and analysis, and personal protective equipment, he teaches a variety of undergraduate courses and serves on administrative, education, and research committees (including the human subject's ethics committee).

The authors have approximately 150 combined scientific publications.

Chapter 1

Introduction to Air Pollution Science

LEARNING OBJECTIVES

By the end of this chapter the reader will be able to:

- discuss natural phenomena that impact air quality
- discuss the impact of humans and their technologies on air quality
- identify three early writers who shaped current thought on the health effects of air pollution
- describe the three great air pollution disasters of the twentieth century and what groups of people were the most affected
- explain how epidemiology, toxicology, and basic laboratory research are all needed to understand the health effects of air pollution

CHAPTER OUTLINE

I. INTRODUCTION: HISTORY

Air pollution has two histories, an early unrecorded one and a more recent recorded one. By examining these histories, one can gain a broad perspective on air pollution, including its trends, and the relationship between the evolution of human technology and air pollution exposures. History also allows us to understand the way our current ideas about air pollution and its hazards might have developed, and how our regulations and controls came about.

Early History and Natural Events

About 4 billion years ago in the *Hadean* era, the surface of the newly-formed Earth went through a violent period characterized by intense bombardment from meteorites, frequent volcanic eruptions, boiling seas, and extreme ultraviolet radiation exposure. These conditions would certainly have precluded the complex and varied plant and animal life as we now know it. During the following *Archaen* era (3.8 to 2.5 billion years ago), the Earth cooled, and life consisted of bacteria that flourished in an atmosphere believed to be devoid of oxygen, and therefore toxic to modern life. **Figure 1–1** depicts these and other geologic eras. Meteorological and geological processes along with any existing life forms have shaped the atmosphere throughout the Earth's history. Long before humans appeared, there were several periods of time that had large changes in the composition of the Earth's atmosphere.

Because early primitive life depended on an environment with little or no oxygen, the eventual rise of early photosynthetic (relating to use of radiant energy to create new compounds) plant life resulted in the emission of large quantities of a highly reactive, and therefore toxic air pollutant, oxygen (**Figure 1–2**). This period (the *Proterozoic* era) would have been catastrophic for many of the established life forms, even producing some total extinctions. Thus the Proterozoic era produced the first, and greatest, air pollution disaster. The new oxygen-rich atmosphere eventually stabilized with an oxygen content of about 20 percent, which led to the flourishing of more of the new forms of life. This life included complex plants and animals. The current oxygen content in the atmosphere is about 20.9 percent at sea level under dry conditions. Should the oxygen content increase to, say 30 percent, extensive uncontrollable fires would result. Combustible materials, such as wood and other organic materials, ignite easily and burn rapidly at high oxygen concentrations. Low oxygen levels, less than 15 percent, would threaten the existence of complex animal life. The abundant life we know today fortunately serves to stabilize our current atmosphere. As a result, atmospheric

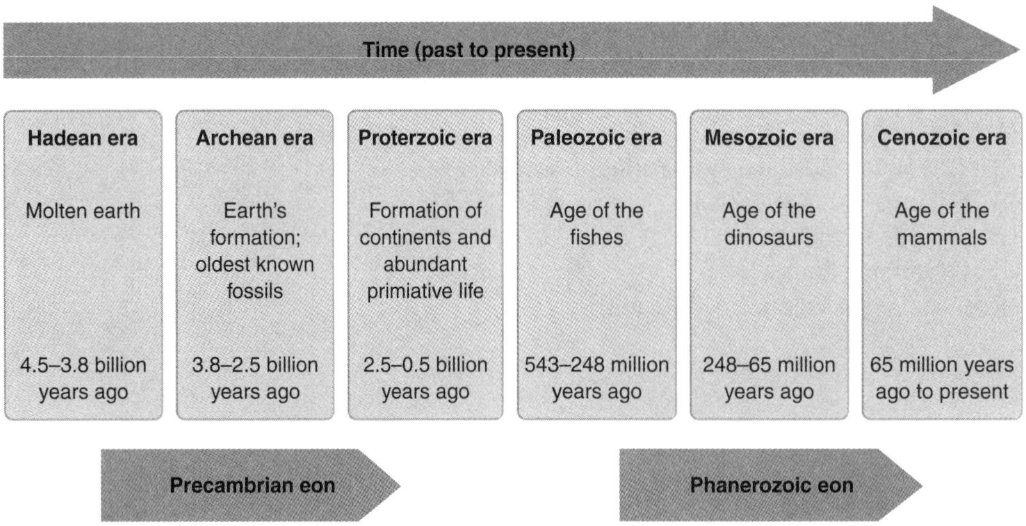

Figure 1–1 Geologic Time
Data from exhibits at the University of California Museum of Paleontology (http:/www.ucmp.berkeley.edu). *Source:* The University of California Air Pollution Health Effects Laboratory, with kind permission.

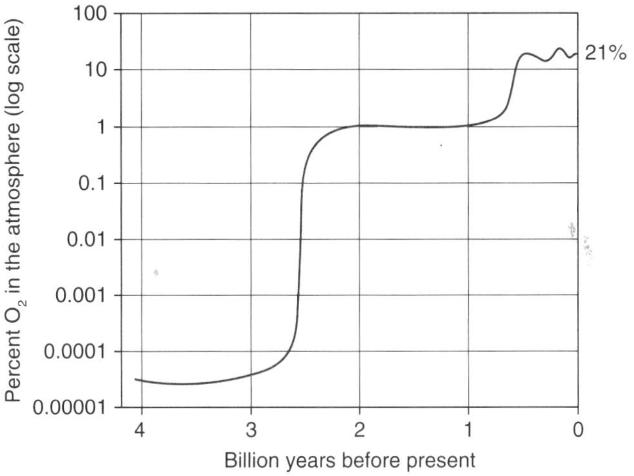

Figure 1–2 Modern view of the Earth's atmospheric oxygen over time. Fortunately, the current level of oxygen appears to be regulated by the interplay of several natural processes.
Source: The University of California Air Pollution Health Effects Laboratory, with kind permission.

Figure 1–3 The eruption of Mount St. Helens in Washington, 1980.

oxygen levels have oscillated around the current level for hundreds of millions of years.

In addition to the impact of such long-term climate changes in the atmosphere, shorter-time events shape the atmosphere. Meteoric impacts and major volcanic eruptions, such as the one that formed Crater Lake in Oregon about 7,700 years ago, have significantly contaminated the global atmosphere periodically and even led to the extinction of some species. More recently, the eruption of Mount St. Helens in 1980 destroyed all nearby life and deposited ash thousands of kilometers (km) downwind (**Figure 1–3**). Natural fires, dust storms, additional meteoric impacts, and sporadic volcanic activity produced significant air pollution episodes. These natural events further shaped life, leading to a continuing series of extinctions and the emergence of new species. Natural changes in climate, including alternating cooling and warming eras, will continue to modify conditions that favor some species and make survival difficult for others. The role of humans and their associated air emissions on the evolution of climate is a topic of active current research (see Chapter 5).

Use of Fuels by Humans

Our human ancestors, who emerged only 4 to 6 million years ago, learned to use and eventually control fire. Archeologists have found hearths and fire-hardened wood spears that date to about 750,000 to 500,000 years BC. It is reasonable to assume that the early burning of organic fuels, such as wood, dried dung, and natural oils, would have generated combustion-related air contaminants in caves and other early dwellings. Evidence from observations of sinus-bone damage on ancient skulls and alterations in mummified Egyptian lung tissue is highly suggestive of the role of early indoor combustion products in producing disease. The acute effects of irritating smokes were certainly evident to the ancients. Whether or not they were able to link air quality to chronic health effects is another matter.

The eventual emergence of large population centers and associated primitive industrial processes would have led to community-level air pollution episodes that resulted from the burning of wood as a primary fuel. However, it was the introduction of a new fuel, coal, in the thirteenth century AD that stimulated several early writers to describe the adverse health effects of air pollutants. Coal usage and the rise of newer industrial activities, such as the smelting of metal ores, produced acidic, odorous, and irritating sulfur-containing pollutants which would have also contained toxic levels of metals such as lead and iron. The success of coal as a fuel and its widespread availability for industrial and domestic uses not only led to increasingly polluted air in outdoor and indoor environments, but it also served as the impetus that would eventually drive regulatory actions.

History of Attitudes and Perceptions

Our modern concepts about environmental contaminants can be traced in the writings of influential thinkers over the past 2,000 years. In ancient Greece, town controllers had the responsibility of maintaining environmental quality, including control of sources of odor such as that generated by rubbish and presumably other sources. Roman courts were involved in civil suits that were designed to protect wealthy suburbs from pollutants generated by a number of industrial processes. Greek and Roman physicians, including Hippocrates (c 460–375 BC), Galen (c 129–200 BC), and Pliny "The Elder" (c 23–79 AD) were prolific writers who helped establish the early foundations of medical practice, including descriptions of diseases (and treatments) related to the effects of natural and anthropogenic (human generated) toxicants. Both Hippocrates and Pliny were interested in occupational diseases because of the often extraordinary levels of industrial exposures and their adverse effects on the health of workers. The health effects of air pollutants were more evident in the most heavily exposed workers, such as those closest to the sources.

Alchemy, the predecessor of the science of chemistry, was practiced for about 1,000 years (c 750–1800 AD). In addition to dealing in secret elixirs and claims of the ability to turn lead into gold, alchemists worked hard to understand the causes of diseases, and to develop the equipment and laboratory methods that allowed modern chemistry to eventually emerge. Paracelsus (the pseudonym of Philippus Aureolus Theophrastus Bombastus von Hohenheim, 1493–1541 AD) was a noted Swiss alchemist and physician who revolutionized medical practice of his time by insisting that it must be based on *observation* and *experience* instead of just time-honored theory. This shift in thinking from relying on theory to drawing conclusions from data was revolutionary in its time. Paracelsus introduced substances such as sulfur, lead, arsenic, and iron into the realm of pharmaceutical chemistry, and he also studied occupational diseases extensively. As a result of his arduous work and fame, he is considered to be the father of the scientific discipline of toxicology, despite his persistent mystical beliefs and teachings. One of his greatest contributions to science was his remarkably astute observation related to the concept of *dose.* Paracelsus is quoted by Gallo (2008) thusly:

> "All substances are poisons; there is none which is not a poison. The right dose differentiates a poison from a remedy."

This proclamation is at the heart of modern toxicology. It is also the basis of many of our current regulations for air pollutants, where the goal is to set acceptable levels of specific air contaminants such that their doses do no significant harm to public health.

Alchemy, and its leading practitioners, not only shaped modern thought, but they also helped medicine and chemistry to become entwined in a manner that helped both to advance and mature. In parallel with these events during the period of alchemy, the discipline of toxicology was emerging from the early use of poisons. Plant extracts and toxic animal venoms were used for hunting, assassinations, and as deadly agents for use in warfare. Over thousands of years humans learned to fear toxic substances and to mistrust those who had the knowledge to use them. The use of poison gas in World War One (1914–1918) heightened any existing fear of air contaminants on the part of the public. Such fear, which generated a mistrust of new technological and chemical applications, persists in much of the population in our time. Although the concept of toxicity is well understood by the public, the role of dose in producing harm is not generally appreciated. This topic is elaborated in Dr. M. Alice Ottoboni's book, *The Dose Makes the Poison* (Ottoboni, 1997).

An early English environmental activist and writer, John Evelyn (1620–1706) courageously adopted a stern moral stance toward the effects of industrial air pollution. As a fellow of the *Royal Society of London* (established by Evelyn and others in 1662), and publisher of an influential booklet, *Fumifugium, or the Smoke of London Dissipated (together with some remedies humbly proposed)* he described, among other things, various means of control of air contaminant emissions. Although Evelyn was mainly concerned with the health of industrial workers, his basic idea of the vulnerability of workers can be seen as also applicable to sensitive groups of individuals in the general population. Evelyn's teachings, which were seen as revolutionary in his time, would fit well in our century. His message was strong, as is evident in a quote from *Fumifugium* (Evelyn, 1661):

> ". . . Inhabitants breathe nothing but an impure and thick Mist accompanied with a fuliginous and filthy vapor, which renders them obnox-

ious to a thousand inconveniences, corrupting the Lungs . . ."

Although several other early thinkers and writers shaped the way in which air pollutants were perceived, two examples serve to demonstrate the evolution of thought. Bernardino Ramazzini (1633–1714), an Italian Professor of Medicine, described the diseases associated with the dangerous trades of his time. As a result of his work and writings, Ramazzini is generally considered to be the father of occupational medicine. A successful famous London surgeon, Percival Pott (1714–1788) is credited with linking chimney-sweep's scrotal cancers to their work; perhaps the first recorded observation of chemical carcinogenesis (the development of cancers).

Impact of the Industrial Revolution

By the time of the *Industrial Revolution,* which was marked by the introduction of steam-powered machinery in the mid-1800s, the linkages between severe air pollution and a variety of human diseases had been recognized. Coal- and oil-fired boilers not only ran power plants, ships, locomotives, and factories, but they also emitted large quantities of smoke that contained ash, partially-burned fuel solids, sulfur, oxides of nitrogen, and a variety of metals and organic gases and vapors (vapors are the gaseous states of volatile liquids). Legislation limiting atmospheric emissions and the establishment of governmental agencies that were intended to enforce regulations soon followed. Great Britain introduced what may be its first Public Health Act in 1848, which was followed by several other attempts to control air-pollutant emissions. In the United States, similar local ordinances were issued in the 1880s that were aimed at controlling smoke and ash emissions. However, the pressure for progress and its many associated benefits largely outweighed the enthusiasm for enforcement. Although the general nuisance and effects on health were recognized, little was done to effectively control air pollutants. It was the great air pollution disasters of the next century that changed the way in which the adverse effects of air pollution were perceived and addressed in our society.

II. THE GREAT AIR POLLUTION DISASTERS

The combined impact of widespread combustion emissions, cold weather, persistent fog, stagnant winds, and low air inversions (see Chapter 2 for a description of air inversions) led to sharp increases in deaths and illnesses in several affected communities. These events drastically changed the relatively tolerant attitudes toward air pollution. The three notable episodes in the first half of the twentieth century that were well documented became known as "*the great air pollution disasters,*" or "*the historic pollution episodes.*" These episodes made world-wide headlines, and they are still widely referred to by air pollution researchers and regulators. There were other air pollution episodes in the twentieth century as well, but they were less well publicized than the three major episodes that occurred in Europe and the United States of America.

Meuse River Valley, 1930

The first of the three historic air pollution episodes occurred in eastern Belgium in a river valley about 2½ km wide and 100 meters (m) deep. The Meuse River Valley was heavily industrialized with a variety of air pollutant sources including several electric power generating plants, over two dozen major factories, substantial railroad, truck and automobile traffic, and the domestic use of coal for heating homes (**Table 1–1**). A six-day period starting on December 1, 1930 had an unprecedented combination of low temperatures, fog, and low wind speeds. The fog droplets facilitated the conditions for a variety of chemical reactions in the air.

Table 1–1 Sources of air pollutants in the Meuse river valley in 1930.

Industry	*Transportation*	*Other*
Five coking operations	Railroads	Use of coal for domestic heating
Four large steel plants	Trucks	
Three metallurgical factories	Automobiles	
A fertilizer plant		
A sulfuric acid plant		
Four electric power plants		
Six glass works		
Three zinc plants		

Data from Clayton and Clayton (1978).

Cold weather increased the burning of coal for home heating. The low wind speed prevented the dispersal of the air pollutants that had accumulated in the valley. The buildup of a variety of gaseous and particulate air pollutants soon produced a large spike in excess human deaths and illnesses, along with a substantial loss of cattle. In a two-day period, December 4 and 5, sixty-three excess deaths (about 10 times the expected number), and 6,000 illnesses were observed. Most of the deaths were in two groups, the elderly and persons with preexisting heart and lung diseases, but others were also affected. Although concurrent air concentration measurements were not made, subsequent estimates by scientists indicated that high levels of particles in the respirable size range, significant sulfur dioxide levels, and associated acidic conditions all occurred. Notably, it was determined that the levels of individual air pollutants were probably not sufficient to produce the health problems; the effects of some unknown combination or combinations of meteorology and several air pollutants were likely causal. Professor J. Firket of the University of Liége was a member of an "enquiry" group that investigated the incident. In his report (Firket, 1936), he made a prophetic statement:

> ". . . the public services of London, e.g., might be faced with the responsibility of 3200 sudden deaths if such a phenomenon occurred there."

This is exactly what happened 16 years later in London (discussed later in this chapter), which probably brought no pleasure to the esteemed Professor Firket.

Donora Pennsylvania, 1948

The second notable incident took place October 25 to 31, 1948 in a river valley that included the communities of Donora and nearby Webster in southwestern Pennsylvania. The heavily industrialized Monongahela river valley, about 120 m deep, used soft coal as the main fuel for domestic and industrial establishments, and several major sources of air pollutants were present (**Table 1–2**). The episode began with persistent cool stagnant winds and heavy fog, described by Ashe (1952) as "unique in intensity as far back as history is available." The fog had the sharply irritating pungent odor of sulfur dioxide, and the ground-level visibility was so low (about 15 m) that it essentially brought traffic to a standstill. While only 1 to 2 deaths were expected during the time of the event, an astonishing 18 to 20 excess deaths were attributed to the episode. Although the exact number is debated, about 40 percent of the 15,000 residents was likely affected by the air pollutants; farm animals, especially chickens, were also apparently vulnerable. As in the 1930 Meuse River episode, the elderly and those with preexisting heart and lung diseases were most affected. The symptoms included eye and respiratory tract irritation, along with coughing and breathing difficulty. No air samples were taken at the time, but subsequent estimates indicated that sulfur dioxide levels as high as 2 ppm (5.5 mg/m^3 of air) and particle levels as high as 30 mg/m^3 (200 times the U.S. EPAs 2010 24-hour limit for particles with diameters under 10 μm in diameter) were present. Several other air pollutants including carbon monoxide, sulfuric acid, oxides of nitrogen, carbon, and several particulate metals were probably present in significantly elevated concentrations. Despite these high levels of individual pollutants, a subsequent U.S. Public Health Service study determined that a combination, rather than any individual pollutant, would be required to produce the adverse health effects.

Table 1–2 Sources of air pollutants affecting Donora, PA in 1948.

Industry	*Transportation*	*Other*
Four steel plants One zinc plant An electric power plant A glass company	Railroad, steamships, and traffic	Use of soft coal for fuel

Data from Clayton and Clayton (1978).

London, 1952

As predicted by Professor Firket in 1936, the most severe air pollution disaster in modern history took 3,000 to 4,000 lives of Londoners during a 4-day period, December 5 to 8, 1952. London lies in a wide valley of the Thames River, and it had a 1952 population of 8.6 million people. Again, meteorological conditions were unusually intense, with cool nearly stagnant air, heavy fog, and an air-pollutant trapping air-inversion layer at about 100 m above ground level. There was a rapid buildup of acidic soot-filled *smog* ("smog" is a compound word originally meaning smoke + fog) that interfered with traffic, and even caused pedestrians to become lost (**Figure 1–4**). Preexisting heart and/or lung disease,

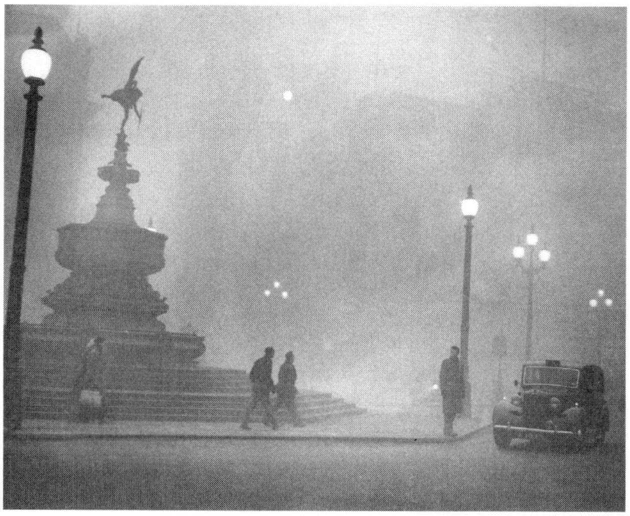

Figure 1–4 Daytime visibility during the 1952 London air pollution episode.

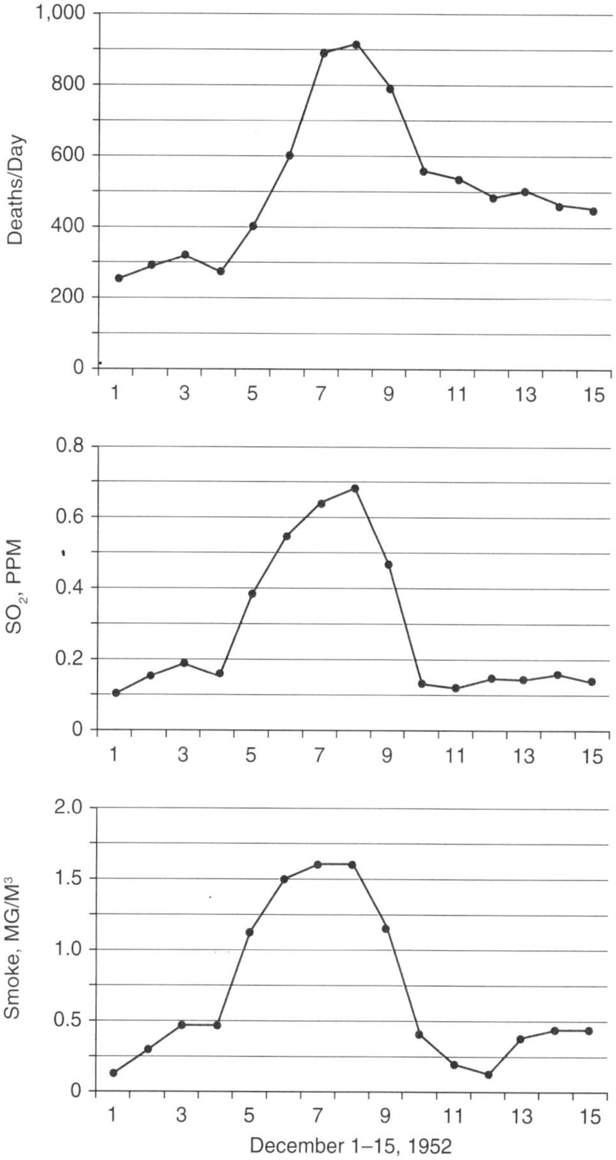

Figure 1–5 Data from the 1952 London air pollution episode; Top: daily deaths including normal deaths; Middle: city average sulfur dioxide concentrations; Bottom: City average smoke concentrations. Data from Wilkins (1954).
Source: The University of California Air Pollution Health Effects Laboratory, with kind permission.

age 45 years and older, and infancy (under 1 year of age) were risk factors in 80 percent of the deaths. The causes of deaths included *pneumonia* (severe deep lung inflammation usually associated with infection), *bronchitis* (inflammation of the bronchial air passages, usually accompanied by fever, cough, and excess mucus production), and heart disease. Most illnesses occurred on the third and fourth days of the episode. The excess acute death rate was estimated to be between 2.6 and 5 times normal by various authors. A contributing factor could have been a prolonged influenza outbreak at the time.

In this case, air-sampling data were available (**Figure 1–5**). Prior to the episode, particle levels averaged a substantial 500 μg/m³ of air, and sulfur dioxide levels averaged 0.15 ppm (which is not generally considered to be excessive). During the episode, particle levels averaged approximately 4,500 μg/m³, and the sulfur dioxide level reached a substantial 1.3 ppm. The *British Smoke Shade* method was used to estimate particle levels based on the dark color of filter samples, so the actual levels of particles could have been higher. In addition to the observed health problems and excess deaths, soiling of metal surfaces and damage to clothing indicated that the smog was strongly acidic. This time, the use of soft coal (which has a high sulfur content) for heating homes was identified as a primary source of the air pollutants, although other sources were also present. As in the Meuse River Valley and Donora episodes, a combination of pollutants, rather than individual components, was the likely causal agent for the excess deaths and other damaging effects. Again, the London episode made world-wide news headlines, but this time the public impact was amplified because a large major modern city was severely afflicted. As a result, this episode triggered the British Clean Air Act of

Table 1–3 Summary of the historic air pollution episodes of the twentieth century, all of which occurred in geographies within a valley.

Location and Period	*Days Excess Deaths Occurred (Increase in Death Rate)*	*Contributing Pollutants Identified*				
		SO_2	*CO*	*Acids*	*Metals*	*Other*
Meuse Valley, Belgium Dec. 1–5, 1930	5–6 (10 fold)	Yes	Yes	Yes	Yes	Yes
Donora, PA, Oct. 27–31, 1948	3 (10 fold)	Yes	Yes	Yes	Yes	Yes
London Dec. 5–8, 1952	3–4 (2–5 fold)	Yes		Yes		Yes

Notes: SO_2 = Sulfur dioxide; CO = carbon monoxide. Data from Clayton and Clayton (1978).
Source: The University of California Air Pollution Health Effects Laboratory, with kind permission.

1956, which limited the use of soft coal for heating homes. It also set up the conditions for other earnest regulatory activities in Britain and other countries to control air pollution. Among these was the U.S. National Clean Air Act of 1963 and its subsequent amendments (http://www.epa.gov/air/caa/, accessed November 11, 2010).

Conclusions from the Three Air Pollution Disasters

All of these air pollution disasters had many factors in common. Severe, even unprecedented, meteorological conditions including persistent nearly stagnant air, intense fog, low-altitude air inversions, and cool to cold temperatures occurred simultaneously. Low temperatures led to increased use of domestic heating. Deaths were also seen to lag the beginning of the highest levels of air pollutants by two or more days. Those with preexisting heart and lung diseases, especially the elderly, were the most severely affected. Infants were also reported as being a susceptible group in the 1952 London episode. However, no single air pollutant could be blamed for the excess deaths and illnesses. An unknown combination of pollutants was more likely to have been responsible for the observed increases in deaths and illnesses. **Table 1–3** summarizes the episodes. Taken together, these incidents generated extraordinary public concern. The governmental responses led to an emphasis on research and legislation directed at both understanding the possible causes and at developing strategies for preventing future similar disasters. As previously noted, there were other episodes that were clearly less disastrous than the three great air pollution episodes, but they were also less influential.

III. MODERN AIR POLLUTION ISSUES

In the immediate period following the London episode of 1952, numerous epidemiologic studies and complimentary laboratory studies, with isolated cells, humans, and animals (see Chapter 9 and 10) were begun. Many of these initial studies were challenged on the basis of the use of unrealistically high concentrations of pollutants in laboratory animal and human clinical studies, or the failure to control for confounding variables (e.g., nonair-pollution factors that could influence health outcomes). However, the initial studies helped to:

- identify potentially harmful combinations and individual components of air pollutants;
- improve air sampling techniques;
- clarify the range of possible health effects associated with air pollutants;
- improve study designs; and
- demonstrate the combined value of *in vitro* (e.g., studies of biochemicals, isolated cells, and cell cultures), laboratory animal, human clinical, and epidemiology studies.

As a result: sampling and analysis methods for particles and gases in the air were improved; effects of pollutants on biochemical events in mammalian cells were delineated; new laboratory animal models and methods for exposing them to particles and gases were introduced; clinical research on resting and exercising human subjects were conducted; and several epidemiological studies that focused on comparing mortality (deaths) and morbidity (illnesses) in cities with different types and levels of air pollutants were conducted. The findings largely supported the conclusions made earlier from studying the great air pollution disasters, especially with respect to the vulnerable population groups, and the likely causal role (in deaths and illnesses) of combinations of air pollutants. Possibly the best way to summarize this early period of intense research is an observation from Dr. David Rall, who was the director of the U.S. National Institute of Environmental Health Sciences from 1971 to 1990. He observed that there is no better way to protect public health from environmental chemicals "than the combination of well conducted animal experiments and well conducted epidemiological experiments" (Rall, 1979). In a similar vein, Dr. Roger McClellan, former President of the Chemical Industry Institute of Toxicology and of the Lovelace Inhalation Toxicology Research Institute (now called the Lovelace Respiratory Research Institute), presented the concept of a three-leg stool (**Figure 1–6**), which illustrated the important role of a combination of mechanistic studies, laboratory animal toxicology studies, and human studies for protecting human health.

The large research effort eventually prompted the U.S. EPA to issue a series of National Ambient Air Quality Standards (NAAQS, pronounced "knacks"), which in 1997 tightened the acceptable limits for airborne particles, and introduced particle size-selective ranges (see Chapter 6). These actions stimulated considerable controversy, which was described in the book, *The Particulate Air Pollution Controversy: A Case Study and Lessons Learned* (Phalen, 2002). The controversy centered around several issues, including:

- the impact of the new standards on the cost of goods and services;
- the use of particle size and mass, rather than chemical composition for setting air standards;
- the use of new sophisticated epidemiologic models to estimate the health effects;
- the lack of confirmatory laboratory studies to establish cause and effect relationships among small fluctuations in particle levels and health; and
- the power of the U.S. EPA to independently establish the new air standards.

After a period of extensive litigation, the U.S. EPA was supported by the U.S. Supreme Court, and the new regulations had the force of law.

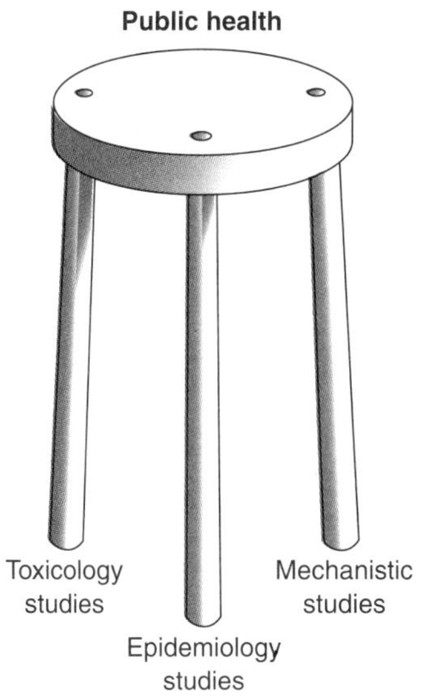

Figure 1–6 Three leg stool representing research on environmental chemical exposures conducted for the purpose of protecting public health.
Source: The University of California Air Pollution Health Effects Laboratory, with kind permission.

IV. RISKS VS. BENEFITS ASSOCIATED WITH AIR POLLUTANT PRODUCING ACTIVITIES

It is safe to say that all human behavior will produce some form of air pollution. In fact, the mere presence of humans, and their routine activities, inevitably contaminates the air (**Table 1–4**). On a larger scale, many essential productive activities such as farming, dairying, electric power generation, manufacturing, construction, spraying (e.g., paints and pesticides), and transportation all have their associated characteristic air contaminants (**Table 1–5**). Similarly, medical procedures, recreation, children's play, entertainment, hobbies, and other valued endeavors produce a large variety of associated environmental air contaminants. Thus, benefits accompany the potential adverse effects of sources of air pollution.

The foregoing examples make it clear that attempts to control air pollution can have counterbalancing effects (also called *tradeoffs*) on human health and welfare that must be seriously considered in regulatory actions. A monograph, *Risk versus Risk: Tradeoffs in Protecting Health and The Environment,* describes the issue in detail (Graham and Wiener, 1995). Accordingly,

Table 1–4 Humans as sources of air contaminants (examples only).

Factor	*Contaminant*	*Possible Effects*
Human presence	Shed skin cells (dander)	Transmission of virus, bacteria, fungi, and allergens
Exhaled air	Organic and inorganic gases, microorganisms	Spread of infectious diseases, and generation of odors
Clothing	Fibers, dyes, preservatives, etc.	Spread of potential airborne allergens
Use of sprays and powders	Cosmetics, disinfectants, cleansers, etc.	Respiratory tract irritation, allergic responses, and chemical poisoning
Cooking, cleaning, and other routine activities	Combustion products and resuspended dusts	Asthma and bronchitis exacerbation, and in rare cases initiation of lung diseases
Tobacco smoking, burning of candles, incense, and wood	Environmental tobacco smoke and other airborne combustion products	Asthma and bronchitis exacerbation, and possible initiation of lung diseases

when any specific activity is heavily regulated it must be modified, or sometimes even banned or replaced. This process can increase the cost of living, and even introduce new, as yet unstudied environmental contaminants. The main point is that, like the targeted air contaminants, regulatory activities can also have their potential hazards. Of course, the regulation of air contaminants is undeniably an essential activity, and some researchers point to the cost-effectiveness of modern regulations (e.g., Hall, et al., 1992). However, as people have to live with all of the consequences of regulations, good and bad, a thorough analysis and a responsible rate of implementation is necessary in order to prevent unacceptable unintended consequences. The modern trend is to balance the monetary costs of controls with the money saved from the expected health benefits. Even today, a thorough analysis of the regulation-related tradeoffs is not usually performed.

V. AGENCIES INVOLVED IN AIR POLLUTION ASSESSMENT AND CONTROL

The list of professional groups involved in air pollution assessment and control is large indeed: **Table 1–6**

Table 1–5 Some essential activities along with their emissions and benefits (examples only).

Activity	*Emissions*	*Benefits*
Agricultural practices (farming, dairying, and animal husbandry)	Sprays, ammonia, pollens, particles, microorganisms, dust, diesel exhaust, etc.	Affordable food and milk combats malnutrition and starvation, ammonia neutralizes air acidity
Electric power generation (except hydroelectric, wind, and nuclear, which have negligible air emissions)	Sulfur, metal-containing particles, and a variety of gases and vapors	Affordable electrical power is essential for food preservation, heating, air conditioning, lighting, and has a variety of other economic benefits
Transportation including cars trucks, ships, aircraft, etc.	Diesel and gasoline engine exhaust, tire and brake dusts, and partially-burned lubricants in exhaust	Personal and commercial transportation are essential for employment and the availability of food and other important products
Manufacturing and construction	A large variety of particles, gases, and vapors	Housing, roads, and numerous products are essential for maintaining a tolerable and healthful lifestyle

Table 1–6 Groups involved in assessing and controlling air pollutants (examples only).

Group	*Roles*
Researchers, including chemists, engineers, industrial hygenists, ecologists, and biomedical scientists	Measure air contaminants; evaluate their effects on humans, wild and domestic animals, and crops; and define which substances are high-priority for control
Engineers including automotive, chemical, electrical, and others	Design machinery and other systems; and measure air emissions
Health professionals, including physicians, public health specialists, and industrial hygienists	Evaluate health effects; develop treatments; solve practical problems; and provide advice to scientists, regulators, and patients
Governmental professionals	Identify issues of concern; provide public education; establish and enforce standards; and advise researchers and regulators
Environmentalists	Identify issues of concern; provide public education; and advise researchers and regulators
Health groups, related associations, and activist organizations	Identify issues of concern; provide public education; monitor problems and progress; raise money for research; and advise researchers and regulators

lists only a few types of such groups and their roles. In the United States, the Environmental Protection Agency (U.S. EPA) is responsible for setting ambient air standards to protect the general public. These standards are called the National Ambient Air Quality Standards (NAAQS). The NAAQS, which are intended to be revised every five years, include sulfur dioxide, particles (in various size ranges), carbon monoxide, ozone, nitrogen dioxide, and lead. Within the EPA the Office of Air and Radiation (OAR) deals with outdoor air quality, indoor air quality, toxic air pollutants (which differ from the NAAQS pollutants in that they expose the population less uniformly), and radiation. The OAR develops national programs, technical policies, and regulations for controlling air pollution and radiation exposure for the general public. The U.S. Centers for Disease Control and Prevention (CDC, http://www.cdc.gov, accessed November 11, 2010) develops and applies information on disease prevention and control, environmental health, health promotion, and health education, in order to improve the health of people in the United States. Within the CDC, the National Institute for Occupational Safety and Health (NIOSH, http://www.cdc.gov/niosh, accessed November 11, 2010) is responsible for research, education, and training geared toward promoting safe and healthful conditions for workers. In this role NIOSH develops information on safe air quality in workplaces. The Occupational Safety and Health Administration (OSHA, http://www.osha.gov, accessed November 11, 2010) is the enforcement agency for workplace air quality, among other factors (such as noise, vibration, and injuries) that relate to workers' safety and health. The Mine Safety and Health Administration (MSHA, http://www.msha.gov, accessed November 11, 2010) enforces compliance with safety and health standards for miners. As is evident, there is considerable overlap in the responsibilities of these governmental agencies, which are only examples of the total number that deal with air pollution in the United States. The European Union (EU), Canada, Japan, China, and other nations have similar regulatory and advisory structures that relate to air quality issues (Olesen, 2004; Chen et al., 2008). An influential professional association, the American Conference of Governmental Industrial Hygenists (ACGIH®), reviews its voluntary workplace air standard recommendations yearly and publishes them in their booklet "TLVs® and BEIs®" (ACGIH®, 2010). These recommendations, based strongly on animal studies and epidemiologic evidence, are not intended as regulations, but they are used worldwide for worker protection. Interestingly, ACGIH® does not always appear to consider the feasibility of meeting a recommended air quality recommendation.

A challenge for the long-term future is to develop uniform (also called *harmonized*) world-wide environmental standards without devastating productivity and prosperity. Such standards would presumably prevent large disparities in the exposures of workers and the general population throughout the world. Harmonized standards could also help prevent disputes over transboundary pollution (the movement across national boundaries), and discourage the practice of locating polluting industries in nations with lax air standards. As one might expect, harmonizing air standards is a complex and politically difficult effort that has large economic consequences.

VI. THE SCOPE OF MODERN AIR POLLUTION SCIENCE

All life that exists and performs its activities on the surface of the Earth is immersed in a finite sea of air. Similarly, non-living things, both natural and anthropogenic (made by humans), can affect and be affected by the air and the many substances contained therein. Even the planet itself, its overall climate, incident surface illumination, and exposure to ultraviolet radiation are influenced by air quality. These factors serve to establish the enormous scope of today's air pollution science, and thus the scope of this textbook. Essentially all of the scientific disciplines that are practiced today have roles to play in understanding, evaluating and, controlling air quality. In the following, a few examples of this broad involvement are listed.

Earth Science, Meteorology, and Climate

Earth science is the discipline that deals with the large picture. The scope includes the oceans and other bodies of water, the atmosphere, and the geologic Earth. The associated atmospheric scientists seek to understand the climate and the atmosphere through the development and use of surveillance, modeling, and data analysis technologies. Included in the Earth scientists' tools and interests are:

- Earth-observation satellites;
- aircraft, and Earth-based instrumentation;
- the use of supercomputers for modeling and understanding climate, weather, and air pollution sources and transport;
- analyses of trends in air quality; and
- defining the challenges, actions, and strategies associated with mitigating adverse effects.

The U.S. National Academies of Science (NAS, established in 1863 under President Abraham Lincoln) currently has several Earth science subdivisions (called "Program Units" or "Boards") that relate to air quality; they include:

- Agriculture and Natural Resources;
- Atmospheric Sciences and Climate;
- Chemical Sciences and Technology (which deals with energy production and use);
- Earth Sciences and Resources (which coordinates several research activities and provides information to policy makers); and
- Environmental Studies and Toxicology (which addresses air and water pollution and their impacts on human health and the environment).

Each subdivision produces reports published by the National Research Council (NRC), which was established in 1916 by the NAS to provide information for critical decision-making, formulating public policy, and important public education services, as well as stimulating the acquisition of new knowledge related to science, technology, and health. Hundreds of reports, many of which are available free online, have been published by the National Academies Press (http://www.nap.edu, accessed November 11, 2010). These reports are an invaluable resource for students as well as practicing professionals.

In addition, there are nearly 50 periodical scientific journals, and countless university research programs that deal with Earth sciences.

Ecology

The discipline of *ecology* deals with topics such as biodiversity, ecosystems, and associated environmental assessments. *Ecosystems* are defined by the plants, animals, and microbes and their interactions in a given defined area. The systems are usually categorized as being *aquatic, terrestrial,* or *wetland.* Ecosystems depend on the local environmental quality, and also modify it. The production of methane (a product of organic decay), carbon dioxide (produced by oxidation), oxygen (a product of photosynthesis), and other atmospheric gases, along with particulate emissions such as pollen and spores, contribute to the local and downwind air quality. In addition, air contaminants from outside sources, both natural and anthropogenic, have their effects on ecosystems. Volcanic eruptions and other geologic processes produce sulfur compounds and other gases and particles that can significantly affect the health of ecosystems. In some cases, anthropogenic air pollutants from cities, farms, factories, and electric power plants impact local and distant ecosystems. Ecologists sample habitats and ecosystems for pollutants, assess the potential consequences, and propose remedial actions. Nearly 200 periodical scientific journals and, perhaps thousands of university programs, cover topics related to ecology and ecosystems.

Epidemiology and Controlled Studies

Many questions arise about the potential adverse effects of air pollution on human and animal health. What chemical and physical (e.g., mass, count, and surface area) characteristics of particles are the most important? What air concentrations of gases and particles are problematic? What are the biologic fates and health consequences of inhaled pollutants? How do effects of *acute* (brief, e.g., hours or a few days) and *chronic* (e.g., months, years, or decades) exposures to air pollutants differ? What segments of the population are particularly susceptible to developing adverse health effects? What are the biological mechanisms that occur within the body that can produce adverse health effects? There are many other important questions as well, but these examples serve to introduce epidemiology and toxicology in air pollution research.

The many roles of epidemiology in protecting public health are described in *Epidemiology for Public Health Practice* (Friis and Sellers, 2008) and *Environmental Epidemiology: Study Methods and Applications* (Baker and Nieuwenhuijsen, 2008). Epidemiology is primarily concerned with the scientific study of the distribution and causes of human diseases and deaths in populations. Epidemiology is essential for showing that a problem exists in the real world. Epidemiology can also establish *associations* between health data (such as deaths and hospital admissions) and environmental data (such as air concentrations of measured pollutants). If the associations are strong, such as those that imply a doubling or larger change (e.g., in normal death rates), there is a reasonable likelihood that the measured environmental parameters are *causes* of the adverse effects. Conversely, weak associations, such as a 1 percent increase in deaths, are more problematic when used as evidence that the measured environmental parameters are the actual causes. The measured pollutants might be *surrogates* for the real causes. For example, if the airborne particle mass is measured and the association is weak, the real culprit could be an unmeasured copollutant, or a combination of the measured particle mass and unmeasured factors. In addition, *confounders* (unexpected factors that produce the outcome of the study), such as lifestyle factors (e.g., smoking, diet, and obesity) and weather extremes, can act to obscure the epidemiology study conclusions. The main point is that epidemiology, although essential, usually cannot establish conclusive scientific causality without outside evidence. Yet strong epidemiologic findings alone can be convincing enough to stimulate regulatory action, as was done for cigarette smoking and the use of lead (Pb) in gasoline.

Sir Bradford Hill (1965) discussed the conditions by which a strong case for causality can be made by epidemiologists. Hill's conditions include the strength of the association, consistency, specificity, temporality, biological gradient, plausibility, coherence, and analogy (for an explanation of these criteria, see Chapter 10). Because epidemiology involves studies of real human populations exposed in real environmental conditions it is given great weight in air pollution research and regulatory decisions.

Controlled studies, such as *in vitro* (literally meaning "in glass"), include investigations of isolated biochemicals, living cells, laboratory animals, and human volunteers (also called clinical studies). These studies seek to control potentially confounding factors and to establish causality. In such well-controlled studies, suspected causal agents can be delivered in known concentrations without the significant interference of other factors (e.g., other exposures or stressors). Thus, causal factors, dose-response relationships, and mechanisms of action can be identified (see Chapter 9). As previously mentioned, it is the combination of well-conducted epidemiology studies and well-conducted controlled studies that are essential for protecting public health. Such controlled studies are performed by specialized scientists who are experts in using their specific experimental systems (e.g., cells, laboratory animals, and human subjects) and interpreting their findings. Because some segments of human populations are more susceptible to the adverse effects of air pollutants, a major challenge lies in developing animal models that respond similarly; such models are often referred to as *compromised animal models*.

About 150 periodical scientific journals cover the areas of epidemiology and toxicology. Many of these journals cover both areas, which allows their complementary nature to be appreciated. For each discipline to be successful, it must be aware of relevant advances in the other one.

Air Chemistry

The Earth's air is actually a complex and dynamic chemical mixture. Our air is a two-component system of particles suspended in a gas. When tiny particles are suspended in a gas, the mixture is called an *aerosol.* Although mostly made of the inert gas nitrogen (N_2), air is chemically active. To appreciate this reactivity, consider the second most abundant component, oxygen (O_2). This reactive gas is involved in both creating and destroying potentially problematic air contaminants. Diatomic oxygen (O_2) is split into two very reactive oxygen atoms (O) by ultraviolet light and by electrical discharges. Among other fates, these oxygen atoms can form ozone, which is triatomic oxygen (O_3). Ozone is chemically reactive, and in sufficient concentrations it can have adverse effects on living things. In the higher regions of the atmosphere (specifically the stratosphere) ozone interacts with ultraviolet (UV) radiation, which serves to reduce the UV levels at the Earth's surface, and thus diminishes the harm to living things. One way to think of this is that ozone is desirable at high altitudes, but undesirable at low altitudes (which is inhabited by living things). The chemistry of ozone is just one small example of the problems of interest to air chemists (see Chapters 3 and 5).

Air chemists do much more than just measure the substances and their reactions in the air. Developing new instrumentation and predictive computer models are important endeavors. Lightweight, inexpensive monitors that can be worn by human volunteers during normal activities are needed by epidemiologists who are trying to find the chemicals and concentrations in the air that pose potential risks. Fast-response biological aerosol monitors that can accurately identify harmful microbes immersed in the air with a variety of similar, but relatively benign ones, are also in need of further development. Collaboration among air chemists, microbiologists, and aerosol scientists is a key for success in improving field bioaerosol monitors. Another important issue, developing computer models that can predict the chemical behavior of the air, should not be overlooked. Without such models, air pollution episodes cannot be reliably predicted, let alone controlled. Attempts to control one air pollutant can even increase levels of other pollutants, and the unintended results could be significant. Also, the many effects of potentially expensive control measures cannot be predicted unless accurate air chemistry models are developed and validated. Air chemistry is a central scientific discipline in the study of air pollution, and it is continually improving the ability to understand and modify air quality.

Dosimetry

There is no question that the *initial doses,* which are defined as *the amounts initially deposited in the respiratory tract,* are critical determinants of the effects of air pollutants. Without a significant dose, there will be no significant toxic effect. *Dosimetry, i.e., the measurement of dose,* deals with several issues including the following examples:

- What descriptors of dose are appropriate?
- What are the doses to specific regions of the respiratory tract?
- What segments of the population receive the largest doses?
- How do age and gender (or probably more importantly, body size) modify doses?
- How does exercise modify dose?
- How do humans and laboratory animals differ with respect to doses?

The methods used by dosimetry researchers include:

- controlled laboratory studies involving human volunteers and laboratory animals;
- use of mathematical computer models to predict population and individual doses;
- use of markers (also called tracers) of inhaled doses; and
- development and validation of new computational dosimetry models, such as computational fluid dynamic (CFD) models and physiologically based pharmacokinetic (PBPK) models.

Scientists involved in dosimetry include mathematicians, physicists, engineers, physiologists, physicians, and veterinarians, to name a few. Chapter 7 includes more information on dosimetry.

Risk Assessment

Risk cannot be completely avoided, but it can usually be managed. If one were attempting to avoid all risk, one might resolve to not get out of bed each morning. But even then, lack of exercise and the absence of the beneficial pressure of life's many challenges would eventually lead to a deterioration of one's health. Nearly everything that people do is associated with risk. Since risk is inevitable, a method is needed for comparing various risks and prioritizing them so that money and effort spent on risk reduction can be properly apportioned. Money and talent are almost always short in supply and long in demand, so they must not be wasted.

In the early 1980s the U.S. Congress asked the National Research Council for help in providing advice on options for characterizing and managing risks. The result was a report, *Risk Assessment in the Federal Government: Managing the Process* (NRC, 1983). This report, also known as the "Red Book," set forth the elements of risk assessment. Risk assessment was described as having four phases:

- Hazard identification
- Dose-response assessment
- Exposure assessment
- Risk characterization

Risk assessment is intended to be followed by *risk management,* which involves using methods that will control risk. Risk assessment has become a mainstay in the methods used to analyze risks and to set priorities for their management. The topic was revisited by the National Research Council (NRC, 2009) by request of the U.S. EPA, for the purpose of improving the process (see Chapter 11). Risk assessment is in a period of modification, mainly to improve its efficiency with respect to its cost, time required, and responsiveness to specific needs.

Regulations

The pressure for industrial, domestic, and other uses of energy understandably leads to the widespread use of the many energy resources. As previously mentioned, the early use of wood and coal as fuels led to problems with air quality, including the associated adverse health effects. Today, the major fuels include coal, burner oil, natural gas, gasoline, diesel fuel, radioactive isotopes, and other, sometimes exotic, substances and technologies. Each of the currently-used fuels and technologies have their associated environmental contaminants. As time passes, the mix of energy sources changes. Advances in methods for energy production are presently being made rapidly. Thus, modern regulatory endeavors are more complex than those of the past.

Dr. Frederick Lipfert (1994) summarized some of the events that shaped the development of air pollution control. Focusing mainly on Great Britain, the United States, and Canada, Dr. Lipfert acknowledged that many other countries had similar histories. Effective regulatory activities probably began in the late 1800s. The Alkali Act in Britain in 1863 established official smoke inspectors. In 1881, the U.S. cities of Chicago and Cincinnati passed local smoke ordinances and California followed much later in 1947. The British Clean Air Act of 1956 (following the 1952 episode in London) regulated smoke, and the first U.S. Clean Air Act of 1963 called for cooperation of state and federal agencies in their enforcement efforts. Shortly afterward, the U.S. Air Quality Act of 1967 and the extended British Clean Air Act of 1968 more firmly set the stage for stern enforceable emission standards. In the United States a milestone event was the creation of the U.S. Environmental Protection Agency in 1970. These events were followed in 1971 by the Canadian Clean Air Act, and in 1980, the U.S. and Canada moved to develop policies to control transboundary air pollution. These and other events marked the basis of effective and evolving air quality regulation.

Several other U.S. agencies also moved to regulate air quality and to insure that funds and effort were dedicated to performing the necessary research efforts. The following organizations are active in their respective arenas: The Bureau of Mines; The Public Health Service; The National Institutes of Health; The National Institute for Occupational Safety and Health; The Occupational Safety and Health Administration; The Consumer Product Safety Commission; and the U.S. EPA. Thus, workers and the public have extensive efforts directed toward understanding their exposures and the health effects of air pollutants. Such efforts permitted defining acceptable levels of air pollutants to which humans are exposed. As previously mentioned, the variety of governmental agencies involved in air quality regulations is extensive. This complex topic is addressed in Chapter 6.

Environmental Justice

Although there are several ways to describe *environmental justice,* it can be defined as *the equal (or at least fair) treatment of those of all incomes, races, genders, cultures, and educational attainments with respect to the social and physical environments in which they work, live, and otherwise spend their time.* Disparities in exposure to violence, drug addiction, health care, and other social environmental factors are known to have strong influences on lifespan and disease rates (U.S. Department of Health and Human Services, 2002). Recently, there has been interest in examining disparities in environmental exposure to potential hazards in water, food, and air. Both outdoor and indoor air quality are generally known to be degraded in many communities that have low incomes and educational levels. Disentangling the combined impacts of income, education, other socioeconomic factors, and air quality on health is a formidable problem. Gee and Payne–Sturges (2004) argue that a major factor that increases the vulnerability to environmental hazards is psychosocial stress in disadvantaged communities. On the other hand, epidemiologists have reported that even when socioeconomic factors are accounted for, exposure to air pollution is associated with excess risks in certain disadvantaged populations (U.S. Department of Health and Human Services, 2002). Many challenges lie ahead in understanding the factors required to promote environmental justice.

VII. SUMMARY OF MAJOR POINTS

This chapter has traced the history of air pollution from before the emergence of humans to the present day. In addition to the pollutant sources and adverse effects, the emergence of awareness and current attitudes toward air contaminants has been described. The continuing tension between progress and the undesirable effects of air contaminants has stimulated large and diverse research efforts. In a sense, this introductory chapter has raised more questions than it has answered.

The story of air pollution and its effects on life predates human existence by billions of years. Natural forces have historically had the greatest impact on life forms, including mass extinctions and the destruction of vulnerable species. With respect to human activities, the conquest of fire and history of fuel usage over time led to observations of air pollution related diseases. Because of their often large exposures, the health and diseases of workers were the focus of several early writers. The introduction of coal as a partial replacement for wood as the fuel of choice changed the air pollution related disease burdens faced by both workers and the general public. Early thinkers and writers, including alchemists, helped to shape attitudes and develop the physical and intellectual tools that benefitted and brought together medical practice and chemistry. Concepts, such as the importance of dose, developed by Paracelsus hundreds of years ago, and John Evelyn's early ideas about vulnerable populations, are still important today.

The industrial revolution, and particularly the great air pollution disasters of the twentieth century, provided the impetus for both large research programs and meaningful regulatory actions related to the adverse effects of air pollution. Research efforts resulted in a realization that epidemiology and controlled studies of cell systems, laboratory animals, and human subjects were all necessary for identifying the important air pollutants and their harmful concentrations. Although it is clear that air pollutants must be regulated, it is also clear that many essential activities, including food production, manufacturing, transportation, electric power generation, and the practice of medicine unavoidably produce air contaminants. Thus, regulating these activities can have unintended effects on health, and these tradeoffs must be taken into account.

Several governmental agencies in the United States and worldwide have been created to study and regulate air quality in workplaces, outdoors, and in dwellings. Accordingly, current air regulations are complex and overlapping. A difficult challenge for the future relates to harmonizing worldwide environmental standards.

Today, the scope of air pollution research is very broad, including Earth sciences, climatology, meteorology, air chemistry, ecology, risk assessment, air quality regulations, and environmental justice. It is clear that we are still in an early phase of understanding and controlling the effects of air pollutants. In fact, it might be said that *air pollution science is in a state where chemistry was before the periodic table of the elements.* Thus, there are many challenges ahead, and therefore, many opportunities for careers in the field.

VIII. QUIZ AND PROBLEMS

Quiz Questions

(select the best answer)

1. The earliest life forms on the Earth:
   a. were much like the dominant plants and animals of today.
   b. were largely microorganisms that could survive in an atmosphere devoid of significant oxygen levels.
   c. did not require substantial concentrations of atmospheric oxygen to survive.
   d. Both b. and c are true.

2. The primary natural forces that have historically determined air quality include:
   a. volcanic eruptions.
   b. meteoric impacts.
   c. the industrial revolution.
   d. Both a. and b are true.
3. Primitive humans and ancient cultures did not experience poor air quality because they did not burn coal.
   a. True
   b. False
4. Modern attitudes toward anthropogenic air pollution were shaped by:
   a. thinkers and writers such as Galen, Pliny, and Paracelsus.
   b. Albert Einstein's theory of relativity.
   c. the great plagues, especially the "black death of the 14th and 15th centuries."
   d. None of the above are true.
5. The three great air pollution disasters were:
   a. caused by strong winds and low humidities.
   b. caused by strong winds and high humidities.
   c. not news at the time, but were recognized decades later.
   d. in the Meuse River Valley (1930), Donora, Pennsylvania (1948), and London (1952).
6. Epidemiologic investigations:
   a. provide the strongest evidence for establishing mechanisms of action of inhaled air pollutants.
   b. are relevant to establishing air pollution regulations because they study real human exposures.
   c. are not relevant to establishing air pollution regulations because of their shortcomings.
   d. use laboratory animals as their primary study subjects.
7. Diesel engines:
   a. emit water vapor and ozone, but not particles.
   b. should be banned because they do not provide essential services.
   c. are important contributors to public health due to their roles in farming, transportation, etc.
   d. Only a. and b. are true.
8. Earth science:
   a. deals with the oceans, the atmosphere, and the Earth's geology.
   b. uses observation satellites and aircraft to study the atmosphere.
   c. is represented in a large number of periodical scientific journals.
   d. All of the above are true.
9. Ozone:
   a. is also called "diatomic oxygen."
   b. is not very reactive chemically.
   c. is formed by the addition of oxygen to a zinc atom.
   d. is beneficial to living things when it is in the stratosphere.
10. Risk assessment:
   a. is a method that is used to eliminate all risks.
   b. is a formal process used to evaluate and prioritize risks by governmental agencies.
   c. is a follow-on process that follows risk management.
   d. was developed by Professor Firket in 1936.

Problems

1. Describe the risk assessment process and explain why it is important for establishing environmental regulations.
2. How did the ideas championed by Paracelsus contribute to our modern attitudes toward air pollution?
3. What were the major conclusions about the health effects of air pollutants that followed the great air pollution disasters?
4. Discuss the roles of governmental groups in understanding and controlling air pollutants.
5. In Table 1–4, which factors are most likely to:
   a. spread infectious diseases?
   b. produce an asthma attack in children?
   c. produce chemically-induced respiratory tract irritation?
6. What adverse health effects would occur if diesel engines were outlawed because of their emissions? List at least six adverse effects.
7. Strict air pollution regulations can both help and harm low socioeconomic status populations. Give an example of a harm.

IX. DISCUSSION TOPICS

1. Photosynthetic plant life utilizes carbon dioxide (CO_2) and produces atmospheric oxygen (O_2). Complex animal life consumes O_2 and produces CO_2. Over the next billion years do we expect the atmospheric O_2 concentration to increase, decrease, or remain about the same at 21 percent?

2. About 600 years ago Paracelsus claimed that all substances are poisons. Is this simple claim still true today? Are there exceptions?
3. In the great air pollution disasters of the twentieth century, deaths lagged the start of each air pollution episode by a few days. Why didn't the deaths show up sooner?
4. Would the immediate banning of all human activities that generate air pollution improve the human lifespan?
5. What are the most important career opportunities that relate to improving future air quality?

References and Recommended Reading

ACGIH®, 2009 TLVs® and BEIs®, American Conference of Governmental Industrial Hygienists, Cincinnati, OH, 2009.

Ashe, W. F., "Acute effects of air pollution in Donora, Pennsylvania," in *Air Pollution,* McCabe, L. C., ed., McGraw–Hill, New York, pp. 455, 458, 1952.

Baker, D. and Nieuwenhuijsen, M. J., eds., *Environmental Epidemiology: Study Methods and Applications,* Oxford University Press, New York, 2008.

Ball, P., *The Devil's Doctor: Paracelsus and the World of Renaissance Magic and Science,* Heinemann, London, 2006.

Brimblecombe, P., "Air pollution and health history," Chapter 2 in *Air Pollution and Health,* Holgate, S. T., Samet, J. M., Koren, H. S., and Maynard, R. L., eds., Academic Press, San Diego, CA, 1999, pp. 5–18.

Chen, Y., Craig, L., and Krewski, D., Air quality risk assessment and management, *J. Toxicol. Environ. Health, Part A—Current Issues,* 71:24–39, 2008.

Clayton, G. D. and Clayton, F. E., eds., *Vol 1, General Principles, Patty's Industrial Hygiene and Toxicology, 3rd Edition,* John Wiley, New York, 1978.

Dockery, D. W., Pope, C. A., Xu, X., Spengler, J. D., Ware, J. H., Fay, M. E., Ferris Jr., B. G., and Speizer, F. E., An association between air pollution and mortality in six U.S. cities, *N. Eng. J. Med.,* 329:1753–1759, 1993.

Dockery, D. W., Schwartz, J., and Spengler, J. D., Air pollution and daily mortality: Associations with particulates and acid aerosols, *Environ. Res.,* 59:362–373, 1992.

Evans, G. W. and Kantrowitz, E., Socioeconomic status and health: The potential role of environmental risk exposure, *Annu. Rev. Public Health,* 23:303–331, 2002.

Evelyn, J., *Fumifugium,* originally published by W. Godbid for Gabriel Bedel and Thomas Collins, London, 1661. Reprinted by Exter: The Rota, 1976. U.S. Library of Congress (www.http://lccn.loc.gov/78301303, accessed June 22, 2011).

Firket, J., Fog along the Meuse Valley, *Trans. Faraday Soc.,* 32:1192–1197, 1936.

Friis, R. H., and Sellers, T. A., *Epidemiology for Public Health Practice, 4th Edition,* Jones and Bartlett, Boston, MA, 2008.

Gallo, M. A., History and scope of toxicology, Chapter 1 in *Casarett & Doull's Toxicology: The Basic Science of Poisons, 7th Edition,* Klaassen, C. D. ed., McGraw–Hill, New York, 2008, pp. 4–5.

Garrison, F. H., *An Introduction to the History of Medicine, 4th Edition,* W.B. Saunders, Philadelphia, 1929, p. 136, 165, 205, 281, 344.

Gee, G. C., and Payne–Sturges, D. C., Environmental health disparities: A framework integrating psychosocial and environmental concepts, *Environ. Health Persp.,* 112:1645–1653, 2004.

Graham, J. D. and Wiener, J. B., eds., *Risk vs. Risk: Tradeoffs in Protecting Health and the Environment,* Harvard University Press, Cambridge, MA, 1995.

Hall, J. V., Winer, A. M., Kleinman, M. T., Lurmann, F. W., Brajer, V., and Colome, S. D., Valuing the health benefits of clean air, *Science,* 255, 14 February: 812–817, 1992.

Hill, A. B., The environment and disease: Association or causation?, *Proc. Roy. Soc. Med., Occup. Med.,* 58:295–300, 1965.

Holgate, S. T., Samet, J. M., Koren, H. S., and Maynard, R. L., eds., *Air Pollution and Health,* Academic Press, San Diego, 1999.

Kump, L. R., The rise of atmospheric oxygen, *Nature,* 451 (7176):277–278, 2008.

Lipfert, F. W., *Air Pollution and Community Health: A Critical Review and Data Sourcebook,* Van Nostrand Reinhold, New York, 1994, pp. 111–141.

Lippmann, M., Cohen, B. S., and Schlesinger, R. B., *Environmental Health Science: Recognition, Evaluation and Control of Chemical and Physical Health Hazards,* Oxford University Press, New York, 2003, pp. 3–12.

NRC (National Research Council), *Risk Assessment in the Federal Government: Managing the Process,*

National Academies Press, Washington, D.C., 1983.

NRC (National Research Council), *Science and Decisions: Advancing Risk Assessment,* National Academies Press, Washington, D.C., 2009.

Olesen, B. W., International standards for the indoor environment, *Indoor Air,* 14:18–26, 2004.

Ottoboni, M. A., *The Dose Makes the Poison: A Plain Language Guide to Toxicology, 2nd Edition,* John Wiley and Sons, New York, 1997.

Phalen, R. F., *The Particulate Air Pollution Controversy: A Case Study and Lessons Learned,* Kluwer Academic Publishers, Boston, MA, 2002, pp. 15–27.

Pope, C. A. III, Review: Epidemiological basis for particulate air pollution health standards, *Aerosol Sci. Technol.,* 32:4–14, 2000.

Rall, D. P., Relevance of animal experiments to humans, *Environ. Health Perspect.* 32:297–300, 1979.

Shrenk, H. H., Heimann, H., Clayton, G. D., Gafafer, W. M., and Wexler, H., *Air Pollution in Donora, PA,* Public Health Bulletin No. 306, Public Health Service, Washington, DC, 1949.

Schwartz, J., Air pollution and daily mortality: A review and meta analysis, *Environ. Res.,* 64:36–52, 1994.

Taubes, G., Epidemiology faces its limits, *Science,* 269:164–169, 1995.

U.S. Consumer Product Safety Commission (http://www.cpsc.gov, accessed March 20, 2010).

U.S. Department of Health and Human Services, *Healthy People 2010: Understanding and Improving Health, 2nd Edition,* U.S. Government Printing Office, Washington, D.C., November, 2002 (http://www.health.gov/healthypeople, accessed April 21, 2009).

Wilkins, E. T., Air pollution and the London fog of December, 1952, *J. Roy. Sanitary Inst.,* 74:1–21, 1954.

Zanobetti, A. and Schwartz, J., Race, gender, and social status as modifiers of the effects of PM10 on mortality, *J. Occup. Environ. Med.,* 42:469–474, 2000.

Chapter 2

Sources and Emissions of Air Pollutants

LEARNING OBJECTIVES

By the end of this chapter the reader will be able to:

- distinguish the "troposphere" from the "stratosphere"
- define "polluted air" in relation to various scientific disciplines
- describe "anthropogenic" sources of air pollutants and distinguish them from "natural" sources
- list 10 sources of indoor air contaminants
- identify three meteorological factors that affect the dispersal of air pollutants

CHAPTER OUTLINE

I. Introduction
II. Measurement Basics
III. Unpolluted vs. Polluted Air
IV. Air Pollutant Sources and Their Emissions
V. Pollutant Transport
VI. Summary of Major Points
VII. Quiz and Problems
VIII. Discussion Topics
References and Recommended Reading

I. INTRODUCTION

Structure of the Earth's Atmosphere

The Earth, along with Mercury, Venus, and Mars, is a terrestrial (as opposed to gaseous) planet with a permanent atmosphere. The Earth is an oblate (slightly flattened) sphere with a mean diameter of 12,700 km (about 8,000 statute miles). The highest surface feature, Mt. Everest, reaches about 8.87 km (5.5 miles) above sea level. Mt. Everest is thus a minute bump on the globe that adds only 0.06 percent to the Earth's diameter.

The Earth's atmosphere consists of several defined layers (**Figure 2–1**). The ***troposphere***, in which all life exists, and from which we breathe, reaches an altitude of about 7–8 km at the poles to just over 13 km at the equator: the mean thickness being 9.1 km (5.7 miles). Thus, the troposphere represents a very thin cover over the Earth, barely higher than Mt. Everest. If the Earth is represented by a ball with a diameter of 2 m, the thickness

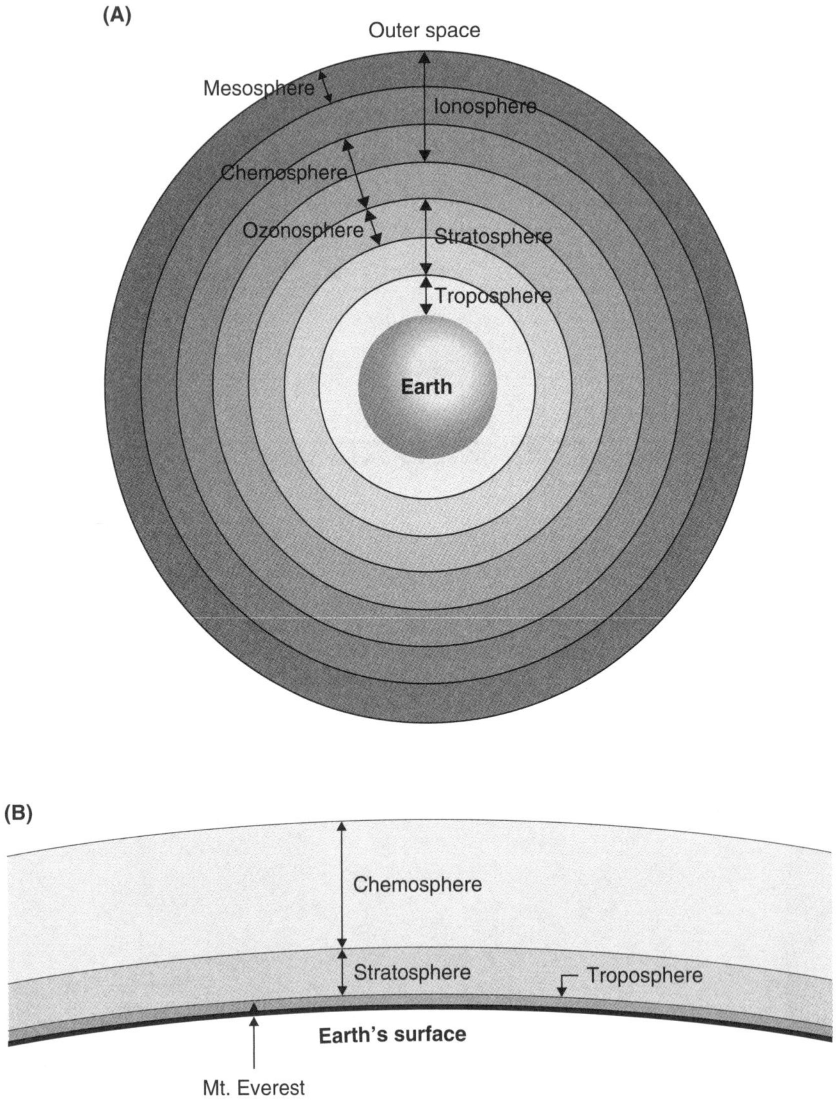

Figure 2–1 A: Layers of the earth's atmosphere (not drawn to scale); B: To-scale depiction of layers of earth's atmosphere. *Source*: The University of California Air Pollution Health Effects Laboratory, with kind permission.

of the troposphere would be about 1.3 mm (**Figure 2–1**). The main point is that surface air pollutants are released into a thin atmospheric skin.

Gravitational attraction binds the atmosphere to the Earth. The outermost traces of the atmosphere reach to over 100 km (60 miles) above the surface. The protective ozone layer is above the troposphere at a variable altitude of about 20 to 30 km.

Vertical Mixing and Inversions

From our knowledge of thermodynamics and hot air balloon observations, we know that heating the air causes it to expand and rise. Because the temperature of the air in the troposphere normally decreases with increasing altitude, warmer surface air rises vertically until it cools sufficiently. This vertical air movement and subsequent mixing dilutes air contaminants generated near the surface, thus preventing their accumulation in the air we normally breathe. However, the situation is changed by *air temperature inversions,* in which vertical air movement is impeded. Inversions are characterized by local parcels of *warm* air over *cooler* air, which prevents the normal vertical mixing. Inversions can trap air pollutants very close to the Earth's surface. Inversions may be low, less than 100 m high, in some cases. In such cases, clear air may be seen above a sharply defined smoggy yellow-brown air layer that extends down to the surface.

Inversions are caused by many phenomena, including (1) movement of a warm front into a region; (2) solar heating of the air above a shaded valley; (3) downward flow of cold mountain air into a warm valley; and (4) nighttime cooling of the Earth's surface by the radiation of surface heat to space. Several types of air contaminants (e.g., black carbon particles) absorb solar energy and heat the surrounding air, thus it is also possible that the trapping of solar heat in the upper layers of a heavily polluted air mass can produce or sustain an air inversion (i.e., a *smog inversion*). When a low-level inversion is coupled with stagnant winds and heavy local emissions of air contaminants, the conditions are right for an air pollution episode. The major historical air pollution episodes all occurred during stagnant inversions in regions that had significant sources of combustion-related air pollutants (see Chapter 1). Although local differences in air temperature and the Earth's rotation normally drive winds that transport and dilute air pollutants, air stagnation and low wind-speed conditions are common in many regions.

Tropospheric vs. Stratospheric Pollutant Effects

Just above the troposphere, the stratosphere reaches to an altitude of about 50 km (30 miles). The boundary between the troposphere and the *stratosphere* is called the *tropopause.* Above the tropopause the temperature rises, creating a natural global inversion that traps pollutants in the troposphere. Large meteorologic events, such as major storms, volcanic eruptions, and nuclear bomb detonations, can penetrate the tropopause and locally mix tropospheric and stratospheric air. Also, the tropopause does not represent an absolute barrier to gaseous molecules that rise through it, which is a factor in the stratospheric *ozone depletion* caused from the rise of chlorofluorocarbons (CFCs) and other reactive substances from the troposphere.

Pollutants in the troposphere can have *direct effects* on living things and physical objects on the surface of the Earth. The physical and chemical properties of the pollutants, the time of contact, and other factors will determine the types and magnitudes of these direct effects. In contrast, pollutants in the stratosphere do not usually contact the surface or interact with life, but they can produce *indirect effects.* These indirect effects relate primarily to absorption and scattering of solar radiation (visible light, infrared, and ultraviolet rays) and cosmic ionizing radiation. The stratosphere provides important protection for life by shielding the surface from such damaging radiations. It also influences and stabilizes the temperature of the Earth, which has significant implications for the sustainability of life on Earth (see Chapter 5). Similarly, tropospheric pollutants can affect surface temperatures by scattering and by absorbing thermal energy from the sun and by trapping nighttime heat that radiates from the surface.

II. MEASUREMENT BASICS

Quantitation and Scale

Quantitative measurements are among the indispensable tools of science. Without such measurements one only has generalities and personal statements, such as "it's too cold"; "the air is clean"; and "visibility is poor." Such generalities convey some information, but they are imprecise and of little use in air pollution science. Quantitative measurements are based on accepted *units of measurement* that are precisely defined. Such units do not vary from time-to-time, place-to-place, or observer-to-observer. They quantify time, size, location, temperature, pressure, altitude, mass (or weight, which is mass times the acceleration due to gravity), humidity, particle count, and many other measurable factors. Ratios such as parts per million (ppm) or milligrams per cubic meter (mg/m^3) are quantitative, and also frequently used in air pollution science.

To demonstrate the use of measurement units, consider sulfur. The total mass of 10 billion, trillion sulfur atoms is only about 0.5 grams. In contrast, the mass of sulfur released in a volcanic eruption can be several billion grams. In order to express such wide ranges of scale, powers of 10 (i.e., exponents) are used. **Table 2–1** provides the scientific nomenclature associated with powers of 10. Such exponents are commonplace in air pollution science.

Powers of 10 are also used for expressing ratios. For example, 1 cubic cm (cm^3) of a gas dispersed in 1 million cubic cm of air is 1 part per million (ppm_v, volume-to-volume ratio). The concentrations of gaseous contaminants in the air are typically reported as ppm_v values (usually without the subscript, v). To avoid confusion, think of ppm as similar to a percentage. Instead of a ratio of 1/100 as with a percent, ppm is a ratio of 1/1,000,000. The scale is different by several powers of 10.

Table 2–1 Powers of ten, using the International System of Units (SI) nomenclature.

	Numerical Equivalent	*Prefix/Name*	*Symbol*
10^{-12}	0.000000000001	pico/trillionth	p
10^{-9}	0.000000001	nano/billionth	n
10^{-6}	0.000001	micro/millionth	μ
10^{-3}	0.001	milli/thousandth	m
10^{-2}	0.01	centi/hundredth	c
10^{-1}	0.1	deci/tenth	d
10^{0}	1	unity	none
10^{1}	10	deca/ten	da
10^{2}	100	hecto/hundred	h
10^{3}	1,000	kilo/thousand	k
10^{6}	1,000,000	mega/million	M
10^{9}	1,000,000,000	giga/billion	G
10^{12}	1,000,000,000,000	tera/trillion	T

Note: In the United Kingdom, 1 billion = 10^{12} (i.e. a million million), and 1 trillion = 10^{18} (i.e. a billion billion).

One could also express ratios on a weight-to-volume basis. For example, 1 cubic cm of sulfur dioxide gas (SO_2) at standard laboratory conditions has a mass of 2.6 mg, thus the measure of mass of sulfur dioxide to a 1 m^3 volume of air would be 2.6 mg per 1 million cubic cm (or 2.6 mg/m^3). Such weight per unit volume measures are commonly reported for solid or liquid contaminants suspended in the air. However, this measure can still be used with gaseous components in air. In some cases the air contaminant may exist in both the gaseous and solid or liquid phases within the air. The weight per volume measure is often used in these instances.

Additional ratios of measure can also be used. For example, as a m^3 of air has a mass of about 1.2 kg, the weight-to-weight ratio for the sulfur dioxide example above would be about 2.2 ppm_w. One could also express the ratio of the *number* of sulfur dioxide molecules to air molecules, but this is seldom done in air pollution science. The weight of an air contaminant in a cubic meter of air (mg/m^3) and the volume ratio (ppm_v, or just ppm) are normally used. **Table 2–2** shows some useful conversion factors for changing mg/m^3 to ppm_v.

Variations in Units of Measurement

Air pollution science involves several disciplines including meteorology, chemistry, physics, ecology, biology, physiology, toxicology, and medicine. Each discipline uses its own measures for pollutant particles, gases, or vapors. Meteorologists are interested in emissions measured in mass or weight units, so concentrations are measured in units such as mg/m^3 of air or parts per million ppm_v. Chemists tend to use units such as moles per cubic meter and atoms or molecules per unit volume. Physicists may use the number, mass, or surface area of particles per cubic centimeter and the partial pressure of a gas. Biologists and physiologists, who are interested in the effects of air pollutants on living systems, use measures such as mass per unit volume of air and *partial pressure* for gases and vapors. Physicians are frequently interested in doses, so the concentrations in organs and tissues (e.g., micrograms per gram of tissue or mass of pollutant delivered to an individual subject) are of interest. Although many more examples could be given, the important points can be made: (1) a number of measures of pollutants are in common use; (2) converting units from one measure to another is often required. This introductory text will use those specific units that best suit the topic at hand.

Table 2–2 When converting mg/m^3 (mass of gas/volume of air) to ppm (volume gas/volume of air) multiply by k. Valid for 25°C and 760 mm Hg pressure.

Pollutant Gas	*Chemical Formula*	*Gas Mol. Wt. (g/mole)*	*k = 24.45/MW (ppm/[mg/m^3])*
Ammonia	NH_3	17.03	1.44
Carbon dioxide	CO_2	44.01	0.556
Carbon monoxide	CO	28.01	0.873
Chlorine	Cl_2	70.91	0.345
Formaldehyde	HCHO	30.03	0.814
Hydrogen sulfide	H_2S	34.08	0.717
Methane	CH_4	16.04	1.52
Nitric oxide	NO	30.01	0.815
Nitrogen	N_2	28.02	0.873
Nitrogen dioxide	NO_2	46.01	0.531
Nitrous oxide	N_2O	44.02	0.555
Oxygen	O_2	32.00	0.764
Ozone	O_3	48.00	0.509
Sulfur dioxide	SO_2	64.07	0.382

III. UNPOLLUTED VS. POLLUTED AIR

Clean Air, Can it be Harmful?

Unpolluted (clean) tropospheric air is defined in **Table 2–3**. Of the two major atmospheric gases, N_2 and O_2, one is considered to be *chemically inert* (N_2), and the other *oxidizing* (O_2). However, N_2 is not always *biologically* inert. In deep-sea diving, where the pressure is several atmospheres, divers breathing compressed air can become disoriented due to the narcotic effects of N_2 dissolved in the fatty tissue of the brain. So N_2 is biologically inert at atmospheric pressure, but toxic at elevated pressures. The use of 80 percent helium gas plus 20 percent O_2 in diving air prevents gas toxicity problems at elevated pressures. Similarly, O_2 is beneficial at normal partial pressure (20.9 percent of 1 atmosphere), but toxic when breathed for a prolonged time at

Table 2–3 The gaseous composition of dry and wet (100% RH) unpolluted air.

Substance	*ppm by Volume, Dry Air*	*µg per m^3, Dry Air*	*ppm by Volume, Wet Air*
Nitrogen	781,000	9×10^8	757,000
Oxygen	209,000	2.7×10^8	203,000
Water	0	0	31,200
Argon	9,300	1.5×10^7	9,000
Carbon dioxide	315	5.7×10^5	305
Neon	18	1.5×10^4	17
Helium	5.2	8.5×10^2	5.0
Methane	1.0–1.2	7.0×10^2	1.0
Krypton	1.0	3.4×10^3	1.0
Nitrous oxide	trace	trace	trace
Hydrogen	trace	trace	trace
Xenon	trace	trace	trace
Organic vapors	trace	trace	trace

100 percent concentration at a pressure of 1 atmosphere. In toxicology (the study of adverse effects of chemicals), not only must gas concentrations be considered, but also the gas pressures (or partial pressures of each gas in a mixture) must be taken into account. Thus, the toxicology of a gas or vapor in a spacecraft (low ambient pressure) or a submarine (high ambient pressure) may not apply to other environments with different atmospheric pressures.

Defining Air Pollutants

Air contaminants (also *air pollutants*) are difficult to concisely define. They exist in several forms including solids, liquids, vapors (vapors are gases formed by volatile liquids/solids), gases, ions, and mixtures of these primary states of matter. Pollutants may persist in the air only briefly (e.g., short-lived reactive chemical species with lifetimes less than 1 second), or for several years (e.g., very small particles and nonreactive gases).

What makes something an air pollutant? From a *regulatory perspective,* an air pollutant is a substance that is, or may be, present at a concentration that exceeds its *safe* concentration. Also, from a regulatory point of view the air is not necessarily polluted if none of its constituents are out of *compliance* with an air quality standard. Note that the air may quickly go from "unpolluted" to "polluted" if an air standard is tightened. Regulatory agencies may refer to *criteria pollutants, air toxics,* or *hazardous air pollutants* (HAPs) that have published standards that are intended to protect susceptible populations or the general public. Criteria air standards are reviewed by the U.S. Environmental Protection Agency (U.S. EPA) about every five years. HAPs are reviewed only when deemed necessary. Workers are usually permitted exposure to higher levels of air pollutants than the public, as workers are healthier (as a group). For the protection of workers from airborne hazards, essentially any substance may added at any time to lists that provide recommended (e.g., ACGIH®, 2010) air concentrations.

From a *practical perspective,* an air pollutant can be defined as any substance, regulated or not, that interferes with a person, process, or an object that has value. Thus, for an analytical chemist a trace of water vapor in the air may be an interfering pollutant. For a deep sea diver, N_2 gas in tank air may be a pollutant. And for a manufacturer of computer chips, very tiny amounts of particles in the air are serious pollutants that may affect product quality. As the practical perspective can include essentially any known substance as a pollutant, it is impractical to consider it in the context of this textbook: specialty publications should be consulted as needed.

An air pollutant is a substance that, at realistic environmental concentrations, is known to have adverse effects on humans, other animals, vegetation, visibility, the climate, or other assets (such as objects of art, machinery, or buildings).

Because regulated pollutants have been subjected to the most intensive research, they are a primary focus in this textbook. If the U.S. EPA, the Occupational Safety and Health Administration (OSHA), or another major regulatory body deems something to be an air pollutant, so will the authors of this book. However, the main point is that both the *substance* and its *concentration,* along with a subject's *exposure* and *sensitivity,* are what actually produce a health problem.

IV. AIR POLLUTANT SOURCES AND THEIR EMISSIONS

Terminology and Pollutant Source Categories

The sources of air pollutants are numerous and varied. Three categories of sources may be defined: (1) *natural* (i.e., those that are not associated with human activities); (2) *anthropogenic* (i.e., those produced by human activities); and (3) *secondary* (i.e., those formed in the atmosphere from natural and anthropogenic air pollutants). Each of these three categories of air pollutants can be further subdivided. For example, anthropogenic pollutant sources can be classified as *stationary,* (e.g., factories, farms, homes, etc.) vs. *mobile* (e.g., cars, trucks, boats, aircraft, etc.). Other subcategories, such as *indoor* vs. *outdoor, organic* vs. *inorganic,* or *local* vs. *regional,* are also used. The following sections will provide an overview of the sources and emissions associated with these and other categories and classes of air pollutants.

Natural vs. Anthropogenic

Table 2–4 lists some major sources and annual national emission estimates (in thousands of tons) for some of the air pollutants that are regulated by the U.S. EPA. The natural sources of selected air pollutants are shown in **Table 2–5**. Several points can be made related to these tables. Natural sources are difficult, if not impossible to

Table 2–4 Some major anthropogenic sources and emissions of ambient air pollutants in the United States in 1993, including primary particle sources and sources of precursor gases (sulfur and nitrogen oxides) and volatile organic compounds (VOCs) (EPA, 1996). PM_{10} refers to particles with diameters under 10 micrometers. SO_x and NO_x are oxides of sulfur and nitrogen, respectively. One ton equals 907 kg.

	Thousands of Tons/Year			
Source	*PM_{10}*	*SO_x*	*NO_x*	*VOCs*
Fuel Combustion				
Utilities	270	15,836	7,782	36
Industrial	219	2,830	3,176	271
Other sources	723	600	732	341
Industrial Processes	553	1,862	905	3,091
Solvent Utilization	305	43	90	10,381
On-road Vehicles	197	438	7,437	6,094
Off-road Vehicles	395	278	2,996	2,207
Fugitive Dust				
Roads	22,568			
Construction/Mining	11,368			
Agriculture	7,236			

control. Natural emissions are often generated at locations that are unpopulated, or sparsely populated by people. For these and other reasons, the primary focus of regulators is on studying and controlling anthropogenic sources. However, a problem arises when secondary pollutants (formed in the air) are considered. For example, anthropogenic (or natural) SO_2 is a precursor of H_2SO_4 (sulfuric acid) and other acidic sulfates. But the natural (and anthropogenic) pollutant, NH_3 (ammonia) reacts with H_2SO_4, and neutralizes its strong acidity. Proposed net reactions include:

Table 2–5 Some major global natural sources and emissions of air pollutants.

	Millions Tons/Year			
Source	*Particles*	*Sulfur*	*NO_x*	*CO*
Dust & Soil	90–450		24	
Sea Spray	800	48		
Biological Action	70–180	108	3–30	40–400
Volcanic Eruptions	*	5.5		
Lightning			24	
Fires & Other Oxidation	*		35	100–5,700

Note: * = extremely large amounts are released.
Data, in short tons (2,000 lbs, 907 kg), are from various sources including: Finlayson–Pitts and Pitts (1986); U.S. EPA (2000).

$$SO_2 + H_2O + \tfrac{1}{2}O_2 \rightarrow H_2SO_4 \quad \text{(Eq. 2–1)}$$

$$H_2SO_4 + NH_3 \leftrightarrow (NH_4)\,HSO_4 \quad \text{(Eq. 2–2)}$$

$$2H_2SO_4 + 3NH_3 \leftrightarrow (NH_4)_3\,H(SO_4)_2 \quad \text{(Eq. 2–3)}$$

$$H_2SO_4 + 2NH_3 \leftrightarrow (NH_4)_2\,SO_4 \quad \text{(Eq. 2–4)}$$

Equations 2–1 to 2–4 represent the formation (and back reactions) of gases/vapors to solid or liquid products, in this case sulfate salts. These salts are typically secondary pollutants formed by the reactions of anthropogenic and natural emissions. Sometimes, it is difficult to isolate the roles of human activities in producing poor air quality.

The foregoing example, which represents only a few of essentially countless atmospheric reactions, demonstrates the importance of air chemistry modeling to support regulations. Air chemistry is covered more completely in Chapter 3.

Stationary vs. Mobile and Point vs. Distributed Sources

Anthropogenic pollutant sources are classified as *stationary* or *mobile*. The major stationary sources as defined by the U.S. EPA (2000) include:

- electric utilities that burn organic fuels (note that nuclear, hydroelectric, and wind-powered utilities usually do not produce air pollutants);
- industrial plants that burn organic fuels;
- chemical manufacturing;
- metals processing;
- petroleum industries;
- other industries (agriculture, textile, wood, rubber, electronic, construction, etc.);
- solvent users (degreasing, graphic arts, dry cleaning, etc.);
- storage (terminals, service stations, chemical storage, etc.);
- waste handling (incinerators, open burning, landfills, etc.); and
- miscellaneous sources (forestry, health services, accidental releases, repair shops, cooling towers, etc.).

Mobile sources, which are vehicles, include:

- on-road gasoline powered trucks, automobiles, and motorcycles;
- on-road diesels; and
- off-road vehicles (gasoline and diesel powered automobiles and trucks, aircraft, marine vehicles, and railroad engines).

For each source, data are gathered on pollutants such as carbon monoxide; sulfur dioxide; ammonia; volatile organic compounds; nitrogen oxides; lead; particulate matter; and hazardous air pollutants. **Figures 2–2 through 2–9** depict 2008 U.S. emissions for several air pollutants. The major source of carbon monoxide is on-road vehicles (about 60 percent of the total). Non-road vehicles emit another 20 percent, and other sources emit the remainder. Vehicles are also major emission sources for nitrogen oxides and volatile organic compounds. Electric utilities are major emitters of sulfur dioxide. Most of the lead emissions are from metals processing, followed by waste disposal/recycling, vehicles, and stationary fuel combustion sources. Most of the ammonia is emitted by livestock, organic decay, and fertilizer-related sources. Sources of anthropogenic particulate matter include stationary-source fuel combustion (including managed burning), mobile sources, and industrial processes (which are responsible for about 10 percent of the total). Sources that are highly-localized, such as electric power plants and industrial plants, are often classified as *point sources,* while agricultural operations and urban centers are *distributed sources.* All sources can have a significant impact on local air quality. The distinction, point vs. distributed, is important for modeling local air concentrations of pollutants. In general, point sources are easier to quantify and to control than are distributed sources.

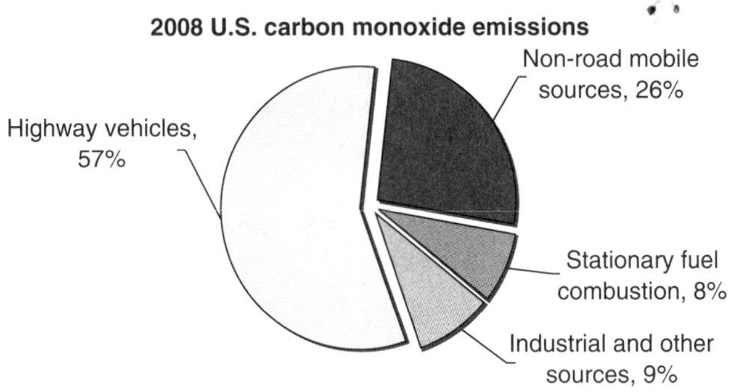

Figure 2–2 U.S. 1998 CO emissions
Source: U.S. EPA (2010).

Enclosed Settings and Workplaces

Overview

Air pollution is not just an outdoor problem. In fact, the air in many *enclosed environments* is more contaminated

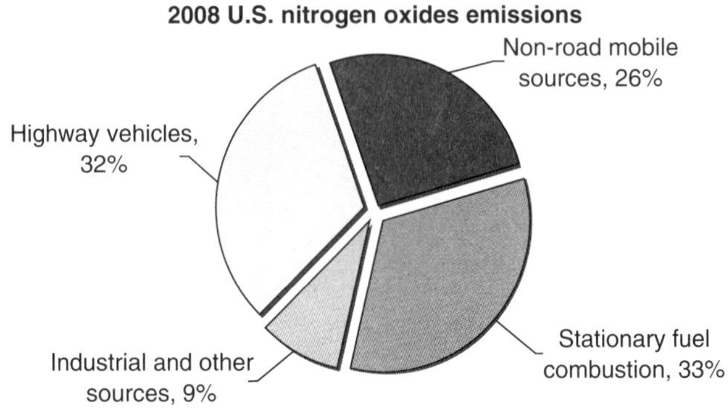

Figure 2–3 U.S. 1998 NO_x emissions
Source: U.S. EPA (2010).

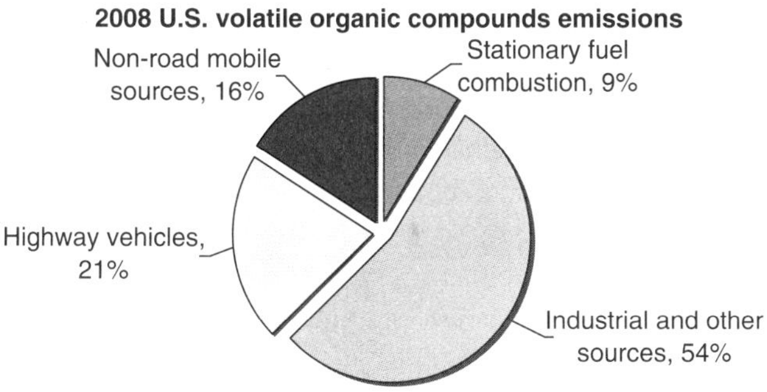

Figure 2–4 U.S. 2008 Volatile organic compound emissions. *Source*: U.S. EPA (2010).

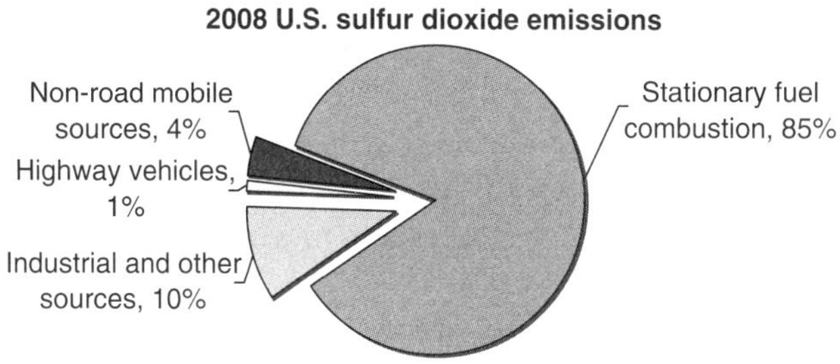

Figure 2–5 U.S. 2008 SO_2 emissions. *Source*: U.S. EPA (2010).

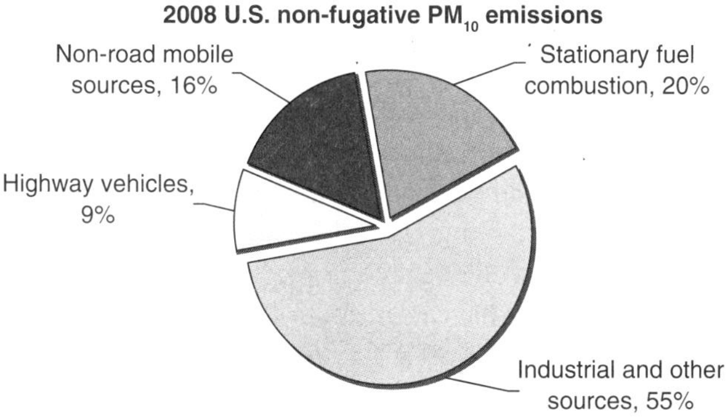

Figure 2–6 U.S. 2008 PM_{10} (particles under 10 μm diameter) emissions from nonfugitive dust sources. *Source*: U.S. EPA (2010).

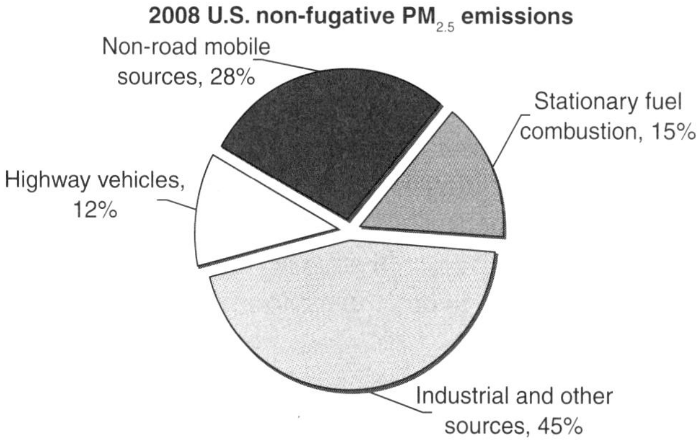

Figure 2–7 U.S. 2008 $PM_{2.5}$ (particles under 2.5 μm diameter) from nonfugitive dust sources. *Source*: U.S. EPA (2010).

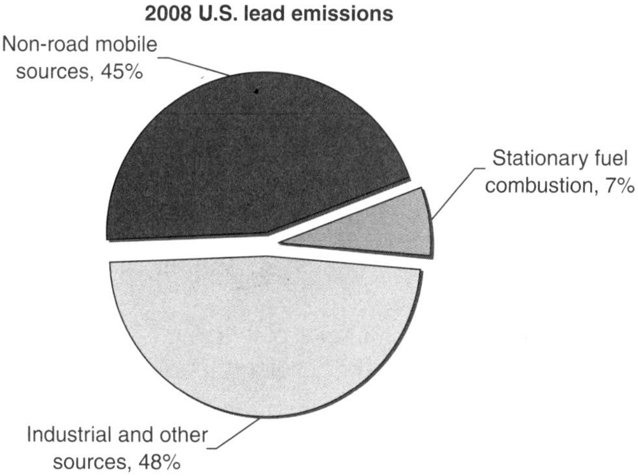

Figure 2–8 U.S. 2008 Lead emissions. *Source*: U.S. EPA (2010).

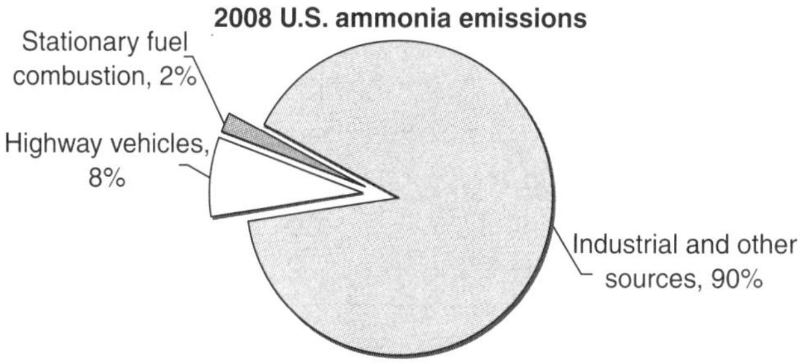

Figure 2–9 U.S. 2008 Ammonia emissions. *Source*: U.S. EPA (2010).

than the local outdoor environment. The reasons for this are (1) the limited volumes of dilution air; (2) the low wind speeds indoors; and (3) the sometimes intense indoor sources (e.g., smoking, cooking, and cleaning). Although one might think of residences as the major occupied enclosed settings, workplaces, public buildings, and vehicles are also places in which indoor exposures occur.

Time-activity patterns (how long and where people are exposed) are used to estimate human exposures. People spend about 90 percent of their time in enclosed environments. Each type of indoor setting has its unique characteristics, and important variations in sources and emissions occur within each. Consider two workplaces, an office and a metal foundry; they have very little in common. Similarly, the air quality for an air conditioned, gas-heated residence will differ considerably from one that has no air conditioning and where the source of heat comes from burning coal or wood indoors. A similar contrast could be drawn between air quality in an upscale automobile and an old "clunker." Because enclosed settings are so variable, this chapter will cover only a few specific examples in order to explore basic principles.

Residences

In contrast to outdoor air pollution, which is both source-dominated and meteorology-dominated, residential air pollution is source-dominated. Pollutants generated in the home are not rapidly dispersed by the wind, and residential air cleaners can seldom overpower uncontrolled pollutant sources. **Table 2–6** lists some of the most common sources of residential air contaminants. **Figure 2–10** illustrates some indoor sources. Whether or not these sources produce health problems depends on the air contaminant concentrations, whether co-stresses (such as excessive heat, cold, or extreme humidity) are present, and the sensitivity and general health status of the occupants.

Allergens and *irritants* (**Table 2–7**) can pose a significant residential problem, especially for sensitive individuals. Such "sensitives" include people with allergies, asthma, bronchitis, emphysema, respiratory-tract infections, and cardiovascular diseases. These, and other health-related conditions, may both impair normal respiratory tract defenses and lead to exaggerated (even life-threatening) responses to modest levels of air contaminants. Such vulnerable individuals may require a very clean air environment. In some cases, medical advice and treatment by a qualified allergist and/or a pulmonary physician is needed.

It is wise to control indoor sources of air contaminants, both for the sake of the healthy residents and any sensitive or otherwise compromised individuals. **Table 2–8** has a checklist that can be used in the home (also the school, office, etc.) for reducing indoor sources. In addition, commercial air cleaners can be effective if the major pollutant sources have been controlled. Such air cleaners should be selected for their *air throughput rate* and their *cleaning efficiency.* An air cleaner that processes only a few cubic feet of air per minute may be totally ineffective in a room with an air volume of 3,000 cubic feet (8.5 cubic meters). Similarly, an air cleaner that is not capable of removing fine particles (e.g., those from smoking or cooking) or organic vapors, may provide little relief to indoor occupants. An uncontrolled source within a poorly ventilated and/or poorly filtered residence

Table 2–6 Common sources of residential air pollutants.

Pollutant	*Sources*
Asbestos and other fibers	Old insulation in/on ceilings, walls, stoves, pipes, floor tiles, heaters, etc.
Lead	Old paint, toys, and other commercial products
Biological aerosols	Mold, pets, insects, rodents, people, humidifiers, indoor plants, etc.
Volatile organic compounds (including formaldehyde)	Air fresheners, spray products, solvents, glues, new building-materials, paints, printers, ink, etc.
Combustion products	Smoking, cooking, wood and gas fireplaces, water heaters, candles, incense, etc.
Dust	Vacuuming/cleaning, plants, dirt, outdoor air, etc.
Gases	Off-gassing from water and various products, electrical discharges, combustion, outdoor air, gas appliances, etc.
Radon	Intrusion from soil, masonry, and underground water

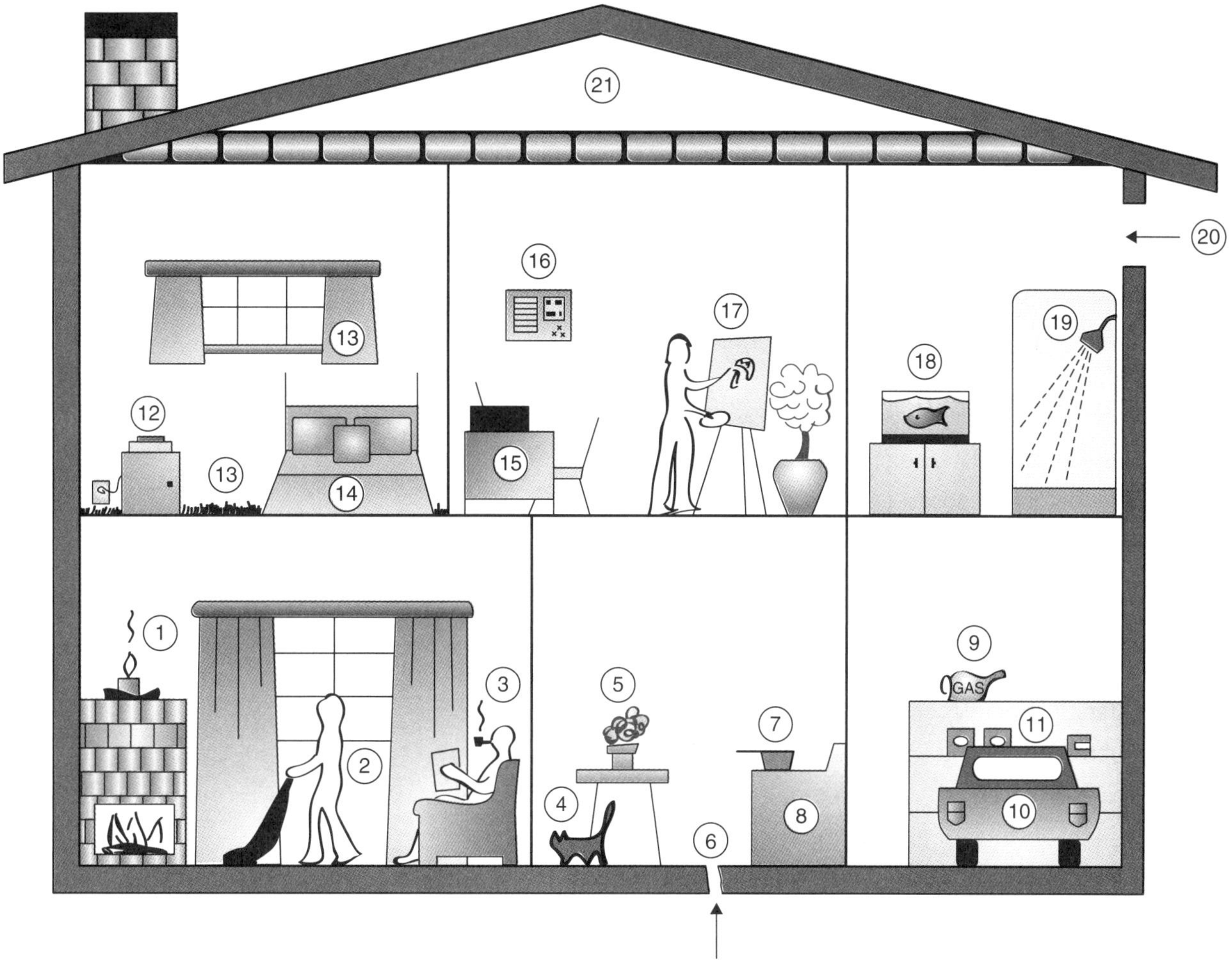

Figure 2–10 Some common sources of residential air contaminants include: (1) fireplace, candles, and incense; (2) vacuuming and cleaning; (3) smoking; (4) indoor pets; (5) potted plants; (6) insects, radon, and methane; (7) food preparation; (8) gas appliances; (9) stored fuel; (10) vehicles; (11) paint, thinner, etc; (12) humidifier; (13) carpet and drapes; (14) bedding and furniture; (15) computer/printer; (16) air conditioner; (17) hobbies; (18) fish tank; (19) shower; (20) outdoor air; and (21) insulation.
Source: The University of California Air Pollution Health Effects Laboratory, with kind permission.

Table 2–7 Irritants and allergens in residences.

A. Irritants (examples only)	*Sources*
Solvents, alcohols, etc.	Cleaning and polishing agents, plastics, liquid fuels, hobbies that involve glue, paint and/or solvents, etc.
Formaldehyde	Air fresheners, other spray products, solvents, glues, new building-materials, paints, printers, ink, etc.
Dust	Vacuuming/cleaning, plants, dirt, outdoor air, etc.
B. Allergens (examples only)	
Insect products (droppings, body parts)	Dust mites, cockroaches, other insect infestations
Pet-related	Cat, dog, and rodent allergens from saliva, fur, droppings, bedding, food, water dishes
Plant-related	Pollens, molds (from dead plant material and soil), volatile fragrances, etc.

Table 2–8 Indoor air pollutant control checklist.

_____ 1. House dust
_____ 2. Deteriorating cloth, rubber, wood, plastic, paper
_____ 3. Indoor animals (birds, dogs, cats, rodents, fish tank bubblers, etc.)
_____ 4. Pet bedding, food, water, and litter boxes
_____ 5. Sprays (air fresheners, paints, insecticides, cleaners, cosmetics)
_____ 6. Smoking (cigarettes, cigars, pipes) and burning incense and candles
_____ 7. Fireplaces, wood, coal, and kerosene stoves
_____ 8. Smoke from cooking food
_____ 9. Poorly-vented or broken ovens, stoves, heaters, clothes dryers, and water heaters
_____10. Dusty or soiled carpets, drapes, blankets, pillows, sheets, or mattresses
_____11. Dirty air filters, soiled or dusty air ducts
_____12. Dirty humidifier and dehumidifier water tanks
_____13. Standing water or evidence of mold
_____14. Damp cloth, paper, wood, carpets, walls, etc.
_____15. Indoor plants with pollen or mold growth on container or soil
_____16. Dust-catching decorations, stuffed toys, and drapery
_____17. Hobby-associated activities and materials
_____18. Paints and solvents that are not tightly sealed
_____19. Insects, including mites, ants, fleas, roaches, spiders, and flies
_____20. Poorly-ventilated rooms
_____21. Garbage and spills
_____22. Odorous building materials, insulation, carpets, drapes or furniture

Note: If completing this checklist does not resolve a residential air problem, air cleaners with sufficient throughput and efficiency should be used. Use of an air cleaner in the bedroom may be particularly effective, if the contaminant sources have been removed. When sensitive or ill residents are present, or when symptoms persist, professional medical help is needed.

Source: The University of California Air Pollution Health Effects Laboratory, with kind permission.

could result in a rapid buildup of contaminants within the living space. Thus, elimination or effective control of a contaminant source is often more critical to air quality than its dilution or mechanical removal from the air.

Workplaces

The variety of air pollutants and their sources in workplaces is great enough to defy any succinct summary. Offices may have solvents, ozone (from printers and copiers), allergens from insect infestations, dust, microorganisms from ducts and air conditioners, carbon monoxide from air intakes near truck loading zones, and a variety of air contaminants from nearby industrial operations. In addition, crowded offices will have human-generated air contaminants from clothing fibers, dander, coughing, and sometimes body-generated odors. Manufacturing operations may produce airborne solvents, organic and inorganic dusts, exotic chemicals, metal fumes, and many other substances. In addition, heat, noise, vibration, and extremes in humidity can exacerbate the direct effects of air pollutants on workers.

Manufacturing and other heavy industrial operations usually require the participation of trained Industrial Hygienists to control the potential risks. It is essential that workplaces conform to modern air quality criteria as well as standards for heat, noise, vibration, and other hazards. In the United States, private professional organizations, such as the *American Conference of Governmental Industrial Hygienists* (ACGIH®) and the *American Industrial Hygiene Association* (AIHA) publish voluntary guidelines that are applicable to workplaces. Also, governmental organizations, such as the *National Institute for Occupational Safety and Health* (NIOSH), the *Occupational Safety and Health Administration* (OSHA), the *Centers for Disease Control and Prevention* (CDC), the *Mine Safety and Health Administration* (MSHA), and numerous state agencies publish important reports, guidelines, and regulations. Although the cost of maintaining a healthful workplace may be substantial, it is usually negligible to the alternative (i.e., significant worker illnesses). A good example of air quality and other workplace recommendations can be found in the ACGIH® publication *TLVs® and BEIs®* (2010). TLVs® are 8-hour averaged threshold limit values (air concentrations) not to be exceeded for air contaminants, and BEIs® are *biological exposure indices,* which are concentrations in blood, urine, etc., that can be measured in workers to determine their actual exposures. The ACGIH® publication also covers other "agents" such as (1) noise; (2) electromagnetic radiation and fields (including lasers, ultraviolet light, and ionizing radiation); (3) thermal (heat and cold) stress; and (4) ergonomics (lifting, vibration, etc.). These other agents, if not controlled, can produce direct harm and increase the risks associated with exposure to air pollutants.

Even if all of the recommendations for workplace air are followed, some workers may not be adequately protected. The TLVs® and BEIs® acknowledge and caution that the recommendations are designed to protect "healthy" working adults, and that they may not protect all workers. Sensitive individuals may require reassignment of duties, stringent local engineering controls, and if all else fails, additional personal protective equipment (e.g., respiratory protection) to reduce exposures.

Criteria Air Pollutants vs. Hazardous Air Pollutants

Criteria air pollutants (CAPs) and *hazardous air pollutants* (HAPs) are specific agents that have been defined by legislation such as the U.S. *Clean Air Act,* or similar international standards. In the United States, there are seven criteria air pollutants (see Exhibit 6–1, Chapter 6). Each one has associated *National Ambient Air Quality Standards* (NAAQS) that are to be revisited and possibly revised by the U.S. EPA at least every five years. The NAAQS include *primary pollutant criteria* for protecting human health and *secondary pollutant criteria* for other effects such as reduced visibility, and/or damage to monuments, crops, or livestock. Primary NAAQS must protect sensitive subpopulations and include a *margin of safety.* Considerations of the cost and feasibility of attainment for primary and secondary NAAQS are currently prohibited by law.

Hazardous air pollutants (HAPs), also designated by the U.S. EPA, currently comprise about 200 substances, including both specific agents (such as asbestos, benzene, and formaldehyde) and chemical classes (such as arsenic compounds, coke oven emissions, nickel compounds, and radionuclides). The distinction between CAPs and HAPs is twofold: (1) CAPs have been determined to present a general risk to the health of the public, and HAPs are believed to cause adverse health and/or environmental effects; and (2) unlike CAPs, HAPs may be added to or deleted from their listings, and they do not have the mandated five-year review cycle.

Similar approaches to listing air pollutants have been established by Canada, the United Kingdom, and the World Health Organization, to name a few. However, nations differ in their standards, often significantly. Two main points are (1) that air standards are constantly evolving; and (2) that different approaches are used throughout the world.

Accidents and Disasters

An accidental release of air pollutants can produce a major health catastrophe for a local community. Explosions, fires, industrial containment failures, railroad car derailments, and trucking accidents can release substantial quantities of fuels, solvents, combustion products, radionuclides, and/or toxic air contaminants. Similarly, natural accidents, such as volcanic eruptions, earthquakes, forest fires, and hurricanes can generate high levels of air pollutants. Such accidents can take hundreds to thousands of lives. Warfare and terrorist attacks must also be included in the list of causes of air pollution disasters. Poison gases (e.g., sulfur and nitrogen mustards, phosgene, chlorine, hydrogen cyanide and nerve agents), biological aerosols (e.g., botulin, ricin, anthrax bacteria, and viruses), and radionuclides (from atomic, hydrogen and so-called "dirty" bombs) are among the potential air pollutants. Warfare agents are sought by terrorists in order to achieve their ends.

Air pollution is not simply a problem caused by industry, transportation, agriculture, and other activities that are often essential to maintaining public health.

An especially-tragic industrial disaster occurred in 1984 in *Bhopal,* India. An explosion at a pesticide facility during an atmospheric inversion released tons of methyl isocyanate into the air of a nearby community, and about 200,000 people were exposed. The highly-irritant chemical most likely produced over 3,000 deaths, including those at the time of the explosion and those over the following four years. Thousands more delayed Bhopal fatalities are projected into the future. Such disasters lead to public fear, and they result in pressure on governments to provide better protection.

V. POLLUTANT TRANSPORT

Overview

The transport of air pollutants is largely driven by weather phenomena. Vertical air motions along with prevailing winds both dilute and disperse particles and gases emitted from any given source. On a global scale, unequal solar heating causes warm air near the equator to rise and then descend as it cools and travels toward the north and south poles. This effect alone would produce wind circulation toward the poles at high altitudes and toward the equator at low altitudes. But several other phenomena complicate the global weather picture. The Earth's rotation has two effects, surface friction that drags surface air in the direction of rotation and a *Coriolis force* that, in general, turns air movement to the right in the Northern hemisphere and towards the left in the Southern hemisphere. The Coriolis force occurs as a consequence of the Earth's rotation. Surface characteristics (e.g., oceans and other large bodies of water vs. land areas) that modify the air temperature and moisture content also influence large air flow patterns. Mountains, forests, and even cities affect wind patterns more locally; **Figure 2–11** depicts the major convective cells and the general climate systems. Although the principles of weather phenomena are understood, long-range weather predictions are highly uncertain due to the effects of small-scale phenomena. Weather is a *chaotic system,* in that small changes in initial conditions can produce large long-term consequences.

If emitted particles and gases strictly followed the winds, concentrations downwind could be accurately predicted over the time scale of a few days. However, such predictions are risky because airborne particles and gases can separate from their air parcels (the volume of air into which they were emitted), due to diffusion and sedimentation. Also, local meteorology can influence the rise and transport of air on a local scale. Such local meteorology includes air inversions and convective phenomena (as produced by local heating effects, eddies, terrain effects, and other phenomena).

Gases and vapors more closely follow wind patterns, but they have diffusion velocities that can cause them to spread beyond the air parcel into which they are emitted.

Particles with large aerodynamic diameters (20 μm and above; see Chapter 3 on aerodynamic diameters) have behaviors strongly influenced by gravity and by the particle's inertia. Therefore, they readily settle out of the air parcels in which they are entrained. Particles in this size regime settle out of the air at velocities that exceed 3 m per hour. Thus, if such particles are emitted

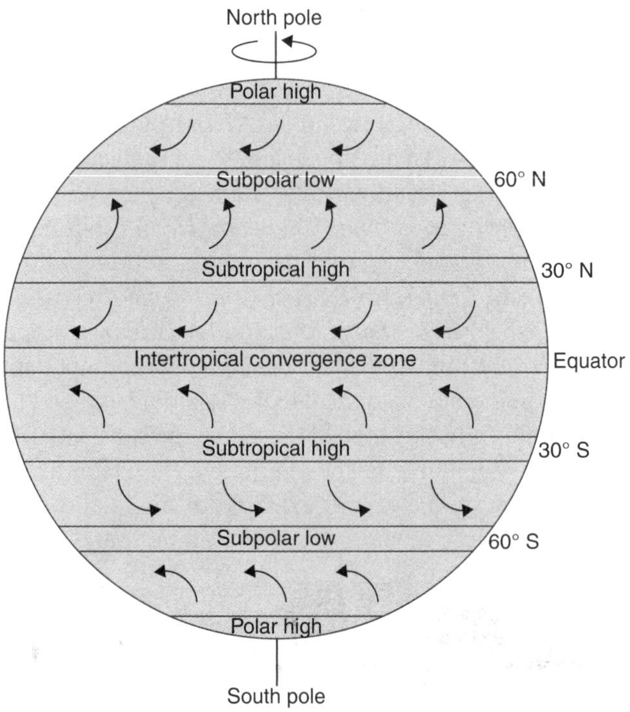

Figure 2–11 The large-scale motion of the Earth's atmosphere. This motion is influenced by thermal gradients, friction between the air and rotating Earth, and the Coriolis force.
Source: The University of California Air Pollution Health Effects Laboratory, with kind permission.

1 m above the surface they will settle to the ground in about 20 minutes, unless they encounter substantial updrafts or descending terrain. Small particles, e.g., those 1 μm or smaller in diameter have settling velocities (see Chapter 3) of less than a few cm per hour. Such particles can travel in the wind for more than several km before settling to the ground. Very small particles can travel thousands of km before settling out of the air.

Other phenomena that produce the loss of contaminant particles, gases, and vapors from the air include:

- evaporation of volatile particles;
- chemical transformations;
- capture by vegetation and other objects (both anthropogenic and natural);
- capture by precipitation (rain, snow, etc.); and
- attachment to particles that settle out.

Gaussian Plume Model

The Gaussian plume model is used for estimating downwind pollutant concentrations emitted from point sources. The model is three-dimensional in that downwind, crosswind, and vertical dispersing components are considered. At the emission point, pollutant dilution occurs in direct proportion with the wind speed. The plume shape near the source is Gaussian (i.e., bell-shaped normal-distributions) in its cross-sections. The standard deviations of the normal distributions increase with distance from the emission point source. The model is useful for predicting air pollutant concentrations downwind from smokestacks. However, the basic model does not include pollutant losses (or formations) in the plume caused by chemical reactions and particle sedimentation. Chemical reactions in plumes can lead to significant acidification and hence downwind acid deposition (see **Exhibit 2–1**).

Plumes and Smokestacks

Plumes from smokestacks present a good case study in which to examine air pollutant transport. Increasing the height of a smokestack will usually decrease the near-stack ground level concentrations of emitted particles and gases and cause emissions to dilute and deposit further downwind, which is why smokestacks tend to be tall. Ambient air velocity has effects on stack emissions. Low winds tend to increase nearby ground level concentrations, but at high wind speeds the additional dilution of the emissions decreases downwind concentrations. As an example, if the pollutant concentration downwind from a point emission source is 10 mg/m^3 of air, then doubling the wind speed should result in a concentration of about 5 mg/m^3, or half the original concentration.

The buoyancy of a plume is a factor in initial plume rise. A warmer or otherwise lower-density plume will rise to a height where it matches the density of the air. Interestingly, humid air has a lower density than dry air. This is because the molecular weight of H_2O vapor (18 atomic mass units) is less than that of both

Exhibit 2–1 Acid deposition (acid rain, snow, and aerosols).

The deposition of acidic air pollution can have adverse effects on soil and fresh water lakes that are downwind from large sources of sulfur dioxide, sulfuric acid, NO, NO_2, and nitric acid. Acid deposition, which refers to the deposition of rain, snow, fog, or acid aerosols, can cause immediate and/or delayed decreases in the pH of water and soil. The delayed effects are mainly due to the melting of acidified snow and the subsequent runoff. The effects, which depend on the buffering capacity of lakes and soil, include decreases (or elimination) of sensitive species of fish and decreased growth of forests. The Eastern United States and Eastern Europe have been most heavily impacted by acid deposition, primarily due to the presence of emissions of acids and acid precursors. Emission controls, e.g., sulfur reductions in coal, are partially directed toward reducing the adverse effects of acid deposition in impacted areas.

The chemistry of acid rain, snow, fog, and aerosols has been extensively studied. The neutral pH of pure water, i.e., neither acidic nor basic, is 7.0. Even pristine rain has a slightly-acidic pH of 5.0 to 5.65, due to the natural CO_2 and other substances in the air. Downwind of some types of industries and electric power plants, the pH of rain can be as low as 4.2. This acidity forms over time in plumes. Interestingly, taller smokestacks can produce more acidic conditions downwind because of the longer times available for acids to form.

N_2 (28 atomic mass units) and O_2 (32 atomic mass units), which make up most of the air; therefore, water vapor displaces these heavier gases.

When the smokestack emits a plume at a low upward velocity and at a temperature and density near ambient conditions one can observe the interaction of the plume with the wind and the vertical air temperature profile. **Figure 2–12** shows two likely vertical temperature profiles in the atmosphere. The solid line shows the standard *lapse rate*, which is the normal decrease in air temperature with altitude in the troposphere. The standard lapse rate is 9.8°C per km, but it can vary. The dashed line represents the temperature profile in a low-level (400 m) inversion condition. Such inversions, which can be lower or higher than 400 m, are not uncommon. **Figure 2–13** shows a top view of a typical smokestack plume carried by a wind moving from left to right. Note that the plume spreads outward and is diluted as it travels with the wind. The concentration of pollutants in the plume is usually greatest along its centerline, tapering to low concentrations at its periphery, as predicted by the Gaussian plume model. Similarly, if viewed from the side, the centerline of the plume will have the highest concentration of stack emissions.

Figure 2–14 shows the effect of wind on a plume in the normal lapse-rate case. With no wind, the typical plume will rise and spread laterally. In a wind, the plume will also spread as it is blown. The spreading plume may touch the nearby ground if the smokestack is short, which can produce high levels of emissions at ground level near the stack.

Figure 2–15 demonstrates plume behavior in an inversion condition when the smokestack is above or below the inversion height. As for the short smokestack, an emission below the inversion layer can produce substantial local ground-level concentrations. The condition of a low inversion height and an even shorter smokestack is called *trapping*.

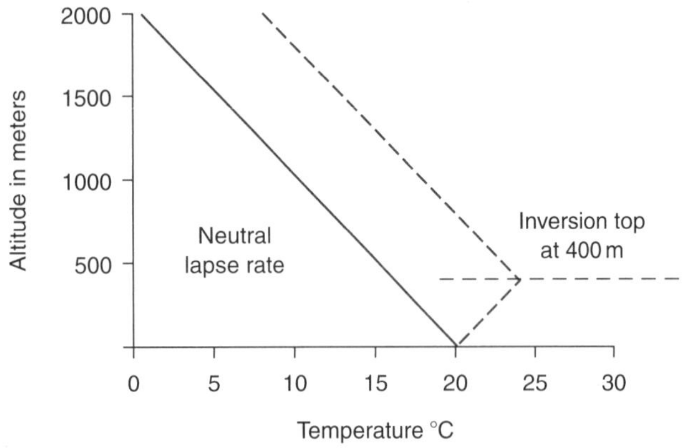

Figure 2–12 Variation of air temperature vs. height above the Earth's surface. The slope of the solid line is the neutral lapse rate of 9.8°C per km. The dashed line depicts a hypothetical inversion layer at 400 m above ground level.
Source: The University of California Air Pollution Health Effects Laboratory, with kind permission.

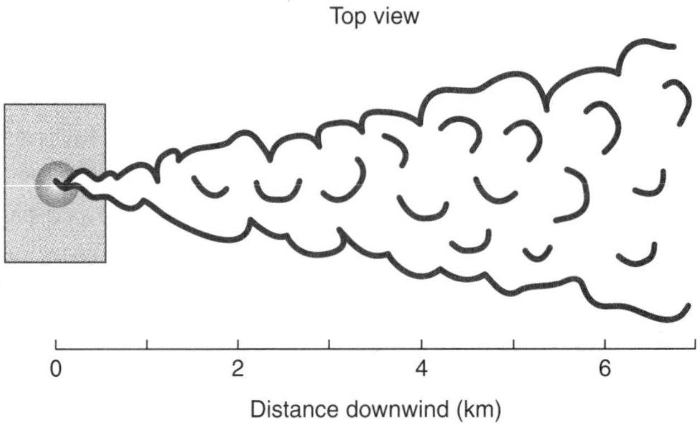

Figure 2–13 Top view of a point-source plume in a light wind.
Source: The University of California Air Pollution Health Effects Laboratory, with kind permission.

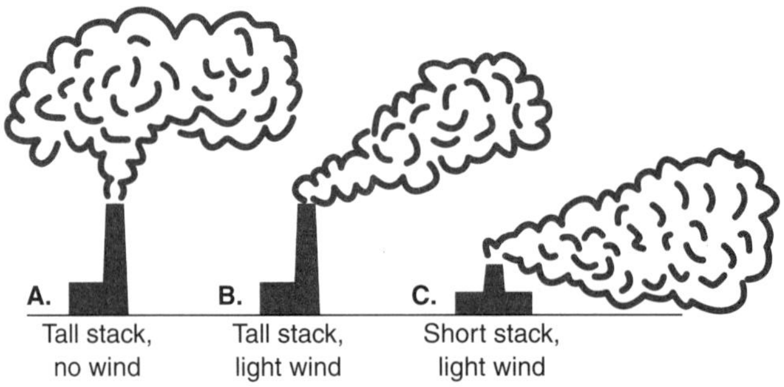

Figure 2–14 Plumes from: A: tall stack with no wind; B: tall stack with a light wind; and C: short stack, with a light wind.
Source: The University of California Air Pollution Health Effects Laboratory, with kind permission.

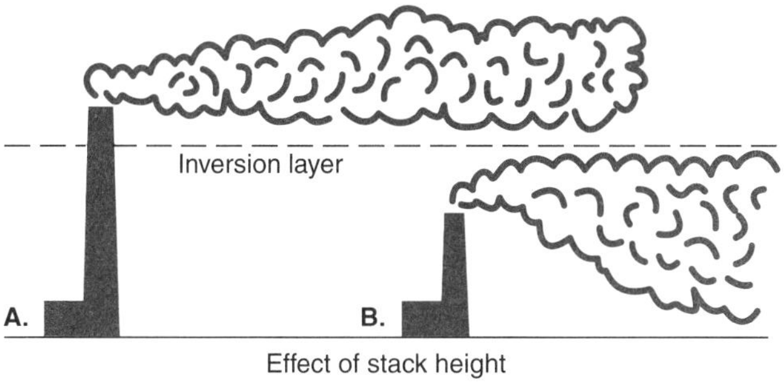

Figure 2–15 Effect of stack height: A: above inversion layer; and B: below inversion layer.
Source: The University of California Air Pollution Health Effects Laboratory, with kind permission.

Figure 2–16 shows the effect of the strength (steepness of the air temperature profile) of the environmental lapse rate. If the lapse rate is strong, *looping* of the plume can occur, which can cause the plume to descend to the surface, rise, and descend again. If the lapse rate is weak, the plume tends to remain aloft.

The most serious conditions, such as a short smokestack, an emission below a low inversion, and a strong lapse rate, have the potential for producing high ground-level concentrations of emissions. In the ideal case the smokestack will be tall enough to allow substantial dilution of the plume before it reaches the surface. The examples given demonstrate just some of the fates of smokestack plumes. The principles of smoke stack plume dispersion also apply to other emission point sources, such as factories, vehicles, and fires.

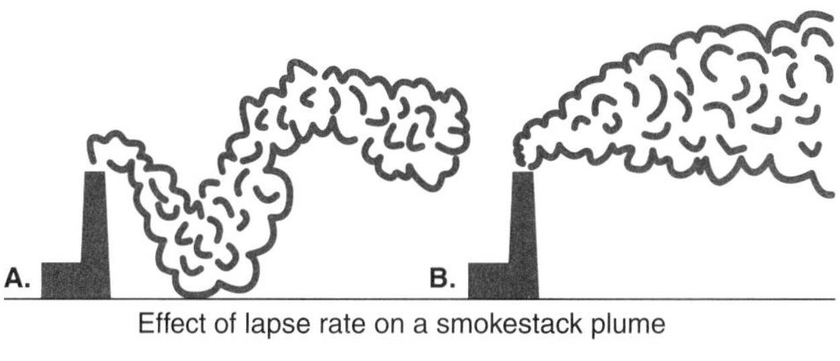

Figure 2–16 Effect of lapse rate on a smokestack plume: A: for a strong lapse rate in which the decrease in temperature vs. altitude is much greater than normal, producing looping of the plume; B: for a normal lapse-rate condition.
Source: The University of California Air Pollution Health Effects Laboratory, with kind permission.

VI. SUMMARY OF MAJOR POINTS

The surface of the Earth has a limited air volume, the *troposphere,* which supports life. Above the troposphere, the *stratosphere* is also essential to life because it shields the surface from harmful radiations and helps to stabilize the temperature of the planet. Several scientific disciplines, each with their own nomenclature, study air pollutants in the atmosphere. Air pollution studies require quantitation and precise units of measurement in order to develop reliable knowledge and to devise acceptable air quality standards. There are several ways to define the term "air pollution" depending on one's perspective. Regulated (e.g., criteria and toxic) air pollutants are a main focus of modern research. Air pollutants are emitted by a variety of sources, which may be "natural" or "anthropogenic," "outdoor" or "indoor," "stationary" or "mobile," and "point" or "distributed." Enclosed spaces and workplaces are highly varied with respect to sources and emissions of air pollutants. However, due to the lack of effective dilution and wind dispersal, strict source control is usually needed in order to protect occupants in these environments. Although governmental regulation has been effective in controlling "normal" emissions, periodic accidents and natural disasters add to the continuing health-related toll of air pollution.

The *transport* of air contaminants must be considered in order to understand their downwind concentrations, and thus exposures of people, animals, and plants. One aspect of transport that has been well studied is the dispersion of plumes. Smokestack plumes are dispersed under the influence of meteorological conditions (e.g., wind speed and vertical air temperature profiles) and smokestack design (specifically smokestack height above the surface).

VII. QUIZ AND PROBLEMS

Quiz Questions

(select the best answer)

1. The "troposphere":
   a. is a term that describes the uppermost region of the Earth's air.

b. is over 1,000 km thick.
c. provides the air in which the Earth's living things exist.
d. consists of 80 percent oxygen and 20 percent ozone.

2. An air inversion is caused by:
   a. stagnant winds.
   b. low atmospheric pressure.
   c. cooler air on top of warmer air.
   d. warmer air on top of cooler air.
3. The three air contaminants, ozone, sulfur dioxide, and carbonaceous fine particles, are:
   a. only natural air contaminants.
   b. only anthropogenic air contaminants.
   c. both natural and anthropogenic air contaminants.
   d. neither natural nor anthropogenic air contaminants.
4. How do air pollutant concentrations of 10^{-3} and 10^{-6} mg/m^3 compare to each other?
   a. 10^{-3} is a thousand times greater than 10^{-6}.
   b. 10^{-6} is two times greater than 10^{-3}.
   c. 10^{-3} is two times greater than 10^{-6}.
   d. 10^{-6} is three times greater than 10^{-3}.
5. What factors can classify a substance as an air pollutant?
   a. presence in the air.
   b. interference with human health or welfare, damage to materials or objects of value, or interference with important scientific processes.
   c. both a and b.
   d. None of the above are true.
6. How would one classify the population of cross-country motorcycles as sources of air pollution?
   a. stationary, distributed, and natural.
   b. mobile, distributed, and anthropogenic.
   c. mobile, point, and natural.
   d. stationary, point, and anthropogenic.
7. The major anthropogenic sources of particulate air pollutants are:
   a. stationary-source fuel combustion.
   b. mobile-source fuel combustion.
   c. volcanic eruptions and natural fires.
   d. both a and b.
8. Indoor air pollution is:
   a. source-dominated.
   b. meteorology dominated.
   c. negligible in comparison to outdoor air pollution.
   d. independent of the activities of occupants.
9. The use of air cleaners indoors:
   a. is the best way to deal with building-related illnesses.
   b. is the primary means of controlling indoor air quality in the home.
   c. is a good method for controlling indoor carbon monoxide, but not indoor allergens.
   d. are effective if the major sources of air pollutants have been removed and the filtering rate and efficiency are sufficient.
10. The U.S. EPA's designated hazardous air pollutants (HAPs):
    a. are also called "secondary NAAQS."
    b. is a list of substances that can change through both additions and deletions by the U.S. EPA.
    c. include classes of air pollutants, but not individual substances.
    d. have been determined to cause adverse environmental effects but not adverse human health effects.
11. How does the height of a smokestack influence the ground-level concentrations of its emitted particles and gases?
    a. It is not an important factor.
    b. It is only important if there is an inversion condition.
    c. It is only important when wind speeds are low.
    d. None of the above are true.
12. Which meteorological factors are important in influencing ground-level air pollutant concentrations?
    a. precipitation.
    b. wind speed.
    c. inversions.
    d. All of the above are true.

Problems

1. Convert the following to ppm.
   10 percent
   100 percent
   1 ppb

2. If the Earth and its atmosphere shrunk to the size of an apple, how thick would the troposphere be?
3. How much time is required for an air cleaner with a throughput air rate of 15 cubic feet per minute (1 cubic foot = 0.0283 m^3) to filter the volume of air in a room 20 feet by 20 feet by 10 feet (1 foot = 0.305 m)?
4. Using Table 2–4, estimate the percent of total PM_{10} emissions that come from vehicles.
5. If an asthmatic child lived in your home, name six sources of air pollutants that should be eliminated or greatly reduced.
6. Recommendations for children at school include allowing them to go outside during recess even in smoggy cities. How can this recommendation be justified?
7. If gasoline-powered cars were all replaced by electric vehicles, what air pollutants would be reduced? What air pollutants, if any, would not be reduced?

VIII. DISCUSSION TOPICS

1. What are the major sources and emissions of air pollutants in your community? Do the sources benefit the community? How would one decide on whether or not a specific pollutant source should be shut down or required to reduce its emissions?
2. What jobs/occupations expose workers to the poorest air quality? Contrast the air quality-related hazards for an elementary school teacher, a farmer, a paramedic, and a construction worker. Which occupation is most likely to develop lung disease?
3. Should indoor potted plants be recommended for treating people with asthma? Discuss the pros and cons associated with your selection.
4. Can the air in a city be too clean?
5. You are charged with the task of designing a smokestack for a new chemical plant that will be built somewhere near a large city. The adverse public health effects are a concern. What factors would you take into account that would influence your design?

References and Recommended Reading

ACGIH® (American Conference of Governmental Industrial Hygienists), *ACGIH®, TLVs®, and BEIs®*, American Conference of Governmental Industrial Hygienists, Cincinnati, OH, 2010.

Costa, D. L., "Air pollution," in *Casarett and Doull's Essentials of Toxicology,* Klaassen, C. D., and Watkins, J. B., III, eds., McGraw–Hill, New York, 2003, Chapter 28, pp. 407–418.

Finlayson–Pitts, B. J. and Pitts, J. N. Jr., *Atmospheric Chemistry: Fundamentals and Experimental Techniques,* John Wiley & Sons, New York, 1986.

Gammage, R. B. and Berven, B. A., eds., *Indoor Air and Human Health, 2nd Edition,* Lewis Publishers, Boca Raton, FL, 1996.

Grant, L. D., Shoaf, C. R., and Davis, J. M., "United States and international approaches to establishing air standards and guidelines," in *Air Pollution and Health,* Holgate, S. T., Samet, J. M., Koren, H. S., and Maynard, R. L., eds., Academic Press, San Diego, CA, 1999, Chapter 42, pp. 947–982.

Hinds W. C., *Aerosol Technology: Properties, Behavior and Measurement of Airborne Particles, 2nd Edition,* Wiley–Interscience, New York, 1999.

Lippmann, M. and Maynard, R. L., "Air quality guidelines and standards," in *Air Pollution and Health,* Holgate, S. T., Samet, J. M., Koren, H. S., and Maynard, R. L., eds., Academic Press, San Diego, CA, 1999, Chapter 43, pp. 983–1017.

(NOAA) National Oceanic and Atmospheric Administration, National Aeronautics and Space Administration and U.S. Air Force, *U.S. Standard Atmosphere,* 1976, NOAA–S/T 76–1562, Washington DC, 1976.

Perkins, J. L., *Modern Industrial Hygiene, Vol. I: Recognition and Evaluation of Chemical Agents, 2nd Edition,* American Conference of Governmental Industrial Hygienists, Cincinnati, OH, 2008.

Rom, W. N. and Markowitz, S., eds., *Environmental and Occupational Medicine, 4th Edition,* Lippincott Williams & Wilkins, Hagerstown, MD, 2006.

Samet, J. M. and Spengler, J. D., eds., *Indoor Air Pollution: A Health Perspective,* Johns Hopkins University Press, Baltimore, MD, 1991.

Scaillet, B., Are volcanic gases serial killers? *Science,* 319:1628–1629, 2008.

Seinfeld, J. H. and Pandis, S. N., *Atmospheric Chemistry and Physics, 2nd Edition,* Wiley–Interscience, New York, 2006.

Sidel, V. W., Onel, E., Geiger, H. J., Leaning, J., and Foege, W. H., "Public health responses to natural and human-made disasters," in *Public Health and Preventive Medicine, 13th Edition,* Last, J. M., and Wallace, R. B., eds., Appleton & Lange, Norwalk, CT, 1992, Chapter 74, pp. 1173–1186.

Stern, A. C., ed., *Air Pollution, Vol. I., 3rd Edition,* Academic Press, New York, 1976.

U.S. EPA (United States Environmental Protection Agency) *Building Air Quality: A Guide for Building Owners and Facility Managers,* EPA/400/1–91/033, U.S. Environmental Protection Agency, Washington DC, 1991.

U.S. EPA (United States Environmental Protection Agency) *National Air Pollutant Emission Trends,* EPA–454/R–00–02, U.S. Environmental Protection Agency, office of Air Quality Planning and Standards, Research Triangle Park, NC, 2000.

U.S. EPA (United States Environmental Protection Agency), *Our Nation's Air: Status and Trends Through 2008,* EPA–454/R–09–002, U.S. Environmental Protection Agency, Research Triangle Park, NC, 2010.

Chapter 3

Important Properties of Air Pollutants

LEARNING OBJECTIVES

By the end of this chapter the reader will be able to:

- define "aerosol," "gas," "vapor," "hydrocarbons," "primary air pollutants," and "secondary air pollutants"
- discuss the ways in which particle size affects inhaled doses and potential health effects
- distinguish between "criteria air pollutants" and "hazardous air pollutants"
- discuss how ozone is formed in the troposphere and why it is important for public health

CHAPTER OUTLINE

I. INTRODUCTION

Understanding Aerosols

In order to understand air pollution and its health and welfare effects, it is necessary to study some physical and chemical characteristics of the atmosphere. The portion of the atmosphere from which we breathe (the troposphere) is a complex mixture of aerosol particles and gases. An aerosol consists of particulate material suspended in a gas. The suspending gas is usually air, plus water vapor and any other gaseous contaminants. The particulate material consists of condensed matter (i.e., particles that are solids, liquids, or a mixture of the two).

The troposphere, which has an average thickness of about 9 km (5.7 miles), is a dynamic reacting aerosol. It can be visualized as a large ocean of air that holds birds, insects, pollens, spores, seeds, microorganisms, dust, and a variety of small particles, gases, and vapors. Just how many different chemicals are present in our air may never be known, but the number must be near the total number of chemicals that exist on Earth. All natural gaseous and volatile substances, and all substances in the soil, lakes, rivers and oceans, will be present. In addition, products of combustion, volcanism, and products of living things, will be there. There are so many chemical substances in the troposphere, and so many physical and chemical reactions among them that it is impossible to even envision all of them, let alone describe a significant fraction. Therefore, the scope of this chapter must be greatly simplified.

International System of Units

The scientific units of measurement accepted worldwide, *The International System of Units (SI)*, are introduced in **Exhibit 3–1**. However, because many scientific specialties still use their own time-honored units of measure, we will sometimes deviate from the SI units in this textbook.

Exhibit 3–1 The SI units.

The International System of Units (SI) are preferred for scientific usage. However, other units are in common use. The base SI units are:

- *length*—meter (m)
- *mass*—kilogram (kg)
- *time*—second (s)
- *electric current*—ampere (A)
- *temperature*—Kelvin (K, or °K)
- *amount of a substance*—mole (mol)
- *luminous intensity*—candela (cd)

Some SI-derived and special units are:

- *area*—square meter (m^2)
- *volume*—cubic meter (m^3)
- *speed*—meter per second ($m \cdot s^{-1}$, or m/s)
- *mass density*—kilogram per cubic meter ($kg \cdot m^3$, or kg/m^3)
- *specific volume*—cubic meter per kilogram (m^3/kg)
- *force*—newton (N, which = $m \cdot kg \cdot s^{-2}$)
- *luminance*—candela per square meter ($cd \cdot m^{-2}$, or cd/m^2)
- *frequency*—hertz (Hz, which = s^{-1})
- *energy, work, quantity of heat*—joule (J, which = $m^2 \cdot kg \cdot s^{-2}$)
- *electric charge, quality of electricity*—coulomb (C, which = $s \cdot A$)
- *celsius temperature*—degree celsius (°C; °C = °K – 273.15)
- *dynamic viscosity*—pascal second ($Pa \cdot s$)

Some alternate units outside of the SI system include:

• *minute*—min	(1 min = 60s)
• *hour*—h	(1 h = 60 min)
• *day*—d	(1 d = 24h)
• *liter*—L	(1 L = 10^{-3} m^3)
• *electron volt*—eV	(1 eV 1.602 × 10^{-19} J)

The foregoing is not a complete list of the SI units. For any given application, it is important to become acquainted with the units being used.

Which Particles and Gases are Important?

Because a major theme of this book is to understand the impacts of air pollution on health and the environment, several simplifications are possible. First, in order to be considered here, an air contaminant must be either (1) present in sufficient concentrations to have health effects, or (2) important to the chemistry of the atmosphere.

These criteria eliminate most substances present in only "trace" amounts (e.g., less than a microgram per m^3 of air, which is about 1 part per billion parts of air). Another simplification is to stress those substances recognized as most worthy of regulating in the interest of protecting public health and welfare, e.g., by the United States Environmental Protection Agency (U.S. EPA). With these simplifications, one can proceed to provide an introduction to the physics and chemistry of the troposphere.

As a beginning, a sample list of notable air pollutants includes:

- ozone (O_3);
- sulfur dioxide (SO_2);
- oxides of nitrogen (NO and NO_2);
- carbon monoxide (CO); and
- particulate material in three diameter classifications: (1) *coarse particles* ≤ 10 μm (PM_{10}), or also PM_{10}–$PM_{2.5}$; (2) *fine particles* ≤ 2.5 μm ($PM_{2.5}$); and (3) *ultrafine particles* ≤ 0.1 μm (UFP).

This list, along with lead (Pb), is close to the U.S. EPA's list of *criteria air pollutants* for which the National Ambient Air Quality Standards (NAAQS) have been established (**Table 3–1**). Another U.S. EPA list to consider is the hazardous air pollutants (HAPs), which are potential health and environmental hazards (**Table 3–2**). HAPs are commonly identified by their CAS numbers, which are described in **Exhibit 3–2**. Although regulated pollutants are one focus of this chapter, other particles (such as ultrafine particles and nanoparticles) and gases will be included as needed to address important issues. Also, many bioaerosols, including infectious and allergenic particles, can have serious health effects, so they will be described. Basic information on particles, gases, and vapors will be presented along with applied topics that apply to understanding the effects of air pollution. Chemical reactions (with an emphasis on photochemistry) and pollutant transport will also be described in this chapter.

Table 3–1 The 2010 criteria air pollutants for which the U.S. Environmental Protection Agency issues the National Ambient Air Quality Standards.

Pollutant	*Type**	*Standard*	*Averaging Time*
Sulfur Dioxide	Primary	0.14 ppm	24 h
	Primary	0.03 ppm	annual
	Secondary	0.5 ppm	3 h
Nitrogen Dioxide	Primary	0.053 ppm	Annual
	Primary	0.1 ppm	1 h
Ozone	Primary	0.075 ppm	8 h
	Primary	0.12 ppm	1 h
Carbon Monoxide	Primary	9 ppm	8 h
	Primary	35 ppm	1 h
Lead	Primary	0.15 μg/m³	3 mo (rolling avg).
	Primary	1.5 μg/m³	quarterly avg.
PM_{10}	Primary	150 μg/m³	24 h
$PM_{2.5}$	Primary	35 μg/m³	24 h
	Primary	15 μg/m³	annual

*Primary standards are intended to protect public health, and secondary standards are set to protect public welfare (e.g., impacts on vegetation, crops, man-made materials, ecosystems, visibility, climate, etc.). When the secondary standard is the same as the primary standard the secondary standard is not listed.

Source: http://epa.gov/air/criteria.html

II. PARTICLE BASICS

Aerosol Terminology

One way to appreciate the importance of aerosols in the atmosphere is to review the large number of English terms used to refer to various aerodisperse systems. Such terms include *air contaminants, air pollutants, Aitken nuclei, aerocolloids, aerosols, ash, bioaerosols, clouds, colloids, condensation nuclei, dispersoids, droplets, dusts, emissions, exhausts, fallout, fine particles, floculates, fly ash, fogs, fumes, granules, hazes, lapilli, mists, motes, nanoparticles, nuclei, particles, particulates, plumes, pollens, powders, smogs, smokes, soots, sprays, suspensions,* and *ultrafines*. Particle dimensions are measured in micrometers (μm) or nanometers (nm). Some of these terms are defined as follows:

- *aerosol*—(1) A disperse system in air. According to Drinker and Hatch (1936), the term *aerosol* was first introduced by Gibbs in 1924. Or (2) a relatively time-stable suspension of small liquid and/or solid particles in a gas. The diameter size range of aerosol particles is about 0.001 to 100 μm. Larger particles do not remain suspended in the air unless they have very low specific gravities (densities).
- *cloud*—A free (not spatially confined) aerosol with a definite overall shape and size. Atmospheric clouds and cigarette smoke rings are examples.
- *condensation nuclei*—Particles that serve as centers for condensation of water vapor. This condensation occurs naturally under supersaturated conditions in

Table 3–2 Selected list from a list of nearly 200 Air Toxics. Also known as hazardous air pollutants (HAPs), which are either suspected or known to cause cancer or other serious health or environmental effects by the U.S. EPA. The EPA works with state, local, and tribal governments to reduce their emissions. The CAS Number is a unique chemical substance identifier (see Exhibit 3–2).

Name	*CAS Number*	*Name*	*CAS Number*
Acetaldehyde	75070	Phenol	108952
Acrolein	107028	Phosgene	75445
Asbestos	1332214	Phosphorus	7723140
Benzene	71432	Quinone	106514
Captan	133062	Styrene	100425
Carbon disulfide	75150	Toluene	108883
Chlorine	7782505	Antimony compounds	none
Formaldehyde	50000	Arsenic compounds	none
Lindane	58899	Beryllium compounds	none
Hydrochloric acid	7647010	Chromium compounds	none
Naphthalene	91203	Lead compounds	none
Parathion	56382	Mercury compounds	none
Epichlorohydrin	106898	Nickel compounds	none
Ethyl chloride	75003	Radionuclides	none

Source: http://www.epa.gov/ttn/atw.

the relative humidity range of about 200 to 300 percent. The nuclei particles are usually in the diameter range of about 0.01 to 0.2 μm.

- *dust*—Dry particles dispersed in a gas as by mechanical disruption of a solid or powder. Windblown dust and airborne sawdust are examples. Most dust particles are greater than 1 μm in diameter.
- *fine particles*—Particles having aerodynamic equivalent diameters from about 0.1 to 2.5 μm. There are many examples, including cigarette smoke particles, internal combustion engine exhaust particles, and coal combustion particles.

Exhibit 3–2 CAS registry numbers.

> CAS (Chemical Abstracts Service) is a team of scientists and a division of the American Chemical Society that produces and provides information on chemicals. The CAS Registry lists over 48 million substances (organic and inorganic), each having a unique CAS number. Approximately 12,000 new substances are added daily. CAS numbers are used internationally by scientists as links to detailed information on each substance. The CAS website, http:/www.cas.org/ provides information on registry numbers and links to additional resources and information.

- *fume*—An agglomerated aerosol consisting of clusters of smaller primary particles. The agglomerates form by condensation and resist disruption into free individual primary particles. Welding fumes are an example. Most fume particles, aside from large agglomerates, range in diameter from 0.001 to 1 μm.
- *mist*—A liquid droplet aerosol with many particles having diameters greater than about 20 μm. The term is also used to describe all liquid aerosols, even those with particles in the submicrometer diameter range. Fine water mists resemble rain, but the droplets are smaller than rain drops. Mists also form above evaporating liquids.
- *nanoparticles*—Particles smaller than ultrafine particles. Dimensions are usually 0.001 to 0.05 μm (1–50 nm) but sometimes larger. Such particles include those produced by combustion, and those specifically generated for nanotechnology applications.
- *particle*—A small piece of matter that may or may not be suspended in a liquid or a gas. Particles are the condensed portion of aerosols; the suspending gases are the other portion.
- *particulate*—This term is usually an adjective, meaning "in the form of separate particles." The

term is also used as a noun meaning "particles" (either solid or liquid).

- *smog*—A mixture of natural and anthropogenic aerosol particles and gases found in the air in or downwind from urban or industrial centers. Smog originally meant "smoke" plus "fog," but it is now associated with air pollution in general, even when fog is not present.
- *smoke*—Any concentrated, visible aerosol formed by condensation of vapors. Smokes usually result from combustion of organic fuels and may contain a variety of solid, liquid, and gaseous components. Due to their high gas and particle concentrations, smokes may also often exhibit "cloud behavior" (i.e., they behave as an ensemble, instead of individual particles). Most smokes, such as tobacco or oil smoke, contain particles in the 0.01 to 1 μm diameter range.
- *ultrafine particles*—Particles that have diameters less than 0.1 μm. Ultrafine particles are produced by condensation of vapors, gas-gas reactions that have solid reaction products, and the decay of some radioactive gases. Such particles are too small to be seen with optical microscopes but are in the size range that can be readily seen with electron microscopes.

There is a general lack of agreement on the exact particle size ranges for the above aerosols. The various special disciplines of atmospheric chemistry, industrial hygiene, engineering, inhalation toxicology, combustion technology, and medical therapy each have their own criteria and terminology. Thus, the size ranges provided above will vary among the disciplines. These discrepancies should not detract from the overall importance of particle size and composition on human health. Several key references on aerosols are given in **Exhibit 3–3**.

Exhibit 3–3 Key references on aerosols.

> A large number of reference books on aerosols have been published. The pioneering theoretical reference is a work by the late Nicholai A. Fuchs (1964) entitled *The Mechanics of Aerosols,* which was translated from Russian into English by R. E. Daisley and Marina Fuchs and edited by C. N. Davies. A modern text by William C. Hinds (1999) entitled *Aerosol Technology: Properties, Behavior, and Measurement of Airborne Particles* includes topics of interest to those dealing with air pollution. A general reference for industrial hygienists is Vincent's *Aerosol Science for Industrial Hygienists* (Vincent, 1995). Engineers use *Smoke Dust and Haze* by S. K. Friedlander (2000) and *Atmospheric Chemistry and Physics* by J. H. Seinfeld and S. N. Pandis (2006).
>
> Bioaerosols, and the larger general topic, *aerobiology,* are both complex topics and important to the health of animals (including humans), plants, and ecosystems. Some useful reference books include *Microbial Evolution and Co-Adaptation* (IOM, 2009), *A Framework for Assessing the Health Hazard Posed by Bioaerosols* (NRC, 2008), *Bioaerosols: Assessment and Control* (Macher, 1999), *Bioaerosols* (Burge, 1995), *Bioaerosols Handbook* (Cox and Wathes, 1995), and *Emerging Infections* (Davis and Lederberg, 2001).

Particle Size

One of the first questions asked about aerosol particles is: "What size are the particles?" **Table 3–3** lists some common particles, their sizes, and potential impacts. Aerosol particles generally have dimensions, such as a diameter, in the range of 0.001 μm to 100 μm. This size range is small compared to familiar objects, as human hair diameters are about 30 to 150 μm. Only the largest aerosol particles, above about 70 μm in diameter, are visually resolvable into individual objects by the unaided healthy human eye. On the other hand, one can see the light scattered from an intense beam by individual particles less than 1 μm in diameter. When a beam of light is made visible by aerosol particles, it is called a *Tyndall beam* after John Tyndall (1820–1893) who investigated light scattering in the late 1800s.

Below a diameter of about 70 μm, microscopes are required to produce visual images of individual particles. The resolution of the optical microscope is limited by the wavelengths of visible light (400 to 700 nm), which makes optical microscopy inadequate for imaging particles under about 0.2 μm in diameter. However, intensely fluorescent particles can still be counted in ultraviolet light microscopes; such particles are seen as pinpoints of light. Electron microscopes, including both transmission and scanning types, are capable of resolving the shapes of the smallest aerosol particles. The reason for this is that the images are made by electrons that have very small wavelengths, and yes, electrons have been shown to exhibit both wave and particle properties. In theory, such microscopes should even be able to

Table 3–3 Common particles and their potential health impacts.

Source	*Particle Type/Agent*	*Typical Diameter Range (μm)*	*Potential Health Impact*
Natural	Bacteria	0.2–30	Infections
	Fern spores	20–60	Unknown
	Fungal spores	2–100	Allergic reactions
	Moss spores	6–30	Unknown
	Plant and insect bits	5–100 and up	Allergic reactions
	Pollen	10–100 and up	Allergic reactions
	Molecular clusters (gaseous ions)[a]	0.001–0.005	Centers for droplet condensation (composition specific; e.g., sulfur and H_2SO_4)
	Natural fog	2–80	Contributes to smog
	Viruses	0.01–0.45	Infections
Anthropogenic	Coal dust	3–30	Lung diseases
	Fly ash	0.5 and up	Unknown
	Metal fumes	0.01–100	Lung diseases
	Molecular clusters (gaseous ions)[a]	0.001–0.005	Centers for droplet condensation (composition specific; e.g., sulfur and H_2SO_4)
	Tobacco smoke	0.05–5	Lung and other diseases

[a] Not true particles; do not persist if uncharged.
Data from Phalen (2009).

image molecules that are much smaller than aerosol particles. However, electron microscopes have several associated artifacts that limit their use in imaging many types of samples. The high vacuum and intense heating (600°C or more) can rapidly evaporate all but the most stable particles. This and other artifacts, such as the coating of specimens by oil and grease vapors, set a practical lower limit of resolution of about 0.001 μm for electron microscopes.

In the foregoing paragraphs, the word "diameter" has been used without being defined. The reason for this apparent oversight is to establish why diameter is important before getting into some of the more abstract aspects of aerosol particle size.

Particle Aerodynamic Equivalent Diameter

The only physical object that has a single unique diameter is a smooth, rigid, perfect sphere. Other objects, including virtually all environmental aerosol particles, have several possible diameters depending on how (and where on the particle) the measurement is made. For example, the largest dimension and the smallest dimension on a particle could be measured and averaged to obtain a diameter. However, such an "average" diameter is poorly associated with its behavior in the air, or its possible health effects. Primarily, it fails to account for particle density and the surface-to-volume ratio, which are important determinants of particle behavior. In addition, many health effects are more closely related to a measure (an "equivalent diameter") that correlates with the particle's mass and its deposition efficiency in the respiratory tract upon inhalation. Such a diameter that has been defined for a group of particles is the *mass median aerodynamic diameter* (MMAD). In order to understand the MMAD, the concept of aerodynamic diameter (D_{ae}) of a single particle must first be introduced.

The aerodynamic diameter of a particle of any shape and any size is the physical diameter of a smooth rigid standard density (1 g/cm^3) spherical particle that has the same terminal settling velocity in still air (under standard laboratory conditions) as the particle in question. For spheres, the aerodynamic diameter can be calculated by standardizing the diameter to that of a sphere with a density of 1g/cm^3 (the density of a water droplet), as:

$$D_{ae} = D_p \left(\frac{\rho_p}{\rho_0} \right)^{1/2}, \qquad \text{(Eq. 3–1)}$$

where D_p is the particle diameter, ρ_p is the particle density in g/cm^3, and ρ_o is 1 g/cm^3.

As an example, the aerodynamic diameter for a spherical particle with a diameter of 1 μm and a density, ρ_p, of 2.5 g/cm^3 is:

$$D_{ae} = 1\,\mu m \left(\frac{2.5\ g/cm^3}{1\ g/cm^3} \right)^{1/2}$$

$$D_{ae} = 1\,\mu m\ (1.58)$$

$$D_{ae} = 1.6\,\mu m.$$

The terminal settling velocity of a particle released from rest into still air is rapidly reached when the particle picks up enough speed to cause the viscous drag (generated by the air on the particles) to exactly counterbalance the downward pull of gravity (with a small correction for buoyancy). The size, shape, and density of a particle all influence its D_{ae} (**Figure 3–1**). The D_{ae} has limited utility for particles with diameters under a few tenths of μm because diffusion, rather than sedimentation, can dominate their motion in still air (**Table 3–4**). Particle motion (i.e., aerosol dynamics) is covered in more depth later in this chapter.

Diameter (μm)	11.2	7.1	5.0	1.58
Density (g/cm^3)	0.2	0.5	1.0	10
Mass (10^{-12} g)	146	92.5	65.4	20.7
Term. velocity (cm/s)	0.078	0.078	0.078	0.078

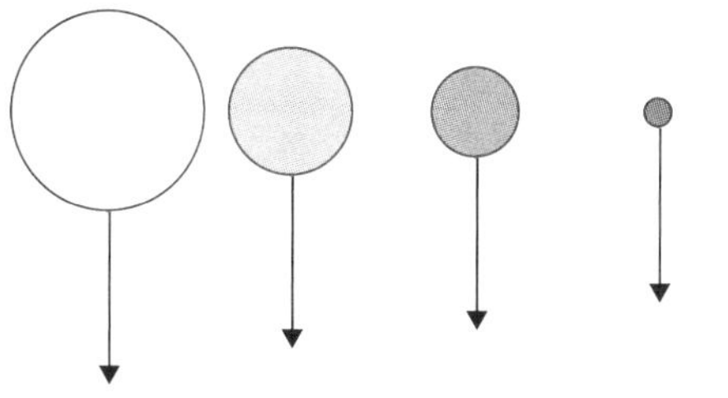

Figure 3–1 Particle aerodynamic diameter, D_{ae}, depends on the particle density, ρ_p, and particle size and shape. For solid spherical particles, the aerodynamic diameter is proportional to the particle density times the square of particle geometric diameter, D_g. V_t is the terminal settling velocity. Each particle in the Figure has the same aerodynamic diameter.
Source: The University of California Air Pollution Health Effects Laboratory, with kind permission.

Table 3–4 The effect of particle diameter of a standard density particle, 1 g/cm^3, on its slip-corrected settling velocity and diffusion displacement speed.

Diameter (μm)	*Terminal Settling Velocity (cm/s)*	*Diffusion Displacement (cm/s)*	*Slip Correction Factor (unitless)*
0.001	6.75×10^{-7}	3.26×10^{-1}	224
0.01	6.92×10^{-6}	3.30×10^{-2}	23.0
0.1	8.82×10^{-5}	3.73×10^{-3}	2.93
1.0	3.48×10^{-3}	7.40×10^{-4}	1.16
10	3.06×10^{-1}	2.20×10^{-4}	1.02
100	2.49×10^{1}	6.88×10^{-5}	1.00

D_{ae} is important in inhalation toxicology because it correlates with the probability of deposition of inhaled particles in the respiratory tract. It also correlates with the time required for a particle to settle out of still air and deposit on a surface (such as the floor of a room or the ground). The U.S. EPA uses the aerodynamic diameter for defining particle size ranges as shown in Table 3–1 and **Table 3–5**. For real-word aerosol particles, the D_{ae} is directly measured using calibrated instruments, such as cascade impactors and other aerodynamic sizing instruments (see Chapter 4 for a description of particle sizing instruments). For a collection of particles, (i.e., a distribution) the MMAD is the aerodynamic diameter of the particle for which half of the total mass is in aerodynamically smaller particles and half of the total mass is in aerodynamically larger particles. Thus, the MMAD can be used to best represent the aerodynamic diameter of the total mass of the group of particles. In toxicology, the total mass is important because mass is usually (but not always) the best measure of the effective dose.

Dose is sometimes assumed to equal the total mass deposited in the respiratory tract resulting from inhaling an aerosol. However, this rather crude measure is often further refined by dividing it by the mass of the tissue it deposits in, or some other normalizing factor (e.g., tissue surface area) meaningful for determining the biological impact (see section III of this chapter). Dose is usually expressed in terms of mass of contaminant per unit body mass or body weight, i.e., mg/kg body weight. Weight and mass are often used interchangeably, but weight requires gravity (or another acceleration), while

Table 3–5 General composition of atmospheric particulate matter in bulk (PM), coarse ($PM_{10\text{-}2.5}$), and fine ($PM_{2.5}$) modes.

PM (Total)	*$PM_{10\text{-}2.5}$ (Coarse Mode)*	*$PM_{2.5}$ (Fine Mode)*
sulfate ions	crustal calcium,	sulfate ions
nitrate ions	aluminum, silicon,	nitrate ions
ammonium ions	magnesium, and iron	ammonium ions
hydrogen ions	pollen	hydrogen ions
particle-bound water	spores	elemental carbon
elemental carbon	plant debris	secondary organics
organic compounds (a large variety)	animal debris	primary organics
crustal material	nitrate ions	certain transition metals
sea salt (in coastal areas)		crustal materials (in some regions)

Data from U.S. EPA (2004).

mass is independent of this force. The MMAD can be used to establish the effective dose.

Brownian Motion (Particle Diffusion)

The uneven bombardment on the surfaces of aerosol particles by gas molecules leads to randomly directed changes in the velocities of tiny particles. The particles are seen to jiggle when viewed using a microscope. Such movement, called *Brownian motion* (the particle motion that produces diffusion), is particularly important for particles less than about 0.5 μm in diameter. Brownian motion (named after the botanist Robert Brown, 1773–1858) produces such phenomena as coagulation, spreading of aerosol clouds, and deposition of particles on nearby surfaces including respiratory tract surfaces. The average particle velocity due to diffusion is greater for higher atmospheric temperatures and for smaller particle diameters. Table 3–4 gives the average displacement speeds (also called "root mean square" velocities) due to diffusion for a variety of particle sizes at 1 atm air pressure and 20°C temperature. The table also provides slip-correction factors used to correct drag forces for different particle sizes (see section in this chapter on *particle motion*). The net effects of diffusion and sedimentation on particle displacements in still air are shown in **Figure 3–2**. Note that the minimum in this curve is at a particle diameter of

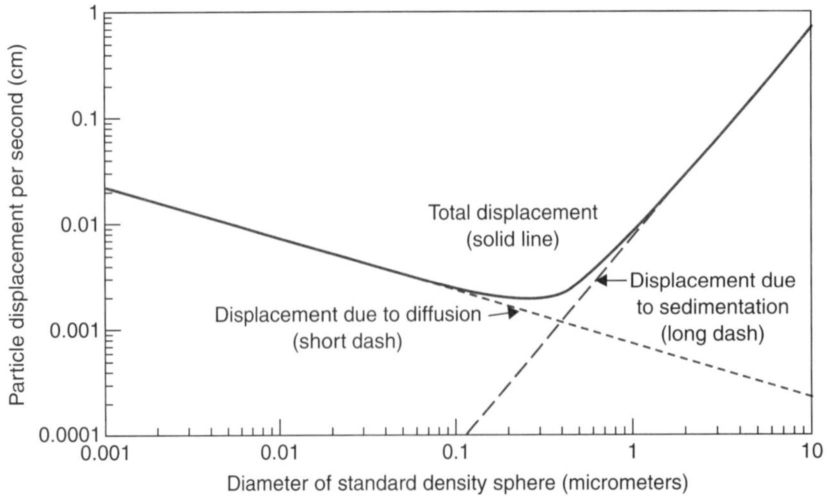

Figure 3–2 The displacement of a standard density particle in 1 second in still air. The total displacement is the sum of that produced by diffusion and by sedimentation.
Source: The University of California Air Pollution Health Effects Laboratory, with kind permission.

about 0.3 µm. Such particles, which are minimally influenced by the combined actions of molecular bombardment and gravity, have relatively long settling times in air, and also low deposition efficiencies in the respiratory tract when they are inhaled. The shape of the total displacement curve in Figure 3–2 is similar to the particle collection curves of fiber filters, sampling lines, and the mammalian respiratory tract, because deposition in these and other devices depends on particle displacement velocities. Deposition of particles occurs when they leave air streams and contact and adhere to a surface.

Distributions of Particle Sizes

Individual particles in an aerosol vary in size and shape, and this variation is represented by mathematical *size distributions.* The most familiar distribution is the bell-shaped normal curve, but a *lognormal distribution function* usually provides a better fit to aerosol samples. The lognormal distribution is skewed and best described by the *median size* and the *geometric standard deviation (GSD or* σ_g*)* (**Figure 3–3a**). These are analogous to the mean and standard deviation measures used to describe the central tendency and spread of a normal distribution. In this case, the median is the measure of central tendency and the σ_g a measure of the distribution spread. From these two values one can use an equation to reconstruct the original distribution. The lognormal distribution has many useful properties. For example, from a count distribution for simple particle shapes, the distributions of volume and surface area can be derived with the aid of some simple equations (the equations are shown later in this section). Also, lognormal distributions graph as straight lines on special, "log-probability," graph paper (**Figure 3–3b**).

For many purposes, lognormal distributions provide the best approximation or representation of commonly encountered particle size distributions. If several different types of sources contribute to generating an aerosol, then the sum of all of the lognormal distributions can be used to describe the particles.

The lognormal distribution is similar in form to the normal distribution except that the log of a particle property, such as the diameter, is normally distributed.

The equations that follow show the similar forms of the normal and the lognormal distributions when used to describe particles with varying diameter, D:

Normal Distribution $f(D)$

$$= \frac{1}{\sigma(2\pi)^{1/2}} \exp\left(-\left[D - \bar{D}\right]^2 / 2\sigma^2\right) \quad \text{(Eq. 3–2)}$$

$$\text{Lognormal Distribution } f(D) = \frac{1}{D \ln \sigma_g (2\pi)^{1/2}}$$

$$\exp\left(-\left[\ln D - \ln D_g\right]^2 / 2 \ln^2 \sigma_g\right), \quad \text{(Eq. 3–3)}$$

where $\bar{D}$ is the mean value of the particle diameter, σ the standard deviation for the normal distribution, D_g the geometric mean (or count median diameter), and σ_g the geometric standard deviation for the lognormal distribution. Two parameters, one measuring the central tendency and the other the spread of the distribution, uniquely describe each distribution.

When particle counts ($C(D)$) vs. diameter are plotted on log-probability graph paper a straight line results. The point at which $C(D)$ equals 0.5 determines the count median diameter, D_g. The geometric standard deviation σ_g is found from the values of D for $C(D)$ = 0.5, 0.16, and 0.84 by the relationships:

$$\sigma_g = \frac{D_{50}}{D_{16}} = \frac{D_{84}}{D_{50}}. \quad \text{(Eq. 3–4)}$$

For convenience, aerosol size distribution parameters are often determined from cumulative plots of sizing data on log-probability graph paper (Figure 3–3b). On this type of graph paper one axis has a log scale and the other a probit scale.

Numerical computational methods are also useful for estimating D_g and σ_g. If the particle-sizing data are organized into pairs of numbers, one the midpoint of a diameter interval D_i and the other the number of particles in that diameter interval N_i, then estimators for D_g (also known as the count median diameter, CMD) and σ_g are:

$$\ln D_g = \ln \text{CMD} = \frac{1}{N} \sum_i N_i \ln D_i, \; \textit{where, } i = 1 \rightarrow k. \quad \text{(Eq. 3–5)}$$

$$(\ln \sigma_g)^2 = \frac{1}{N-1} \sum_i N_i \left(\ln D_i - \ln \text{CMD}\right)^2, \; \textit{where, } i = 1 \rightarrow k. \quad \text{(Eq. 3–6)}$$

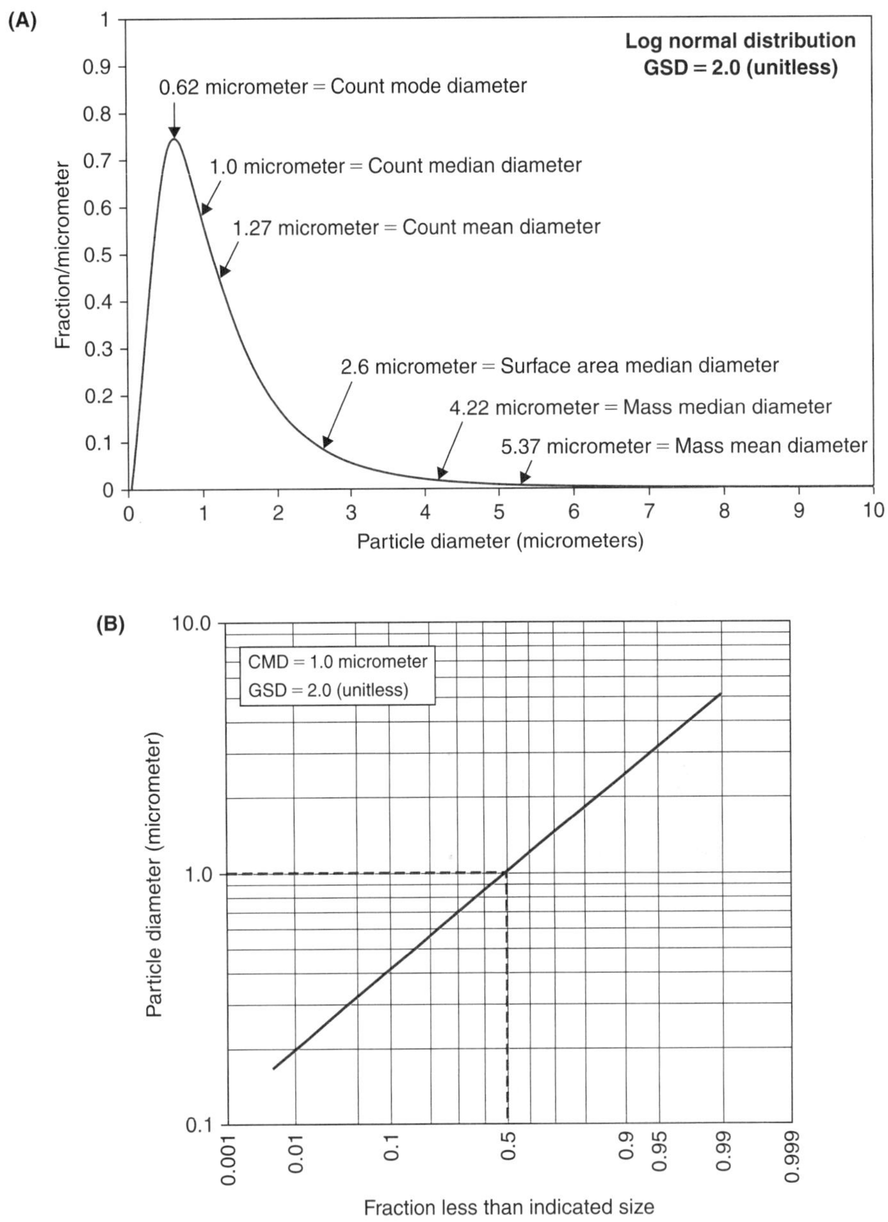

Figure 3–3 A: A normalized lognormal distribution for particles with a CMD of 1 μm and a σ_g of 2.0.; B: The use of "log-probit" graph paper to linearize a lognormal particle size distribution.
Source: The University of California Air Pollution Health Effects Laboratory, with kind permission.

And k is the number of intervals and N is the total number of particles such that:

$$N = \sum N_i. \quad \text{(Eq. 3–7)}$$

The equations of Hatch and Choate (1929) allow one to estimate the volume median diameter, VMD, and the surface median diameter, SMD, for a particle population that is log-normally distributed with a known D_g and σ_g:

$$\ln \text{VMD} = \ln \text{CMD} + 3 \ln^2 \sigma_g \quad \text{(Eq. 3–8)}$$

$$\ln \text{SMD} = \ln \text{CMD} + 2 \ln^2 \sigma_g. \quad \text{(Eq. 3–9)}$$

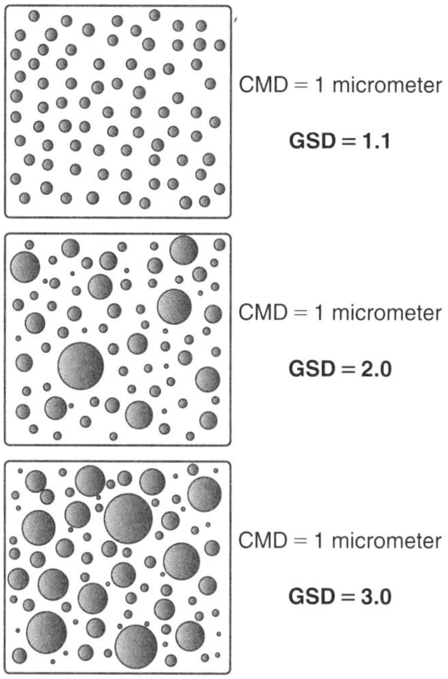

Figure 3–4 The effect of GSD (σ_g) on particles with a CMD of 1.
Source: The University of California Air Pollution Health Effects Laboratory, with kind permission.

Both the volume and surface distributions have the same σ_g as the count distribution. In many applications the VMD, CMD, and SMD are noted as D_V, D_N, and D_S, respectively.

In order to aid in understanding the geometric standard deviation, it is useful to consider some collections of hypothetical particles, each collection with the same count median diameter but having different geometric standard deviations. The collections of dark circles depicted in **Figure 3–4** represents geometric standard deviations of 1.1 (essentially monodisperse), 2.0, and 3.0 (polydisperse). These values span those typically encountered for aerosols generated in the laboratory or by environmental sources, such as coal-fired power plants, motor vehicles, and many industrial processes.

Particle Shape

Atmospheric particles exist in a large variety of shapes. The four types of shapes adequate for describing the majority of particles of interest are (1) globular (nearly spherical), (2) plate-like, (3) fibers, and (4) irregular. Some *designer nanoparticles* can deviate from these four types as they may have odd geometric shapes.

If the particle boundary extends to about the same distance along any direction, then the particle can be classified as *globular* or roughly spherical. Liquid particles are normally spherical (due to the effect of surface tension), as are many particles formed by the condensation of vapors or by the evaporation of solute containing droplets.

If one axis of the particle is much shorter than the other two, then the shape is considered *plate-like* and flat. Examples of this shape include particles of graphite, talc, dander, diatoms, and plant or insect scales. Plate-like particles are usually formed by mechanical means, such as the wear or abrasion of a surface. For example, human skin sheds plate-like particles consisting of dead skin cells.

Fibers are particles for which the length along one axis is three or more times greater than along the other two axes. Examples of fibers are asbestos, glass fibers, mineral wool, and hair fragments. These particles tend to orient in moving air with their long axis parallel to their direction of air flow. Therefore, fibers can have aerodynamic diameters more influenced by their width diameters rather than their lengths. As a result, inhaled fibers can travel and deposit more deeply in the respiratory tract than globular particles that have the same volumes. Also, long insoluble fibers resist engulfment by mobile cells (macrophages) in the lungs. Fibers can remain in the body for long periods and may even kill the macrophages that try to engulf them. The release of digestive enzymes and reactive chemicals from the macrophages into the lungs can cause diseases, such as emphysema (tissue destruction) and asbestosis (scar formation).

A fourth class of particles includes those whose shapes are *irregular*. Examples include crystalline particles of zinc oxide smoke and clusters of small primary particles including soot, fly-ash, and metal fumes. Irregular particles that consist of hundreds of small primary units are commonly encountered (**Figure 3–5**). Their shapes often approximate those of distorted spiders, having long, branched, and chain-like limbs. Such particles may have very large specific surface areas (surface area per unit mass) and low values of effective particle density (as small as 10 percent of the parent material). The large surface and low densities of branched-chain agglomerates permit them to follow airstreams and remain suspended in air for long periods of time. They can also travel deeply into the lungs

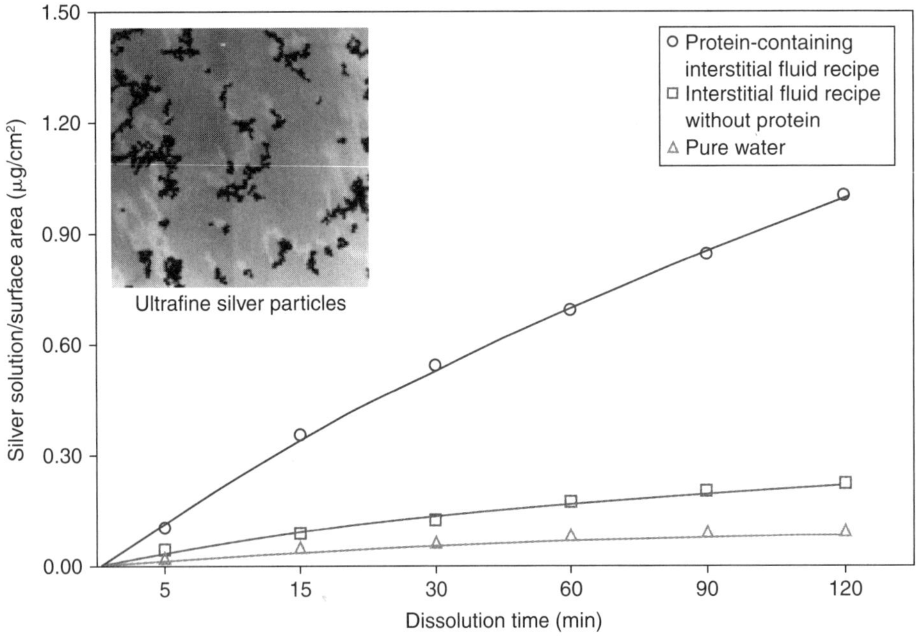

Figure 3–5 Dissolution of ultrafine silver particles shown in (inset) in various solvents.
Source: The University of California Air Pollution Health Effects Laboratory, with kind permission.

where they deposit and expose the alveolar cells (cells that line the terminal air sacs).

The persistence of such particles after depositing in the lungs is often brief since a large surface is available for dissolution. Metal fumes of silver, lead, or inhaled welding fumes may dissolve within a few hours in the lungs, and appear rapidly in the bloodstream (Figure 3–5). The distinction between *primary particles* and *clusters,* or *agglomerates,* of such particles should be noted. Agglomerates that do not dissolve are typically quite stable, being held together by electrical and molecular forces. In summary, the aerodynamic behavior and toxicity of agglomerates can differ considerably from those of their individual primary particles.

Particle Surface Area

The surface area of a particle is important both before and after it deposits in the respiratory tract. In the air, the particle's surface area relates to its interaction with the surrounding gases; the greater the surface area, the more gas may be carried on it and the more rapidly it can react chemically. If a given amount of a material is dispersed in very small particles, the total surface area will be larger than if it is dispersed in fewer larger particles. Reactions that normally proceed slowly in the air, such as oxidation, can proceed explosively when the particle size is very small. This is a fundamental reason for the extreme nature of industrial *dust explosions.* In addition, it is the reason that gasoline and diesel fuels are atomized before being ignited in the combustion chambers of engines. If such liquid fuels were not converted into a fine aerosol, they would burn slowly, which would not produce enough power to make the engines run.

A large surface area, such as exists on *activated charcoal,* can also adsorb large quantities of gases and liquids. Think of adhesion to a surface when you see the "ad" in "adsorb." One gram of activated charcoal can have a surface area of 800 m^2, which is three times the playing area of a doubles tennis court. Charcoal is activated by heating it to high temperatures to drive adsorbed gases from its surface. Therefore, activated charcoal is used in air cleaning to adsorb large quantities of gases and vapors. Similarly, other types of high surface area aerosol particles, such as diesel soot, can adsorb pollutant gases and deliver them in concentrated forms to the body upon inhalation.

The surface areas of inhaled particles can influence their fates after they deposit in the body. Substances that are quite insoluble in pure water (such as lead, silver, and other metals) can dissolve within hours or minutes in the lungs. The dissolved forms of such metals can enter the bloodstream and expose distant organs. In most cases, the amounts of such dissolved material is small and of little consequence. But highly-toxic aerosols inhaled in sufficient concentrations can dissolve and affect organs other than the lungs. A dedicated issue of the scientific journal *Inhalation Toxicology* (vol. 16, nos. 6–7, 2004) has 11 papers on the effects of particulate air pollutants on organs other than the lungs. In some of these cases it is probable that dissolved material has accumulated in distant tissues.

In summary, the following points can be made regarding particle surface area:

- Larger individual particles have larger surface areas than do individual smaller particles.
- A given amount of material divided into smaller particles has a larger total surface area than the same amount of material divided into larger particles.
- Greater surface area leads to more rapid chemical reaction rates.
- Large particulate surface areas can adsorb gases and carry them into the respiratory tract.
- Greater surface areas can cause more rapid dissolution of material deposited in the body.

Particle surface areas can be crudely estimated by performing calculations, and they can be more precisely measured in the laboratory by weighing a particle sample as it adsorbs a gas on its surface. **Tables 3–6** and **3–7** provide information on the surface areas of particles.

Table 3–6 Specific surface areas of particles of various types and sizes.

Particle	*Diameter (μm)*	*Specific Surface (m^2/g)*	*Method*
Droplets of water	100	0.06	Theoretical
	10	0.6	Theoretical
	1	6	Theoretical
	0.1	60	Theoretical
	0.01	600	Theoretical
Quartz	100	0.05	Experimental
	10	0.4	Experimental
	1	3.5	Experimental
Metal fume	0.03	16	Experimental
Charcoal	—	800	Experimental

Data from Phalen (2009).

Particle Density

Particle density, defined as the amount of mass per unit volume of a particle, influences both toxicity and particle motion in several ways. Consider two spherical particles that have the same diameter but differ in their densities. They will have different masses and different terminal settling velocities in still air. If their geometrical diameters are in the micrometer range, their settling velocities will be directly proportional to their densities (over a wide range). Conversely, if two spherical particles have the same settling velocity, the one with lower density will have the greater mass (Figure 3–1), and thus can deliver a greater amount of toxicant after it has deposited in the respiratory tract. Also, if two spherical particles have equal masses, the one with greater density will be geometrically smaller and its motion in air influenced more by Brownian bombardment.

Published values for the densities of various bulk materials found in the literature are seldom applicable to most real-world aerosol particles. Most aerosol particles

Table 3–7 Relationships between diameter, mass, surface area, surface to mass ratio, and number of particles in a one microgram sample. Applicable to smooth spheres that have a standard specific gravity of 1g/cm^3.

Diam. (μm)	*Mass (g)*	*Surface Area (cm^2)*	*Surface to Mass Ratio (cm^2/g)*	*Particle number/μg*
0.001	5.24×10^{-22}	3.14×10^{-14}	6.00×10^{7}	1.9×10^{15}
0.01	5.24×10^{-19}	3.14×10^{-12}	6.00×10^{6}	1.9×10^{12}
0.1	5.24×10^{-16}	3.14×10^{-10}	6.00×10^{5}	1.9×10^{9}
1.0	5.24×10^{-13}	3.14×10^{-8}	6.00×10^{4}	1.9×10^{6}
10.0	5.24×10^{-10}	3.14×10^{-6}	6.00×10^{3}	1.9×10^{3}
100.0	5.24×10^{-7}	3.14×10^{-4}	6.00×10^{2}	1.9

Source: The University of California Air Pollution Health Effects Laboratory, with kind permission.

Table 3–8 Aerosol particle densities and densities of bulk material.

Material	Aerosol Source	Particle Density (g/cm^3)	Bulk Density (g/cm^3)
Au	Electric arc	0.2–8	19.3
Ag	Electric arc	0.64–4.22	10.5
Hg	Heating	0.07–10.8	13.6
MgO	Burning Mg	0.24–3.48	3.6
$HgCl_2$	Heating	0.62–4.3	5.4
CdO	Electric arc	0.17–2.7	6.5

Data from Fuchs (1964).

will have *effective densities* less than the densities of their corresponding bulk materials. The reason for the lower density is due to *porosity,* i.e., the presence of voids within the particles. It is not uncommon for up to 90 percent of a particle volume to be void. **Table 3–8** lists typical density values for some aerosol particles. The determination of particle density is not a simple matter; several methods are used, including:

- Flotation tests of the particles in fluids that have different densities, both smaller and greater than the particle's density;
- The simultaneous determination of geometrical diameter and aerodynamic diameter followed by a calculation of density; and
- Calculations based on particle mass, morphology (size and shape), and surface area.

Electrical Charges on Particles

Under ordinary conditions a significant fraction of the particles in any aerosol will be electrically charged (even if an attempt has been made to discharge them). The amount of electrical charge on aerosol particles influences their behavior in several ways, including the following:

- Rates of coagulation in the air;
- Deposition rates on surfaces (including airway surfaces); and
- Motion in the air in the presence of external electromagnetic fields.

Unless the particles are highly charged, or the external field is strong, these effects are minor. However, freshly-generated aerosols can carry significant charge levels that can increase their rates of deposition on surfaces, including the respiratory tract when they are inhaled. Large particles can carry more net charge than smaller particles. Charge phenomena are used to enhance the collection efficiency of particles in industry, homes, and laboratories. Devices that exploit electrical charge in order to enhance particle collection are called *electrostatic precipitators* (see Chapters 4 and 6). In addition, some particle sizing instruments operate by first charging particles and then directing their motion within the device. Such instruments include *electrical size analyzers.*

Attraction of a Charged Particle to a Nearby Conductor

When a charged particle in free space is at a distance, *d,* away from an electrically conductive surface, the requirement that the conductor's surface have a uniform electrical potential leads to an interesting effect. Charges within the conductor redistribute to produce a counter-field that is equivalent to the electrical field that would be produced by a second oppositely-charged particle placed behind the conductive surface at a distance, *d.* This second, apparent charge is called an *image charge*. The two oppositely-charged particles then produce fields that just cancel at the surface of the conductor. Curiously, the real charged particle experiences a force, *F,* that attracts it to its image. The magnitude of this attractive force, which attracts the particle to the conductive surface, is given by *Coulomb's law,* which in cgs units is:

$$F = \frac{q^2}{d^2}, \qquad \text{(Eq. 3–10)}$$

where F is in dynes, q in statculombs, and d is in cm.

Coulomb's law can also be used to calculate attractive and/or repulsive forces between charged particles, and between a charged particle and a surface that is grounded or has a net charge with respect to the particle. Examples where Coulomb attraction is relevant include the deposition of inhaled particles and losses of particles in sampling lines and instruments.

Charge Distributions

The atmosphere has numerous gaseous ions that are largely produced by natural ionizing radiation. Radioelements in the soil and the air and cosmic rays are such natural sources. Charged aerosol particles will attract oppositely-charged ions in the air and lose their charges. Uncharged particles can likewise collect ions in the air and become charged. Under usual environmental conditions, an equilibrium state will eventually be achieved. The equilibrium charge distribution on an aerosol in the presence of uniform bipolar ions in air (i.e., equal positive and negative conductivities) is symmetrical with a central peak where the charge per particle is zero. In this case, *Boltzmann equilibrium* is said to be established, and the average number of charges, positive or negative, q (in election charge units) on individual particles is a function of particle diameter, D (in µm):

$$q = 2.37\,D^{1/2}. \qquad \text{(Eq. 3–11)}$$

Light Scattering by Particles

Aerosol particles will absorb and scatter incident light. Extinction, which is a reduction in intensity in the direction of the incident light, is due to both absorption and scattering.

Absorption converts the radiant energy into heat and other forms of potential or kinetic energy including excitation of chemical bonds. Scattering involves the re-radiation of incident light without a change in wavelength. For a homogeneous body, the intensity of the original light beam, I_o, is related to the intensity, I, behind the body by the relationship:

$$I = I_o e^{-kx}, \qquad \text{(Eq. 3–12)}$$

where k is the extinction coefficient and x the thickness of the body. The light intensity is decreased by the presence of the body. The ratio of transmitted light intensity to the original intensity, I/I_o, is called the *transmittance*. The product of the extinction coefficient and the thickness, kx, is called the *turbidity*. For several particles, N, per unit volume of air, each having a projected area, a, the area that intercepts the light beam, k, equals NaE, where E is the single particle extinction coefficient, and

$$I = I_o e^{-NaEx}. \qquad \text{(Eq. 3–13)}$$

When several particles are present in the air, such light scattering can reduce the visibility. Reductions in visibility are considered by the U.S. EPA to be a *welfare effect* (an adverse effect that is not a direct human health effect). Chapter 5 covers visibility effects and the effects of aerosols on climate, which can both be related to the interactions of sunlight with aerosol particles.

Hygroscopic Growth of Particles

Hygroscopic particles are those soluble in water and that grow in a high humidity environment. Examples include sea salt, sulfuric acid droplets, tobacco smoke, some inhaled medicinals, and several spray-generated aerosols. Since the relative humidity in the respiratory tract is high (near 100 percent) most water-soluble aerosol particles will grow when they are inhaled. Whether or not this growth will alter the deposition of these particles in the respiratory tract is an important question. The Task Group on Lung Dynamics of the International Commission on Radiological Protection recommended that the equilibrium, or final, size of hygroscopic particles be used for calculating inhalation doses. Computational schemes that take growth rate into account have also been proposed. These more sophisticated approaches appear to be warranted since many hygroscopic particles grow significantly during passage through the airways. As an example, the calculated growth-time curve for a 1 µm initial diameter dry sodium chloride particle inhaled into the respiratory tract indicates that the diameter increases at a significant rate during the first 1 to 5 seconds and reaches a final diameter of 4 µm in approximately 10 seconds. Therefore, more proximal (closer to the nose) deposition is expected due to the larger size. Many materials of interest to inhalation toxicologists are hygroscopic. However, even small quantities of non-hygroscopic substances in particles can coat the particles' surfaces and alter growth rates, so caution must be used when applying simple theoretical growth models to chemically-complex real-world aerosols. Most real-world hygroscopic particles also contain non-hygroscopic substances.

Particle Motion in the Air

An important property of aerosol particles is their prolonged suspension times in the air. In order to understand this stability and to make quantitative estimates of aerosol deposition in the respiratory tract (and in the environment), it is necessary to understand particle motion, i.e., *aerosol dynamics.* Three types of forces are used to predict the motion of aerosol particles:

- *External forces*—including gravitational, buoyant, and those produced by electrical and/or magnetic fields.
- *Resistance forces*—that arise from the surrounding gaseous medium and depend on particle size and particle velocity relative to the air.
- *Interaction forces*—among particles (e.g., electrical attraction or repulsion).

Aerosol dynamics usually considers only the first two types of forces; the third is usually ignored, or added as a correction. A brief summary is presented as follows.

Gravitational and Buoyant Forces

Gravity pulls airborne particles toward the center of mass of the Earth with a downward force, F_G, which is known as the particle weight. The weight is equal to the product of the particle mass, M_p, and the gravitational acceleration constant, g (980 cm/s^2; 0.98 m/s^2):

$$F_G = M_p g. \qquad \text{(Eq. 3–14)}$$

Since the particle mass is its volume ($\pi D_p^3/6$) times its density, ρ_p, Eq. 3–14 becomes

$$F_G = g\pi D_p^3 \rho_p / 6. \qquad \text{(Eq. 3–15)}$$

This gravitational force is offset by an opposite buoyant force, F_B. The buoyant force is an upward acting force (opposite to that of gravity) caused by the fluid pressure of the surrounding air. This force, given by Archimedes' principle, equals the weight of the fluid displaced by the particle, i.e., the particle volume times the fluid density, ρ_f:

$$F_B = g\pi D_p^3 \rho_f / 6. \qquad \text{(Eq. 3–16)}$$

The net force, F_{GB}, is simply the difference of downward gravitational and upward buoyant forces:

$$F_{GB} = g\pi D_p^3 (\rho_p - \rho_f)/6. \qquad \text{(Eq. 3–17)}$$

The magnitude of F_{GB} in air is quite small compared to F_G, so it is usually ignored. The density of most particles is more than a thousand times the density of air. In contrast, in liquids the buoyant force can more readily exceed the gravitational force, producing flotation or buoyancy.

The Resistance or Drag Force

The resistance, or drag force, is caused by the interaction of air with a moving particle. It depends on the size of the particle, D_P, its velocity relative to the air, U, and the viscosity of the air, η. Calculating this force is simple when the particle is large with respect to the molecular mean-free path length (i.e., the average distance between collisions among air molecules). In this case, the particle "experiences" the surrounding medium as a continuous fluid rather than a collection of rapidly moving individual gas molecules. Also, if the particle is solid and small enough so that its movement does not greatly disturb the air fluid, only viscous forces need be considered. When these conditions apply, one is dealing with Stokes' (or continuum) mechanics, and the drag force, F_D, is opposite to the direction of motion:

$$F_D = 3\pi\eta D_p U. \qquad \text{(Eq. 3–18)}$$

Stokes' law (Eq. 3–18) applies to particles in the air that have diameters between about 1 and 100 μm. Stokes' law can be extended downward to particles with diameters as small as about 0.05 μm by application of a correction factor (the *Cunningham slip correction).* The slip correction for a 10 μm diameter particle is 1.015, and that of a 0.1 μm diameter particle is 2.93. Some slip corrections are also shown in Table 3–4.

Terminal Settling Velocities

A particle falling due to the effect of gravity will accelerate until the air drag, F_D, increases and becomes equal to the net accelerating force, F_{GB}. At this point, the net force on the particle is zero, and the particle will continue moving downward at a constant velocity. This constant velocity, called the *terminal settling velocity,* U_t, is found by setting the gravity-buoyant force equal to the drag force at velocity, U_t, and solving for U_t:

$$F_{GB} = F_D \qquad \text{(Eq. 3–19)}$$

$$g\pi D_p^3 (\rho_p - \rho_f)/6 = 3\pi\eta D_p U_t \qquad \text{(Eq. 3–20)}$$

$$U_t = \frac{g D_p^2 (\rho_p - \rho_f)}{18\eta}. \qquad \text{(Eq. 3–21)}$$

In the Stokes' realm, the terminal settling velocities can be very small. The terminal velocity for a standard density 1 μm diameter sphere in only 35 μm/s. So, a 1 μm diameter standard density particle has a remarkably slow settling rate: 12.5 cm per hour. Such particles stay suspended for long times. In contrast, the time required for particles at rest to reach their terminal settling velocity, the *relaxation time*, is short: For a 100 μm diameter particle it is 0.03 seconds, and for a 1 μm diameter particle 3.5×10^{-6} seconds.

The strong dependence of settling velocity, U_t, on particle diameter can be seen in Table 3–4. The terminal settling velocity is also directly proportional to particle density, so one can multiply the settling velocity value of standard density spheres by the density of the given particles to estimate their settling velocities. So a 1 μm diameter lead (Pb) particle having a density of 11.3 g/cm^3 will settle in the air at a velocity of 1.4 m per hour. Even a low wind speed of 1 m/s can carry this airborne Pb particle (if released 1 m above the ground) further than 2 km before it settles to the ground. If the Pb particle was 100 μm in diameter, it would settle within about 20 cm of the source under the same wind conditions. Under these conditions the 1μm diameter standard density particle would settle out about 30 km downwind.

Coagulation

Airborne particles that contact each other in the air will usually stick together and form a single agglomerate particle. This phenomenon, called *coagulation* or *agglomeration*, leads to a decrease in the number of particles and an increase in the average particle size. Coagulation leads to the production of secondary aerosol particles that can be chemically heterogeneous.

Coagulation is caused by phenomena that produce net differences in the velocities of nearby aerosol particles. Such phenomena include diffusion (due to Brownian motion), sedimentation due to differences in particle aerodynamic diameter (leading to differences in settling velocities), velocity gradients in the air during laminar or turbulent flow, and particle motion produced by external forces (e.g., thermal, magnetic, etc.). Aerosol coagulation can lead to enhanced losses of particles from the air to nearby surfaces due to increased sedimentation. However, the losses due to diffusion will be lower. Forced coagulation, as produced by acoustic or ultrasonic vibrations, can be used to increase the collection efficiency of fine particles.

Nanoparticles

Nanoparticles are those solid particles that have one or more of their physical dimensions under 0.1 μm. Nanoparticle aerosols unquestionably predate humans, as such particles are produced by high-temperature processes and gas-to-particle reactions. What is new today is the large increase in the design, production, and use of novel nanomaterials.

Nanoparticles have some unique properties related to their small size that can be exploited to produce useful products. Nanoscale materials have a variety of potent characteristics not found at the macro- or even the micrometer scales. Examples include:

- access to microscale environments such as the interiors of living cells, organelles within cells, and pores and crevices in living and non-living materials (e.g., fabrics, plastics, and ceramics);
- precise specification of surface chemical composition;
- dense ordering of several chemical entities, which permits rapid reactions, specific detection, and control of biological distribution (e.g., targeting cancer cells);
- encapsulation, and thus protection of fragile nanosize structures including tiny mechanical devices;
- availability of large specific surface (surface to mass ratio) for adsorbing gases and controlling microscale chemistry; and
- self-assembly, based on two steps: strong affinity, and predictable structure formation (e.g., the assembly of macromolecules such as DNA and hemoglobin).

These and other properties permit the construction and use of agents previously not even imaginable. There are literally thousands of nanoparticle applications in medicine, electronics, materials science, chemistry, physics, biology, and engineering that are currently under rapid development. The following list highlights only a few in the medical arena:

- delivery of anticancer, antifungal, and antibacterial agents to targeted cells and tissues, without unintended toxicity;
- targeted delivery of labels and markers to specific cells (e.g., cancer cells) and/or disease sites in order to image their locations in the body;

- bypassing normal barriers, such as the blood-brain barrier and the vascular epithelium barrier to access organs and tissues that have previously been inaccessible to medical treatments;
- delivering tiny magnetic structures to specific cells so that subsequent displacements by variable magnetic fields can destroy the cells;
- supplementing essential biochemicals such as hormones (e.g., insulin), enzymes, cytokines (cell-signaling molecules), antioxidants, antibodies, etc., at sites where they are needed; and
- assembling biological structures such as microtubules, collagen, and membranes where they are needed.

Researchers are well aware of the potential toxicity and environmental hazards of nanoengineered particles, and several agencies and working groups have defined the research needed to reduce the potentially harmful effects to acceptable levels. The new areas of nanotechnology, nanochemistry, nanomedicine, nanoengineering, and nanotoxicology, present opportunities for scientists. Incidentally, nanotechnology can be a *green science* in that tiny quantities can be used to replace large amounts of many materials now used for several applications.

Bioaerosols

Although considerable money, time, and talent are expended on controlling chemical air pollutants, bioaerosols have the potential for far greater impacts on human death, disease, and misery. The 1918 influenza pandemic (widespread epidemic) killed over 20 million people worldwide in less than 1 year. The World Health Organization reported that 8 million new cases of clinical tuberculosis are diagnosed annually. Asthmatic attacks, measles, anthrax, and pneumonia, which can all be fatal, as well as the common cold, have bioaerosol connections. Of the more than 50 emerging infectious microorganisms identified by the Institute of Medicine, eight are clearly spread in the air. Respiratory infections (e.g., tuberculosis, pneumonia, and influenza) and measles combined are responsible for several million annual deaths worldwide. In addition, the threat of *bioterrorism,* and even the emergence of large-scale green technologies (e.g., waste recycling, water reclamation, and composting) point to the importance of understanding and controlling bioaerosols. What are bioaerosols?

Bioaerosols are defined as aerosols in which the particulate phase is either a living organism, or a piece of, or a product of a living organism.

The bioaerosols include:

- *viroids*—also called naked viruses (which are short strands of uncoated RNA);
- *viruses*—their components and metabolic products;
- *bacteria*—their components and metabolic products;
- *spores* and spore fragments;
- *yeasts*—their components and metabolic products;
- *molds*—other fungi and their components and products;
- *pollen* and pollen fragments;
- *insect* body parts and excretions;
- *plant fragments;*
- *animal dander* (dead skin cells)—hair, excretions, and secretions (e.g., saliva and mucus); and
- *biotechnology products*—such as enzymes.

From a health perspective, the important properties of bioaerosols include their aerodynamic sizes, infectivity, toxicity, and allergenicity.

The sizes of bioaerosol particles can range from a few nanometers to 100 µm or larger (Table 3–3). Size distributions can be monodisperse (e.g., spores and pollens) or polydisperse (e.g., dander and plant fragments). The shapes are similarly variable. Other important properties are their abilities to replicate and their survival in the air. As an introduction to bioaerosols, a few types will be described.

Viruses

Viruses and viroids, the simplest infectious agents, can be described as potential components of living cells. They consist of nucleic acids (DNA or RNA) usually enclosed in a protein-containing shell. The amount of lipid in the virus shell relates to its vulnerability to temperature tolerance. The influenza virus has a lipid in its coat that favors its low-temperature tolerance; hence influenza epidemics are more common in cold weather. The polio virus, which has a lipid-free coat, is associated with epidemics in warm weather. Some viruses are inactivated by desiccation, or by oxygen exposure when airborne. Viruses differ greatly in their resistance to destruction by ozone, formaldehyde, and other disinfectants. Some of the most difficult to treat diseases, such as smallpox, polio, measles, and influenza are viral. Traditional antibiotics are ineffective against viruses, so widespread immunization is necessary for disease prevention and control. Control of virus-borne illnesses falls into three general categories:

- *Immunity*—which may be naturally-acquired, or induced by vaccination;
- *Hygiene*—hand-washing, covering sneezes and coughs, and isolation of infected subjects; and
- *Engineering*—ventilation, air filtration, and disinfection.

Bacteria and Endotoxins

Bacteria are complete cells without a nuclear membrane (i.e., prokaryotic cells). Their diameters are typically 1 to 5 μm, but they can also form long chains that are simple or branched. Thus, their shapes can be spherical, globular, rod-like, spiral, chain-like, or branched. While some are infectious or toxin-producing, others are benign, or even beneficial to complex life forms (including humans). Bacteria are ever present in the troposphere and in the body.

A few of the major infection-producing bacteria that have the potential for airborne transmission include the following:

- *Mycobacterium tuberculosis*—tuberculosis;
- *Bordetella pertussis*—whooping cough;
- *Staphylococcus aureus*—wound infections;
- *Mycoplasma pneumoniae*—pneumonia;
- *Legionella pneumophila*—legionellosis;
- *Bacillus anthracis*—anthrax infections; and
- *Pseudomonas aeruginosa*—respiratory tract infections.

Control techniques for infectious bacteria include those used for viruses, plus the use of antibiotics to kill infectious cells in infected subjects and on surfaces.

Endotoxins, which are lipid components of bacterial cell walls, are commonly found in air samples, and they are associated primarily with acute health problems (e.g., asthma attacks, fatigue, and mucous membrane irritation). They may also have chronic effects on health. Massive airborne exposures, as occur in cotton mills and around the agricultural use of fecal materials, can even be life threatening to healthy people. Endotoxins have some beneficial properties as well. They can stimulate the immune system (i.e., have *adjuvant* properties), and they have anti-tumor and anti-cancer activity. Thus, they are valuable in biomedical research and in some medical treatments.

Fungi

Fungi are neither plants nor animals. They may be microscopic (e.g., yeasts, molds, and mildews) or macroscopic (e.g., mushrooms, puffballs and bracket fungi). They may be single-celled or multi-celled, often with an extensive thread-like structure (hyphae). Fungal cells have thick walls (mainly consisting of polysaccharides) and membrane-enclosed cell nuclei (i.e., eukaryotic cells). Fungi primarily reproduce through spores that have diameters ranging from under 2 μm to over 100 μm. Most spores are dispersed by air currents, but some are spread by animals (including insects) or by raindrops and water flows.

Fungi are chemically active, secreting digestive enzymes that prepare their food for absorption, along with a large number of other secreted and excreted metabolic products. The self-protective metabolic products of fungi include antibiotics, poisons, and odorants. Many of these products are antigens that are capable of producing allergies. Among the thousands of fungal products, some are potent liver, kidney, gastrointestinal, and nervous system toxins, and several are carcinogens. Fungi can infect their hosts, which include humans (e.g., athlete's foot, and lethal aspergillosis), animals, plants, other fungi, and microorganisms.

Fungal spores are ubiquitous in outdoor air, and sometimes found in substantial concentrations indoors, especially in the presence of excess water or humidity. Water-soaked building materials, furniture, and carpeting as well as indoor potted plant surfaces and damp soils can produce large numbers of airborne fungal spores. Hospitals and some homes can require fungal control to protect the health of occupants.

Indoor control measures for fungi include:

- adjusting environmental conditions, especially keeping surfaces dry and maintaining low air humidity (< 60 percent relative humidity);
- increasing ventilation to dilute and remove spores and other fungal products;
- filtering the air to remove fungal products;
- removing damp/wet substrates such as plants, soils, deteriorating wood, cloth, and other materials that can be colonized;
- applying fungicides or sanitizing surfaces (e.g., bleach is often used); and
- removing fungi by physical means.

As a cautionary note, remedial actions that disturb fungi can temporarily produce high airborne fungal concentrations. Therefore, use of effective masks and ventilation must be considered during removal and treatments.

Review of Particle Basics

The following is a brief summary of the important points to consider regarding particle morphology and characteristics:

- Aerosol particles are conveniently described by "equivalent" diameters that relate to their behavior in the air. An important equivalent size, the aerodynamic diameter (D_{ae}), is used to describe the rate at which particles settle out of the air, and the efficiencies with which they deposit in the respiratory tract when inhaled.
- Particle size distributions (mass, diameter, and surface area) typically follow a log-normal distribution that is best approximated by a median diameter and a geometric standard deviation.
- Particle shapes vary and are classified as (1) globular, (2) plate-like, (3) fibers, or (4) irregular.
- Additional particle properties of significance in air pollution science include composition, density, surface area, electrical charge, and hygroscopicity.
- Particle motion in air is governed primarily by (1) gravitational and buoyant forces, (2) drag forces, (3) terminal settling velocities, and (4) interactions among particles (e.g., coagulation).
- Bioaerosols and engineered nanoparticles pose special challenges in the recognition, evaluation, and control of particle air pollutants.

III. PARTICLE MORPHOLOGY AND TOXICITY

Overview

Independent of composition, particle size is an important modifier of toxicity. The particle's mass, aerodynamic diameter, surface area, physical size, and state of agglomeration must all be considered when estimating the potential effects on health.

Particle Mass

As Paracelsus pronounced (see Chapter 1), the health effects of all substances are dose dependent, i.e., the magnitude and type of response are associated with a greater dose.

Dose is commonly expressed in terms of the amount of a given substance present in a given mass of target tissue. The target tissue may be an organ, a tissue within an organ, or even a specific cell type in a tissue:

$$\text{Dose} = \frac{\text{amount of toxicant}}{\text{mass of target tissue}}. \quad \text{(Eq. 3–22)}$$

Although the mass of a toxicant is usually the appropriate measure of particulate dose, this is not always true. The particle number, surface area, chemistry, or some properties other than mass have also been shown to drive the effects. A list of some of the potentially toxic characteristics (other than mass) of particulate urban air pollution is presented in **Table 3–9**. Ultrafine particles in particular can be more toxic per unit of inhaled deposited mass than are larger particles. Therefore, the simple definition of dose (Eq. 3–22) must be modified for many types of particles (see Chapter 9).

Fibers and Metal Fumes

Fibers have aerodynamic diameters nearly independent of their length up to a length-to-diameter ratio of about 20. This influences the deposition of inhaled fibers. For this reason, asbestos fibers with significant mass can behave aerodynamically similar to smaller particles, and thus deposit deeply in the respiratory tract. Also, if the particles are insoluble they may not be efficiently removed by phagocytic cells in the lungs. When long insoluble fibers are partly ingested by macrophages they can kill the cells by protruding outside of the cell walls. These effects on lung cells are emphasized by the cases of pulmonary disease in individuals who inhale large amounts of toxic forms of asbestos. A proposed mechanism of asbestosis-induced lung fibrosis (a widespread buildup of scar tissue) is that killed cells release toxic chemicals that stimulate a local, often progressive, tissue reaction.

An interesting aerosol in inhalation toxicology, the *metal fume,* consists of chain-like clusters (agglomerates) of large numbers of primary particles smaller than 0.1 µm (Figure 3–5). The agglomerates can have hundreds, or even thousands, of small particles. Pulmonary injury is known to be associated with the inhalation of

Table 3–9 Some properties of inhaled urban particulate matter, other than particle mass, that have been proposed as driving the health impact.

Property	*Mode of Injury*	*Pollutant Sources*
Oxidative capacity	Direct cell damage	Combustion, soil, industrial processes, photochemistry
Silica or asbestos content	Inflammation, macrophage toxicity, cell injury, cell death, fibrosis	Soil, brake lining, insulation, sandblasting, ore processing
Infectivity, allergenicity	Inflammation, cell killing, bronchoconstriction, immunologic cascade	Viruses, bacteria, pollen, spores, plants, animals
Acidity	Mucus changes, impaired particle clearance, cell killing, inflammation, bronchoconstriction	Sulfur containing fuel combustion, industrial processes, photochemistry
Ultrafine size	Inflammation, tissue permeability, cell damage	Combustion, high temperature processes, industrial processes
Metallic content	Catalytic effects, inflammation, cell killing/injury	Soil, fuel combustion, industrial processes, welding

Data from Phalen (2002).

significant quantities of agglomerated metal fumes (as by welders). The large surface area associated with a given mass of metal fume aerosol has interesting effects. The aerodynamic drag on large fume particles can be significant, which allows them to follow airstreams and escape impaction in the nose, mouth, and other portions of the upper respiratory tract. As with asbestos fibers, the metal fume particles can penetrate to the deep lung and deposit in the alveoli. Once deposited, they can produce inflammation in the deep lung. This undoubtedly contributes to their hazard, which includes the sometimes fatal condition in welders called *metal fume fever.* Fortunately, most patients fully recover without serious complications beyond flu-like symptoms. Dose is a primary contributing factor in the severity of the illness.

Surface Area and Dissolution Rate

Inhalation toxicologists define two categories of particulate materials: those that must dissolve in order to produce harmful effects and those that can be harmful even if they do not dissolve and enter the blood. When inhaled, materials such as asbestos and quartz are hazardous because of their shape or surface characteristics. Other materials, for example Pb and Mn, are known to require dissolution in order to be toxic. For both soluble and insoluble particulate materials, the *specific surface,* or surface-to-mass ratio, will affect their toxicity. The surface-to-mass ratio for smooth spherical particles is given by the equation, $6/\rho D$. In this relationship, ρ is the physical density and D is the geometric diameter. Standard density particles of 1 μm diameter have a specific surface of 6 m^2/g, while 0.01 μm diameter particles have a larger specific surface of 600 m^2/g. As an example, finely-divided silica has increased toxicity, which may relate to increased surface area. The mechanism for this toxicity has been attributed to a tissue contact reaction due to the particle surface character. Research on ultrafine particles indicates that the particle surface area is an important factor in their toxicity. Surface characteristics are more important than particle mass in these cases.

The rate of dissolution of particles deposited in the lungs has been mathematically modeled. The model proposed by Thomas Mercer (1967) correlates well with experimental data on the persistence of particles deposited in the deep lung. Mercer's model assumes that the rate of dissolution of a particle is proportional to the available surface area of the particle. For materials that are toxic only when they dissolve, increased surface area will enhance their toxicity. As an example, silver particles (mass median diameter = 0.04 μm) studied in various aqueous media show that even a so-called *insoluble* material can rapidly dissolve in body fluids when in a finely divided state (Figure 3–5). Using the dissolution rate of the silver particles in protein-containing fluid, Mercer's model predicts that they should essentially completely dissolve in the lungs in about two days. Proteins attach to dissolved silver atoms and prevent them from re-condensing on the particles.

Other Size Dependent Factors and Toxicity

Particle size influences the magnitude and distribution of dose, deposition pattern, and the dissolution rate of particles in the body. But other size-related factors also bear on the toxicity of inhaled particles. An optimal particle diameter of 1.5 μm for efficient uptake of tiny spheres by macrophages was introduced by Bo Holma (1967). Holma suggested an upper particle diameter limit of 8 μm for phagocytic uptake. The size range for efficient uptake is the size range of bacteria. Wolfgang Kreyling et al. (2006) later provided evidence that ultrafine particles smaller than 0.1 μm in diameter are not efficiently phagocytized by lung cells. This effect is most likely due to their small chemical and/or physical signal. The macrophages must have a particle of a certain size in order for it to be recognized and thus initiate engulfment. Because macrophages are geared to engulf and destroy bacteria that have diameters in the range of a few μm, they appear to be most efficient for attacking particles in this range.

The permeability of alveolar membranes to penetration by bare particles was described by Gross and Westrick in 1954 and by Tucker et al. in 1973. In the Gross and Westrick study, rats were exposed to carbon particles (< 0.2 μm in diameter) by intratracheal injection (direct placement in the trachea). The particles were found in extracellular interstitial spaces (spaces between cells) 19 hours later. The authors considered this to be proof of the penetration of lung membranes by intact particles. In Tucker's experiments, carmine particles (ground-up insect shells) ranging from about 5 μm down to below 0.05 μm in diameter were inhaled by rats. Three hours after the inhalation, microscopic examination revealed "small aggregates, up to cell size" in the interstitial spaces (spaces outside of cells). Particles found in these spaces could (1) remain intact, (2) dissolve, (3) undergo transport to the lymph fluid or blood, or (4) be moved up the respiratory airway via mucocilliary transport (by mucus and cilia) to subsequently be swallowed. The role of particle size on such membrane penetration is important for understanding their distribution beyond the respiratory tract, and thus their potential toxicity.

As a result of the foregoing information on particle size and toxicity, it is useful to examine Tables 3–6 and 3–7. Values for particle mass, surface area, surface to mass ratio, and number of particles in a microgram sample are shown for unit specific gravity spheres in Table 3–7. Potentially toxic characteristics of particles are shown in Table 3–9. Note that particle mass varies over 15 orders of magnitude as the diameter varies from 0.001 to 100 μm! Also, note the large number of particle attributes that must be considered by toxicologists. Although much is known about how particle size and other properties affect their toxicity, there is still much to be discovered.

IV. GASES AND VAPORS

What Are Gases and Vapors?

A given chemical substance can exist in several physical forms called *states of matter.* The commonly recognized states of matter include the following:

- *solids*—Solids have a definite shape and volume, and may have either a crystalline or amorphous structure. Diamonds and table salt are crystalline, and ordinary glass and plastics are amorphous. Elemental carbon (EC) has three forms: crystalline diamonds, crystalline graphite, and amorphous graphite. Ionic bonds provide rigidity to solids. Metals are solids that have freely-shared valence electrons, which make them good conductors of electricity.
- *liquids*—Liquids have a definite volume but variable shape, depending on the shape of their containment. When unconfined, or in a microgravity environment, liquids tend to take spherical shapes due to their surface tension. Liquids are held together by intermolecular forces.
- *gases*—Gases are free molecular dispersions that fill a space in which they are confined. When unconfined, their molecules diffuse and may adhere to bodies, solid or liquid, that they encounter. Gases can dissolve in solids and liquids. If there are no forces between the gas molecules (or atoms), the dispersion is called an *ideal gas.* In reality, some forces will exist between the molecules of a gas, but such forces become negligible as the gas expands. Real, i.e., *non-ideal gases,* are sufficiently concentrated such that intermolecular forces also influence their behavior (e.g., produce deviations in Boyle's law).

- *other states—Ions* are molecules or atoms that have gained or lost electrons, or are free charges (such as electrons). Their behaviors are influenced by their electrical charge. *Plasmas* are collections of ions and electrons (usually at very high temperatures) in which the motions of their individual components are dominated by electromagnetic interactions.

Most, but not all substances, can exist in solid, liquid, or gaseous states. For example, at a pressure of 1 atmosphere, water is a solid at or below 0°C, a liquid from zero to 100°C, and a gas from 100 to over 2,000°C (at very high temperatures, water decomposes into hydrogen and oxygen atoms). Some substances, such as paper, wood, and coal decompose at a given temperature before they change to liquid or gaseous states.

Vapors, which are free molecular dispersions, are the gaseous phases of substances that are normally liquids or solids at normal laboratory temperature and pressure (e.g., 25°C and 1 atm). As a result, vapors coexist with their condensed phase sources. One common example is water vapor, but many other substances also produce vapors. The *vapor pressure* is the pressure of the vapor in equilibrium with the parent liquid or solid. When the pressure of the vapor has reached the equilibrium point, no more net evaporation occurs to produce more vapor. It is important to keep in mind that the vapor pressure of a liquid increases exponentially with temperature, and that not all vapors behave as ideal gases at even moderate concentrations above the ppm range. However, most environmental air pollution occurs at significantly lower concentrations. To simplify, a vapor can usually be considered to be no different than a gas.

Ideal Gas Laws

An ideal gas is one in which the molecules do not attract one another and the collisions are elastic. Relationships that describe ideal gases are:

$$\text{Boyle's law: } \frac{P_1}{P_2} = \frac{V_2}{V_1}, \text{ at constant } T \qquad \text{(Eq. 3–23)}$$

$$\text{Charles' law: } \frac{V_1}{T_1} = \frac{V_2}{T_2}, \text{ at constant } P \qquad \text{(Eq. 3–24)}$$

$$\text{Ideal gas law: } PV = nRT, \qquad \text{(Eq. 3–25)}$$

where P is the pressure (kilopascal, kPa), V the volume (L), T the absolute temperature (K), n the number of moles, and R is the gas constant, (8.31 LkPa/Kmol). The foregoing units are SI (International System of Units), but alternates, such as atmospheres (atm, for pressure) and liters (L, for volume) are sometimes used. Dilute gases follow these ideal gas laws but concentrated gases do not.

Vapor Pressure

Vapor pressures for a few liquids at various temperatures are shown in **Table 3–10**. Liquids with higher vapor pressures evaporate more rapidly at a given temperature than those with lower vapor pressures. When the vapor pressure of a liquid at normal laboratory pressure reaches 760 mm Hg (1 atm), the liquid will boil to produce a gas.

Partial Pressure

The *partial pressure* of a gas in the atmosphere can be used to express its concentration relative to other gases that are present. The partial pressure of a gas relates to its behavior in the atmosphere, to its rate of dissolution in liquids, and to its condensation on surfaces. The *partial pressure* of a gas dissolved in a liquid (e.g., oxygen in water) is used to predict its behavior with respect to entering or leaving the liquid, as well as its potential importance to the liquid chemistry and toxicity. Thus, the interest in gas partial pressures is driven by its roles in chemical and biological phenomena.

Henry Cavendish (1731–1810) and Sir John Dalton (1766–1844) each performed early work on respiration

Table 3–10 Vapor pressures in mm Hg for various liquids as a function of temperature.

°C	*Water*	*Carbon Tetrachloride*	*Ethyl Ether*	*Ethyl Alcohol*	*n-Octane*
0	4.58	33	185	12	3
20	17.5	91	442	44	10
40	55.3	216	921	135	31
60	149	451	g	353	78
80	355	843	g	813	175
100	760	1,463	g	g	354

Notes: 1. To convert mm Hg to Pa (pascal) multiply by 133.32.
2. g = gas

Data from Barrow (1966).

and the composition of air. Their efforts led to *Dalton's Law* of partial pressures, which is stated thusly:

> The total pressure exerted by a mixture of gases in a vessel is equal to the sum of the individual pressures that each gas exerts if it is placed individually into the containing vessel.

So, for several coexisting ideal gases, 1, 2, 3, etc., the total pressure, *P*, is:

$$P = P_1 + P_2 + P_3 + \ldots P_n. \quad \text{(Eq. 3–26)}$$

This simple empirical law, which holds for non-reacting gases, applies to gases dissolved in the air or in liquids. Stated another way, where n_i is the number of moles of gas, i, Dalton's law is:

$$P = (n_1 + n_2 + n_3 + \ldots n_i)\frac{RT}{V}. \quad \text{(Eq. 3–27)}$$

In air at standard conditions the total pressure, $P =$ 101.3 kilopascal (1 atm).

Similarly, when gases are dissolved in a liquid, e.g., water, the equilibrium partial pressure of each in the liquid is defined thusly.

The partial pressure of the dissolved gas (in a closed container that holds the liquid and the gas above the liquid) is equal to the partial pressure of the gas above the liquid when equilibrium has been reached.

When the net movement of gas is neither entering nor exiting the liquid (the definition of equilibrium), its partial pressure in the liquid is equal to the partial pressure above the liquid. If the partial pressures are unequal (the non-equilibrium condition) the gas will either move into or out of the liquid until equilibrium is reached. When equilibrium is reached, the gas has reached *saturation* for the existing conditions in the liquid. As one might speculate, the amount of gas in the liquid at saturation is related to its solubility in the liquid. Gases highly soluble in water include carbon dioxide, ammonia, and sulfur dioxide. Gases poorly soluble in water include nitrogen, helium, carbon monoxide, oxygen, and ozone.

A non-reacting gas will diffuse from a region of higher concentration (or higher partial pressure) to a region of lower concentration (or lower partial pressure). This process has important implications for the movement of gaseous air contaminants into and out of water droplets in the air. It is also important in relation to the uptake of gases by the body upon inhalation and the movement of dissolved gases within the body. For example, oxygen has a partial pressure of 152 mm Hg in the atmosphere, 105 mm Hg in the alveoli, 90 mm Hg in the blood, and 0 mm Hg in the tissues. Thus, oxygen moves from the air into the tissues.

Physiologic Implications of Gas Partitioning

Thus far, the focus of gas solubility has been in pure water. Although the mammalian body is mostly water, many tissues also contain other substances, such as *lipids* (i.e., fats). In the same way that the gases have equilibrium partition concentration ratios in air vs. water, they also have equilibrium concentration ratios in water vs. lipids. That is, some gases are more soluble in water than in lipids, and for some it is the other way around. **Table 3–11** shows some equilibrium distribution ratios, also called *partition coefficients* in air vs. water and in water vs. oil (oil represents body lipids). Because the brain has high lipid content (in the insulating myelin sheaths around nerves), it avidly absorbs gases with large oil/water partition coefficients (such as nitrogen and chloroform) from the blood. Incidentally, nitrogen at high pressure, as in deep-sea diving, is readily taken up by the brain. If the partial pressure of nitrogen is 4 to 5 atm, it becomes an anesthetic that can produce disorientation and death. Substitution of helium for nitrogen in diving air can prevent this toxicity.

When the air concentration of an inhaled gas drops to zero, the partial pressure gradients are reversed and the gas begins to desorb from the body. The rates of accumulation and desorption of inhaled gases in specific organs

Table 3–11 Equilibrium distribution coefficients for some non-reacting gases in water/air and water/oil. Biologically, water represents blood serum and other body liquids, and oil represents lipids in blood, cells, tissues, and organs.

		Partition Coefficients	
Substance	*MW*	*Water/Air*	*Oil/Water*
Helium	4.0	0.0097	1.7
Nitrogen	28.0	0.0144	5.2
Ethylene	28.1	0.089	14.4
Chloroform	119.4	4.6	110
Ethyl ether	74.1	15.5	3.2

Data from Phalen (2009).

depends on (1) the gas partition coefficients; (2) the blood flow to the organ; (3) the makeup of the organ (e.g., amount of lipid and water); and (4) metabolic and other factors. Note that the body's fat storage deposits are poorly perfused by blood, so accumulated fat-soluble substances may require long elimination times to deplete storage in fat deposits after exposure ceases. However, the brain, kidneys, and liver are well perfused with blood, so the wash-in and wash-out rates of dissolved gases are rapid. Toxicologists are interested in the partition coefficients of air pollutant gases in tissues of the body, along with associated blood flows.

Inhaled Gases

In order to simplify and focus on basic principles, what follows will primarily pertain to gases that do not avidly react with biological tissue. The properties of molecular weight and relative equilibrium concentrations in air, water, and oil media will largely account for the behavior of these gases in the body after inhalation.

The movement of a gas from air into tissue involves a series of steps. Inhalation initially brings the gas into the respiratory tract. Gases very soluble in water will be avidly absorbed into the wet lining layer of the nose, mouth, and major airways. Gases not trapped in the upper airways will then flow into the small airways, and some will diffuse into the alveoli deep in the lungs. This final diffusion step is driven by concentration gradients. If the atmospheric partial pressure of a gas is greater than that in the alveoli, the gas will undergo net movement into the alveolar region. In the alveolus, the gas will move toward equilibration with the cellular tissue and blood. The initial rate of approach toward equilibrium is dependent upon the appropriate partition coefficients that are the ratios of volumes of gas in air to those dissolved in tissues at equilibrium (Table 3–11):

$$\text{Partition Coefficient} = \text{Distribution Coefficient} = \frac{\text{Volume of gas in medium A}}{\text{Volume of gas in medium B}} \text{(at equilibrium).} \quad \text{(Eq. 3–28)}$$

In the case of the alveolus, medium A can be thought of as air and medium B as water (representing tissue). Dissolved gas will distribute within the nearby tissue, including the blood, and will then partition among aqueous and non-aqueous (e.g., fat or lipid) tissue compartments. Gas dissolved in the blood will rapidly distribute throughout the body and deposit in various organs according to their relative blood flows and their compositions.

If a subject breathes a gas at a fixed concentration in air for a sufficient time, their tissues will effectively reach equilibrium, and the rate of movement of molecules into and out of any tissue compartment will balance. This assumes that the gas does not react or chemically bind to any tissue components, with simple dissolution being the only mechanism for movement.

Sophisticated physiologically-based pharmacokinetic (PBPK) models are available for investigating inhaled gases and vapors in various mammalian species.

Expressing Gas Concentrations

The concentration of a gas in air can be expressed many ways including the following:

- percent composition, parts per million (ppm), parts per billion (ppb), etc.;
- $\mu g/m^3$ micrograms per cubic meter of air; and
- partial pressure in mm of Hg or other pressure units.

In most toxicologic research, gas concentrations are expressed as a ratio or percent volume of the gas per volume of air (v/v), or on a weight to volume basis. In either case, unless the atmospheric pressure is stated, this means of expressing concentration is problematic. Recall the previous nitrogen in diving air example. At 79 percent (790,000 ppm), this component of air is without significant harmful effects. However, at elevated pressures of 4 to 5 atm, as occur in deep-sea diving, 79 percent nitrogen becomes an anesthetic gas that can produce disorientation and death. Although this is an extreme example, reduced or elevated atmospheric pressures can alter the driving force for uptake of gas molecules by tissues. Thus, use of volume percent, mass percent, or related units should be used with caution. The direct use of toxicity data obtained at a given pressure to environmental conditions with significantly different pressures should generally be avoided.

Use of the weight of a gas per unit volume of air (w/v), e.g., mg/m^3 to express concentration is reasonable in that this unit is proportional to the number of molecules present regardless of the ambient pressure. Similarly, the partial pressure of a gas relates to the number of molecules present per unit volume of air and to the driving pressure for uptake into tissues. It is frequently useful to convert gas concentrations from ppm (v/v) to mg/m^3(w/v) and vice versa. Assuming that the ideal gas law is sufficiently accurate under the prevailing conditions, the conversion factors in Table 2–2 can be used. For convenience, several conversion factors are listed in the table.

Gas Solubility and the Role of Particles in Transporting Inhaled Gases

As previously mentioned, the degree of solubility of a gas in water is an important determinant of where the gas will deposit within the respiratory system. An important exception to this generalization occurs in the presence of significant amounts of aerosol particles. Liquid droplets can carry dissolved gases into the respiratory tract and thus increase exposure of the deep lung to materials that would normally be trapped in the upper airways. Even dry particles that have large surface areas may carry significant amounts of gases adsorbed on their surfaces.

V. IMPORTANT PHOTOCHEMICAL AND OTHER REACTIONS

Gas Spectroscopy and Photochemistry

Absorption of Radiant Energy

Visible light and adjacent portions of the electromagnetic spectrum, e.g., ultraviolet (UV) and infrared emissions, can be absorbed by atmospheric gases. The absorption of such radiant energy drives important chemical reactions. It is also the basis of spectroscopy, a tool for detecting atmospheric chemical components. Spectroscopy involves measuring the forms and concentrations of gases by analyzing their electromagnetic wave absorption characteristics. Spectroscopy is one of the important analytical tools in atmospheric chemistry.

When gases absorb solar electromagnetic energy (especially UV wavelengths) they can enter into reactions that would not otherwise occur. Such reactions are the basis of photochemistry. Photochemistry and spectroscopy are related, in that it is the absorption of electromagnetic radiation that underlies both.

The Beer–Lambert law describes the absorption of electromagnetic radiation by molecules, including atmospheric gases. The incident energy is I_o and the energy that passes unabsorbed through the absorber is I:

$$\frac{I}{I_o} = e^{-\sigma N l} \text{ or, } \ln\left(\frac{I}{I_o}\right) = -\sigma N l. \qquad \text{(Eq. 3–29)}$$

The units commonly used by gas spectroscopists are I and I_o, light intensity in quanta or energy per unit time; σ, the gas phase absorption coefficient (also called the absorption cross section), in cm^2 $molecule^{-1}$; N, the number of molecules per cm^3, in cm^{-3}; and l, the path length in cm. However, other self-consistent units can be used.

Quantization of Absorbed Energy

When radiant energy is absorbed by a gas molecule, the added energy can create several "excitation" forms including (1) vibrational, (2) rotational, and (3) electronic transitions. Vibration of the bonds between atoms is quantized, i.e., only specific vibrational energies are permitted. Not all molecules can convert radiant energy to rotational energy. For those that can, the rotational energy states are also quantized, in that only specific energies are permitted. Electronic transitions relate to changes in the state of orbital electrons in the molecule. Such states are also quantized.

The quantization of absorbed energy means that only specific wavelengths of the incident radiant energy can be absorbed; other wavelengths will pass through the gas without being absorbed. Because different molecules have different permitted energy states, the *absorption spectrum* is unique to each type of gas molecule. Trained spectroscopists can use absorption spectra to measure the molecules present, their energy states, and their concentrations in the air. Incidentally, it is the absorption of visible wavelengths that give gases their colors.

Photochemistry

Photochemistry in the troposphere involves the chemical reactions produced by solar radiant energy in the lower portion of the Earth's atmosphere. However, important photochemical processes, notably those related to ozone, also occur in the stratosphere. The conversion of radiant energy into molecular energy, which was described above in relation to spectroscopy, also drives chemical and physical processes in the atmosphere. As one might expect, such reactions are responsible for many of the differences between the chemical compositions of air pollutants in day vs. night air.

Photochemistry of Ozone and Nitrogen Dioxide

Among the numerous reactions driven by solar radiant energy, the production of ozone (O_3) and nitrogen dioxide (NO_2) are perhaps the most significant with respect to the effects of air pollution on human health and plant life. Free radical production and reactions involving formaldehyde (HCHO), peroxyacetyl nitrate (PAN), and hydrogen peroxide (H_2O_2) are also notable examples. Some of these and other reactions occur in water droplets.

When solar electromagnetic energy is involved in a chemical formula, it is indicated by hυ. An excited molecule is indicated by a superscript "*."

The photoactivation of NO_2 can lead to several processes including photodissociation, production of NO_3 and singlet oxygen (O), and fluorescence (the prompt emission of energy by an excited atom or molecule):

$$NO_2 + h\upsilon \rightarrow NO_2^* \quad \text{(Eq. 3–30)}$$

$$NO_2^* \rightarrow NO + O \quad \text{(Eq. 3–31)}$$

$$NO_2^* + O_2 \rightarrow NO_3 + O \quad \text{(Eq. 3–32)}$$

$$NO_2^* \rightarrow NO_2 + h\upsilon. \quad \text{(Eq. 3–33)}$$

These reactions have been simplified in that many intermediate chemical forms and third body interactions are not shown. Intermediates are often short-lived chemical species, and third bodies are molecules that properly position the reactants, but that are not transformed themselves (e.g., a catalyst).

If only NO_2 and O_2 are present (as in a laboratory experiment), the photochemical formation and depletion of ozone involves atomic oxygen and NO_2:

$$NO_2 + h\upsilon \rightarrow NO + O \quad \text{(Eq. 3–34)}$$

$$O + O_2 \rightarrow O_3 \quad \text{(Eq. 3–35)}$$

$$O_3 + NO \rightarrow NO_2 + O_2. \quad \text{(Eq. 3–36)}$$

Note that the products of these reactions are also the starting gases (NO_2 and O_2). Thus, the reaction process is *cyclic* with O_3 and NO as intermediate molecules. Such cyclic reactions will reach a steady state in which the amounts of all of the chemicals are constant if the conditions (temperature, pressure, and level of solar irradiation) are also constant. Although such constant conditions do not occur outside of the laboratory, the reactions themselves do. When the sun rises on an atmosphere containing nighttime NO, NO_2 is initially formed. The solar irradiation then produces O_3 from the NO_2 throughout the late morning and afternoon. As the solar radiation intensity decreases in late afternoon, the O_3 level falls.

The reactions involving NO_2, NO, and O_3 in the atmosphere are simulated in the laboratory in large inert bags or chambers with solar simulator irradiators. For such simulations to be environmentally realistic, hydrocarbons must also be present. One such simulation, using propene (C_3H_6; CH_3–CH=CH_2) as the hydrocarbon, is shown in **Figure 3–6**. In this simulation, not only does the O_3 concentration rise, but the propene concentration declines as secondary formaldehyde (HCHO), acetaldehyde (CH_3CHO), and peroxyacetyl nitrate (PAN) concentrations increase. Such simulations are not only important for understanding atmospheric chemistry, but they are also useful in explaining the health and environmental effects of air pollution mixtures, and in providing guidance on controlling air pollutants.

Not all important atmospheric reactions are photochemically driven: the reactions involving sulfur, especially sulfur dioxide (SO_2), ammonia (NH_3), and a large variety of hydrocarbons, including oxygenated hydrocarbons, are examples.

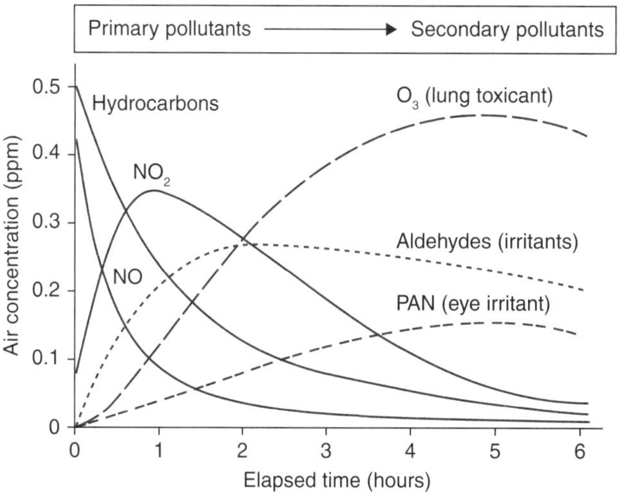

Figure 3–6 Time-course photochemistry of irradiated propene and nitrogen dioxide in a chamber. Data from Finlayson-Pitts and Pitts (1986).

Hydrocarbons and Their Derivatives

Hydrocarbons (molecules that contain only C and H) are among the simplest and most abundant organic molecules. The large number of hydrocarbons in the troposphere implies that they can have major roles in air pollution chemistry. Hydrocarbons can be arranged either by *class* or in *homologous series.* The hydrocarbon classes include:

- *alkanes*—carbon chains (straight, branched, or closed) connected by single bonds and having the structural formula C_nH_{2n+2};
- *alkenes*—planar and linear carbon chains with unsaturated double carbon to carbon bonds, are more reactive than alkanes, and have the structural formula C_nH_{2n};
- *alkynes*—planar and linear carbon chains with unsaturated triple carbon to carbon bonds are more reactive than alkanes and alkenes and have the structural formula C_nH_{2n-2}; and
- *cyclic hydrocarbons*—closed carbon chains including *cycloalkanes* and *aromatics* that are benzene-like molecules and less reactive than alkenes and many alkanes.

The homologous series classification relates to the ratio of carbon and hydrogen atoms in a molecule. Three such series are:

- *methane series*—general formula, C_nH_{2n+2};
- *ethylene series*—general formula, C_nH_{2n} (which has one carbon-carbon double bond); and
- *acetylene series*—general formula C_nH_{2n-2} (which has one carbon–carbon triple bond).

Among the more familiar hydrocarbons are:

- *methane*—CH_4;
- *propane*—$CH_3–CH_2–CH_3$;
- *butane*—$CH_3–CH_2–CH_2–CH_3$;
- *ethylene* or *ethene*—$CH_2=CH_2$;
- *propylene* or *propene*—$CH_3–CH=CH_2$;
- *acetylene*—$CH \equiv CH$;
- *benzene*—hexagonal, C_6H_6; and
- *toluene*—hexagonal, $C_6H_5CH_3$.

Saturated bonds are indicated by a "–," and unsaturated double and triple bonds by a "=" or a "≡," respectively. The double and triple bonds are more easily broken than single bonds, and are thus points of reactivity.

Simple hydrocarbons are primarily found in natural gas, petroleum, and coal tar. When one or more of the hydrogen atoms are replaced by another kind of atom or group of atoms the molecule belongs to the class of *derivative hydrocarbons.* Several types of derivatives are recognized, primarily by the functional group(s) attached to the hydrocarbon structure. The symbol "R" stands for an alkyl unit, with general formula C_nH_{2n+1}:

- *alcohols*—contain a hydroxyl group (OH) of the general form R–OH;
- *aldehydes*—have the general form R–CHO;
- *ketones*—have a –CO– group, R–COR;
- *carboxylic acids*—have the general formula R-COOH and can release a hydrogen ion in water solution;
- *esters*—have a general formula of R–COOR; and
- *ethers*—have a general formula of R–O–R.

As evident in the foregoing basic introduction to hydrocarbons and their derivatives, they have a large variety of forms, properties, and chemical reactivities. Hydrocarbons and their derivatives have important roles in the chemistry of the troposphere and the effects of air pollutants on living things, ecosystems, and the environment. Understanding these roles is an enormous endeavor that is still in an early stage.

Sulfur

Sulfur (S) ranks fourteenth in abundance of the elements found in the Earth's crust and fifth in abundance in seawater. Sulfur is also a required element for all higher organisms. Similar to carbon, sulfur forms a large number of compounds with linear, branched, and cyclic molecular structures. Sulfur chemistry is therefore extensive in terms of the variety of reactions and molecular forms available.

From the standpoint of air pollution science, sulfur dioxide (SO_2), sulfur trioxide (SO_3), sulfuric acid (H_2SO_4), hydrogen sulfide (H_2S), metal-containing and other sulfates and sulfides, and mercaptans (R–SH) are important air contaminants. In order to narrow the focus, we will consider the sources and formation of SO_2, SO_3, H_2SO_4, and other sulfates.

Sulfur dioxide is a colorless gas with a sharp, irritating odor. Sulfur dioxide is added to the atmosphere primarily as a byproduct of (1) combustion of coal, fuel oil, diesel engine fuel; (2) biological processes; (3) sea spray; and (4) to a lesser extent, the combustion of wood. Coal combustion is the major source of electrical power in the United States and the major source of anthropogenic sulfur. The sulfur content of coal mined in

Table 3–12 The sulfur content of selected coal deposits mined in the United States.

County	*U.S. States*	*BTU/lb*	*% Fixed Carbon*	*% Sulfur*
Schuykill	Pennsylvania	14,280	84.1	0.77
Lackawanna	Pennsylvania	14,880	79.4	0.60
McDowell	W. Virginia	15,600	77.3	0.74
Indiana	Pennsylvania	15,580	64.9	2.20
Pike	Kentucky	15,040	57.5	0.70
Belmont	Ohio	14,380	47.3	4.00
Emery	Utah	13,560	50.2	0.90
Musselshell	Montana	12,075	46.7	0.43
Sheridan	Wyoming	9,745	40.8	0.30
Mercer	N. Dakota	7,610	32.2	0.40

Data from Considine (1989).

the United States varies from less than 0.3 to 4.0 percent of the total mass, as shown in **Table 3–12**. Because sulfur compounds are biologically active, efforts are made to control sulfur emissions. Sulfur control can involve selection of low-sulfur fuels, pre-cleaning of fuels, or scrubbing of sulfur emissions post-combustion. SO_2 is both regulated and controlled because of its potential direct effects on living systems and its conversion in the atmosphere to other sulfur species.

Sulfur trioxide is a colorless gas that rapidly forms sulfuric acid when in contact with water or water vapor, in the presence of a catalyst (M) such as a metallic oxide. Thus, under the right conditions SO_3 can lead to the formation of sulfuric acid, a strong irritant and corrosive agent. Ultimately, sulfuric acid is formed in water droplets from the interaction of SO_2 and hydroxyl radicals (OH•), in the presence of water and a catalyst:

$$SO_2 + OH\bullet \rightarrow HOSO_2 \qquad \text{(Eq. 3–37)}$$

$$HOSO_2 + O_2 \rightarrow HO_2\bullet + SO_3 \qquad \text{(Eq. 3–38)}$$

$$SO_3 + H_2O + M \rightarrow H_2SO_4 + M. \qquad \text{(Eq. 3–39)}$$

The rapid reaction of SO_3 and H_2O produces sulfuric acid plus heat (i.e., the net reaction is exothermic):

$$SO_3 + H_2O + M \rightarrow H_2SO_4 + M + 21{,}300 \text{ cal.} \qquad \text{(Eq. 3–40)}$$

In combustion-generated plumes, SO_2 and the above reaction products are surrounded by a large number of other substances, organic and inorganic. The result is the formation of particulate sulfates other than sulfuric acid (e.g., $(NH_4)_2SO_4$ and metal-containing sulfates). Sulfates contribute significantly to the fine particulate mass, $PM_{2.5}$, in regions that use fuels containing sulfur.

Nitrogen

Most of the nitrogen in the atmosphere is the diatomic form, N_2, which is biologically inert at normal atmospheric temperature and pressure. The major inorganic and organic tropospheric nitrogen compounds of interest in air pollution science include:

- NO_2—nitrogen dioxide
- NO—nitric oxide
- NH_3—ammonia
- HNO_3—nitric acid
- NH_4NO_3—ammonium nitrate (there are also other important nitrates)
- peroxyacetyl nitrate (PAN)
- organic compounds that contain nitrogen (e.g., amines)

Given the large variety and complexity of nitrogen compounds in the atmosphere, only a few will be described here.

The sources of NO and NO_2 (together, NO_x) are dominated by fuel combustion (**Table 3–13**). Nitric oxide, a colorless gas, is formed during combustion from atmospheric N_2:

$$N_2 + O_2 \leftrightarrow 2NO. \qquad \text{(Eq. 3–41)}$$

Table 3–13 Average annual NO_x emissions in the United States in 2005 in million metric tons (1 metric ton = 103 kg = 2,205 lb).

Source	*NO_x Emissions*
On-road vehicles	6.5
Electricity-generating units	3.8
Non-road mobile sources	4.2
Industrial, commercial, and residential fuels	2.4
Industrial processes	1.2
Waste disposal	0.2
Fertilizers and livestock	0.02
All other	0.2 (approx.)

Data from U.S. EPA Emissions Inventories (http://www.epa.gov/ttn/chief/eiinformation.html).

At high temperatures the reaction moves to the right, but at low temperatures it shifts to the left. Rapid cooling with dilution in the atmosphere (as occurs with engine exhausts) favors high NO yields. Some NO_2 is directly emitted from combustion sources, but NO is the dominant primary emission product.

Nitrogen dioxide is an orange to reddish brown gas with its own characteristic sharp irritating odor. NO_2 is more corrosive and more toxic than NO, as it is more reactive. Since it is a colored gas, it absorbs light and is photochemically active. Some atmospheric NO is oxidized to NO_2. Some NO_2 also forms NO and singlet oxygen in the presence of sunlight and nitric acid (HNO_3) in the presence of water:

$$2NO + O_2 \rightarrow 2NO_2 \quad \text{(Eq. 3–42)}$$

$$NO_2 + h\upsilon \rightarrow NO + O \quad \text{(Eq. 3–43)}$$

$$2NO_2 + H_2O \leftrightarrow HONO + HNO_3. \quad \text{(Eq. 3–44)}$$

Because the above reaction products are biologically and/or chemically reactive, they are targeted for control, largely by improvements in internal combustion engine design and by post-combustion scrubbers (see Chapter 6).

Ammonia, NH_3, is generated primarily by biological decay. As it is used in industrial processes and as a refrigerant, it also has other anthropogenic sources. Ammonia does not photolyze, but it can be destroyed by reacting with OH• to form the NH_2 radical. The fate of NH_2 is uncertain, but its oxidation has been postulated:

$$NH_3 + OH\bullet \rightarrow NH_2 + H_2O \quad \text{(Eq. 3–45)}$$

$$NH_2 + O_2 \rightarrow OH\bullet + HNO. \quad \text{(Eq. 3–46)}$$

Ammonia is readily absorbed by water and may also deposit on many surfaces. Therefore, its residence time in the atmosphere is short, but it is continually renewed by various sources. Ammonia's main roles in air pollution are the neutralization of atmospheric acidity and the formation of particulate nitrates and sulfates. Ammonium nitrate, NH_4NO_3, and ammonium sulfate, $(NH_4)_2SO_4$, are examples of components of fine particulate air pollution. Ammonium sulfate is a common component of regional haze.

VI. PRIMARY AND SECONDARY AIR POLLUTANTS

Overview

In air pollution science, the terms "primary air pollutants" and "secondary air pollutants" have specific meanings. Primary air pollutants are those in the same chemical form that was emitted from a source (the source can be either natural or anthropogenic). Secondary air pollutants are those formed in the atmosphere by chemical reactions or physical processes. The reactants could come from natural, anthropogenic, or both types of sources.

There are two main reasons for making the distinction of primary or secondary. First, in order to perform and interpret atmospheric chemistry research, one must distinguish the chemical reactants (primary pollutants) from the chemical products (secondary pollutants). The second reason is that emission controls can only be effectively directed at primary anthropogenic air pollutants. In order to control the air concentrations of secondary air pollutants, their formation process must be understood and somehow interrupted. Controlling the air concentrations of primary anthropogenic pollutants is much easier than controlling the concentrations of secondary pollutants, or of primary natural pollutants.

The distinction between primary and secondary air contaminants is not always clear, as the same chemical can either be directly emitted into, or formed by reactions in the air. Some examples include sulfuric acid, other sulfate compounds, nitrates, and gases, such as carbon dioxide, nitrogen dioxide, and formaldehyde.

As secondary air contaminants are mainly formed by chemical reactions, and chemical reactions usually produce products that are less reactive than their reactants, it would be convenient to assume that secondary pollutants are more inert than primary pollutants. In many cases this is true. However, as sunlight drives many atmospheric reactions, additional energy (and thus, greater reactivity) can be found in some secondary pollutants. Ozone is a primary example of a very reactive (and relatively toxic) secondary air contaminant. Thus, *photoactivation* can produce highly-reactive products. The peak concentrations of many such secondary contaminants are built late in the day, after the sun has had time to drive their formation.

Reaction Intermediates

A third class of air contaminants includes the *reaction intermediates* (also called chemical intermediates). These are short-lived substances that exist only briefly as unstable (and therefore, short-lived) chemicals that disappear after a stepwise reaction is complete. The study of such stepwise reactions falls into the topic of *chemical kinetics*. Short-lived chemical intermediates include free radicals, which have unpaired orbital electrons, and electrically-charged atoms or molecules (ions) that rapidly react and disappear. Thus, when chemical reactions are occurring, the intermediates exist only for a short time, and they are not usually detected in routine chemical analyses of sampled air contaminants. Because chemical intermediates are usually transient, they are not considered in evaluating air quality for the purpose of determining health effects.

Primary Particulate Matter

The main primary particulate air pollutants can be classified chemically, or by the type of source that emits them. One type of chemical classification is:

- inorganic elements (e.g., metals)
- inorganic compounds
- organic compounds
- radioactive substances

Another classification for primary particles is by origin:

- natural (e.g., soil, sea spray, microbial phenomena, fires, plants, etc.)
- stationary combustion source (e.g., coal, and oil)
- industrial (e.g., metallurgical and manufacturing processes)
- transportation-related (e.g., exhausts, resuspended particles, tire and brake wear)
- other (e.g., tobacco burning, road paving, spraying, cooking, etc.).

A third classification of primary particles used by the U.S. EPA is:

- sulfate
- nitrate
- ammonium
- organic carbon
- elemental carbon
- metals
- bioaerosols

Figure 3–7 depicts the major particle modes in the atmosphere (although several other modes have been seen). **Table 3–14** shows some of the primary particles and their sources. Each locale will have its own mix of particles depending on the climate, sources of electrical power (e.g., coal, oil, gas, solar, wind, hydroelectric, or nuclear), the industries present, traffic, and the population.

Secondary Particulate Matter

The major mechanisms that form secondary particles include:

- conversion of gases and vapors via condensation or chemical reactions to the particulate (solid or liquid) state;
- chemical reactions of primary particles with gases and/or vapors that lead to new chemical forms of the particles;
- coagulation of various types of primary particles to form larger complex particles; and
- acquisition of surface coatings on solid particles that alter their composition and potentially their biological activity.

Formation of Secondary Particles

Water is an important medium in which secondary pollutants form. Gases including sulfur dioxide and oxides of nitrogen that are absorbed by water droplets produce sulfates and nitrates in solution. When the relative humidity is lowered, water evaporates and residue particles (in the $PM_{2.5}$ range) are formed. Similar gas reactions in condensed water on the surfaces of solid particles can also produce new chemical species without changing the particle size. Such reactions in water droplets and in condensed water on solid particles are modified by the presence of reactive species, such as hydroxyl radicals, hydrogen peroxide, ammonia, and organic species. Solar radiation, elevated temperature, and catalysts (e.g., iron and manganese) can accelerate the formation of secondary particulate chemical species. Thus, the particle composition varies from day to night. Of note, most of the acidity in the air is in the form of secondary particles.

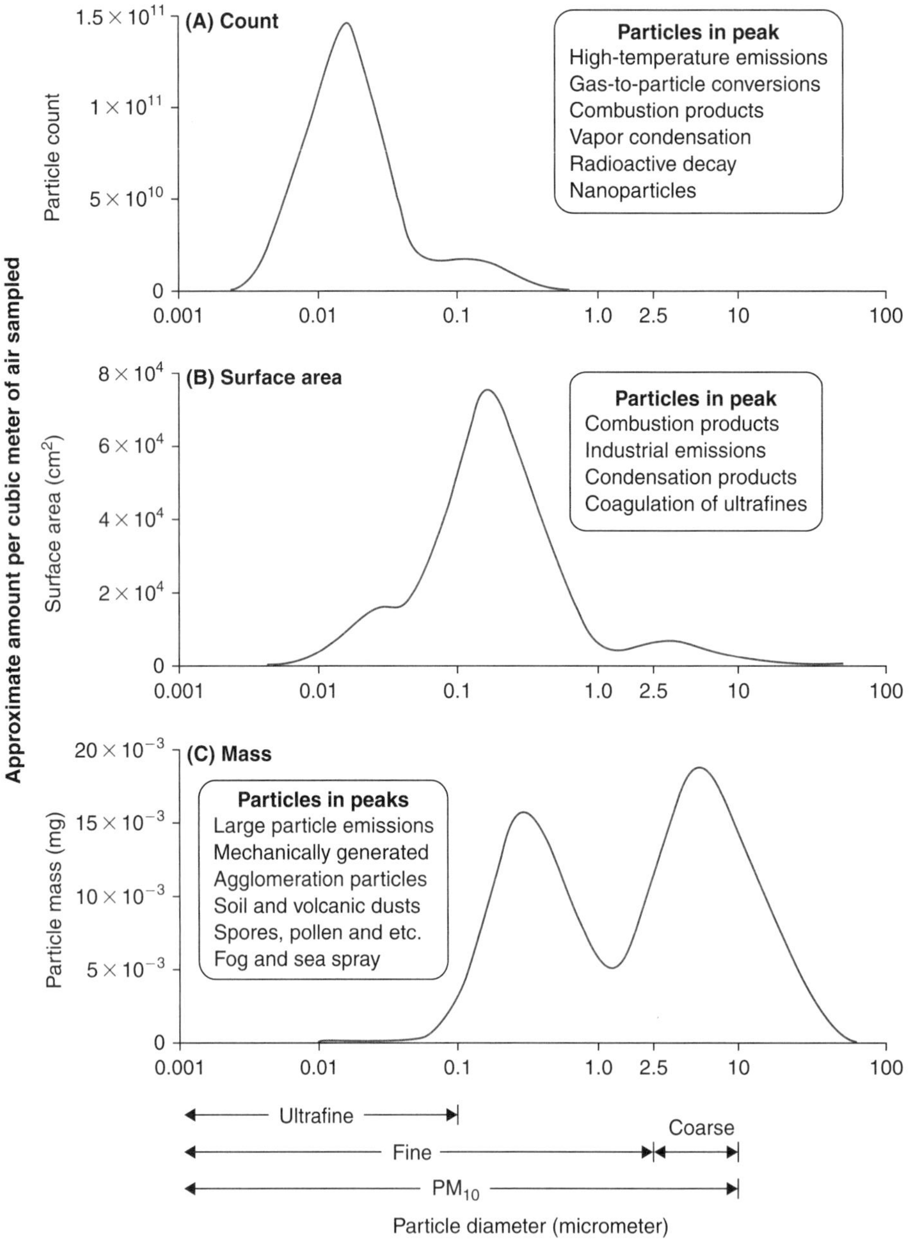

Figure 3–7 Sample urban aerosol plotted three ways by: (A) count; (B) surface areas; and (C) mass. Data from Wilson (1978) and Whitby (1978).
Source: The University of California Air Pollution Health Effects Laboratory, with kind permission.

The formation of secondary agglomerated particles by the coagulation of fine particles is favored when particle concentrations are high (as in automobile exhaust). Examination of atmospheric particles by electron and optical microscopy confirms that such complex secondary particles exist.

Secondary organic particulate matter (SOPM) contributes significantly to photochemical air pollution. Volatile organic compounds that contain at least seven carbon atoms tend to condense to form new particles, or to attach to existing particles. Such volatile organic compounds are produced by both natural and anthropo-

Table 3–14 Atmospheric primary particles and some of their major sources.

	PM < 2.5 μm in diameter		*PM > 2.5 μm* in diameter	
Particle Species	*Natural*	*Anthropogenic*	*Natural*	*Anthropogenic*
Sulfates	Sea spray	Fossil fuels	Sea spray	—
Minerals	Soil erosion	Roads, construction, agriculture	Soil	Roads, construction, agriculture
Organic carbon	Wildfires	Wood burning, motor vehicles, cooking	Soil, wildfires	Tire wear, paved road dust, asphalt wear
Elemental carbon	Wildfires	Motor vehicles, wood burning, cooking	Wildfires	Tire wear, paved road dust, asphalt wear
Metals	Volcano activity	Fuel burning, smelting, brake wear	Erosion, organic debris, re-entrainment	—
Bioaerosols	Viruses, bacteria	—	Plant and insect fragments, pollens, spores, bacterial agglomerates	—

Note: "—" means either minor source or no known source of component.
Data from U.S. EPA (2004).

genic sources. Reactions of such compounds with ozone, OH radicals, HO_2 radicals, and NO_3 radicals participate in the formation of SOPM. As for inorganic secondary particles, solar radiation can initiate reactions that form SOPM.

Particle surfaces are also significant sites for the formation of secondary organic compounds. This process is enhanced when the particle surface is acidic. Research on the particulate surface chemistry is both complex, and in an early phase. Such research is also important for understanding air chemistry and the effects of secondary organic particles on people, livestock, crops, ecosystems, visibility, and climate.

Secondary Gases and Vapors

The most notable secondary gases are ozone (O_3) and nitrogen dioxide (NO_2). In the case of O_3, essentially all of the gas is secondary, although electrical arcs and ultraviolet sources produce primary O_3 locally. Some NO_2 is a primary gas produced by combustion sources, but the conversion of NO to NO_2 produces the secondary gas. Most secondary organic compounds have low volatility and are thus not present in the air as gases or vapors.

VII. UNCERTAINTIES RELATED TO PUBLIC HEALTH ISSUES

Much is known about air pollutant particles and gases, including their basic properties, origins, physical and chemical reactions, transport, and removal from the troposphere. As with scientific research in general, each new discovery reveals a multitude of new possibilities and questions. More simply, scientific research often raises more questions than it answers.

With respect to pollutants in the troposphere several problems are still not solved, including the following examples:

- What are the proper ways to describe the surface area of "spongy" particles that have surface cracks and internal voids (different methods of measuring surface areas give different results for such particles)? The surface area is an important factor in the toxicity of many types of particles.
- What are the rates of dissolution of realistic environmental particles in biological media such as mucus, lung surfactant, and the media inside of living cells?

- What are the fates of inhaled ultrafine particles and nanoparticles within the body?
- What are the fates of inhaled volatile particles, including the rates of deposition of their components on airway surfaces?
- Information on the viability of bioaerosols under various environmental conditions is needed.
- More research is needed on organic species.
  - How do their interactions with airborne inorganic particles influence their deposition in the respiratory tract?
  - More information is needed on the solubilities of non-reacting gases in the many biological substances in the body, such as components of mucus, pulmonary surfactants, and cellular components (e.g., DNA, proteins, lipids, etc.).
  - What are the important properties of solids, gases, and vapors with respect to toxicity?
  - The secondary organics and their reactions in the air and in the body are in need of additional research.

VIII. SUMMARY OF MAJOR POINTS

Humans and many other organisms are immersed in an ocean of air (the *troposphere*) that extends to an average altitude of about 9 km. This "ocean" is actually a two-component system that consists of particulate material suspended in the air. The tropospheric aerosol is chemically reactive because it contains a vast number and variety of gases, vapors, and particles. Solar energy promotes some important reactions. For the purpose of understanding the basic principles related to the impacts of the air on health and the environment, it is helpful to focus on a few of the atmospheric components. This focus is mainly on those particles and gases regulated by the U.S. EPA. These include some of the substances listed in the National Ambient Air Quality Standards (NAAQS) but others such as *bioaerosols* and *nanoparticles* are important as well.

Aerosol particles are a class of matter that has a large range of sizes, shapes, and compositions. Most aerosol particles are not directly visible to the unaided eye, so optical and electron microscopy must be used to image them. Aerosol particles are conveniently described by "equivalent" diameters that relate to their behavior in the air. For example, an important equivalent size, the *aerodynamic diameter* (D_{ae}) is used to describe the rate at which particles settle out of the air, and the efficiencies with which they deposit in the respiratory tract when inhaled. D_{ae} is used by the U.S. EPA to define size classes, such as PM_{10} and $PM_{2.5}$, that relate to where particles deposit in the respiratory tract, and to some extent the particle source. In general, PM_{10} is associated more with natural sources and $PM_{2.5}$ with combustion sources. Aerosol particle size distributions are seldom symmetric, and *lognormal* distribution equations are used to summarize them.

Aerosol particles have many important properties, including their chemical composition, surface areas, material densities, electrical charges, optical parameters, hygroscopicities, and rates of diffusion and coagulation. Such properties are used to describe their chemical and physical behavior in the atmosphere, and in the body after they are inhaled. *Nanoparticles,* because of their small size, have unique properties that are the basis of the introduction of new products with new uses. *Nanotoxicology* is a relatively new specialty aimed at defining the potential adverse effects of many types of nanoparticles.

Gases and vapors that are confined have temperature-volume-pressure relationships described by the *ideal gas laws.* However, intermolecular forces in concentrated gases can produce deviations from this ideal scenario. Individual gases in a mixture exert their pressure independent of one another, as described by the *law of partial pressures.* Gases dissolved in liquids or solids are described by their partial pressures as well. The *solubilities* of different gases in various media, such as water and oils, are used to model their behavior in biological systems. The solubility of a non-reactive gas in water influences where it will initially deposit in the respiratory tract upon inhalation, as well as its uptake in the blood. The lipid solubility of a gas dissolved in the blood is used to model its uptake by lipids in the body, specifically in the brain and fat deposits. Inhaled gases tend to approach equilibrium concentrations throughout the body until their external concentrations drop and their movement reverses to wash them out of the body. *PBPK* models are used to describe such uptake, internal distribution, and wash-out kinetics.

The absorption of energy by gases illuminated with visible, infrared, and ultraviolet radiation is quantized. Such quantization relates to allowable molecular energies in their vibration, rotational, and excitation states.

The uptake of energy by gas molecules is important to their *photochemical* reactions, and for their detection using *spectroscopy* techniques. Photochemical reactions are important in atmospheric chemistry. Examples include the production of ozone and nitrogen dioxide, two gases that have relevance to public health.

Organic chemicals, which are present in the atmosphere in gaseous, liquid, and solid states, play prominent roles in air chemistry and effects on living systems. Given the great variety of organic chemicals, research on understanding their roles in atmospheric science is still in an early stage.

Air contaminants are conveniently classified according to whether they are *primary* (in physical and chemical forms as emitted from sources) or *secondary* (formed in the atmosphere). Primary anthropogenic contaminants are easier to regulate, as their sources can be controlled. Primary natural air contaminants are typically not regulated nor controlled. Secondary air contaminants can only be controlled by understanding how they are formed and by regulating concentrations of their precursors when feasible.

Although much is known about the aerosol particles, gases, and vapors in the atmosphere, several important public-health-related uncertainties remain. Thus, many opportunities exist for productive careers related to atmospheric chemistry and the transport of air contaminants.

IX. QUIZ AND PROBLEMS

Quiz Questions

(select the best answer)

1. Primary air pollutants, as opposed to secondary pollutants, are those that:
   a. are present in levels that can be weighed on filter samples.
   b. are in forms directly emitted from identified sources.
   c. have undergone chemical transformations in the air after they are emitted.
   d. are designated as Hazardous Air Pollutants.
   e. All of the above are true.
2. Which statement about hydrocarbons is true?
   a. They consist only of hydrogen and carbon atoms.
   b. If they have more than six carbon atoms they are more likely to be liquids or solids than gases.
   c. If they are "unsaturated" they are more chemically reactive than if they are "saturated."
   d. All of the above are true.
   e. Only a. and b. are true.
3. Secondary air pollutants:
   a. are unimportant with respect to human health effects.
   b. have undergone chemical and/or physical transformations after they are emitted into the air.
   c. do not include organic species.
   d. None of the above is true.
4. How do gases and vapors differ?
   a. Vapors are actually liquid aerosols.
   b. Vapors are like gases, but they are evaporation products of liquids or solids.
   c. Vapors are toxic, but gases are not.
   d. Vapors are photochemically excited forms of gases.
5. Inhaled non-reactive gases:
   a. do not deposit in the body.
   b. deposit only in the nose and large airways of the tracheobronchial tree.
   c. deposit in the respiratory tract regions according to their solubility in biological fluids.
   d. are removed from the body primarily by macrophage cells in the lungs.
6. Inhaled particles:
   a. deposit in the respiratory tract with 100 percent efficiency independent of their size.
   b. deposit with efficiencies that depend on their size.
   c. remain in the respiratory tract without undergoing any clearance.
   d. are only deposited in the body if a person is breath-holding after an inhalation.
7. Which of the following are anthropogenic pollutants?
   a. Bacteria and viruses
   b. Diesel engine exhaust
   c. Smokestack emissions
   d. Volcano plumes
   e. Both b and c
8. Which particles have the largest aerodynamic diameters?
   a. Nanoparticles
   b. Diesel exhaust particles
   c. Fog droplets.
   d. 1.0 micrometer diameter salt crystals

Problems

1. Calculate the aerodynamic diameter for a spherical particle with a diameter of 10 micrometers and a density, ρ_p, of 1.15 g/cm^3.
2. How much faster does a 2-micrometer aerodynamic diameter particle settle than a 1-micrometer particle of the same shape and density?
3. What is the surface area of a 1-micrometer diameter smooth spherical particle?
4. How long does it take for a 100-micrometer aerodynamic diameter particle to settle out of still air if it is released 1 meter above the ground?
5. The wind blows past an air pollutant source at 10 kilometers per hour and the downwind SO_2 concentration is measured as 1 part per million. What could the concentration be if the wind speed is only 1 kilometer per hour? Ignore factors other than dilution.
6. Calculate the volume median diameter of a log-normal distribution of particles that has a count median diameter of 1.0 micrometers and a geometric standard deviation of 3.0.
7. Image forces on inhaled charged particles can increase their deposition in the upper airways. What is the image force between a particle with a charge of 10 statculombs and conductive surface when the separation distance is 1.0 cm? Use Eq. 3–10.
8. Using Table 3–3 and Figure 3–2, compare the particle displacement in 1 second for a virus particle 0.01 micrometers in diameter and a pollen grain that is 10 micrometers in diameter. Which particle has the greatest displacement per second?
9. Explain why the dissolution rate of silver particles in interstitial fluid with protein differs from the rate in water. Use Figure 3–5 for reference.

X. DISCUSSION TOPICS

1. Current U.S. EPA National Air Quality Air Standards are based on particle mass in two size ranges, $PM_{2.5}$ and PM_{10}. Should the air standards take chemical composition and/or other factors into account? What particle properties other than mass and size should be considered, and why?
2. Secondary air contaminants are those formed in the atmosphere after they are emitted. What steps could be taken to control the secondary pollutant ozone?
3. Review the uncertainties in section VII. What research areas do you believe will have the greatest impact on future regulations?
4. Why is it important to study the chemistry and physics of air pollutants, rather than only their health effects?

References and Recommended Reading

Ansborlo, E., Hengé–Napoli, M. H., Chazel, V., Gibert, R., and Guilmette, R. A., Review and critical analysis of available *in vitro* dissolution tests, *Health Phys.* 77:638–645, 1999.

Barrow, G. M., *Physical Chemistry, 2nd Edition,* Mc Graw–Hill, New York, 1966, p. 521.

Bell, K. A. and Ho, A. T., Growth rate measurements of hygroscopic aerosols under conditions simulating the respiratory tract, *J. Aerosol Sci.,* 12:247–254, 1981.

Brunauer, S., Emmett, P. H., and Teller, E., Adsorption of gases in multimolecular layers, *J. Am. Chem. Soc.,* 60:309–319, 1938.

Burge, H. A., "Airborne contagious disease," in *Bioaerosols,* Burge, H. A. ed., Lewis Publishers, Boca Raton, FL, 1995, pp. 25–47.

Cohen, B. S., Xiong, J. Q., and Li, W., "The influence of charge on the deposition behavior of aerosol particles with emphasis on singly charged nanometer sized particles," in *Aerosol Inhalation: Recent Research Fontiers,* Marijnissen, J. C. M., and Gradon, L., eds., Kluwer Academic Publishers, Dordrecht, 1996, pp. 153–164.

Considine, D. M., ed., *Van Nostrand's Scientific Encyclopedia, 7th Edition,* Van Nostrand Reinhold, New York, 1989, p. 664.

Cox, C. S. and Wathes, C. M., eds., *Bioaerosols Handbook,* Lewis Publishers (CRC Press), Boca Raton, FL, 1995.

Davis, J. R. and Lederberg, J., eds., *Emerging Infectious Diseases From the Global to the Local Perspective,* National Academy Press, Washington, DC, 2001.

Dennison, J. E., Andersen, M. E., and Yang, R. S. H., Pitfalls and related improvements of *in vivo* gas uptake pharmacokinetic experimental systems, *Inhal. Toxicol.,* 17:539–548, 2005.

Drinker, P. and Hatch, T., *Industrial Dust,* McGraw–Hill, New York, 1936, pp. 1–2.

Ferron, G. A. and Busch, B., Deposition of hygroscopic aerosol particles in the lungs, in *Aerosol Inhalation: Recent Research Frontiers,* Marijnissen J. C. M. and Gradón, L. eds., Kluwer Academic Publishers, Dordrecht, Netherlands, 1996, pp. 143–152.

Finlay, W. H., *The Mechanics of Inhaled Pharmaceutical Aerosols: An Introduction,* Academic Press, New York, 2001.

Finlayson–Pitts, B. J. and Pitts, J. N. Jr., *Atmospheric Chemistry: Fundamentals and Techniques,* John Wiley & Sons, New York, 1986.

Friedlander, S. K., *Smoke, Dust and Haze, 2nd Edition,* Oxford University Press, New York, 2000.

Fuchs, N. A., *The Mechanics of Aerosols: Revised and Enlarged Edition,* Dover Publications, Inc. New York, 1964.

Gross, P. and Westrick, M., The permeability of lung parenchyma to particulate matter, *Am. J. Pathol.,* 30:195–213, 1954.

Hatch, T. and Choate, S. P., Statistical description of the size properties of non-uniform particulate substances, *J. Franklin Inst.,* 207:369–387, 1929.

Heber, A. J., Bioaerosol particle statistics, in *Bioaerosols Handbook,* Cox, C. S., and Wathes, C. M., eds., Lewis Publishers, Boca Raton, FL, 1995, pp. 55–75.

Hinds W. C., *Aerosol Technology: Properties, Behavior and Measurement of Airborne Particles, 2nd Edition,* Wiley–Interscience, New York, 1999.

Hinds W. C. and Kennedy N. J., An ion generator for neutralizing concentrated aerosols, *Aerosol Sci. Technol.,* 32:214–220, 2000.

Holma, B., Lung clearance of mono- and di-disperse aerosols determined by profile scanning and whole-body counting—a study on normal and SO_2 exposed rabbits, *Acta Med. Scand.,* 5 (Suppl. 473):1–102, 1967.

IOM (Institute of Medicine of the National Academies), *Microbial Evolution and Coadaptation,* The National Academies Press, Washington, DC, 2009.

Kreyling, W. G., Semmler–Behnke, M., and Möller, W., Ultrafine particle-lung interactions: Does size matter? *J. Aerosol Med.,* 19:74–83, 2006.

Kreyling, W. G., Möller, W., Semmler-Behnke, M., and Oberdörster, G., Particle dosimetry: Deposition and clearance from the respiratory tract and translocation towards extra-pulmonary sites, in *Particle Toxicology,* Donaldson, K., and Borm, P., eds., CRC Press, Boca Raton, FL, 2007, pp. 47–74.

Lederberg, J., Shope, R. E., and Oaks, S. C. Jr., eds., *Emerging Infections: Microbial Threats to Health in the United States,* National Academy Press, Washington, DC, 1992.

Levetin, E., Fungi, in *Bioaerosols,* Burge, H. A., Ed., CRC Press, Boca Raton, FL, 1995, pp. 87–120.

Macher, J., Ammann, H. A., Burge, H. A., Milton, D. K., and Morey, P. R., eds., *Bioaerosols: Assessment and Control,* American Conference of Governmental Industrial Hygienists, Cincinnati, OH, 1999.

McNeil, S. E., Nanotechnology for the biologist, *J. Leucocyte Biol.,* 78:585–594, 2005.

Mercer, T. T., On the role of particle size in the dissolution of lung burdens, *Health Phys.,* 13:1211–1221, 1967.

NRC (National Research Council), *A Framework for Assessing the Health Hazard Posed by Bioaerosols,* National Academies Press, Washington, DC, 2008.

Oberdörster, G., Pulmonary effects of inhaled ultrafine particles, *Internat. Arch. Occup. Environ. Health,* 74:1–8, 2001.

Phalen, R. F., Evaluation of an exploded-wire aerosol generator for use in inhalation studies, *Aerosol Sci.,* 3:395–406, 1972.

Phalen, R. F., *The Particulate Air Pollution Controversy: A Case Study and Lessons Learned,* Kluwer Academic Publishers, Boston, MA, 2002, pp. 39–53.

Phalen, R. F., *Inhalation Studies, Foundations and Techniques, 2nd Edition,* Informa Healthcare, New York, 2009.

Phalen, R. F. and Morrow, P. E., Experimental inhalation of metallic silver, *Health Phys.,* 24:509–518, 1973.

Raabe, O. G., Generation and characterization of aerosols, in *Inhalation Carcinogenesis,* Hanna, M. G. Jr., Nettesheim, P., and Gilbert, J. R., eds., U.S. Atomic Energy Commission, AEC Symposium, Springfield, VA, 1970.

Schum, G. M. and Phalen, R. F., Modeling hygroscopic particle growth in human lung airways, *Ann. Occup. Hygiene,* 41 (Suppl. 1):60–64, 1997.

Seinfeld, J. H. and Pandis, S. N., *Atmospheric Chemistry and Physics, 2nd Edition,* Wiley–Interscience, New York, 2006.

Stefaniak, A. B., Guilmette, R. A., Day, G. A., Hoover, M. D., Breysse, P. N., and Scripsick, R. C., Characterization of phagolysosomal simulant fluid for study of beryllium aerosol particle dissolution, *Toxicol. In–Vitro,* 19:123–134, 2005.

Stern, A. C., Ed., *Air Pollution, Vol. I., 3rd Edition,* Academic Press, New York, 1976.

Stern, S. T. and McNeil, S. E., Nanotechnology safety concerns revisited, *Toxicol. Sci.,* 101:4–21, 2008.

TGLD (Task Group on Lung Dynamics), ICRP Committee II, Deposition and retention models for internal dosimetry of the human respiratory tract, *Health Phys.,* 12:173–207, 1966.

Tucker, A. D., Wyatt, J. H., and Undery, D., Clearance of inhaled particles from alveoli by normal interstitial drainage pathways, *J. Appl. Physiol.,* 35:719–732, 1973.

U.S. EPA (United States Environmental Protection Agency), *Air Quality Criteria for Particulate Matter, Vol. 1,* EPA/600/P-99/002aF, United States Environmental Protection Agency, Washington, DC, 2004.

U.S. EPA (United States Environmental Protection Agency), *Technology Transfer Air Toxic Website* (http://www.epa.gov/ttn/atw/allabout.html, accessed March 22, 2011).

U.S. EPA (United States Environmental Protection Agency), *National Ambient Air Quality Standards (NAAQS)* (http://www.epa.gov/air/criteria.html, accessed September 5, 2010).

Vincent, J. H., *Aerosol Science for Industrial Hygienists,* Elsevier Science, Inc. Tarrytown, NY, 1995.

von Smoluchowski, M., A mathematical theory of the kinetics of coagulation of colloidal solutions, *Z. Phys. Chem.,* 92:129–168. 1917.

Warheit, D. B., Borm, P. J. A., Hennes, C., and Lademann, J., Testing strategies to establish the safety of nanomaterials: Conclusions of an ECETOC workshop, *Inhal. Toxicol.,* 19:631–643, 2007.

Whitby, K. T., The physical characteristics of sulfur aerosols, *Atmos. Environ.,* 12:135–159, 1978.

Wilson, W. E., Sulfates in the atmosphere: A progress report on project MISTT*, *Atmos. Environ.,* 12:537–547, 1978.

Chapter 4

Sampling and Analysis for Health Assessments

LEARNING OBJECTIVES

By the end of this chapter the reader will be able to:

- define "sample" and identify different types of representative air pollution samples
- define and differentiate between qualitative and quantitative analysis
- discuss statistical measures used to ensure the quality of sampling and analytical results, as well as how analytical results are reported
- describe in what ways air sampling techniques for particles and gases are designed to mimic human exposures
- discuss sampling and analysis methods for particles, gases, and vapors

CHAPTER OUTLINE

I. INTRODUCTION

Overview

In order to evaluate air pollutants in our environment we must first obtain a sample of what is in the air or what enters the exposed populations. A *sample* is a representative part of a larger group or whole. For regulatory compliance, samples representative of the ambient environment are required. For risk assessment, pollutant samples that represent what is inhaled or what exposes relevant target tissues within humans or other organisms are preferred. Obtaining a representative sample is not trivial, because concentrations of air pollutants vary in space (spatial) and time (temporal). For example, some pollutants have transient lifetimes and may change or disappear within minutes of their formation. In addition, a local sample adjacent to a major pollution source may not be reflective of the surrounding population's exposure. Science in general relies on the development of well-reasoned and defined sampling strategies. The sampling methodology and sample size are critical for testing hypotheses.

Once a sample is acquired it is important to preserve its integrity for *qualitative* and/or *quantitative* analysis. Qualitative analysis ensures the proper pollutant has been identified and collected from the environment. Quantitative analysis specifies *how much* of the pollutant is in the sample for purposes of evaluating compliance or risk. The final product or data must be *organized, summarized, validated,* and *presented* in a form that is relevant to the original goal. Good laboratory practices (GLPs) and quality assurance protocols are often used to ensure that each step in this process has been carefully monitored, documented, and verified. When performed correctly the results are valuable for determining regulatory compliance, performing risk assessments, and for use in air pollution studies, such as epidemiological investigations or climate modeling.

Sampling

The importance of sampling is often overlooked. Without an adequate sample the analysis and conclusions drawn from the data will be questionable. In air pollution science, an *air sample* involves the collection of a known volume of air for later analysis. An air sample should be representative of one or more of the following (**Table 4–1**):

- A *source emission:* For example, sulfur dioxide from a coal-fired plant or carbon monoxide from an automobile tailpipe.
- An *area sample:* For example, particulate sulfate and nitrate air concentrations in a national park for visibility (see Chapter 5 on *Visibility and Climate*) or fine particulate matter downwind from a major freeway.
- A *population exposure sample:* For example, ozone levels within the Los Angeles air basin (i.e., the

Table 4–1 Types of environmental air samples used in air pollution studies.

Air Sample Type	*Description*	*Purposes*
Source	Samples collected at an emission source before it is released to the environment	Emission control studies and activities; regulatory monitoring, permits, and controls; air pollution modeling and predictions
Area	Samples collected adjacent or in proximity to an identified source	Identification and characterization of contributing sources within a specified region; monitoring for the effectiveness of emission controls and model predictions
Population	Randomized samples collected within a community or air basin to represent an identified population	Epidemiology studies comparing the health effects of air pollution on exposed and non-exposed populations; risk assessment
Personal	Samples collected within the breathing zone of a human subject	Air pollution health effects studies for susceptible individuals (e.g., asthmatics) within a population; epidemiology studies; risk assessment

Southern California air basin) or airborne allergens within schoolrooms.

- A *personal sample:* For example, coarse particle concentrations in the breathing zone of an asthmatic or diesel exhaust exposures to a local resident living near a major freeway.

For regulatory compliance, both area samples and population exposure samples (within an air basin) are often collected using a *monitoring platform* with multiple samplers for select pollutants. Monitoring for the National Ambient Air Quality Standards (NAAQS) is often performed using a series of stationary monitoring stations within a defined air basin. Some platforms are mobile to allow researchers the flexibility to study new sources (e.g., new roads and freeways).

Remote sensing technologies are also gaining popularity, as they allow the collection of air pollution data using either (1) real-time wireless sensors, (2) remote data logging sensors, or (3) satellite-based instrumentation. Remote sensors allow pollutant measurements in environments not readily accessible.

In practice, a given sample may be used for multiple purposes. For instance, source emission sample data are often used to model and estimate area concentrations and population exposures. Conversely, a series of personal samples or area samples can be used to identify and estimate the strength of upwind emission sources. However, caution must be exercised in these cases. The environment is complex and unpredictable. Scientists are still learning about the chemistry and interactions that occur from the time an air pollutant is created to the eventual exposure. In short, localized or personal exposures may not be representative of the surrounding population, and vice versa.

The *number* of samples collected is equally important. The sample size must be large enough to represent the exposure group or population. However, if the sample size is too large it can overwhelm the analytical capability and available resources (e.g., time and money). There have been many cases where the collected and stored samples were never analyzed. The bottom line is that sampling must be tied to a worthy purpose and matched to the resources available for subsequent analysis.

Analysis

If there were unlimited resources, expertise, and time, then it would be possible to analyze every imaginable property of a sample. However, as with sampling, analysis must be tied to a worthy and well-defined objective. In air pollution science, the analytical objectives of identification and measurement often relate to (1) compliance with regulatory standards, (2) a formal risk assessment, (3) population health effects, (4) ecological effects, or (5) visibility and/or climate impacts. For example, an epidemiologist may require an analysis of the air concentrations of ozone, a criteria pollutant associated with asthma attacks. A toxicologist may require an analysis of the specific properties (e.g., oxidative) of pollutants that may interact with and/or damage target tissues. An air pollution chemist may need to know the specific concentrations of hydrocarbon and NO_x, as both are precursors to ozone formation. A climatologist may require measurement of the indices of refraction of fine and coarse particles in order to estimate the impacts on visibility and cloud formation. Therefore, *analysis must be performed in a manner that supports the original question or hypothesis.*

II. QUALITY ASSURANCE AND STATISTICAL CONSIDERATIONS

The characteristics of a quality-controlled sampling and analysis method are well established. *Quality control* is the acceptance or rejection of sampling and analytical data using specified performance standards. For example, any data collected by methods not meeting the minimum quality standards for accuracy and precision are disregarded. *Quality assurance* is the set of standardized procedures used to establish performance standards and assure data quality. The following *quality assurance/quality control* (QA/QC) concepts must be considered when developing, establishing, or using an analytical protocol:

- *Accuracy*—closeness of measured values to the true value;
- *Bias*—systematic error in an analytical method, due to sampling, instrumental, operator, or methodological errors;
- *Efficiency*—relative time, energy, and cost requirement per sample, which controls the number of samples that can be collected;
- *Interferences*—compounds or properties of the air (e.g., humidity) that may interfere with sampling and/or analytical methods;
- *Precision*—repeatability and reproducibility of analytical results;

- *Recovery*—ability of the sampling and analytical method to reliably recover a known percent of a pollutant within an environmental sample;
- *Selectivity/specificity*—ability of an analytical method to reliably identify and measure a pollutant in the presence of interferences;
- *Sensitivity*—ability of an analytical method to differentiate between small changes in pollutant concentration; and
- *Stability*—freedom from chemical or physical changes that may occur after sample collection, but prior to analysis.

Table 4–2 illustrates some of the common QA/QC measures applied in air pollution sampling and analysis methods. The overall importance of conducting a well-reasoned and quality sampling and analysis plan can be summarized as "questionable methods lead to questionable results."

Accuracy and Precision

Accuracy and precision are used to establish whether a sample concentration or value is representative of the true value. In environmental sampling, accuracy is a measure of the ability of the analytical method to provide a correct answer. Accuracy can also be defined as a measure of the level of agreement between the observed analytical result and true value.

Most environmental methods must meet a minimum accuracy criterion of ± 25 percent before they can be used in the field. Most laboratory-based analytical instruments must meet a minimum accuracy criterion of ± 10 percent when in operation and analyzing samples and standards.

Accuracy is a measure of the ability to provide a correct answer, whereas precision is a measure of the reliability and consistency of the method to continually provide correct answers. Precision can be defined as a measure of the closeness between independently-measured replicate values under the same analytical conditions.

The majority of environmental methods must meet a minimum precision criterion of ± 10 percent before they can be used in the field. Most laboratory-based analytical instruments must meet a minimum precision criterion of ± 5 percent when in operation and analyzing samples and standards.

Table 4–2 Common quality control measures.

Measure	*Description*	*Common Mathematical Expression*
Accuracy	The correctness of an experimental result or agreement of a measured value (*Obs*) to the true value (*True*); measured as relative error or percent recovery	$\text{Relative error} = \frac{Obs - True}{True}(100\%)$ or $\text{Percent recovery} = \frac{Spike - Blank}{Spike\ Amount}(100\%)$
Bias	Systematic error due to an assignable cause, such as instrumental, personal, or methodological errors	$bias = \bar{x}_{experimental} - \bar{x}_{true}$
Precision	The reproducibility of experimental results; the closeness of replicate samples; measured as the *coefficient of variation (CV)*	$CV = \frac{\sigma_{n-1}}{\bar{x}}(100\%)$
Sensitivity	Analytical sensitivity (*S*) is the ability of an analytical method to differentiate between small changes in analyte concentration	$S = \frac{\text{calibration curve slope}}{\sigma_{n-1}\text{ of slope}}$

$\bar{x}$ = mean
σ_{n-1} = standard deviation
Spike = blank or sample with known amount of analyte added

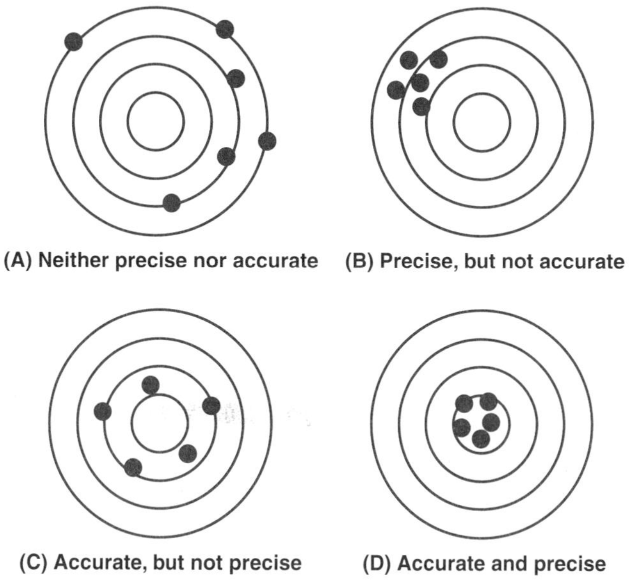

Figure 4–1 Accuracy and precision.
Source: The University of California Air Pollution Health Effects Laboratory, with kind permission.

The differences between accuracy and precision are illustrated in **Figure 4–1**. If the center "bull's eye" represents the true value, then (A) represents results that are neither accurate nor precise. The results for (B) are more reproducible and show greater precision, but the results are less accurate and farther from the true value. The average result for (C) is closer to the true value and accurate, but the replicates are not reproducible and less precise. The results for (D) are both accurate and precise. The primary objective of sampling and analysis methods is to provide accurate and precise results. The greater the accuracy and precision, the more representative the samples will be of the study population.

With sampling and analytical methods accuracy is often measured one of two ways, as a *relative error* or as *percent recovery*. Relative error compares the agreement of a measured value (*Obs*) to the known value (*True*), which must obviously be available for comparison:

$$\text{Relative error} = \frac{Obs - True}{True}(100\%). \qquad \text{(Eq. 4–1)}$$

If the true value is not known, then sample or blank media can be *spiked* by addition of a known amount of the chemical contaminant to determine the percent recovery of the analytical method:

$$\text{Percent recovery} = \frac{Spike - Blank}{Spike\ Amount}(100\%). \quad \text{(Eq. 4–2)}$$

The *Spike* is the analytically-derived contaminant concentration within a spiked blank, the *Blank* is the analytical result for a media blank, and the *Spike Amount* is the known contaminant concentration added to the spiked sample. The resulting percent recovery provides a measure of how accurately the analytical method can measure the contaminant within a sample.

Percent recovery is also used to determine the ability of the sampling and analytical method to recover a known amount of a pollutant within an environmental sample. Environmental samples consist of a complex *matrix* that often contains hundreds of potential interfering constituents, which may either increase or decrease analytical sensitivity. In contrast, analytical standards are mostly prepared in the laboratory without the presence of contaminants. Drastic differences in complexity between samples and standards could result in erroneous analytical data. The simplest way to test for and account for *matrix effects* is to spike one or more of the environmental samples with a known amount of the contaminant and then determine how much of the spike can be recovered using the analytical method. The sample is split into two portions and a spike is added to one half. The calculation of percent recovery is similar to that shown in Eq. 4–2, with the exception that the unspiked, original sample (*Sample*) takes the place of the blank:

$$\text{Percent recovery} = \frac{Spike - Sample}{Spike\ Amount}(100\%). \quad \text{(Eq. 4–3)}$$

The general criterion for most environmental sampling and analytical methods is that the relative error or percent recovery should be within ± 25 percent of the true value. Thus, a sampling and analysis method should be capable of providing an accurate result within 75 percent to 125 percent of the true environmental value. Fortunately, many environmental sampling and analysis methods exceed this minimum criterion.

Precision, a measure of the reliability of the analytical method over time, is measured by comparing the closeness of replicate samples to the average or mean value. The mean ($\bar{x}$) is the sum of all individual sample values ($\bar{x}$) divided by the total number of samples (N) as represented in:

$$\bar{x} = \frac{\sum_{i=1}^{n} x_i}{N}. \qquad \text{(Eq. 4–4)}$$

The sample standard deviation (σ_{n-1}) is a measure of the closeness or spread of the sample data in comparison with the mean ($\bar{x}$):

$$\sigma_{n-1} = \sqrt{\frac{\sum_{i=1}^{n} (x_i - \bar{x})^2}{N - 1}}, \qquad \text{(Eq. 4–5)}$$

where x_i are the individual measured values and $N - 1$ is the degrees of freedom or total number of samples (N) minus one. The resulting σ_{n-1} gives an approximation of how far away the values are from the true mean. If the data are normally distributed and follow a normal bell-shaped curve, then about 68 percent of the data are within one standard deviation from the mean. In addition, about 95 percent of the data would be within two standard deviations from the mean. Because we still need a measure of closeness relative to the mean, the *relative standard deviation* or *coefficient of variation (CV)* is used as the measure of precision:

$$CV = \frac{\sigma_{n-1}}{\bar{x}}(100\%). \qquad \text{(Eq. 4–6)}$$

The general criterion for most environmental sampling and analytical methods is that the coefficient of variation should be within ± 10 percent.

Most environmental instruments and analytical methods specify the precision and accuracy, which aids in the initial selection process. However, it must be noted that most of these instruments and methods were verified under ideal laboratory conditions. Proper QA/QC requires that accuracy and precision are evaluated and maintained for all environmental monitoring activities.

Field Blanks

To ensure the proper handling and analysis of samples *field blanks* are required with sampling and analysis methods. The goal is to ensure samples are free from potential contamination before, during, and after sampling. In general, at least two field blanks are required for every 10 samples. The field blanks accompany the samples in transit and are handled and analyzed as samples. In some cases a spiked blank can also be carried into the field to determine if sample degradation may have occurred during the sampling process. If no contamination or degradation occurred during transit or sample handling, then the average field blank measure is often subtracted from the samples, as a matrix effect of the sample media. The field blanks can also be used to determine the method detection and quantification limits discussed as follows.

Detection and Quantification Limits

The *method detection limit* (MDL) is a measure of the ability of the analytical method to distinguish a sample from the background or blank samples. It must first be understood that currently scientists cannot measure chemical concentrations, or even distances, down to a precise zero. The subdivisions of measurement can always be broken down into smaller subdivisions until eventually one reaches the limits of the instrumentation or powers of observation. Thus, "zero" does not have a practical application in environmental science. Identifying the limits of our observation and measurement system is more practical. The MDL is a measure of the analytical limit of an observation. Samples with concentrations below the MDL are considered to be "non-detectable" or below the limit of the analytical capabilities.

The *method quantification limit* (MQL) is a minimum criterion used to establish the ability of an analytical method to quantify the contaminant concentration in relation to the background signal, produced by an analytical instrument. While the MDL establishes a limit of detection, the MQL establishes the limit of measurement. The MDL and MQL can be calculated using the standard deviation of the background signal, also known as the "signal" noise, from the analytical instrument. The methods differ between disciplines, government agencies, and scientists, but generally include the following steps:

1. Collect at least seven field or method blanks.
2. Analyze the blanks using the analytical method for the desired contaminant.

3. Calculate the σ_{n-1} of the blank concentration from the instrument response or "signal."
4. The MDL is three times the σ_{n-1}.
5. The MQL is 7 to 10 times the σ_{n-1}.

Calibration

The accuracy of environmental monitoring results is dependent on proper *calibration* of sampling equipment (e.g., flow rates) and analytical instruments. Calibration is the act of checking and adjusting the accuracy of a measurement device by comparison with a known standard(s). The general rule is that calibration must be checked and adjusted prior to sampling and analysis, and then checked again at the conclusion of the process. In some cases, the calibration is also checked at regular intervals during the sampling and analysis method. The goal is to ensure the proper functioning and accuracy of equipment during the entire process. Data must be disregarded if the instrumentation was not functioning properly at any stage of the process. Because most environmental data can be used as evidence in a court of law, it is critical to ensure proper calibrations and calibration checks have been performed and documented. It must be established that the sampling equipment and analytical instrumentation were functioning properly.

Actual calibration of the analytical instrumentation is often termed *standardization* and involves the preparation of a *standard curve,* which is used to convert the instrument's response (e.g., a voltage or peak area) to a concentration. Known standards are prepared at various concentrations (including blanks), analyzed in at least triplicates, and then used to define the standard curve. At least one blank and five concentrations are required to establish a standard curve. The resulting curve must be accurate, precise, unambiguous, and *correlated* such that the concentration can be predicted from the instrument signal. For linear correlations, which are most desired, the *working linear range* concept can be applied to define the standard curve. The criteria vary, but the following is an example of the general criteria used to define the lower and upper concentration limits of a working linear range:

- The lower limit must be greater than the defined MQL.
- The instrument must accurately measure all standards within ± 10 percent of the true value (relative error).
- The instrument must precisely measure all standards within ± 5 percent of the mean value (coefficient of variation).
- The resulting range must be linear with a significant ($p \leq 0.05$) Pearson correlation coefficient (*r*). (Please refer to *Biostatistics: The Bare Essentials* by Norman and Streiner (1994) or any of several general statistics texts on linear regression and tests for significance of the Pearson correlation coefficient.)

All environmental samples are then concentrated or diluted for analysis within the working linear range. Ultimately, this ensures that all samples will be analyzed within the QA/QC limits of the analytical method.

Reporting Analytical Results and Errors

Analytical results must be reported in a manner that reflects the analytical limitations of the method. The following classifications are often used in reporting environmental concentrations:

- "Below detectable level" or "Below detection limit"—measurements below the MDL.
- "Trace level"—measurements between the MDL and MQL.
- A numerical concentration can be reported when the measurement is above the MQL and within the working linear range of the standard curve.

Reporting both sampling and analytical errors is equally important. The resulting data from air sampling will exhibit variability due to random errors throughout the sampling and analysis process. The sources of random errors include:

- variability in sample media and between batches/lots;
- variability in sample volumes and flow rates between air sampling devices;
- variability in the handling of samples (e.g., storage temperature and transportation times);
- variability in sample preparation; and
- variability in analytical methods and instrumentation.

The combined error associated with all of the potential sources of variability will have an effect on one's overall confidence in the analytical data. It is the responsibility of the scientist to diligently work to reduce errors and increase confidence in the data. Nonetheless, random errors must be evaluated and reported in the final analytical results.

The propagation of error in science is complex and beyond the scope of this text. However, when the variability or errors are said to be random and normally distributed, then the individual standard deviations can be combined to represent the *sampling and analysis variability* (σ_{sa}). Equation 4–7 represents the additive nature of independent random errors and how they are commonly handled in the environmental sciences:

$$\sigma_{sa}^2 = \sigma_s^2 + \sigma_a^2, \qquad \text{(Eq. 4–7)}$$

where σ_s is the sampling variability (e.g., standard deviation of the air sampling device flow rates) and σ_a is the analytical variability (e.g., standard deviation of replicate samples analyzed under the same conditions). When the variability between samples (σ_{n-1}) is found or transformed into a normal distribution, then the total variability σ_t can be determined using these principles of additive error propagation:

$$\sigma_t^2 = \sigma_{n-1}^2 + \sigma_s^2 + \sigma_a^2. \qquad \text{(Eq. 4–8)}$$

The resulting data are then reported as a mean and standard deviation, as $\bar{x} \pm \sigma_t$, to represent the central tendency and spread of the sample data. The mean and standard deviation provide a better representation of the sample population and its variability. When comparing samples to regulatory standards or criteria both the mean and standard deviation are considered using a calculated 95th percentile confidence interval, which is equivalent to about two standard deviations on either side of the mean (see Chapter 11 on *Risk Assessment*). Any overlap of the 95th percentile confidence interval with the criterion level is an indication that the sampling results exceed the established standard.

III. THE HUMAN AS AN AIR SAMPLER

Inhalation is the primary route of exposure associated with air pollution (see Table 7–1). Inhalation exposures to air pollutants are governed by a number of factors, namely (1) anatomy and physiology of the respiratory system, (2) chemical properties (e.g., solubility) of the air pollutant, (3) contaminant concentrations, (4) physical properties (e.g., particle size) of the pollutant, and (5) environmental conditions. In order to conduct air sampling and analysis representative of human inhalation exposures, one must understand and attempt to mimic human exposure mechanisms (see Chapter 7 on *Human Exposures* for more details).

Human Respiratory Tract

The human respiratory tract (**Figure 4–2**) can be broken down into the nasopharyngeal (upper respiratory tract), tracheobronchial, and pulmonary regions. The anatomy and physiology of each of these regions are unique and pollutant deposition (collection) mechanisms vary among them. In addition, many respiratory diseases are specific to one or more of these regions (e.g., nasal cancer associated with nickel exposure, or emphysema in the pulmonary region associated with cigarette smoking). Modern air sampling methods have been devised to represent exposures of these regions.

In Figure 4–2, the nasopharyngeal region, or naso-oro-pharyngo-laryngeal (NOPL) region, includes the head airways region (nose and mouth) descending to the larynx (voice box). The inhaled air is humidified and warmed as it comes in contact with the moist, mucous-laden tissues of the nose, mouth, and back of the throat

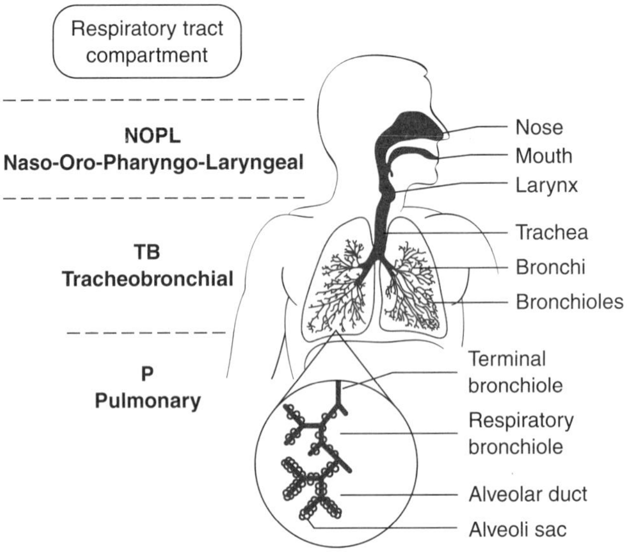

Figure 4–2 Human respiratory tract.
Source: The University of California Air Pollution Health Effects Laboratory, with kind permission.

(pharynx). Gaseous pollutants and a large number of particles less than 100 μm in *aerodynamic diameter* (defined in Chapter 3) enter and pass through the nasopharyngeal region. Water-soluble gases are readily removed from the inhaled air stream. In addition, a significant portion of the larger diameter (1 to 100 μm) and ultrafine (0.001 to 0.01 μm) particles are collected in the nasopharyngeal region.

The tracheobronchial (TB) region is comprised of the main conducting airways of the lungs, including the trachea, bronchi, and bronchioles (Figure 4–2). As the TB airways branch out (16 total generations in the average person) the air velocity decreases and the airflow becomes less turbulent (i.e., more laminar). The airways are covered in fine hair-like cilia and mucus, which are components of the *mucociliary-escalator.* Once again, water-soluble gases and a wide range of aerosol sizes (namely 0.001 to several μm) are collected in the mucous-covered linings of the tracheobronchial region.

The terminal region of the respiratory tract, the pulmonary region (P), includes the terminal bronchioles and alveoli (Figure 4–2). Gas exchange with blood from the right side of the heart (i.e., the pulmonary circulation) occurs in this region where there are no cilia or mucus to interfere with gas diffusion. A thin layer of surfactant coats the inner surfaces of the alveoli, but the purpose is to prevent collapse of the alveoli during exhalation. Particles in the range from about 0.01 to 5 μm in aerodynamic diameter are selectively deposited in the pulmonary region of the average adult. In addition, less water-soluble, non-polar, and low molecular weight gaseous pollutants can interact with alveolar tissues and some may even cross the gas-exchange region into the pulmonary bloodstream.

It should be evident that the anatomical and physiological properties of these three regions of the respiratory tract are important determinants in the deposition of air pollutants. As a result, air pollution sampling methods are designed to represent either the whole respiratory system (termed inhalability) or one or more regions of the respiratory tract.

Gases and Vapors

For gaseous pollutants, chemical properties such as water solubility, polarity, ionic charge, and molecular weight are important considerations for inhalation exposures. Water-soluble, polar, and ionic (e.g., HCl) compounds are more likely to interact with mucous membranes but less likely to cross cell membranes. Conversely, non-polar hydrocarbons (e.g., benzene or hexane) and some low molecular weight compounds (e.g., methanol) are less likely to interact with mucous membranes and more likely to cross cell membranes and enter the circulatory system. Sampling and analysis methods for gaseous pollutants use these chemical properties of solubility, polarity, ionic charge, and molecular weight to collect, identify, and measure pollutants in the air humans breathe.

Aerosol Particles

For aerosol particles, *aerodynamic diameter* is a critical property governing deposition within the regions of the respiratory tract. Several occupational and environmental sampling schemes have been developed to represent the size-selective exposures occurring within the respiratory tract. The American Conference of Governmental Industrial Hygienists (ACGIH®) developed size-selective sampling methods based on the relative collection efficiencies for particles in the inhalable, thoracic, and respirable fractions of the respiratory system (**Figure 4–3**). Samplers for inhalable fractions (0 to 100 μm collection efficiency curve), thoracic fractions (10 μm cut-point), and respirable fractions (4 μm cut-point) are in common use in occupational health.

The cut-point refers to the aerodynamic diameter where the mass fraction collection efficiency is 50 percent, and is representative of the mass median diameter (MMD) of the particles collected.

Inhalable samples are collected for select inorganic compounds (e.g., nickel and thallium), wood dusts, and a large number of pesticides. *Respirable samples* are collected for welding fumes (e.g., zinc and iron oxides), inorganic dusts (e.g., Portland cement and crystalline silica), and coal dust. Samplers for the thoracic fraction (10 μm cut-point) are used for sulfuric acid (H_2SO_4) aerosols and cotton dust, both known to affect pulmonary function. Cotton dust exposures during processing are also known to cause a form of chronic bronchitis called *byssinosis.* In general, occupational health aerosol sampling is most often conducted in association with a known source and identified disease or critical health effect (e.g., bronchitis, asthma, cancer, or irritation).

Currently, the U.S. Environmental Protection Agency (U.S. EPA) uses size-selective sampling methods for PM_{10} (10 μm or less diameter cut-point), $PM_{2.5}$ (2.5 μm or less diameter cut-point), and ultrafine particles

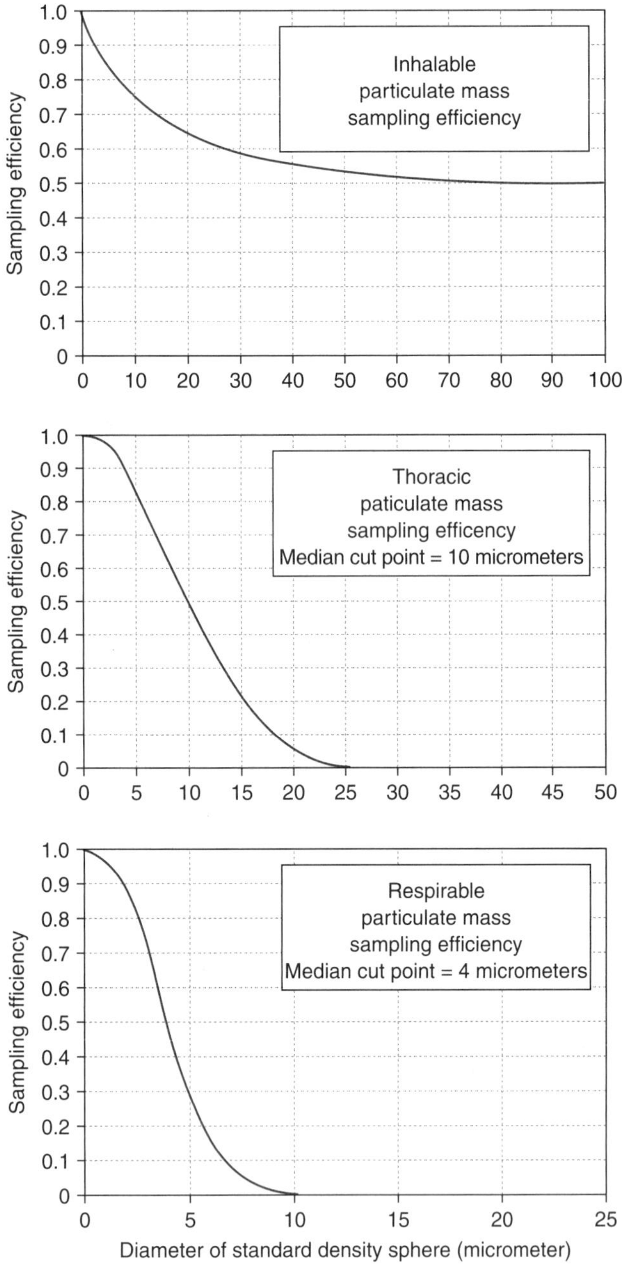

Figure 4–3 American Conference of Governmental Industrial Hygienists (ACGIH®) aerosol mass fractions recommended for particle size-selective sampling.
Source: The University of California Air Pollution Health Effects Laboratory, with kind permission.

(0.1 μm or less diameter). It must be noted that $PM_{2.5}$ was not exclusively selected to measure the respirable fraction, as it was also intended to help separate natural from human-made air pollution sources. However, without identification of the specific pollutants, sources, and critical health effects, reliable distinctions between $PM_{2.5}$ and PM_{10} cannot be easily made. Because $PM_{2.5}$ is part of PM_{10}, and ultrafine particles are gaining more attention, many air pollution studies have expanded the environmental-related particulate classifications into:

- *coarse particulate matter* ($PM_{10-2.5}$)—mass fractions between 10 μm and 2.5μm diameter cut-points;
- *fine particulate matter* ($PM_{2.5}$)—mass fractions below a 2.5 μm diameter cut-point; and
- *ultrafine particulate matter* (UFP)—particle counts below a 0.1 μm cut-point.

The remaining focus of aerosol sampling in this chapter will be on environmental air pollution and the U.S. EPA classifications for PM_{10}, $PM_{2.5}$, and UFP. It must be noted that generalized mass-based measures of air pollutants (e.g., PM_{10} and $PM_{2.5}$ specifically) are useful for regulatory control but do not directly identify sources or protect the population from specific health hazards. The reliance on mass-based measures in air pollution regulation and control is primarily due to the complexity of the ambient environment in comparison with the simpler occupational setting. Numerous occupational standards and sampling and analytical methods exist for specific aerosol hazards, such as metal compounds and oxides (e.g., iron oxide), inorganic salts and compounds, dusts (e.g., coal and wood), pesticides, mineral contaminants (e.g., asbestos and crystalline silica), carbon black, and even aerosol mists (e.g., sulfuric acid and mineral oil). The future success of non-occupational, environmental air pollution controls and improved human health will in all probability rely on a more rigorous treatment of pollutant identification and quantification, much like has occurred the workplace setting.

Deposition Mechanisms for Aerosol Particles

Aerodynamic diameter plays an important role in the deposition of air pollutants within the respiratory tract, as well as in air sampling devices. Deposition occurs when a particle moves out of the inhaled air stream and touches an airway wall. The basic mechanisms of deposition are diffusion, sedimentation, and inertial impaction. A fourth basic mechanism, *interception*, occurs when particles follow an air stream but contact and adhere to a nearby surface. Interception occurs with larger particles and is a function of particle diameter or shape in relation to airway diameter. As particle size increases, or airway diameter decreases, particles are more likely to contact and adhere to nearby surfaces. *Diffusion* is the spontaneous movement and mixing of particles within an air stream across a gradient, from high to low concentrations. Diffusion occurs predominantly with small particles less than 0.3 μm in diameter, whose motion is strongly affected by collisions with gas molecules (i.e., Brownian motion). *Sedimentation* occurs with larger particles under the influence of gravity. According to Newton's Second Law (force equals mass times acceleration) the gravitational force increases with particle mass. In an air stream, sedimentation occurs when the gravitational force overcomes opposing drag and buoyancy (directed upward) forces. The resulting sedimentation velocity is a function of particle diameter, shape and density, air density, and air velocity. In general, particles with larger aerodynamic diameters have larger sedimentation velocities. *Inertial impaction* occurs as a function of particle size and density, air stream velocity, proximity of a collection surface, and change in air flow direction. Particle inertia causes larger particles to resist abrupt changes in direction and leave an air stream with significant velocity. Essentially, the particle is unable to follow the air stream due to its inertia and it contacts a nearby surface. **Figure 4–4** illustrates the relative particle sizes and deposition efficiencies for these four mechanisms.

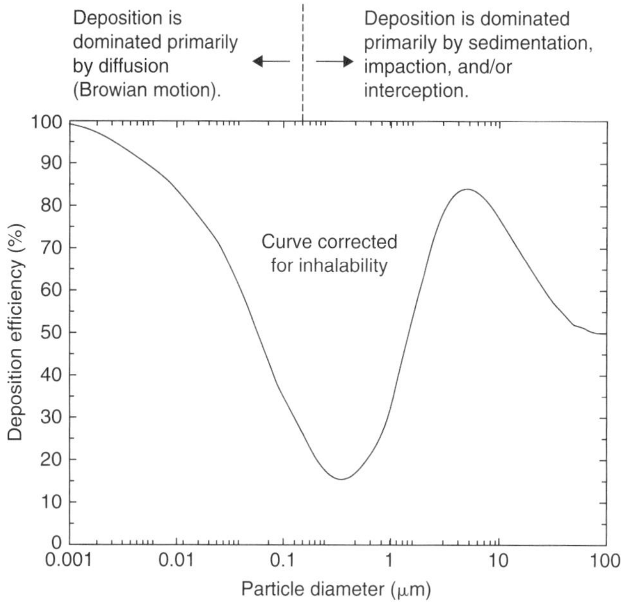

Figure 4–4 Particle deposition efficiency in the human respiratory tract during normal respiration, according to particle diameter. In practice, air samplers should be corrected for inhalability, as shown in the figure.
Source: The University of California Air Pollution Health Effects Laboratory, with kind permission.

IV. PARTICLE SAMPLING

Particle sampling involves the collection of suspended particulate matter from the air using similar mechanisms of deposition (e.g., inertial impaction, interception, sedimentation, and diffusion) as those seen in the human respiratory tract. Two primary methods of particle collection are filtration and inertial collection. The two methods are often applied in tandem to obtain a size-selective sample representative of human respiratory tract exposures. Additional sampling techniques are mentioned briefly in this section. A summary of particle sampling devices is provided in **Table 4–3**.

Filtration

For particles suspended in air, filtration is the process by which the air is passed through a filter medium and the particles are collected on the surfaces of the filter. As with the human respiratory system, the primary mechanisms of collection include diffusion, sedimentation, interception, and inertial impaction. As air is drawn through the filter media the air is forced to travel an often tortuous path through compacted fibers or porous membranes. Collectively, these four deposition mechanisms aid in the efficient collection of a wide variety of particle sizes (Figure 4–4). As can be seen in Figure 4–4 collection efficiency is lowest for particles with an aerodynamic diameter of about 0.3 μm where none of the collection mechanisms are dominant. It is for this reason that the collection efficiencies for filters are often optimized against a challenge aerosol of dioctyl phthalate (DOP) spheres with an aerodynamic diameter of 0.3 μm.

Table 4–3 Particle sampling collection devices for subsequent analysis of particle shape, size, or composition.

Sampler	*General Principle*	*Common Diameter Range*	*Artifacts and Limitations*
Filters	Particles are collected on filter surfaces by interception, impaction, diffusion, sedimentation, or electrostatic interaction	0.01–100 μm	Efficiency changes as the filter loads; filters are fragile; leakage and/or non-isokinetic sampling may occur; and sample may degrade on filter during or after sampling
Electrostatic precipitators	Electrical charges are added to particles prior to collection on an oppositely charged surface	0.01–60 μm	Small particles may lose charge prior to collection; deposits may be non-random and/or overlapping
Thermal precipitators	A thermal gradient is used to move particles towards a collection surface	0.01–5 μm	Temperature related artifacts may occur (e.g. vaporization of sample)
Elutriators	Gravitational sedimentation is used to collect particles	1–200 μm	Very small particles <1μm have inefficient collection efficiencies
Impingers	Particles in air are collected in a liquid after collection by inertial impaction	1–100 μm	Collection efficiency diminishes for particles <1 μm in diameter; agglomerated or fragile particles may break up; soluble components may be re-aerosolized
Impactors	Air is accelerated through an opening and particles are collected on a nearby surface due to inertial impaction; cascade impactors can collect multiple size fractions	0.05–100 μm	Particle loss and re-entrainment may occur due to particle bounce, overloading, or improper sampler preparation; petroleum or silicone collection media may interfere with analysis
Cyclones	Centrifugal motion is used to move particles to a collection surface; particles are collected by inertial impaction	0.05–5 μm	Collection efficiency diminishes for very small particles <0.5 μm in diameter; non-conductive cyclones can develop charges and alter size distributions

Data from Phalen (2009); Lodge and Chan (1986).

Following sample collection, the filters can be analyzed *gravimetrically* (e.g., by weighing the filter) to determine collected mass, chemically for composition (e.g., acidity or lead content) or microscopically for particle identification and classification (e.g., ultrafine particles or asbestos fibers).

Gravimetric analysis is the measurement of a weight change of the substrate before and after air sampling. The initial weight is subtracted from the final weight to provide the mass collected during the sampling period.

In air pollution studies, the mass collected is divided by the air volume sampled to give the mass concentration:

$$\text{Air concentration} = \frac{\mu g}{m^3} = \frac{\textit{mass collected}(\mu g)}{\textit{air volume sampled}(m^3)}. \quad \text{(Eq. 4–9)}$$

Common filter media include glass fiber, quartz, cellulose, and porous organic membranes (e.g., polyvinyl chloride). **Table 4–4** summarizes the properties and uses

Table 4–4 Common filter media used in air sampling for particulates.

Filter Media	*Properties*	*Common Use*
Cellulose	Fibrous; hydrophobic; cellulose can be burned off in a furnace with little to no residual ash; inexpensive	Elemental/metals analysis
Glass Fiber	Fibrous; slightly hydrophobic; withstands high temperatures; efficient particulate retention	Gravimetric analysis for dust and particulates
Mixed Cellulose Ester (MCE)	Membrane filter; hydrophilic; soluble in aqueous solvents for metals analysis; can be made transparent (e.g., cleared with acetone solvent) for light microscopy analysis	Asbestos and mineral fibers; elemental/metals analysis
Polycarbonate	Membrane filter; hydrophobic; transparent	Asbestos and mineral fibers; scanning electron microscopy
Polyvinyl Chloride (PVC)	Membrane filter; hydrophobic; low silica content; non-oxidizing properties; PVC can be burned off in a furnace with little to no residual ash; inexpensive	Gravimetric analysis for dust and particulates; common for respirable dust; use in higher humidity conditions
Polytetrafluoroethylene (PTFE)	Membrane filter; hydrophobic; inert properties (e.g., resistant to acids, bases and many organic solvents)	Gravimetric analysis for corrosive (acids and bases) aerosols; reactive/sensitive organic aerosols or vapors (e.g., pesticides and isocyanates); polynuclear aromatics
Quartz	Fibrous; slightly hydrophobic; withstands high temperatures; efficient particulate retention; low metals content; can be heat-treated to remove organic impurities	Gravimetric analysis for dust and particulates (e.g., PM_{10}); total suspended particulates (TSP); diesel particulates; organic pollutants

Source: University of California Air Pollution Health Effects Laboratory, with kind permission.

of common filter media. *Glass fiber filters* are composed of finely spun borosilicate fibers and are used when desired particle diameters exceed 0.3 μm or in the presence of high temperatures and corrosive atmospheres (e.g., industrial stack effluents). Glass fiber filters are inert, hydrophobic (i.e., do not collect water), inexpensive, and a popular choice for particulate matter sampling. High-purity quartz fiber filters are used in U.S. EPA methods for total suspended particulates (TSP) and PM_{10} using a high volume air sampler, as well as particulate emissions from stationary sources (e.g., industrial stack or power-plant emissions). High-purity *quartz filters* have low metal contents and are heat-treated to remove trace organic contaminants. *Cellulose fiber filters* are often used for metal analyses, because they are low in trace metals and can be reduced to ash in a furnace to improve analytical sensitivity. Porous organic *membrane filters* include mixed cellulose ester (MCE), polytetrafluoroethylene (PTFE), polycarbonate, and polyvinyl chloride (PVC). The general properties and uses of these filter media are provided in Table 4–4.

Most filter media, regardless of reported pore size, efficiently collect ultrafine (<0.1 μm diameter), fine, and coarse particles. In addition, because multiple deposition mechanisms are involved in the filtration process, research-quality filters collect a wide range of particle sizes. However, *home air-filters* can have low collection efficiencies and are often inadequate for environmental sampling. Filters alone are not suited for size-selective sampling, unless they are behind a size-selecting device.

Therefore, filters are often used in tandem with inertial pre-collectors to collect a size-selected subset of the original air sample.

Inertial Collection

As discussed earlier, inertial impaction occurs as a function of (1) particle size, (2) particle shape and mass, (3) air stream velocity, (4) proximity of a collection surface, and (5) change in air flow direction. Particle inertia can cause larger particles to leave an air stream with high velocities and contact a nearby surface. That is, the air stream goes around an obstacle, but the particle crashes into the obstacle, and is collected. In *inertial collection* the air stream velocity and change in direction are specifically designed to remove larger particles from the air stream by impaction on a filter, hard surface (e.g., cyclone collector), or a solid surface often coated with a grease or oil. Softer surfaces (e.g., coated surfaces) have an advantage as they can absorb energy and prevent particle bounce. As with the human respiratory system, these inertial collectors do not effectively remove all particles of a given aerodynamic diameter, and they are defined by their median collection efficiency or 50 percent cut-point. The cut-point is often used to describe the upper limit of particle diameters collected by the device.

Inertial collectors are used to (1) directly collect a size-selective sample for analysis, or (2) to remove larger particles from the air stream prior to subsequent collection and analysis of the uncollected air sample. In the first case, a filter or grease-coated substrate may be used to collect the impacted particles. In this case, the filter is used as a collection surface but does not actually filter the air, as air does not pass through the filter. Liquid aerosol particles do not bounce, and they can be collected on uncoated substrates. The filter or coated substrate can then be analyzed *gravimetrically* to determine collected mass or *chemically* analyzed for aerosol composition (e.g., for pH, sulfur, or metal content). **Figure 4–5** shows a PM_{10} impactor, with an internal filter-based collection substrate, used for personal exposure monitoring. Cascade impactors (**Figure 4–6**) contain several impactor stages in series, which are designed to selectively collect multiple size fractions simultaneously on filter substrates, coated substrates, or uncoated substrates. Size-selective impaction is achieved by modifying the nozzle (air jet) diameter and the distance of the impaction plate from the nozzle (Figure 4–6A).

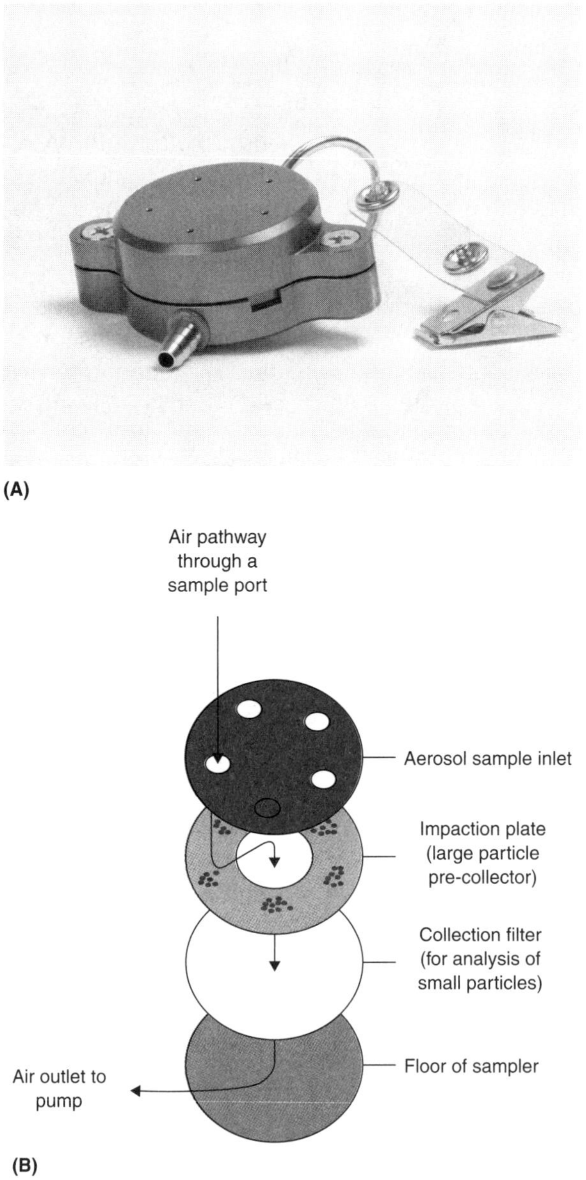

Figure 4–5 (A) Single-stage PM_{10} impactor with internal filter collection device (Personal Environmental Monitor; PEM™). (B) Simplified diagram of single-stage impactor design showing: (1) the air inlets towards the outer edges of the sampler; (2) the impaction plate for collection of larger particles; (3) the collection filter for collection of smaller particles after pre-collection; and (4) the air outlet to a vacuum pump.
Courtesy of MSP Corporation, Shoreview, MN, USA.
Source: The University of California Air Pollution Health Effects Laboratory, with kind permission.

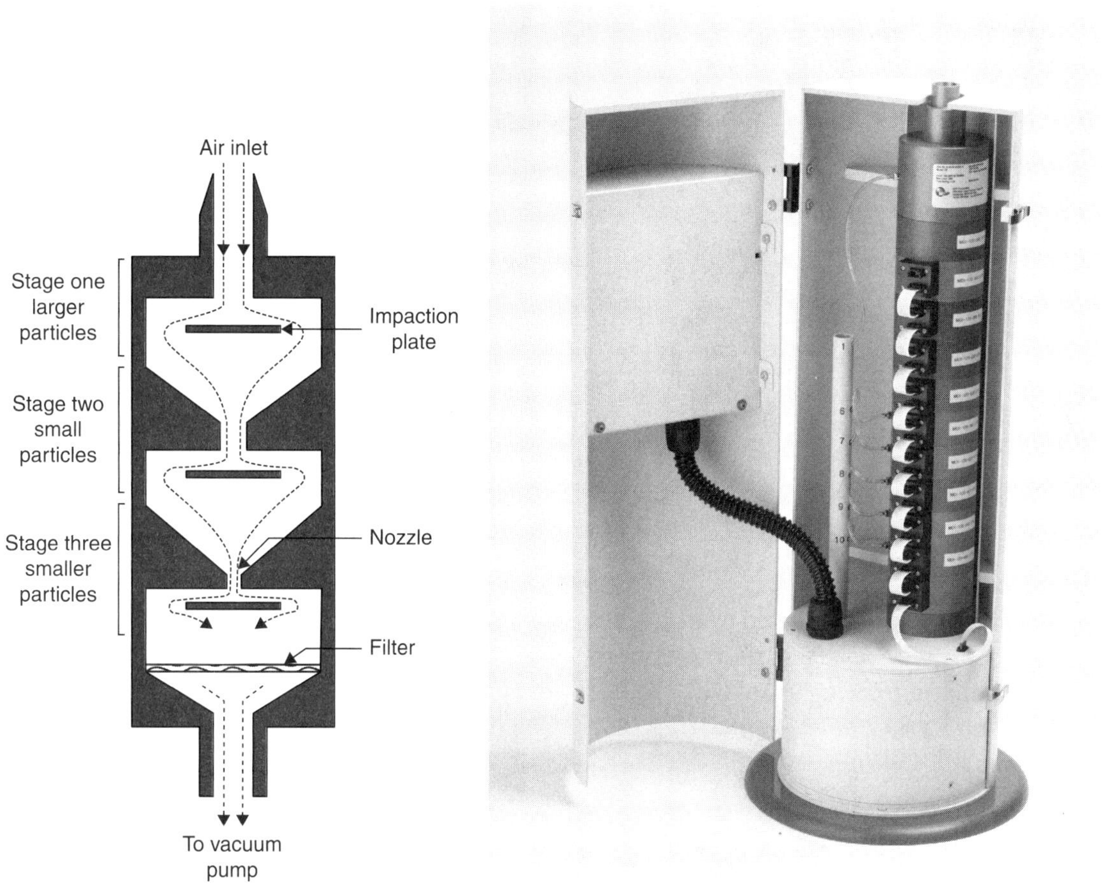

(A) (B)

Figure 4–6 (A) Illustration of a three stage cascade impactor design showing the changes in air flow currents at each stage. Notice the narrowing of the nozzle (air jet) diameter and closeness of the impaction plate at each successive stage. (B) A 10-stage MOUDI™ (Micro-Orifice, Uniform Deposit Impactor) used to collect particle size ranges from about 0.01 to 18 μm.
Courtesy of MSP Corporation, Shoreview, MN, USA.
Source: The University of California Air Pollution Health Effects Laboratory, with kind permission.

In general, particles with increasingly smaller aerodynamic diameters are collected as the nozzle diameter and impaction plate distances are effectively reduced. Larger particles are collected in the initial stages and smaller particles in the final stages.

In the latter case, larger particles are removed from the air stream to allow size-selective sampling and analysis of the remaining air. The air stream may pass through a filter media to collect the remaining particles for analysis. For example, **Figure 4–7** shows a typical PM_{10} sampler with size-selective inlet, impaction plate, and internal filter. **Figure 4–8** shows a size-selective cyclone pre-collector design, which is used to remove larger particles from the air stream before the air is delivered to a filtering device. With common designs the larger particles, mostly above 4 μm in aerodynamic diameter, are collected by the cyclone as a result of inertial impaction. The remaining particles, mostly below 4 μm in diameter, are collected by the filter. The filter sample represents the respirable fraction of particulates, which are able to reach the pulmonary region of human lungs. In addition, the size-selective air stream may be sent to a *direct-reading monitor* for continuous, real-time analysis. Size-selective, direct-reading instruments (**Figure 4–9**) are becoming more and more popular due to their versatility and ability to provide real-time data relevant to changes in pollution sources (e.g., traffic) and meteorological conditions.

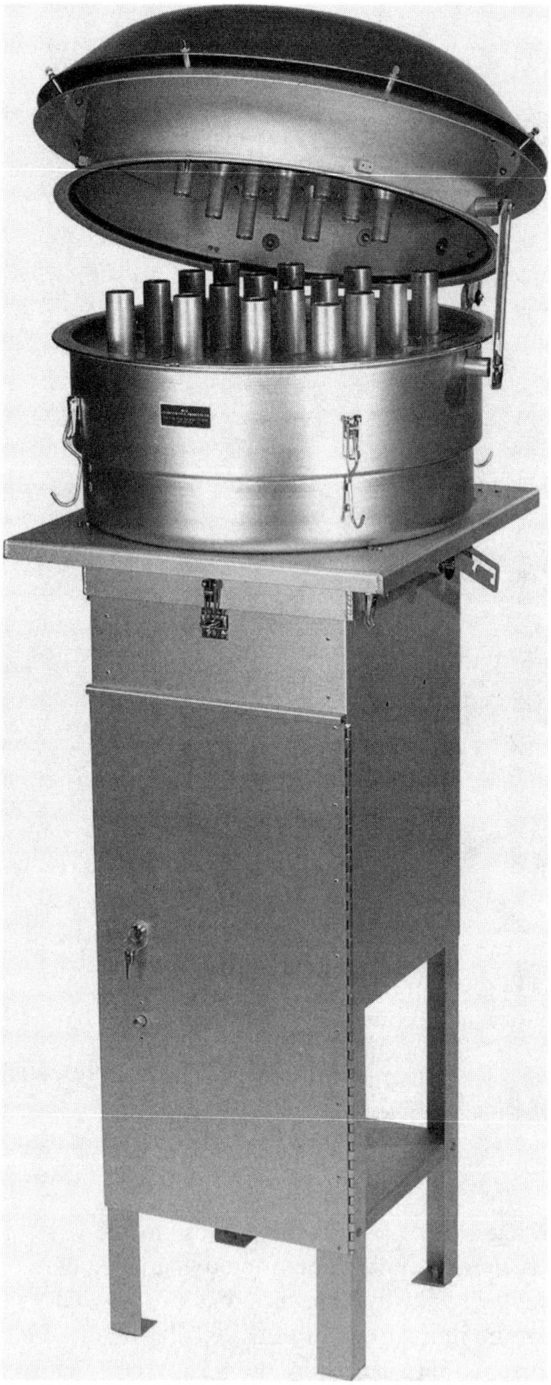

Figure 4–7 Size-selective high volume air sampler with PM_{10} impaction air jets (shown in the illustration) and plate. Larger particles are impacted and collected within the open section of the sampler (shown). An internal filter (just beneath the bottom jets) collects non-impacted particles in the PM_{10} and smaller range.
Courtesy of HI-Q Environmental Products Company, Inc. (San Diego, CA; www.HI-Q.net).

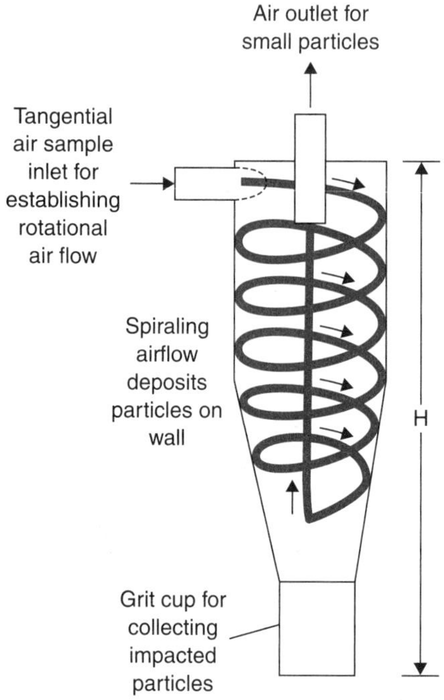

Figure 4–8 Size-selective cyclone pre-collector design. *Source:* The University of California Air Pollution Health Effects Laboratory, with kind permission.

Additional Sampling Methods

Additional sampling methods are listed in Table 4–3. *Electrostatic precipitators* have been used in indoor air quality studies, because of their portability, low flow rates, and quiet operation. Larger systems have been widely used in industry as air cleaning devices (Chapter 6). However, electrostatic precipitators cannot be used in explosive atmospheres due to the electrical discharge (arc) used to charge particles. In contrast, *thermal precipitation*, a sampling technique previously used in coal mines, can be used in explosive atmospheres and is effective in collecting small particles. Air travels through a narrow passage with opposing hot and cold surfaces. The heated gas molecules are more energetic and exert a net force on the particle, which results in movement towards the colder surface. Smaller and less dense particles are affected more by the temperature gradient (interaction with gas molecules). It is anticipated that use of thermal precipitation will become more common for sampling as nanoparticles become a more important air pollution

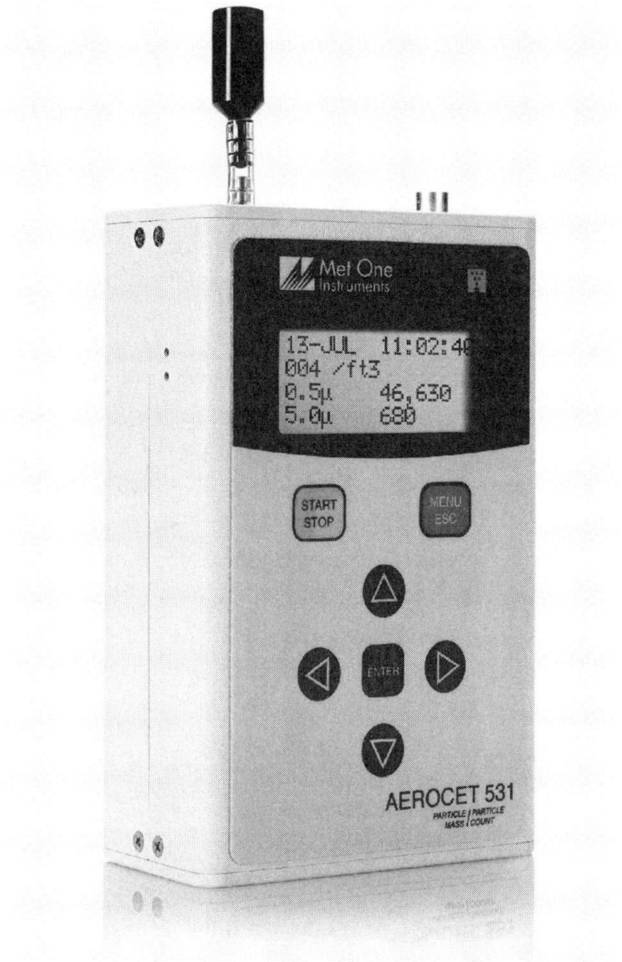

Figure 4–9 A portable, size-selective and direct reading particulate mass profiler and counter used to measure both particle count and particle mass. Courtesy of Met One Instruments, Inc.

issue. *Elutriators* (see Chapter 6) use the force of gravity to aid in the collection of particles by exploiting their terminal settling velocities. Larger and higher density simple-shaped particles will settle faster than smaller, lighter complex-shaped particles. Smaller particles under an aerodynamic diameter of about 1 μm do not settle rapidly. *Elutriation* allows for the separation of particles based on settling velocity and resulting particle trajectory in a moving air stream. Air sampling methods using elutriation, as well as some of the other methods mentioned in this section, are relatively rare compared to filtration and inertial collection for general environmental applications. However, each method has distinct advantages for specialized applications in air sampling and analysis, as well as laboratory-based research and emission controls.

Isokinetic Sampling

The airflow patterns entering sampling devices can be complex and lead to collection of a *non-representative* sample, due to the under sampling or over sampling of different particle sizes (especially large particles). Also, the sampling probe should be thin-walled in order to minimize disturbing the external airflow.

Isokinetic sampling is a sampling technique in which the inlet air velocity is matched and parallel to the air stream flowing around the probe, and where the probe inlet does not disturb the airflow. Isokinetic sampling is applied most often in emission stack sampling, with the stack air velocity and temperatures both matched to reduce sampling errors. Examples of isokinetic and non-isokinetic sampling conditions are illustrated in **Figure 4–10**. In general, isokinetic sampling (A) occurs when the air flow direction and velocity are equivalent outside and inside the sampling probe. Oversampling of large particles (B) can occur when the air velocity within the probe is less than outside the probe. Under-sampling of large particles can occur when the air velocity is greater than outside the probe (C) or with probe misalignment (D) where larger particles may impact on the inner surfaces of the probe inlet. If larger particles are oversampled, then smaller particles are under-represented in the air sample. Likewise, if larger particles are undersampled, then smaller particles are over-represented in the air sample.

Experimental work and theoretical treatments published by Fuchs (1975), Mercer (1973), and Hinds (1999) describe conditions under which particles of various sizes may be sampled isokinetically. It is difficult to isokinetically sample particles with diameters greater than 20 μm, especially when the sampler inlet is tubular and the flow rates are less than about 10,000 cm^3/s or ~17 L/min.

V. PARTICLE ANALYSIS

Instruments for particle analysis (**Table 4–5**) either measure a size-dependent property of particles or exploit a size-dependent property to separate particles prior to collection and analysis. In some cases, instruments will provide both size-dependent and mass-based concentrations. Such is the case with current U.S. EPA regulatory

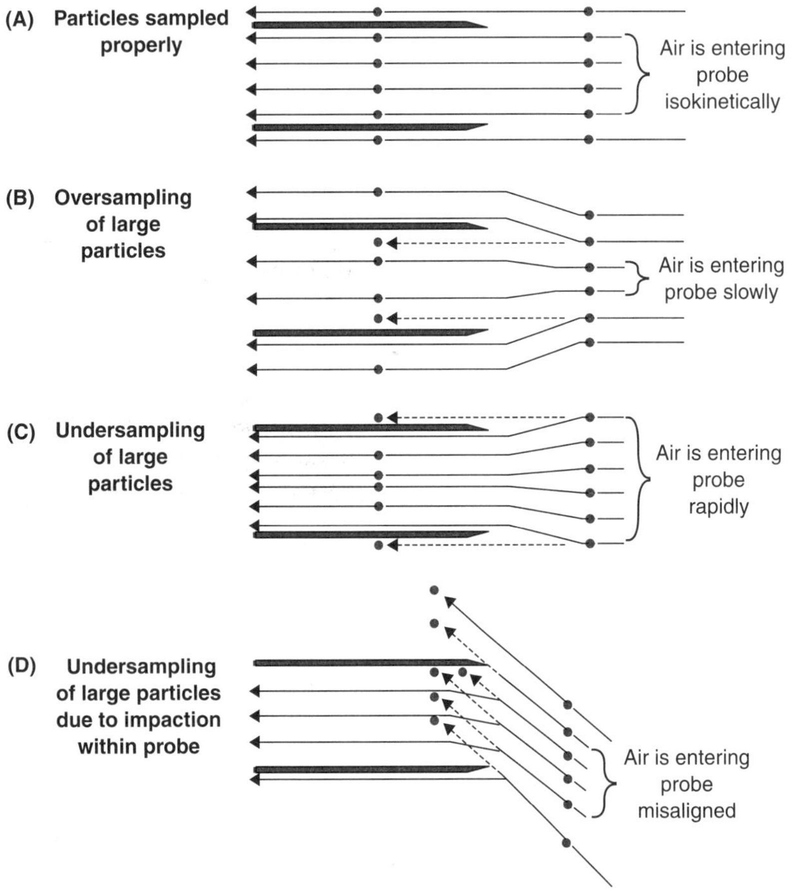

Figure 4–10 Isokinetic and non-isokinetic sampling conditions. Isokinetic sampling (A) occurs when the air flow direction and velocity are equivalent (matched) outside and inside the sampling probe. Oversampling of large particles (B) can occur when the air velocity within the probe is less than outside the probe. Under-sampling of large particles can occur when the air velocity is greater inside the probe (C) or with probe misalignment (D) where larger particles may impact on the inner surfaces of the probe inlet.
Source: The University of California Air Pollution Health Effects Laboratory, with kind permission.

sampling and analysis requirements for PM_{10}, and $PM_{2.5}$. Some of the methods measure distinct size fractions of total particles, using instruments calibrated with a common reference standard (i.e., Arizona road dust). In contrast, health-based air pollution research and activities are increasingly more concerned with particle composition, size, number, and surface area. The rationale behind this is centered on the notion that smaller particles, which have larger surface area per unit mass, are more likely to be toxic according to the surface area that is in direct contact with respiratory tissues. Mass-based collection and analysis is dominated by larger particles, which are often fewer in number when compared to fine and ultrafine particles. Nonetheless, both mass-based and other particle characterization methods of analysis are critical in the recognition, evaluation, and control of air pollutants.

Introduction to Particle Sizing Instrumentation

Particle sizing instruments rely on a number of particle properties and general mechanisms for detection and measurement of particles. The common size-dependent particle properties exploited include:

- particle diffusion coefficient (Brownian movement);
- inertia (e.g., inertial impaction and centrifuge collectors);

Table 4–5 Particle measuring instrumentation for air pollution.

Instrument	*General Principle*	*Common Diameter Range*	*Problems and Limitations*
Optical microscopes	Imaging using glass optics	0.5–500 μm	Limit of resolution about 0.5 μm, which can lead to erroneous sizing data
Electron microscopes	Imaging using electron beam and magnetic "optics"	0.001–100 μm	Vacuum and local heating can alter samples; erroneous sizing data due to shallow depth of focus, small field of view, and coating of small particles with column contaminants
Optical particle counters	Air passes through a light chamber and particles scatter light; the scattered light intensity is indicative of particle number; pre-collectors are used for size-selective analysis; this technology may also be used for mass-based estimates	0.3–100 μm	Interferences from particles that absorb light and high particle concentrations; optics may require frequent cleaning; factory calibration (e.g., Arizona road dust) may not apply to local air pollutants
Radiation mass detectors	Particles collected on a filter tape attenuate a radioactive source (e.g., beta particles); particle mass is correlated with reduction in radioactive intensity; pre-collectors are used for size-selective analysis	1–100 μm (mass-based)	Provides generalized measure of particle mass collected on a filter; suitable only for particles larger than about 1 μm or heavy filter loadings; typical calibration using Arizona road dust must be adjusted for local sources and characteristics
Piezoelectric mass balance monitors	Particles are collected on a piezoelectric, vibrating crystal, or glass element, and change in the resonant frequency is related to collected particle mass; pre-collectors are used for size-selective analysis	1–100 μm (mass-based)	Larger particles greater than a few μm may not effectively adhere to the vibrating crystal; tapered element oscillating microbalance (TEOM) uses a filter to reduce losses, but sensitivity is reduced
Cascade impactor	Multiple impaction stages, arranged in series from large to small collection efficiencies, are used to obtained aerodynamic size distributions	0.05–100 μm	Particle bounce, re-entrainment, and overloading can produce biases; internal losses, low pressures, and electrical charges can distort data
Condensation nuclei counters	Partical number determined by light absorption after particle growth in a supersaturated vapor	0.002–5 μm	Calibration difficult; coincidence errors occur at high concentrations (about 106 to 107 particles/cc); and at the lower limit of detection near 0.002 μm particle counts may be unreliable

(continued on next page)

Table 4–5 Particle measuring instrumentation for air pollution. *(Continued)*

Instrument	*General Principle*	*Common Diameter Range*	*Problems and Limitations*
Centrifuge Spectrometers	Sedimentation across a clean air volume, for deposition on a tape, is enhanced by high rotational speed	0.1–15 μm (varies)	Gases, thermal effects, and inlet losses can distort particle size distributions
Diffusion batteries	Devices measure particle fractions that penetrate through an array of fine screens or channels at various air flow rates; data provide size distributions	0.001–0.1 μm	Data analysis becomes less certain as the size distribution broadens
Charge spectrometers	Particles are drawn between oppositely charged plates and penetration beyond plates is analyzed to yield distribution of charges on particles	1–100 μm	Stability of flow and voltage difficult to maintain; best for monodisperse aerosols; and arcing can invalidate data
Mobility analyzers	After being given known charge levels, particles are selectively deposited according to size	0.002–1 μm	Large internal losses may be particle-size dependent; and calibration dependent on environmental factors
Surface area measurement devices	Degassed particles are allowed to absorb gas at various partial pressures and gravimetric absorption curves provide surface area estimates	0.01–1.0μm	Surface character, such as cracks and voids, make surface area ambiguous conceptually; different gases result in different surface area estimates; low pressures and high temperatures may alter particle properties
Elutriators	Gravitational sedimentation causes particles to deposit, according to their aerodynamic diameter, as a function of the device dimensions and air flow rate	1–200 μm	Requires precise flow control; and electrical charges on particles and diffusion interfere with deposition

Sources: Data from Phalen (2009); Lodge and Chan (1986); and Vincent (1995).

- terminal settling velocity;
- light-scattering properties and patterns;
- electrical charge;
- mobility in electrical fields;
- ability to serve as a nuclei for condensation of supersaturated vapors;
- attenuation of ionizing radiation (e.g., alpha or beta); and
- particle mass effects (e.g., on a balance or an oscillator).

Common particle sizing instrumentation, such as cascade impactors, condensation nuclei counters, centrifuge spectrometers, diffusion batteries, charge spectrometers, mobility analyzers, surface area measurement devices, and elutriators, are briefly described in Table 4–5. In addition, common particle size ranges for available instrumentation are provided, as an indication of the general applications for each methodology. As technologies advance, these size ranges tend to widen and improve in both accuracy and precision.

Particle Distributions

The important particle size distribution parameters reported by particle sizing instruments include measures of central tendency and dispersion, such as (1) the mass median aerodynamic diameter (*MMAD*); (2) count median diameter (*CMD*); and (3) the geometric standard deviation (*GSD*). These important concepts were described in Chapter 3.

Mass-Based Instrumentation

The first two mass-based technologies described in this section are founded on a general principle that changes or perturbations can be observed and quantitatively measured on a collected sample to determine the particle mass (e.g., that has been collected on a filter). The perturbation may be in the absorption of radioactive particles passing through a filter, or changes in the physical properties (e.g., the frequency of oscillation) of a detector attached to the filter. The primary limitations of these methods are:

- the particles must have appreciable mass, which means that they are usually greater than about 1 μm in diameter;
- the methods require a substantial loading of particles on the filter in order to elicit a detectable change; and
- a correction factor, requiring calibration by gravimetric analysis, may be necessary to account for interferences (e.g., high humidity).

The primary advantages of these instrumental methods are their reliability, durability in the field, and low cost of operation. In addition, the U.S. EPA uses mass-based measures for regulatory monitoring and control of regional air pollution. The National Ambient Air Quality Standards (NAAQS) for PM_{10} and $PM_{2.5}$ are both mass-based regulatory standards; however, there is an obvious size-selective component to each one.

Lastly, optical particle counters can also be used for making mass-based determinations of particle concentrations in air. *Nephelometers* measure light scattering or reflected light from particles suspended in air. In contrast, *aethelometers* measure light absorption or the attenuation of a light source by particles suspended in air. The use and application of optical technologies is highly dependent on the optical properties (e.g., index of refraction) of the particles and gases in the air. Thus, a calibration curve or a correction factor based on concurrent gravimetric analysis is often required with optical technologies.

Radiation Attenuation Mass Detectors

With radiation mass detectors, such as *beta attenuation monitors* (BAM), particles collected on a filter-tape attenuate a radioactive source (**Figure 4–11**). With BAM, the radioactive source is a beta particle emitter. The particle mass collected on the filter-tape (**Figure 4–11B**) is correlated with a reduction in measured radioactive intensity. Radiation attenuation mass detectors provide a reliable measure of particle mass collected on a filter. Pre-collectors are used for size-selective analysis of particle mass, with most of the current applications for PM_{10}, $PM_{2.5}$, and $PM_{1.0}$ size fractions. As an aside, some home smoke detectors are based on monitoring the attenuation of alpha radiation by particles flowing between a radiation source and a radiation detector.

Piezoelectric Mass Balance Monitors

Tapered electrical oscillating monitors (TEOM) are one of the more common piezoelectric mass balance technologies used in air pollution monitoring. These monitors collect particles on a filter attached to a vibrating glass element. The resulting change in vibration-related resonant energy is correlated with the collected particle mass. Monitors using piezoelectric vibrating crystals (no filter) provide greater sensitivity; however, particle loss associated with vibration is greatly reduced. Thus, TEOM is often preferred for regulatory applications where sensitivity is currently not an issue of concern. As with BAM, pre-collectors are used with TEOM for size-selective analysis of particle mass.

Optical Particle Monitors

The same technology used with optical particle counters (Table 4–5) can be applied to making mass-based determinations of particle concentrations in air. With light-scattering detectors, known as nephelometers, air passes through a light chamber and particles scatter light. The detection of scattered light can be proportional to total particle mass concentration as compared with a reference standard (e.g., Arizona road dust). However, caution must be applied when particles have indices of refraction that differ (e.g., carbon black and diesel soot) from Arizona road dust because these particles can absorb more light, scatter less light, and not be effectively detected. In contrast, light transmission technologies,

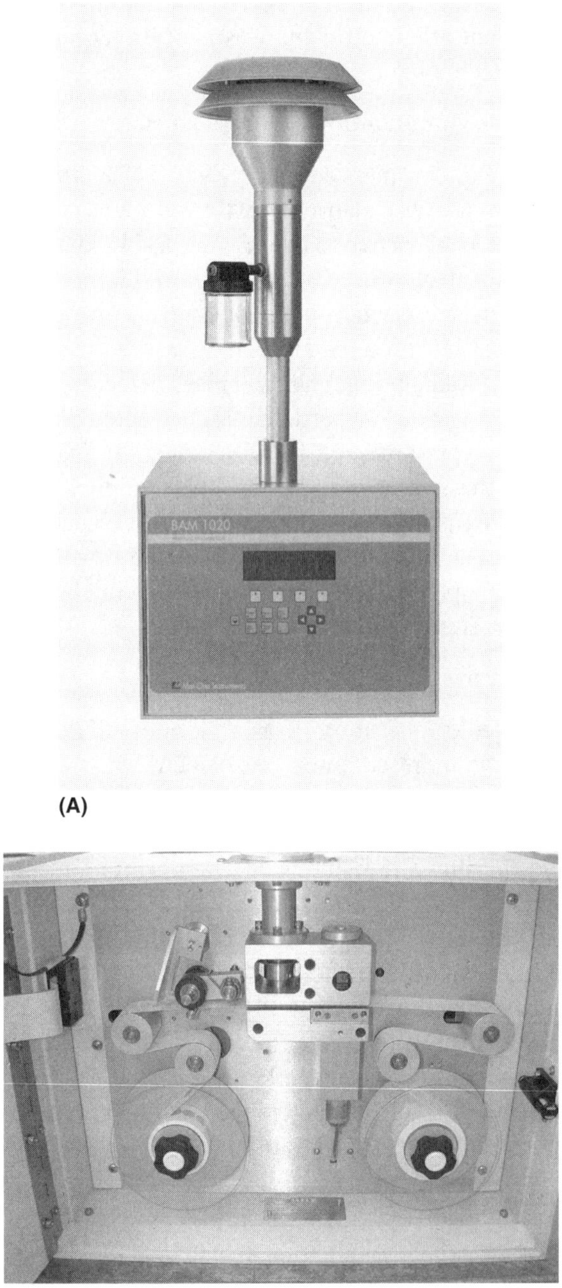

(A)

(B)

Figure 4–11 (A) An external photo of a beta attenuation monitor (BAM) showing a size-selective inlet at the top. (B) Inside of a BAM. Particles collected on a moving filter-tape attenuate a radioactive source (middle). The radioactive source is a beta particle emitter. The particle mass collected on the filter-tape is correlated with a reduction in measured radioactive intensity across the detector.
Courtesy of Met One Instruments, Inc.

also known as aethelometers, are used when particle light absorption is suspected. In both cases, calibration using the local particle sources (e.g., diesel exhaust particles) and environmental conditions (e.g., environmental humidity) must be applied.

Calibration and Correction Factors

Instrument calibration should always be performed in accordance with the air sampling and analysis method (e.g., U.S. EPA method), as well as the manufacturer's specifications. Typical instrumental calibration for mass-based monitors is performed using Arizona road dust as a reference standard, which may require a correction for local sources and environmental conditions. For example, high relative humidity (e.g., greater than 90 percent) or condensing water vapor (e.g., mists or fogs) could collect on the filter media and bias monitoring results. Particle sources significantly different from the calibration source may also require a correction. Mass correction can be made by performing concurrent gravimetric analysis (mass collected on a filter) under comparable sampling conditions. Gravimetric analysis requires preconditioning and post-conditioning of the filters in desiccators for 24 hours, which helps remove unwanted water prior to analysis. For example, PM_{10} samples can be collected using a high volume-filtering device (Figure 4–7) over an equivalent 24-hour sampling period. Gravimetric analysis of the filter will provide the average mass concentration (e.g., in $\mu g/m^3$) during the sampling period. A correction factor for concurrent PM_{10} monitoring by BAM or TEOM can then be determined by dividing the actual 24-hour mass concentration by the instrumental 24-hour mass concentration:

$$\text{Correction factor} = \frac{\textit{Average gravimetric mass concentration}}{\textit{Average instrumental mass concentration}} \quad \text{(Eq. 4–10)}$$

The sampling periods and methods (e.g., PM_{10}) should be similar for both the gravimetric analysis and BAM or TEOM. The real-time data obtained from the PM_{10} monitor can then be corrected by multiplying all results by the correction factor.

Particle Microscopy

Optical, transmission electron (TEM), and scanning electron microscopy (SEM) are useful in the identification and sizing of collected airborne particles. They allow for direct observation of particle size, shape, structure (e.g., crystalline versus amorphous), and even particle composition as with the use of energy dispersive x-ray analyzers (EDXRA) with SEM and TEM. EDXRA provides an analysis of elemental content (e.g., Ca, Na, P, Mg, Si, Fe, etc.), which aids in proper identification of morphologically similar particles.

As an example of possible application, all three microscopy methods have been used in the identification, sizing, and compositional analysis of mineral asbestos fibers. With optical microscopy, asbestos fibers can be collected on mixed-cellulose filters, cleared with acetone, and then identified and counted using phase-contrast microscopy. For proper identification of asbestos type (e.g., amosite versus chrysotile) bulk samples are typically analyzed using SEM/TEM with EDXRA elemental analysis of the Ca, Fe, Mg, and Si content. The general applications and limitations of optical and electrical microscopes are presented in Table 4–5.

All microscopy methods require a high degree of operator skill and judgment regarding identification and sizing. Thus, the skill and training of the operator must never be overlooked. The primary disadvantages with these methods are the large amount of time required for analysis, analytical cost, and potential operator bias. The primary advantages of microscopy are that it can provide evidence of a particle identity or source, as well as the presence of unexpected particles and sources. Essentially, microscopy can provide necessary data regarding particle identification and sizing. It is a powerful tool when coupled with the above particle sizing and mass-based methodologies.

VI. GAS SAMPLING

Introduction to Gas Sampling and Analysis

As with sampling and analysis for particles, *direct-reading instruments* for gaseous pollutants have continued to improve and have increased in popularity within the air pollution field. However, many laboratory-based analytical methods still provide greater sensitivity, as well as the ability to separate, identify, and quantify a multitude of chemical air pollutants within a single sample. *Gas chromatography-mass spectrometry* (GC-MS) can identify hundreds of organic chemicals at a time and measure picogram (10^{-12}) quantities. The same can be said for elemental analyses performed using *inductively-coupled plasma mass spectrometry* (ICP-MS). Because these powerful and sensitive analytical methods are best suited within a laboratory setting, ambient air sampling is often conducted separately and the samples are then delivered to the laboratory. The resulting air concentration is a ratio of the pollutant mass in the sample, determined by laboratory analysis, and the air volume collected during air sampling:

$$\text{Air concentration} = \frac{\mu g}{m^3} = \frac{\text{mass in sample } (\mu g)}{\text{sample air volume } (m^3)}. \quad \text{(Eq. 4–11)}$$

The resulting mass-per-volume (m/v) concentration is a direct product of the sampling and analysis methods. The analytical method provides the pollutant mass and the sampling method provides the air volume. However, the concentrations for most gaseous pollutants are expressed as volume-per-volume (v/v) measures in either parts per million (ppm) or parts per billion (ppb). This requires a conversion of the pollutant mass into a volume, which is quite straightforward when dealing with small concentrations (e.g., less than 100 ppm) of substances that behave like ideal gases. The mass is converted to moles using the molecular weight (*MW*) of the substance, then the moles are converted to a volume using the molar volume or *V/n*, which is derived from the ideal gas law $PV = nRT$:

$$\text{Molar volume} = \frac{V}{n} = \frac{RT}{P}, \quad \text{(Eq. 4–12)}$$

where *R* is the gas constant ($R = 0.0821$ L atm/K mole), *T* is the temperature in Kelvin (K = °C + 273.15), and *P* is the atmospheric pressure in atm.

The molar volume is the volume occupied by one mole of a substance. At normal temperature and pressure (NPT = 1 atm and 25°C) the molar volume is 24.45 L/mole for an ideal gas.

Equation 4–13 is used to convert a μg/m^3 concentration into a *ppb* concentration at NTP:

$$ppb = \frac{(\mu g/m^3)(24.45)}{MW}, \quad \text{(Eq. 4–13)}$$

where *MW* is the gram molecular weight in g/mole. When atmospheric conditions deviate from NTP, then the corrected molar volume ($V/n = RT/P$) is used in the calculation of the concentration:

$$ppb = \frac{(\mu g/m^3)(RT/P)}{MW}. \qquad \text{(Eq. 4–14)}$$

Most air pollution standards for gaseous pollutants are listed as either *ppb* or *ppm* concentration units. This helps distinguish between aerosol/particle pollutants (often w/v) and gaseous pollutants (v/v), but these general rules are not without exception. Ultimately, atmospheric temperature and pressure must be recorded during air sampling and used when converting sampling and analysis concentrations (w/v) into volume-per-volume concentrations.

Figure 4–12 Air monitoring canisters.
Reproduced by permission of Restek Corporation.

Air Sampling Methods

When sampling it is critical to obtain a representative sample that is not significantly altered by the sampling process, as well as minimize potential degradation or sample loss prior to analysis.

The two general categories of air sampling methods used for gaseous contaminants are (1) collecting an unadulterated sample of the air within a container (also called "whole air sampling"), and (2) *removing the pollutants* from a known volume of air. Air sampling may also be either *active* or *passive,* which is discussed in greater detail later in this section.

Whole air sampling involves the use of evacuated glass or stainless steel containers (e.g., evacuated canisters; **Figure 4–12**), inert sample bags, or gas-tight syringes to collect a representative air sample. The ease and simplicity of the sample collection are distinct advantages for these methods. However, they are not well suited for collecting reactive, unstable, or short-lived compounds, or when analytical sensitivity is an issue.

As analytical instruments have continued to improve and become more sensitive, the use of evacuated containers has increased in popularity. The air is evacuated from the containers and either held under vacuum, or the air is partially replaced with an inert gas. During sampling, ambient air is allowed to enter the container at a controlled flow rate. Sampling can be carried out over a short span of several minutes (grab sample) or over longer intervals for *time-weighted average* determinations.

A time-weighted average (TWA) is an average concentration for a specified time period. In many air pollution studies, ambient air sampling is collected over a 24-hour period and subsequent analysis of the sample provides only a single, average concentration.

Three primary methods are used for removing pollutants from a known volume of air:

1. *Cryogenic sampling* uses a very low temperature trap to condense, collect, and concentrate volatile compounds.
2. *Gas absorption methods* use either physical or chemical absorption as the process for collecting a gaseous compound within a liquid or solid matrix.
3. *Gas adsorption methods* use either physical or chemical adsorption as a mechanism to collect gaseous compounds on the surface of a porous solid sorbent.

The advantages and disadvantages of each method are listed in **Table 4–6**. Selection of an appropriate method is primarily dependent on the boiling point (volatility), concentration, polarity, and stability of the pollutant.

Cryogenic Sampling

Cryogenic sampling typically uses a U-shaped tube filled with glass beads as a cryogenic trap. Liquid argon or oxygen can be used to cool the trap to temperatures well below 0°C. As air passes through the trap, volatiles will condense and collect within the trap. Cryogenic traps are best suited for volatile organic compounds at low

Table 4–6 Air sampling methods for gases and vapors.

Method	*General Principle*	*Advantages*	*Disadvantages*
Whole air	Air is collected through a metered device into an evacuated chamber, bag, or syringe	Simplicity of sample collection; air sample representative of ambient air; reduced sampling error	Not suitable for reactive or unstable pollutants; oxygen (~21%) and water (0–2%) in air may result in sample degradation over time; shorter shelf-life of samples
Cryogenic	Air is passed through a cooled chamber (cold trap) and volatile compounds with higher boiling points condense and are collected as a liquid	Allows for pre-concentration of volatile organic compounds; can increase detection of pollutants at extremely low concentrations	Water in air is also concentrated, which can result in unwanted analytical interferences and hydrolysis reactions
Gas absorption	Air is passed through liquid or solid media and gaseous compounds either dissolve in or chemically react with the media	Chemical absorption can produce a stable compound (derivative) for analysis; used for unstable and/or reactive pollutants (e.g., ozone)	Collection efficiencies vary; liquid media losses can occur due to leaks and evaporation; glass bubblers are fragile and difficult to clean
Gas adsorption	Air is passed through a solid sorbent with a high surface area, and gaseous pollutants are attracted to and adhere to the surface molecules; similar to condensation	Physical adsorption is reversible by desorption with a solvent or by heating; sorbents are available for selective absorption of polar (e.g., silica gel) and non-polar (e.g., activated charcoal) gaseous pollutants	Not suitable for many pollutants with low boiling points and compounds that are gases at ambient conditions; high temperatures and humidity extremes reduce absorption efficiency; interfering gases and vapors may compete for adsorption sites

ambient concentrations, which require pre-concentration prior to analysis. One drawback is that water vapor in air may also be collected and concentrated, which can result in unwanted analytical interferences, such as hydrolysis of the sample.

Physical Gas Absorption

With physical gas absorption the pollutant is dissolved in an absorbent medium, typically a liquid or liquid-coated filter. Most liquid absorption devices (**Figure 4–13**) are glass containers with a submerged impinger tube that introduces the air into the liquid at a high velocity. Impinger tubes are narrow-bore in order to accelerate the air stream and increase the rate of collisions of gas molecules with the liquid media. Fritted glass ends with hundreds of small pores are also used to increase air contact with the liquid. Common liquid absorption device designs include *impingers* or *gas bubblers.* Collection efficiencies vary for these absorption devices, and many U.S. EPA methods use two to five bubblers in series to reliably collect gaseous pollutants. Note that the glass containers are fragile and the fritted glass ends can be difficult to clean if they become clogged with solid contaminants (e.g., metal fumes or dusts) from the air.

Chemical Gas Absorption

Chemical gas absorption has the advantage of reacting gaseous pollutants with a liquid media to produce stable, non-volatile derivatives for subsequent analysis. Methods exist for unstable and reactive gases such as

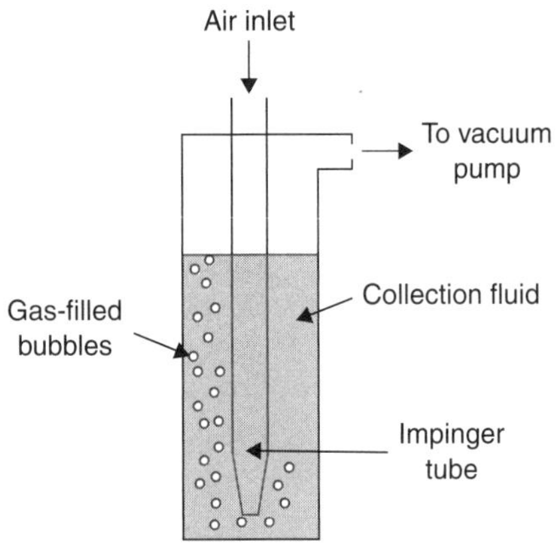

Figure 4–13 Gas bubbler design.
Source: The University of California Air Pollution Health Effects Laboratory, with kind permission.

ozone (O_3), phosgene ($COCl_2$), ammonia (NH_3), and hydrogen chloride (HCl). Acidic and alkaline pollutants can undergo an acid-base reaction to produce a stable salt compound for pH analysis. Methods also exist for reactive volatile organic compounds, namely amines, aldehydes, and ketones. When a pollutant is reacted with another compound to produce a stable analyte for analysis the process is called *derivatization.*

Gas Adsorption (think "ad" for adhesion)

With gas adsorption, air is typically passed through a solid sorbent with a high surface area, and gaseous pollutants deposit on and adhere to the surface molecules.

Adsorption is a process by which compounds adhere to and collect on the surface of a solid or liquid medium. Physical adsorption is a reversible process involving weak forces of attraction such as van der Waals forces, whereas chemical adsorption is used less often due to the irreversible nature of most chemical reactions.

The adsorption forces are dependent on the structure and polarity of the sorbent material. Most adsorbents are porous solid materials, with small nanopores (1 to 5 nm in diameter), and a high surface area to mass ratio. *Activated charcoal,* which is activated by removing adsorbed matter with vacuum and/or heat, attracts primarily low-polarity, volatile organic compounds with higher boiling points (i.e., lower volatility). Activated charcoal is a common sorbent for petroleum hydrocarbons, such as hexane, benzene, toluene, and xylenes. It also has wide applications for many organic compounds, especially as polarity, water solubility, and volatility are all decreased. Additional organic polymers (e.g., Tenax® and Amberlite® XAD resins) with non-polar and hydrophobic properties are in common use in place of activated charcoal. There are also several special grades of activated charcoal (e.g., Anasorb®) in common use. Proper selection of the appropriate non-polar sorbent is primarily based on pollutant (i.e., analyte) volatility and the maximum temperature, which are important for sampling and analysis. In contrast, *silica gel* efficiently collects more-polar and water-soluble compounds, thus it is used for sampling pollutants such as methanol (CH_3OH), many amines ($R–NH_2$), and even water vapor itself. One of the more common and familiar applications of silica gel is as a desiccant that is used to protect leather clothing (e.g., shoes) and dried foods (e.g., beef jerky) from the effects of excess humidity. Additionally, Anasorb, a microporous beaded activated charcoal, has been effectively used for collecting both polar and non-polar organic compounds.

Polyurethane foam (PUF), which is housed in glass sampling cartridges, is another sorbent commonly used in air pollution sampling of organic compounds with low volatility. Common applications include sampling and analysis methods for polynuclear aromatic hydrocarbons (PNAs), polychlorinated aromatic hydrocarbons (PAHs) such as benzo(a)pyrene, polychlorinated biphenyls (PCBs), and organic pesticides. The collected pollutants are extracted from the PUF with an organic solvent (e.g., 5 percent diethyl ether in hexane).

The primary advantage of physical adsorption is that it is reversible by desorption with a solvent or by heating. In addition, similar volatile organic compounds can be sampled simultaneously and then separated and analyzed using chromatographic methods. Thermal desorption can deliver the pollutant directly to a chromatographic column for analysis, which can eliminate additional sample preparation.

With chemical adsorption, or *chemisorption,* the forces of attraction are strong and a chemical reaction or some interaction between valence electrons occurs. Chemical adsorption requires large amounts of energy or effort for desorption. Chemically-treated silica gel sorbents are used for acid gases (e.g., HCl and HF) and reactive organic compounds (e.g., aldehydes and ketones).

Examples of select U.S. EPA sampling and analysis methods for organic gases and vapors are provided in **Table 4–7**. Most sampling methods use a sorbent or molecular sieve. Thermal desorption is commonly used to introduce collected organic compounds into an analytical

Table 4–7 Select U.S. EPA ambient air sampling methods for organic compounds.

EPA Method ID	Air Pollutant(s)	Sample Collection	Sample Recovery
TO-1	Volatile organic compounds	Tenax sorbent tube	Thermal desorption
TO-2	Volatile organic compounds	Carbon molecular sieve	Thermal desorption
TO-5	Aldehydes and ketones	Impinger with DNPH	Solvent extraction
TO-10A	Organochlorine pesticides	PUF tube	Solvent extraction
TO-11A	Formaldehyde	DNPH-coated silica gel	Solvent extraction
TO-12	Non-methane organic compounds	Cryogenic trap (argon)	Heating
TO-13A	Polynuclear Aromatic Hydrocarbons (PNAs)	PUF/XAD sorbent (for gases) and quartz filter (for aerosols)	Soxhlet solvent extraction
TO-15	Volatile organic compounds	Evacuated canister	Delivery to a GC/MS
TO-17	Volatile organic compounds	Various sorbent tubes	Thermal desorption

Source: U.S. EPA (1999)
DNPH = 2,4-dinitrophenylhydrazine reagent
GC/MS = gas chromatography with mass spectrometer
PUF = polyurethane foam
XAD = Amberlite XAD®-2 polystyrene copolymer resin

instrument. Solvent extraction is also used to remove the collected organic from the sample media.

Collection Efficiency

The *collection efficiency* of the adsorbent for a specific pollutant is an important factor in the selection of adsorption media. Collection efficiency is the fraction of the air pollutant deposited on the sorbent media, which is primarily dependent on the sorbent and pollutant properties. A number of environmental and sampling factors can also affect the collection efficiency of adsorption methods, including:

- higher temperature reduces adsorption and increases reactivity, especially with activated charcoal;
- higher humidity reduces adsorption capacity for organic vapors, especially with silica gel;
- higher flow rates can result in *channeling* (uneven airflow across the sorbent bed) and reduce adsorption;
- large concentrations of background contaminants can compete for adsorption spaces and reduce collection efficiency for trace compounds (e.g., trace amounts of benzene < 1 percent in gasoline may be undetectable in the presence of gasoline vapors); and
- highly volatile solvents (e.g., methylene chloride; vapor pressure = 350 mm Hg) are more likely to be desorbed and escape during sampling and transport, resulting in sample losses and reduced collection efficiency.

Active vs. Passive Sampling

Active Sampling

With most of the previously mentioned methods, air is delivered to the sampling device by means of a vacuum pump or via the movement of air into an evacuated chamber, which is termed *active sampling.* In both cases, a pressure drop is used to move air into the sampler. Active air sampling is the continuous collection of airborne contaminants, within a specified air volume, by means of forcing air into or through a sampling device.

In order to provide an accurate air concentration (see Eq. 4–11), an accurate measure of the volume of air sampled is required. The air volume is calculated from the sampling flow rate and time:

$$\textit{Air volume} = \textit{flow rate} \times \textit{time} \qquad \text{(Eq. 4–15)}$$

or

$$m^3 = \left(\frac{m^3}{\text{min}}\right) \times \text{min}. \qquad \text{(Eq. 4–16)}$$

Prior to sampling, flow rate calibration and adjustment must be performed under identical atmospheric conditions as used for the air sample. For convenience, most often a *secondary standard,* such as a rotameter, critical orifice, or mass flow meter is used in the field to adjust and measure sampling flow rates. Secondary standards must be traceable to a *primary standard.*

A primary standard for flow rate is one based on known physical parameters of volume and time, which has been shown not to change under normal operating conditions. Both proper precision and accuracy will provide a valid measurement of flow rate.

Conversely, a *secondary standard* for flow rate is one *not* based on known physical parameters (e.g., a fixed volume). Calibration and validation must be traceable back to a primary standard, and calibration must be routinely conducted to ensure valid sampling. In order to measure the accuracy and precision at least triplicate measurements must be collected before and after sampling. For most air sampling applications the minimum QA/QC requirements are:

- flow rate accuracy (relative error) ± 10 percent for both before and after measurements;
- flow rate precision (*CV*) ± 5 percent.

Accuracy and precision can be calculated using the equations provided in section II or Table 4–2. Deviations outside the established criteria are grounds to disregard the sample, and an indication that the sampling equipment needs service. Overall, active sampling requires specialized technical training and expertise in air sampling methods and sample pump calibration.

Passive Sampling

Passive sampling does not use an air pump or vacuum to move air into the sampling device. Instead, the collection and movement of airborne contaminants into the sampling device is controlled either by diffusion in air, or permeation (i.e., molecular movement) through a permeable membrane.

Passive air sampling is the continuous collection of airborne contaminants based on principles of diffusion, the movement and spread of molecules across a concentration gradient, e.g., without the use of a mechanical sampling device or vacuum pressure.

Passive sampling is best defined by *Fick's first law of diffusion,* which states that the flux (J) of a diffusing species is proportional to the concentration gradient (dC/dx), or change in concentration with distance:

$$J = -D\left(\frac{dC}{dx}\right), \qquad \text{(Eq. 4–17)}$$

where D is the diffusion coefficient (cm^2/s), which is specific for each molecule in air. Flux is the movement of molecules across a unit area, expressed in units of $moles/cm^2 \cdot s$. Theoretically, the compound-specific movement of pollutant molecules by diffusion is similar to that of active sampling. The theoretical sampling rate (Q_T) in cm^3/s, which is equivalent to an air flow sampling rate, is determined as follows:

$$Q_T = \left(\frac{AD}{L}\right), \qquad \text{(Eq. 4–18)}$$

where A is the cross-sectional surface area of the diffusion path (cm^2), D is the compound-specific diffusion coefficient (cm^2/s), and L is the diffusion path length or thickness (cm). The sampling rate can be increased or decreased by adjusting the cross-sectional area (A) and diffusion path (L) (**Figure 4–14**). However, caution must be applied when using passive sampling methods because the sampled air volume is theoretically-derived and the methods are prone to sample bias. Passive

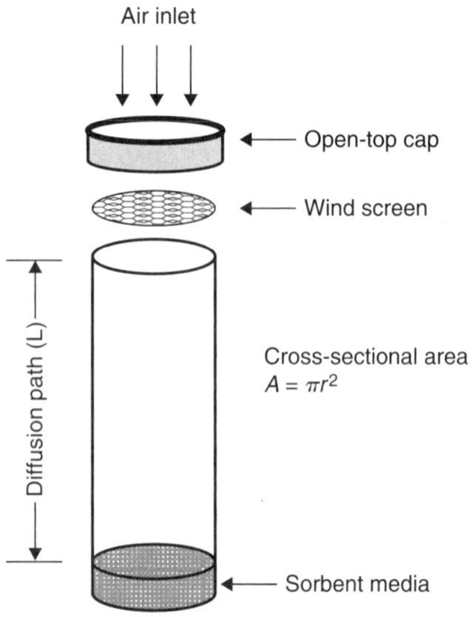

Figure 4–14 Passive diffusion cell design.
Source: The University of California Air Pollution Health Effects Laboratory, with kind permission.

sampling methods are susceptible to environmental conditions of:

- stagnant air with face velocities less than about 0.13 to 0.25 m/s (25–50 feet/min);
- high face velocities (or winds), which vary with sampler design;
- sampler orientation, especially when oriented perpendicular to the wind direction;
- temperature extremes; and
- humidity extremes.

Therefore, passive sampling methods must be tested and validated for the environmental conditions for which they will be used. For more information and guidance on the uses and limitations of passive sampling refer to *Passive Sampling and/or Extraction Techniques in Environmental Analysis: A Review* (Namiesnik et al., 2005). A comparison of the advantages and disadvantages of both active and passive sampling is provided in **Table 4–8**.

VII. GAS ANALYSIS

As stated earlier, analysis of gaseous pollutants can be performed using either *direct-reading* (continuous monitoring) or *sampling and analysis protocols.* Each has distinct advantages and disadvantages associated with its use (**Table 4–9**). For example, direct-reading methods have the advantage of providing continuous data over long periods of time, which allows scientists to identify peak exposures, fluctuations, and possible emission sources. However, these techniques typically monitor only one or two contaminants at a time. In addition, direct-reading instruments are more costly than sampling equipment, which usually limits the number of locations that can be monitored at any given time. Direct-reading technologies are best suited for evaluating and monitoring regional air pollution trends. Portable systems can also be used to identify suspected local sources. In contrast, sampling and analysis methods have the advantage of centralizing the analysis within a controlled laboratory environment. In general, laboratory-based analytical methods provide better sensitivity than direct-reading technologies, as well as the ability to separate, identify, and quantify a multitude of chemical air pollutants within a single sample. The ability to sample from a greater number of locations also increases the statistical power and representation of the study population. The primary disadvantages of sampling and analysis methods include (1) the inability to provide continuous data, and (2) a greater potential for errors

Table 4–8 Advantages and disadvantages of active and passive sampling methods.

Sampling Method	*General Principle*	*Advantages*	*Disadvantages*
Active	Low pressure used to move a specified volume of air into or through a sampling device	– Numerous validated methods exist – Large sampling volumes improve sensitivity – Sampling trains with multiple samplers in series reduce sample losses – Isokinetic (matched-flow) sampling can be used to minimize sampling errors	– Requires frequent calibration of flow meters and pumps – Specialized training required – Bulky equipment and sampling trains are cumbersome and often fragile – Often requires a reliable power source to operate in the field
Passive	Diffusion in air or through a permeable membrane is used to collect an air contaminant on a sorbent, without the use of mechanical means or low pressure	– Low cost – Compact size – Do not require a power source – Ease of use – Minimal training – Large number of samples can be collected	– Theoretical sampling rate must be validated for field conditions – Low sensitivity – Temperature, humidity, and air velocity extremes adversely affect collection efficiency – Sensitive to wind direction or sampler orientation

Table 4–9 Direct-reading vs. sampling and analysis methods.

Method	*Advantages*	*Disadvantages*
Direct-reading	– Provides continuous data with identified peak exposures – Portable instruments can help identify sources – Portable instruments can monitor personal exposures and distinguish between indoor, outdoor, and in-transit exposures – Useful for measuring transient/reactive coumpounds (e.g., ozone)	– Often less sensitive than laboratory instrumentation – Analysis limited to fewer contaminants – Cost limits number of locations that can be monitored – Instrumentation may be affected by environmental conditions – Often require a power source to operate in the field
Sampling and Analysis	– Numerous validated methods exist – Large sampling volumes can increase analytical sensitivity – Large number of samples can be collected – Ability to separate, identify, and quantify multiple contaminants in the air	– Provides average exposure data (non-continuous) – Operator errors associated with collection, handling, and preparation of samples – Requires frequent calibration of flow meters and pumps – Specialized training required

associated with the additional collection, handling, and preparation of samples. Nonetheless, similar analytical technologies are used in both direct-reading and sampling and analysis methods; with the exceptions of gas and liquid chromatography methods for the separation of multiple contaminants and mass spectrometry analyses. Instrument cost, complexity, and longer analysis times are the major limiting factors that restrict use of these separation technologies with direct-reading monitors. Portable gas chromatographs and mass spectrometers are available, but are mostly used for the qualitative (identification) analysis of contaminants associated with hazardous chemical releases and emergency response incidents.

The analytical technologies used for both direct-reading and laboratory instrumentation all exploit some chemical or physical property of the analyte, such as molecular weight, boiling point, polarity, molecular structure, bond energies, the presence of functional groups (e.g., R–OH alcohol group), selective reactivity, or spectral properties (e.g., absorption or emission of electromagnetic radiation). The following sections on detection and separation techniques briefly describe some of the common technologies used in air pollution science. Distinctions between direct-reading technologies and sampling and analysis methods are provided in **Tables 4–10** and **4–11**.

Common Detection Techniques

The common detection techniques used with direct-reading instruments are provided in Table 4–10. Air sampling and analysis methods also use these technologies, with the addition of mass spectrometry. A brief description of each detection technology is provided in this section. As a review, the important considerations and criteria for selection of an appropriate analytical method are as follows:

- *detection limit*—The ability of the analytical method to detect the pollutant in the environment. The MDL is often defined as three times the standard deviation (σ_{n-1}) of the background instrument response or signal.
- *quantification limit*—The ability of the analytical method to measure the pollutant in the environment. The MQL is often defined as 7 to 10 times the standard deviation (σ_{n-1}) of the background instrument response or signal.
- *linearity*—The concentration range of the analytical method that provides a linear association between concentration and instrument response. For example, flame-ionization detectors provide a wide linear response over several orders of magnitude (e.g., ppb to ppm concentrations).

Table 4–10 Representative direct-reading methods for analysis of gases and vapors.

Operating Principle	*Application and Remarks*	*Range*	*Detection Limit*
Chemiluminescence (chemically simulated light emission)	Specific measurement of NO in ambient air and NO_2 following conversion to NO by hot catalyst. Specific measurement of O_3. Not affected by atmospheric interferences.	0-10,000 ppm	Varies: 0.01 ppb to 0.1 ppm
Electrochemical sensors	Continuous monitoring of NO, NO_2, O_3, and SO_2 in ambient air. Chemicals with similar chemical structures and reactive functional groups can interfere with chemical reactions and accuracy.	From parts per billion to percent	Varies: 0.01 – 10 ppm
Flame ionization detector (FID)	Non-specific detection and measurement of volatile organic compounds (VOCs). Useful for surveys to identify VOC exposures and sources.	1–100,000 ppm	Varies: 0.1 ppm (methane)
Infrared spectrometry	Continuous determination of a given component in ambient air by measuring the amount of infrared energy absorbed. Wide variety of applications including CO, CO_2, hydrocarbons, nitrous oxide, NH_3, SO_2, and water vapor.	From parts per million to percent	Varies: 0.5 - 1 ppm (NH_3)
Photoionization detector (PID)	Non-specific detection and measurement of volatile organic compounds (VOCs) with double-bonds and aromatic structures. Useful for surveys to identify VOC exposures and sources.	Parts per billion to parts per million	Varies: 1 ppb to 1 ppm (select VOCs)
Ultraviolet and visible (UV/VIS) absorption spectrophotometry	Continuous determination of a given component in ambient air by measuring the amount of UV/VIS energy absorbed. Wide variety of applications including NO_2, NO_x, SO_2, total oxidants, H_2S, HF, NH_3, Cl_2, mercury vapor and aldehydes in ambient air. Interferences are common.	From parts per million to percent	Varies: 0.01 ppb (mercury) to 10 ppb (NO_2, SO_2)

Data from ACGIH® (1988); Cohen and McCammon (2001); DiNardi (2003); Phalen (2009); and Vincent (1995).

Table 4–11 Representative sampling and analysis methods for gases and vapors.

Gas/Vapor	*Sorption Medium*	*Analysis*	*Interferences*
Ammonia	Sulfuric acid treated silica gel	Visible absorption spectrophotometry	None identified
Benzene	Activated charcoal	GC-FID or –MS	Carbon disulfide; similar aromatic hydrocarbons can be separated
Formaldehyde	Silica gel treated with DNPH	HPLC-UV	Ozone; aldehydes and ketones will be sampled but can be separated
Nitrogen dioxide	Molecular sieve or passive diffusion tube, both treated with triethanolamine	Visible absorption spectrophotometry	Compounds reacting with triethanolamine; excessive dust loading diffusion sampler
Polycyclic aromatic hydrocarbons (PAHs)	Polyurethane foam (PUF)	GC-FID, GC-MS, or HPLC-UV	Excessive heat, UV light, ozone, or NO_2 may degrade samples
Sulfur dioxide	Na_2CO_3 treated filter	Ion chromatography	Sulfur trioxide
Volatile organic compounds (VOCs)	Sorbent tube, evacuated canister, or sample bag	GC-MS	Similar compounds and isomers may be difficult to separate

Data from Phalen (2009); U.S. EPA (1999) and NIOSH (2003).
DNPH = 2,4-dinitrophenylhydrazine reagent
GC-FID = gas chromatography with flame ionization detector
GC-MS = gas chromatography with mass spectrometer
HPLC-UV = high-performance liquid chromatography with ultraviolet detector

- *sensitivity*—The ability of an analytical method to differentiate between small changes in pollutant concentration. Sensitivity is associated with the slope of the instrument calibration or standard curve and its standard deviation (see Table 4–2). In general, sensitivity increases with a steeper slope and lower standard deviation.
- *selectivity/specificity*—The ability of an analytical method to reliably identify and measure a pollutant in the presence of interferences, such as chemical compounds or properties of the air that may interfere with sampling and/or analytical methods.
- *accuracy*—The closeness of measured values to the true value. Accuracy is measured as a relative error or percent recovery (Table 4–2). The minimum requirements are typically ± 25 percent for direct-reading instruments and ± 10 percent for laboratory instrumentation.
- *precision*—The reproducibility of analytical results, measured as the coefficient of variation (*CV*) or relative standard deviation (Table 4–2). The minimum requirements are typically ± 10 percent for direct-reading instruments and ± 5 percent for laboratory instrumentation.

In addition, cost and efficiency are equally important considerations when selecting an analytical method. The relative time, energy, and cost per sample control the number of samples that can be collected. In short, the analytical method must meet the specific design requirements of the study. In some cases, such as clinical trials, the requirements for greater accuracy, precision, and selectivity may outweigh cost considerations. In other cases, such as an investigative survey, accuracy and precision may be less important. Every study requires careful planning and design. Overall, selecting the analytical method of detection and measurement is a critical component of the design process, as well as the final product, which should be quality environmental data representative of the study population.

Chemiluminescence

In *chemiluminescence*, a chemical reaction is used to produce an intermediate in an excited electron state. When the excited electron returns to a more stable state, the energy is released as a photon of light with a wavelength determined by the energy released. A sensitive photon detector, e.g., a photomultiplier tube, can be used to selectively measure the pollutant energy. Chemiluminescent analyzers are typically used for direct-reading monitoring of nitrogen oxides (NO and NO_x) and ozone (O_3). Common instruments have detection limits down to about 0.1 ppb and are linear up to about 50,000 ppb.

Electrochemical Sensors

Electrochemical sensors typically measure some aspect of electrical charge or conductivity associated with a chemical reaction with the measured pollutant. For example, some analyzers use the change in electrical conductance across a gold film to selectively measure the amount of mercury vapor and/or hydrogen sulfide adsorbed on the surface. Other technologies use the change in conductance within a reaction chamber (containing a solid or liquid) to selectively measure a pollutant. Electrochemical sensors are commonly used for continuous monitoring of NO, NO_2, O_3, and SO_2 in ambient air. However, they are less sensitive than chemiluminescence detectors and have detection limits down to about 0.1 ppm.

Infrared Spectrometry

Infrared spectrometry has the ability to measure the infrared absorption and vibrational energy of specific molecular bonds and interactions. In this case "absorption" has a specific meaning, i.e., the alteration (e.g., stretching, bending, etc.) of molecular structures. Absorption is a measure of the quantity of electromagnetic radiation absorbed by a sample.

Nearly all molecules absorb infrared radiation and the resulting infrared spectrum will be unique to that molecule, given that the sample is pure or free from interferences. The absorption of infrared radiation is governed by a common derivation of the Beer–Lambert Law:

$$A = abc, \qquad \text{(Eq. 4–19)}$$

where

A = absorbance
a = molar absorptivity
b = path length
c = concentration

The molar absorptivity (a) is constant for a specific absorption wavelength and molecule. Thus, as long as the path length is constant and unchanging, then absorbance (A) can be directly proportional to concentration (c). Application of the Beer–Lambert Law and its linear relationship between absorbance and concentration is generally limited to a few orders of magnitude.

Infrared spectrometry is commonly used in direct-reading instruments for measuring CO, CO_2, hydrocarbons, nitrous oxide (N_2O), NH_3, SO_2, and water vapor. However, low sensitivity in the presence of interfering compounds present in the air (e.g., water vapor and other organics) is a key limiting factor when using infrared spectroscopy. The detection limits for common infrared monitors range from about 0.5 to 1 ppm. Most absorption technologies are linear over at least two orders of magnitude (i.e., 10^2).

Ionization Detectors

Ionization is a process by which molecular bonds are broken and charged ions are produced. Ionization detectors are designed to measure the presence or absence of the charged ions produced within the detector. The three common detectors are *flame ionization detectors (FID), photoionization detectors (PID),* and *electron capture detectors (ECD).*

Flame ionization detectors use a hydrogen/air flame to ionize organic compounds in air. Changes in electrical current across an electric circuit are used to measure the amount of charged ions produced within the detector. The change in potential across the circuit is proportional to the concentration. The ionization energy of FIDs is about 15.4 eV. Therefore, an FID is capable of breaking molecular bonds with bond strength energies (ionization potential, or IP) less than about 15.4 eV, which means an FID can detect most hydrocarbons (e.g., methane; IP = 12.61 eV) and volatile organic compounds (VOCs) in the air. However, FIDs are non-specific hydrocarbon detectors, which mean they will detect and measure potentially any hydrocarbon with a C-H bond. For direct-reading purposes they are useful for source surveys and leak detection (e.g., in fugitive emissions monitoring) down to about 0.1 ppm. In the laboratory, FIDs are commonly used with gas chromatography separation methods, because they can detect nanogram quantities (ppb levels) and have a wide linear range of six to seven orders of magnitude (10^6 to 10^7). Once compounds are separated and isolated, the selectivity of the detector is of less concern.

Photoionization detectors (PIDs) use an ultraviolet (UV) lamp to ionize molecules in the air. The ionization chamber and ion detection principles are similar to those used by FIDs. However, PIDs are generally less sensitive and less linear than flame ionization detectors. For example, a common and rugged UV lamp on the market has an ionization energy of 10.6 eV. Alternative lamps are available in the range of about 9 to 11.7 eV. Therefore, a PID cannot ionize as many hydrocarbons or VOCs as the FID. For future reference and application, ionization energies (also termed IP or ionization potentials) of common air pollutants can be found within:

- *the NIOSH Pocket Guide to Chemical Hazards* (available at http://www.cdc.gov/niosh/npg);
- *the Hazardous Substance Data Bank (HSDB)* (available at http://toxnet.nlm.nih.gov); and
- *the NIST Chemistry WebBook* (available at http://webbook.nist.gov/chemistry).

The PID ionization energy must be greater than the pollutant's ionization potential. Ionization potentials and common detectors used for select air pollutants are provided in **Table 4–12**. In contrast to an FID, the primary advantage of a PID is that its operation does not require oxygen (in air) and hydrogen gas, which is highly flammable. The field portable PID instruments can detect VOCs down to about 1 ppb. Photoionization detectors are also used with gas chromatography methods for a range of VOCs, especially aromatic hydrocarbons such as benzene.

Electron capture detectors use a radioactive source (e.g., Ni^{63}) to generate ions within the detector, namely free electrons. Certain compounds, namely halogenated compounds (e.g., chlorinated pesticides), nitrates, and organometallic species, have a high electron affinity and "capture" the free electrons. The change in potential across the detector circuit is proportional to the concentration. Electron capture detectors are commonly used, following gas chromatography separation, in the analysis of pesticides down to ppb concentrations. However, the linear range is narrow and only about two to three orders of magnitude (10^2 to 10^3).

Mass Spectrometry

Mass spectrometers use a high-energy electron beam to fragment molecules (break bonds) into ions, which are then separated based on the mass-to-charge ratio (m/z). The instruments use a scanning magnetic field and radiofrequency to sequentially separate and focus the ion fragments onto an electron multiplier detector. Every compound is fragmented in a unique pattern, referred to as the mass spectrum. **Figure 4–15** is a mass spectrum

Table 4–12 Ionization potential of selected air pollutants.

Compound	*Ionization Potential*	*Common Detectors*
Benzene	9.24 eV	FID; PID
Carbon dioxide	13.77 eV	IR
Carbon monoxide	14.01 eV	IR; Electrochemical
Chloroform	11.42 eV	ECD; FID
Ethanol	10.47 eV	FID; PID (low sensitivity)
Methane	12.61 eV	FID
Methanol	10.84 eV	FID
Octane (component in gasoline)	9.82 eV	FID; PID
Ozone	12.52 eV	Chemiluminescence
Phenol	8.50 eV	FID; PID
Styrene	8.40 eV	FID; PID
Toluene	8.82 eV	FID; PID
Trichlorethylene	9.45 eV	ECD; FID; PID
Vinyl chloride	9.99 eV	ECD; FID; PID

Sources: NIOSH Pocket Guide to Chemical Hazards (http://www.cdc.gov/niosh/npg); Hazardous Substance Data Bank (http://www.toxnet.nlm.nih.gov); and NIST Chemistry WebBook (http://webbook.nist.gov/chemistry).

ECD = electron capture detector; FID = flame ionization detector; IR = infrared detector;
PID = photoionization detector

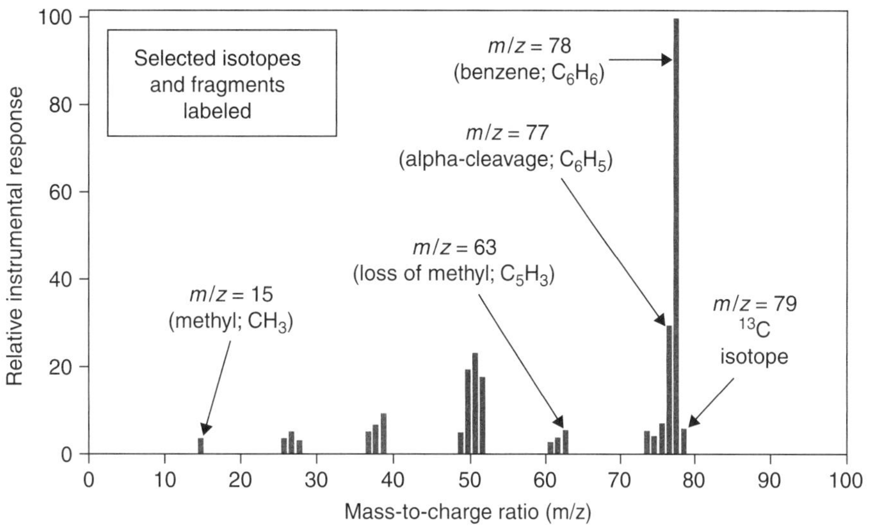

Figure 4–15 Mass spectrum for benzene.
Source: The University of California Air Pollution Health Effects Laboratory, with kind permission.

for benzene, where m/z 78 corresponds to the molecular weight of benzene (the base peak) and m/z 15 corresponds to a CH_3 fragment. Thus, m/z 63 on the spectrum represents the corresponding fragment to the CH_3 fragment from the base peak for benzene. Mass spectrometers are used mostly with gas chromatography (GC–MS) or liquid chromatography (LC–MS) separation techniques to isolate compounds for proper identification. Mass spectrometer detectors are powerful and sensitive tools in the analysis of VOCs and pesticides. They are capable of detecting and measuring chemical compounds down to ppt (part-per-trillion) concentrations and are linear over three to four orders of magnitude (10^3 to 10^4).

Ultraviolet and Visible Absorption Spectrophotometry

Ultraviolet (UV) and visible (VIS) absorption spectrophotometry operate under absorption principles similar to infrared spectroscopy and are also governed by the Beer–Lambert Law. A UV/VIS light photometer measures the amount of light passing through the sample (transmittance), but the relationship with concentration is logarithmic rather than linear.

Transmittance is the ratio of the amount of electromagnetic radiation passing through the sample to the total electromagnetic radiation incident upon it. The transmittance (T) can be converted to the amount of electromagnetic radiation absorbed, or absorbance (A):

$$A = -\log T. \qquad \text{(Eq. 4–20)}$$

Based on principles of the Beer–Lambert Law ($A = abc$), as long as the path length is constant and interferences are minimized, the wavelength-specific peak absorbance will be proportional to the concentration. UV/VIS is commonly used to monitor NH_3, nitrogen oxides (NO_x), ozone, sulfur dioxide, and mercury vapor. The absorption spectra for UV/VIS are broader than those for IR, and thus more susceptible to interferences from similar chemicals. However, mercury vapor absorbs UV radiation strongly at 253.7 nm, which makes UV a sensitive (ppb levels) and selective choice for monitoring purposes. For other pollutants the sensitivity is lower and in the ppm range. Lastly, UV detectors are commonly used with lab-based high performance liquid chromatography (HPLC) methods. Because the HPLC separates compounds prior to their detection, selectivity is not typically an issue.

Common Separation Techniques

In general, the common analytical separation techniques applied in air pollution science are classified as chromatography methods. *Chromatography* is the separation of components within a mixture based on the differential partitioning and interaction of the components within two different material phases.

In most cases, one material phase is stationary, called the *stationary phase,* and the other is moving, called the *mobile phase.* For example, in *capillary gas chromatography,* the stationary phase is often a gel coating on the inner surface of thin capillary tubing, while the mobile phase is generally an inert gas (e.g., helium or nitrogen). The capillary tubing, referred to as the *column,* is generally about 30 to 60 meters long depending on the degree of separation required. Complex samples with multiple pollutants require longer columns. The column is located in an oven, which is used to control column temperature. The sample is introduced into the column at a low temperature and low flow rate. As the temperature is slowly increased, compounds will volatize out of the stationary phase and be carried in the mobile phase to the detector. This simple design allows chemicals to be separated based primarily on their boiling point temperature and molecular weight. Thus, temperature control and programming is essential in gas chromatography methods. A simplified diagram for a gas chromatography system is illustrated in **Figure 4–16**. In contrast, *high performance liquid chromatography* (HPLC) uses a sorbent-packed column as the stationary phase, and a liquid mobile phase under high-pressure. Instead of varying temperature, the polarity of the mobile phase is varied during the analytical run. Therefore, chemicals are separated and delivered to the detector based on their degree of polarity. *Ion chromatography* (IC) is similar to HPLC, but uses an ionic stationary phase to separate the components based on ionic charge.

Figure 4–17 illustrates the chemical separation in a typical chromatogram. Provided that there is complete separation (also termed resolution) between individual compounds, the analysis time for each peak is termed the retention time (t_r), which is specific to each chemical. The detector signal for each chemical peak is triangular in shape and the peak height and width are dependent on the amount of chemical present in the sample. The area of the peak is calculated from measurements of the width and height. The peak area is then compared to a standard curve of peak area versus concentration to determine the concentration. Chromatography methods have the ability to separate and analyze hundreds of chemicals. Qualitative identification, which is based on the retention time, is extremely useful, especially when using a mass spectrometer detector to positively identify each separated compound. Unfortunately, quantitative analysis requires a separate standard curve for each chemical, in addition to all of the statistical considerations discussed earlier.

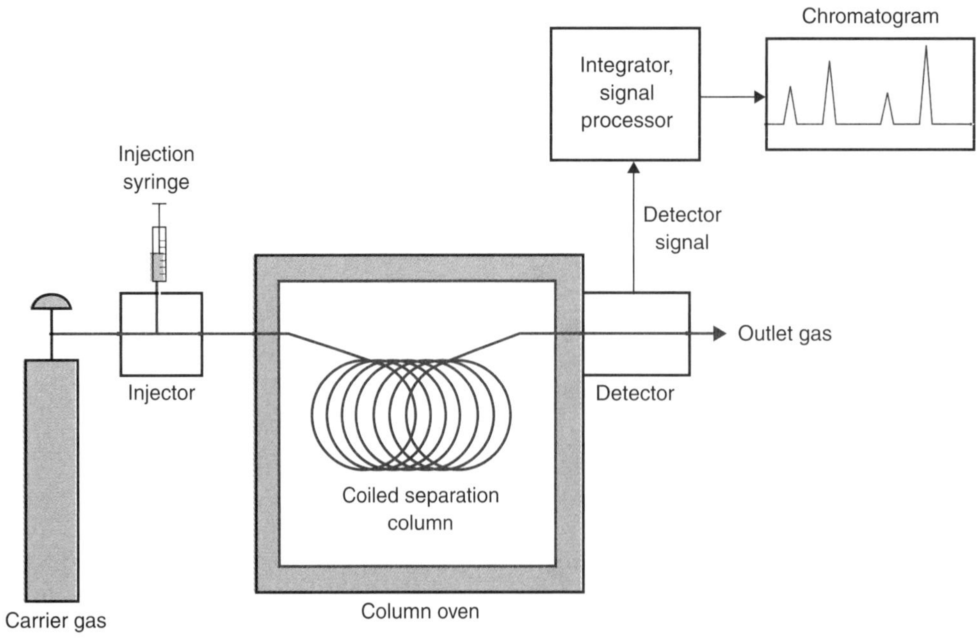

Figure 4–16 Diagram of a gas chromatography system.
Source: The University of California Air Pollution Health Effects Laboratory, with kind permission.

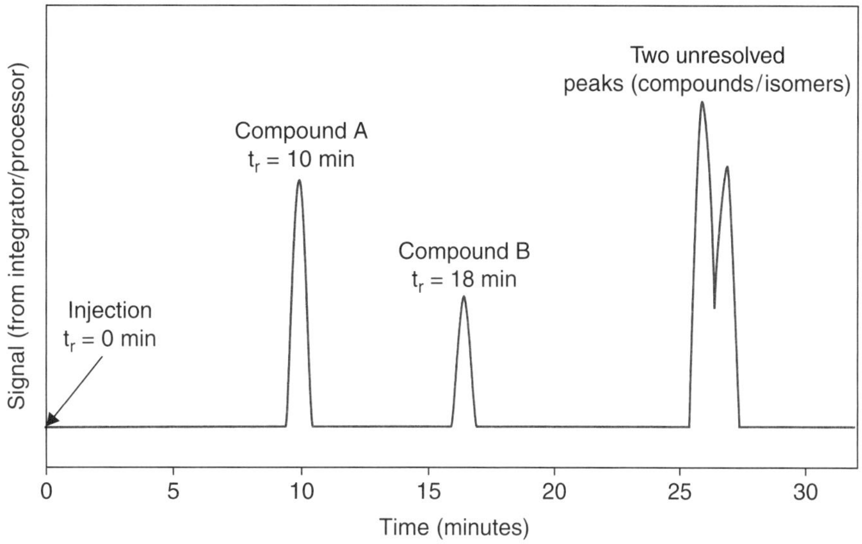

Figure 4–17 Chromatogram showing retention times (t_r) and resolution (separation) of two compounds. *Source:* The University of California Air Pollution Health Effects Laboratory, with kind permission.

The following sections briefly describe the applications of three common analytical separation techniques used in air pollution science. Examples of the common detectors are also provided.

Gas Chromatography

Most analytical methods for gases and vapors use gas chromatography. The chemicals must be thermally stable and have a low enough vapor pressure to volatilize into the mobile phase. *Gas chromatography* is commonly used in the analysis of hydrocarbons, volatile organic compounds, some polycyclic aromatic hydrocarbons, and volatile and semi-volatile pesticides. Examples of common detectors used with GC include:

- electron capture (ECD)
- flame ionization (FID)
- mass spectrometry (MS)
- photoionization (PID)

Additional detectors that are not described in this chapter include flame photometric for sulfur and phosphorus containing compounds; nitrogen-phosphorus for nitrogen (e.g., amines) and phosphorus containing compounds; and thermal conductivity for low molecular weight gases. However, due to its high degree of sensitivity and specificity to positively identify chemicals, GC–MS methods have steadily increased in popularity. A major reason for this increased use is due to reductions in the cost and complexity of the newer systems.

High-Performance Liquid Chromatography

High-performance liquid chromatography (HPLC) is primarily used for thermally unstable compounds that have high boiling points. In essence, HPLC can almost be thought of as the antithesis to GC. In actuality this means that for any given application one method is almost always better suited than the other. The choice between GC and HPLC is often predetermined and an easy one to make. Major analytical uses of HPLC include non-volatile and semi-volatile pesticides, structurally complex carcinogenic compounds (e.g., nitrosamines and polynuclear aromatic hydrocarbons), and reactive organic compounds (e.g., isocyanates and formaldehyde). Examples of common detectors used with HPLC include:

- electrochemical detectors
- fluorescence detectors
- mass spectrometry (MS)
- ultraviolet detectors (UV)

Ultraviolet detectors are by far the most common detectors used with HPLC, but they are semi-selective in nature. Fluorescence detectors measure the emission of specific wavelengths of light following the excitation of electrons, which provides additional selectivity.

However, fluorescence does not occur in all chemicals. Once again, mass spectrometry detectors are increasing in use and popularity as availability increases. Most mass spectrometry systems are simply termed *liquid chromatography-mass spectrometry* (LC–MS).

Ion Chromatography

Ion chromatography (IC) is similar to HPLC with the exception of an ion-exchange column, which is used to separate compounds based on ionic charge. Various IC configurations exist for separation of either anion (negative net charge) or cation (positive net charge) species. Anionic applications include the analysis of chloride (Cl^{-1}), chromate (CrO^{-2}), fluoride (F^{-1}), nitrate (NO_3^{-1}), nitrite (NO_2^{-1}), phosphate (PO_4^{-3}), sulfate (SO_4^{-2}), and sulfite (SO_3^{-2}) anions in solution. Ion chromatography has a special application in air pollution science for the analysis of sulfur dioxide (SO_2). Cationic applications include the analysis of amine salts, ammonium (NH_4^{+1}), calcium (Ca^{+2}), magnesium (Mg^{+2}), potassium (K^{+1}), and sodium (Na^{+1}) cations in solution. Electrical conductivity is used as the primary detector with IC, because it is best suited to detect ionic charge. In rare cases UV, fluorescence, and MS detectors are also used.

VIII. SUMMARY OF MAJOR POINTS

Sampling and analysis are combined as a rigorous process that allows scientists to investigate the composition and complexity of contaminated air. The primary goal of a sampling and analysis method is to obtain an accurate representation of exposure of a study population, or characterization of a pollution source. Different types of air samples are generally representative of one or more of the following: a source emission; an area sample; a population sample; or a personal sample. Each type of sample can benefit air pollution science in different ways. For example, source emission samples are used for *regulatory compliance* and *exposure modeling*. A personal sample can be used to link clinical measures of health to individual exposures. Therefore, a proper sampling technique is more complicated than simply going out into the environment and capturing a sample of the air. Careful planning and design, including statistical considerations, must be performed to ensure the accuracy, precision, and validity of samples. Quality control and quality assurance protocols, which are established performance standards and procedures, are used to assure the validity of data. Two of the most important statistical considerations are *accuracy* (closeness of measured values to the true value) and *precision* (reproducibility of the analytical results). Instrument *calibration* and the establishment of a *working linear range* are critical for ensuring that both samples and the reference standards are evaluated and analyzed under the same conditions. The working linear range can be defined using the following criteria:

1. the lower concentration limit is greater than or equal to the MQL;
2. all standards are accurately measured within at least ± 10 percent of the true value;
3. precision is within ± 5 percent of the mean values; and
4. the resulting standard curve is linear with a significant ($p \leq 0.05$) Pearson correlation coefficient.

Once the accuracy and precision of the analytical method have been established, then samples can be analyzed. The analytical results must then be reported in a manner reflective of the analytical limitations. In some cases, data are not reported with a numerical value but as either "*below detectable level*" if below the MDL or as "*trace level*" if between the MDL and MQL. Only data meeting the statistical criteria for accuracy and precision should be reported as numerical values. In the end, the data must reflect the original goal and provide an accurate representation of the study population exposure or source emissions.

Particle sampling uses mechanisms of deposition, similar to those in the human body, to collect suspended particulate matter from the air. The primary mechanisms are inertial impaction, interception, sedimentation, and diffusion. *Filtration* is a common collection method that uses all four of these mechanisms to capture particles on a filter. In contrast, inertial collection instruments, such as *impactors* and *cyclones,* use particle inertia for size-selective particle collection. Lastly, *isokinetic sampling* is an important consideration in particle sampling. Failure to properly position the sampling probe and operate under isokinetic conditions can result in significant sampling errors for large particles.

Particle analysis uses any number of size-dependent particle properties to count and size particles. Examples of these particle properties include the diffusion coefficient, inertia, settling velocity, light-scattering properties, electrical charge, mobility in electrical fields, and ability to serve as nuclei for condensation. Particle size is often reported as the *mass median aerodynamic diameter* or *count median diameter.* While air pollution scientists are primarily interested in measuring particle count, size, and surface properties, regulatory emissions and controls are primarily aimed at measuring particle mass. Common mass-based instrumentation include *radiation mass detectors, piezoelectric mass balance monitors,* and *optical monitors.* Because most mass-based instruments are calibrated against a reference standard (e.g., Arizona road dust) that is likely to be quite different from the source, a correction (i.e., calibration) factor must often be applied to correct readings for the proper mass.

Gas sampling is aimed at collecting a representative sample of the environment that is unaltered by the sampling process. In addition, the sampling method must minimize degradation and potential loss of the sample prior to analysis. The two primary gas sampling methods involve either (1) whole air sampling or (2) removing the pollutant from a known volume of air. Examples of *whole air sampling* include use of inert containers or evacuated chambers. The primary methods used for the removal of pollutants from a known volume of air include cryogenic sampling, gas absorption, and gas adsorption. Lastly, sampling technologies have continued to progress in the development of both active and passive sampling methods. *Active sampling* uses a pressure drop to move air into the sampling device. Passive sampling uses principles of diffusion and Fick's first law of diffusion to theoretically predict the movement of gaseous chemicals in the air. Both methods have distinct advantages and disadvantages that must be examined when selecting the appropriate sampling method for a project.

Gas analysis can be performed using either direct-reading (continuous monitoring) or sampling and analysis protocols. Each has distinct advantages and disadvantages, which must be considered during the early project design stages. The analytical technologies for gases and vapors all exploit some chemical or physical property (e.g., boiling point or polarity) of the analyte to separate, identify, and measure the pollutant. The primary methods of analytical separation used in air pollution science include *gas chromatography, high performance liquid chromatography,* and *ion chromatography.* Similar to separation techniques, the methods of detection vary depending on the physical and chemical properties of the pollutant. Examples of common detectors used for the analysis of gaseous pollutants include *chemiluminescence, electrochemical, ionization, infrared, ultraviolet/visible,* and *mass spectrometry* detectors. Selection of the appropriate analytical method is dependent on the project design requirements and statistical criteria.

It should be apparent that specialized education and training in chemistry, physics, statistics, sampling, and analytical chemistry are required to conduct sound sampling and analysis procedures within the field of air pollution. Air pollution scientists who perform sampling and analysis must be highly trained and proficient in these relevant skills. In addition, a common assumption that should be made during the sampling and analysis process is that any and all data could come under the scrutiny of a court of law. Inconsistencies and discrepancies can invalidate even the most telling and important data. The overall importance of conducting a well-reasoned and high-quality sampling and analysis plan can be summarized as, "questionable methods lead to questionable results."

IX. QUIZ AND PROBLEMS

Quiz Questions

(select the best answer)

1. Which of the following is an example of a population sample?
   a. Daily particulate air concentrations in a national park.
   b. Hourly traffic-related carbon monoxide air concentrations within an air basin.
   c. Automobile exhaust emissions.
   d. All of the above are true.
2. How is the measure of standard deviation used in sampling and analysis methods?
   a. It is used to determine the dispersion or spread of the sample data from a mean.
   b. It can be used to establish the method detection limit.
   c. It can be used to measure the coefficient of variation, as a measure of precision.
   d. All of the above are true.
3. Most environmental sampling and analysis methods must meet a minimum accuracy criterion of ± _____ percent, as well as a precision criterion of ± _____ percent.
   a. 25 : 10
   b. 10 : 25
   c. 10 : 5
   d. 5 : 10

4. The concentration of acrylonitrile from a sampling and analysis procedure is 0.01 ppb. The method detection limit is 0.005 ppb and method quantification limit is 0.017 ppb. How is the acrylonitrile concentration reported?
   a. < 0.005 ppb (detectable level)
   b. 0.01 ppb
   c. < 0.017 ppb
   d. Trace level
5. The four basic mechanisms that lead to deposition of aerosol particles in the human respiratory tract and in size-selective air samplers meant to represent the human respiratory tract include:
   a. Fick's first law; sedimentation; gravitational settling; and impaction.
   b. Fick's first law; diffusion; sedimentation; and impaction.
   c. Diffusion; sedimentation; impaction; and inertial impaction.
   d. Diffusion; sedimentation; interception; and impaction.
6. Over-sampling or under-sampling of particles due to differences in the air velocities between the sampler and emission source is corrected by:
   a. atmospheric pressure differences.
   b. mathematical equations.
   c. ensuring a high sample flow rate.
   d. isokinetic sampling methods.
7. With the mass median aerodynamic diameter, why are a median and geometric standard deviation used to define the distribution, even though they tend to both under-represent and over-represent the mass distribution? (see Chapter 3)
   a. The mean (average) and arithmetic standard deviation are *more likely* to either under-represent or over-represent the mass distribution than the median and geometric standard deviation.
   b. The average is influenced more by large particles with extreme masses.
   c. Both the median and mean are used interchangeably, as they are approximately the same.
   d. Both a and b are true.
8. Which of the following is NOT an advantage of passive sampling over active sampling?
   a. The theoretical sampling rate accurately predicts the sampling rate under environmental conditions and use in the field.
   b. Passive samplers do not require a power source during the sampling process.
   c. Due to lower cost and compact size a larger number of samples can be collected in many cases.
   d. Passive sampling technology is simple, easy to use, and requires less training.
9. What type of particle collector uses narrow passages and abrupt air current changes to collect particles on a perpendicular surface?
   a. Cyclones
   b. Elutriators
   c. Impactors
   d. Electrostatic precipitators
10. For air sampling gases and vapors, which of the following is NOT an advantage of adsorption in comparison with absorption?
   a. Adsorption can produce stable derivatives for analysis.
   b. Adsorption is mostly reversible by desorption.
   c. Heat can often be used to deliver the collected analyte to an analytical instrument.
   d. Liquid absorption devices can exhibit media losses due to evaporation and leaks.

Problems

1. A reference standard with a concentration of 10 μg/m^3 is analyzed five times on an analytical instrument. The analytical results are 8.5, 9.3, 9.8, 10.1, and 10.9 μg/m^3.
   a. Calculate the mean and standard deviation of the analytical data.
   b. Calculate the precision of the analytical device at 10 μg/m^3.
   c. Calculate the accuracy of the analytical device at 10 μg/m^3.
2. A method blank is analyzed seven times on an analytical instrument. The analytical results are 0.005, 0.01, 0.012, 0.015, 0.02, 0.022, and 0.023 μg/m^3.
   a. Calculate the mean and standard deviation of the analytical data.
   b. Calculate the method detection limit.
   c. Calculate the method quantification limit.
   d. How should a value of 0.015 μg/m^3 be reported?
   e. How should a value of 0.04 μg/m^3 be reported?
   f. How should a value of 0.06 μg/m^3 be reported?
3. An optical particle monitor with an internal filtering device placed after the optical sensor is used

for a 24-hour study. The average instrument reading during the study is 71.4 μg/m^3. The air flow rate through the instrument and internal filter was 0.05 m^3/h. The filter was weighed before and after the sampling period. The weight gain on the filter was 98.9 μg.

a. Calculate the air concentration for the 24-hour filter sample.
b. Calculate the correction factor for the instrument.
c. The peak reading on the instrument was 195 μg/m^3. What is the corrected particulate concentration?

4. An air sampling procedure for benzene (MW = 78.1) vapor in air is conducted using activated charcoal and a calibrated vacuum pump. The air flow rate was 0.001 m^3/min and the total sampling time was 24 hours. GC–MS analysis of the activated charcoal indicated that a total of 315 μg of benzene was collected on the sorbent.
   a. Calculate the average air concentration (in μg/m^3) of benzene.
   b. Assuming normal temperature and pressure (25 °C; 1 atm) convert the air concentration in μg/m^3 to ppb.
5. Methyl isocyanate (MW = 57.1) is analyzed by HPLC–UV at 254 nm. The transmittance is 0.93. What is the absorbance?
6. An air pollution scientist needs to conduct air sampling and analysis for methyl isocyanate associated with the production of polyurethane foam. During the project, she decides to conduct a survey to identify potential leaks and emission sources using a portable direct-reading instrument.
   a. Find the ionization potential for methyl isocyanate.
   b. Which is better suited a FID (15.4 eV) or a PID (10.6 eV) for the detection and monitoring of isocyanate?

X. DISCUSSION TOPICS

1. Why are sampling and analysis considered together in air pollution science?
2. Why is the cost of an instrument sometimes important and at other times not considered?
3. Instead of just reporting the numerical value, why is it important to report analytical results as "below detectable limit" or "trace" when values fall below the MDL or MQL?
4. What are some possible reasons why both sampling and analysis methods and direct-reading (continuous monitoring) methods should be used side-by-side in air pollution studies?
5. Looking to the future, is it feasible to establish chemical-specific (e.g., nickel particulate) or source-specific (e.g., cement dust) air pollution standards, as has been done in the occupational health field? What are the advantages and disadvantages?

References and Recommended Reading

ACGIH® (American Conference of Governmental Industrial Hygienists), *Advances in Air Sampling: Industrial Hygiene Science Series,* Lewis Publishers, Chelsea, MI, 1988.

Cohen, B. S. and McCammon, C. S., Jr., Technical Editors, *Air Sampling Instruments: For Evaluation of Atmospheric Contaminants, 9th Edition,* ACGIH, Cincinnati, OH, 2001.

Davies, C. N., The entry of aerosols into sampling tubes and heads, *Brit. J. Appl. Phys. Ser. 2,* 1968, 1:921–932.

DiNardi, S., *The Occupational Environment: Its Evaluation, Control, and Management, 2nd Edition,* AIHA Press, Fairfax, VA, 2003.

Eller, P. M. and Cassinelli, M. E., eds., *NIOSH Manual of Analytical Methods, 4th Edition (3rd Supplement),* DHHS Publication Number 2003–154, U.S. Department of Health and Human Services, Cincinnati, OH, 2003 (http://www.cdc.gov/niosh/docs/2003-154/, accessed August 15, 2010).

EPA (U.S. Environmental Protection Agency), *EPA Compendium of Methods for the Determination of toxic Organic Compounds in Ambient Air: 2nd Edition,* EPA/625/R–96/010b, Center for Environmental Information, Cincinnati, OH, January 1999.

Fuchs, N. A., Sampling of aerosols, *Atmos. Environ.,* 1975, 9:697–707.

Hinds, W. C., *Aerosol Technology: Properties, Behavior; and Measurement of Airborne Particles, 2nd Edition,* John Wiley & Sons, Inc., New York, 1999.

Hori, H. and Tanaka, I., Effect of face velocity on performance of diffusive samplers, *Am. Occup. Hyg.,* 1996, 40:467–476.

Lodge, J. P., Jr. and Chan, T. L., eds., *Cascade Impactor: Sampling and Data Analysis,* AIHA Press, Akron, OH, 1986.

Mercer, T. T., *Aerosol Technology in Hazard Evaluation,* Academic Press, New York, 1973.

Namieśnik, J., Zabiegała, B., Kot–Wasik, A., Partyka, M., and Wasik, A., Passive sampling and/or extraction techniques in environmental analysis: a review, *Anal. Bioanal. Chem.,* 2005, 381:279–301.

Nelson, P., *Index to EPA Test Methods,* U.S. Environmental Protection Agency New England Region 1 Library, Boston, MA, 2003.

NIOSH (National Institute for Occupational Safety and Health), *NIOSH Manual of Analytical Methods, 2nd Edition (3rd Supplement),* DHHS (NIOSH Publication 2003–154), National Institute for Occupational Safety and Health, Cincinnati, OH, 2003 (http://www.cdc.gov/niosh/nmam/, accessed August 1, 2010).

Norman, G. R. and Streiner, D. L., *Biostatistics: The Bare Essentials,* Mosby, St. Louis, MO, 1994, pp. 106.

Park, C. O., Fergus, J. W., Miura, N., Park, J., and Choi, A., Solid-state electrochemical gas sensors, *Ionics*, 2009. 15:261–284.

Phalen, R. F., *Inhalation Studies: Foundations and Techniques, 2nd Edition,* Informa, New York, 2009.

Skoog, D. A., Holler, F. J., and Crouch, S. R., *Principles of Instrumental Analysis, 6th Edition,* Thomson Brooks/Cole, Belmont, CA, 2007.

Vincent, J. H., *Aerosol Sampling: Science and Practice,* John Wiley & Sons, New York, 1989.

Vincent, J. H., *Aerosol Science for Industrial Hygienists,* Elsevier Science, Inc., Tarrytown, NY, 1995.

Wright, G. D., *Fundamentals of Air Sampling,* Lewis Publishers, New York, 1994.

Chapter 5

Visibility, Climate, and the Ozone Layer

LEARNING OBJECTIVES

By the end of this chapter the reader will be able to:

- discuss the roles of visibility, climate, and stratospheric ozone on human health and welfare
- explain why even pristine air does not permit unlimited visibility
- describe the roles that anthropogenic air pollutants might have on modifying visibility and climate
- differentiate "good ozone" from "bad ozone"

CHAPTER OUTLINE

I. Introduction: Visibility, Climate, and the Ozone Layer
II. Visibility and Air Pollution
III. Climate and Air Pollution
IV. Stratospheric Ozone
V. Summary of Major Points
VI. Quiz and Problems
VII. Discussion Topics
References and Recommended Reading

I. INTRODUCTION: VISIBILITY, CLIMATE, AND THE OZONE LAYER

Some Basic Concepts

Overview

Local visibility and worldwide climate are both influenced by air quality. The effects of climate on public health can be substantial, and visibility can impact human welfare. Low-altitude (tropospheric) air pollutants primarily impact visibility, but tropospheric and stratospheric pollutants also affect climate and the protective ozone layer. The influence of air pollutants is apparent with respect to visibility, but much more subtle with respect to their impacts on climate and stratospheric ozone. Yet, the basic physical mechanisms by which air pollutants affect visibility are similar to those that influence climate.

Aerosol particles and gas molecules can both absorb and scatter light, heat, and other electromagnetic radiation (**Table 5–1**, **Figure 5–1**). Such absorption and scattering impairs visibility by decreasing both the contrast and illumination (i.e., brightness) of distant objects. Under pristine (pollution-free) conditions, the colorless atmospheric gas molecules (N_2, O_2, CO_2, etc.) scatter visible light (Rayleigh scattering), which obscures distant objects (e.g., mountains). Rayleigh scattering also decreases perceived contrast by scattering extraneous light into the visual field. Colored gases (especially NO_2) and many types of particles absorb specific wavelengths of light. Pollutant gases and particles can modify climate by absorbing and scattering solar radiation, retaining terrestrial heat, and modifying clouds, precipitation, and other weather phenomena. Some anthropogenic (human-generated) gases, such as chlorofluorocarbons (CFCs) and chlorine (Cl_2), can enter cyclic reactions that deplete stratospheric ozone.

Visibility

The visibility of large distant objects depends on illumination, contrast, color, haze, and biological factors, such as training, expectation, retinal accommodation, and ocular health. Measures of the obscuration of light beams are used by researchers and regulators as a surrogate for visibility, which is a *subjective* perception. The human eye is most sensitive to yellow-green light, which has a wavelength of about 550×10^{-9} m, so this or a nearby wavelength is typically used in laboratory and field studies to assess visibility. With no pollutants present (i.e., in *pristine air*), the intensity of a light beam at *sea level* will be decreased by 50 percent when viewed at a horizontal distance of about 58 km (36 miles) and

Table 5–1 The electromagnetic spectrum. In reality, the visible band is a tiny portion of the entire spectrum. The types of radiation are not unique, in that a given frequency may be described in various ways depending on the source.

Terminology	*Representative Frequency (Hz)*	*Representative Wavelength (cm)*
Electric waves including utility power	$60–10^5$	$5 \times 10^8–3 \times 10^5$
Radio bands and microwaves	$10^5–10^{11}$	$3 \times 10^5–0.1$
Infrared (heat radiation)	$10^{11}–10^{14}$	$0.1–10^{-4}$
Visible		
Red	4.3×10^{14}	7×10^{-5}
Orange	4.64×10^{14}	6.47×10^{-5}
Yellow	5.13×10^{14}	5.85×10^{-5}
Green	5.22×10^{14}	5.75×10^{-5}
Blue	6.11×10^{14}	4.91×10^{-5}
Violet	7.07×10^{14}	4.24×10^{-5}
Ultraviolet	3×10^{16}	1×10^{-6}
X-Rays	$10^{16}–3 \times 10^{20}$	$3 \times 10^{-6}–10^{-10}$
Gamma and cosmic rays	$10^{18}–10^{23}$	$3 \times 10^{-8}–10^{-13}$

Note: Hz = 1 cycle/second.

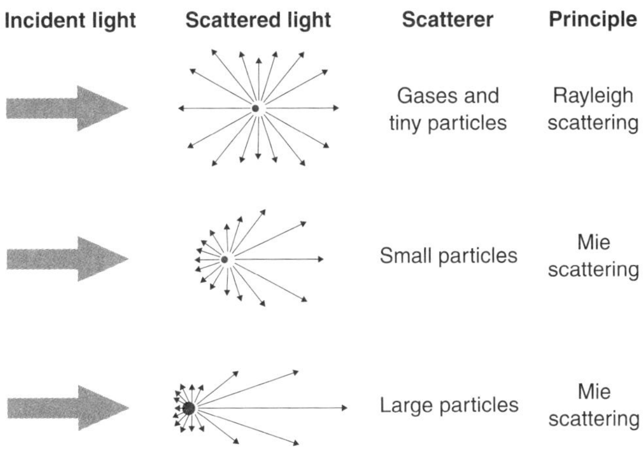

Figure 5–1 Light scattering by particles and gases. Symmetric Rayleigh scattering is characteristic for gases and particles with diameters below 0.05 micrometers. Mie scattering applies to larger particles and it is more asymmetric in the forward direction as particle size increases.
Source: The University of California Air Pollution Health Effects Laboratory, with kind permission.

by 90 percent at a distance of about 190 km (120 miles). Under such ideal conditions, 300 km is just about as far as a non-luminous, but high-contrast, distant object is visible at sea level in pollutant-free air. As altitude increases so does visibility. As will be seen later, the U.S. EPA assumed a greater (than that at sea level) visibility range in the definition of clean air. Because differences in illumination also affect visibility, brightly-lit objects can be seen at greater distances. Even dimly-lit objects can be seen against a dark background, as the rod cells of the eye can adapt so that they are extremely sensitive to light. Planets and faint stars can be seen at night even though they are hundreds of millions, or more, km distant. In severe conditions, such as intense fogs, hazes, and dust storms, visibility can be reduced to less than a meter.

Essentially all natural air components and anthropogenic pollutants can reduce visibility. Colorless particles and gases, both natural and anthropogenic, scatter but do not absorb light. The significant pollutants that both scatter and absorb a light include the brownish colored gas, NO_2, and a large variety of particles. An important question for regulators is: "What level of degradation of visibility from anthropogenic pollutants should be permitted?" The answer to this unsettled question can have significant implications for the economy, because current transportation, power generation, food production, manufacturing, and other valued practices generate air pollutants that impact visibility. Also, visibility in areas such as national parks and near observatories should be more protected than that in industrial manufacturing and other areas.

Climate

The Earth has undergone numerous natural climate changes in its long history, the causes of which are still poorly-understood (see Chapter 1). The influences of air pollutants on climate are less certain than those on visibility. Climate is both complex and difficult to model, and some air pollutants produce both cooling and heating effects.

Climate and other weather phenomena are largely driven by heating of the Earth. The Earth has two sources of heat, solar radiation and the molten core. Solar radiation is the dominant factor. The Earth both *absorbs* and *radiates* heat, and, thus, a *heat balance* must be maintained to preserve the present climate. **Figure 5–2** depicts the primary fates of solar energy that reaches the Earth. When solar energy is absorbed in the atmosphere the air itself is warmed, which can modify cloud formation, cloud persistence, and the types and levels of precipitation. Atmospheric absorption of solar radiation also decreases the heat reaching the Earth's surface, producing surface cooling. The scattering of solar energy by surface features, clouds, and atmospheric particles and gases reflects some (nearly 50 percent) of the solar energy back into space, which also affects the temperature of the Earth. The fraction of energy that is reflected back into space is called the Earth's *albedo.*

The albedo, as applied to astronomy, is the fraction of incident energy (especially light) that is reflected back into deep space by a planet or satellite. The albedo, more simply, is the apparent brightness of a non-self-luminous planetary object. The albedo of the barren moon in the visible electromagnetic spectrum is 0.07, which means that it reflects about 7 percent of the sunlight and earthlight that strikes it. The albedo of the Earth is about 0.4, which makes it a much brighter object than the moon. This additional brightness is mainly due to clouds, snow, and ice (features not found on the moon).

Superimposed on the effects of the albedo is the influence of the intensity of solar energy at the distance of the Earth from the sun (150 million km, 93 million miles). Energy from the sun comes from *nuclear fusion*, in which hydrogen and other atoms are transmuted into helium and other heavier elements; the fusion process releases

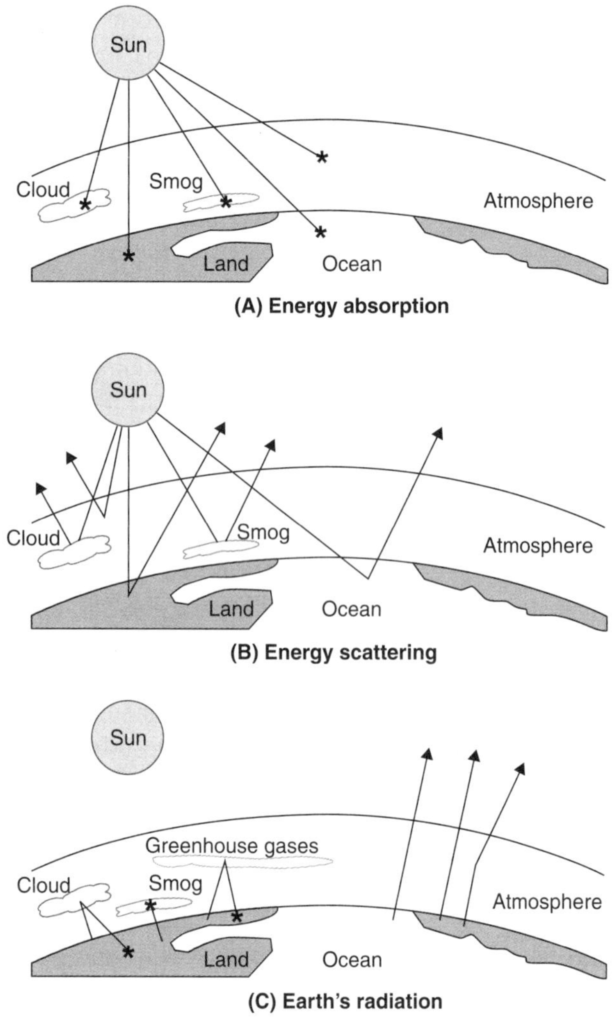

Figure 5–2 Examples of fates of solar energy that contribute to the Earth's heat balance. Short-wave solar energy can: (A) be absorbed in the atmosphere or surface regions; (B) be scattered back into deep space; or (C) be re-radiated as long-wave (heat) energy from the Earth. Not shown is the role of heat from the Earth's core and mantle, which serves to heat the surface from below.
Source: The University of California Air Pollution Health Effects Laboratory, with kind permission.

large amounts of energy. This released energy (E) is at the expense of a loss of solar mass (M), which can be calculated using Einstein's famous equation $E = Mc^2$, c being the speed of light in empty space.

The amount of energy from the sun that reaches the orbit of the Earth per square meter per unit time is termed the *solar constant.* The solar constant varies as a result of still poorly-understood "weather" phenomena on the sun, including sunspots, prominences, and solar flares. The average solar constant, measured at the top of the Earth's atmosphere, is about 1.9 calories per cm^2 per second (1.37 kilowatts per m^2). This value is not really constant, as small variations are observed. The solar energy that reaches the Earth's *surface* is less than 50 percent of that reaching the top of the atmosphere. This energy reaching the Earth's surface, which is seasonal, is much more variable than is the solar constant.

The albedo, which can be modified by air pollutants, in conjunction with the variable solar constant, which is independent of air pollutants, together have effects on the Earth's temperature and climate. The term *pollutant radiative forcing* describes a *change* in the heat balance of the Earth that is produced by the addition of tropospheric air pollutants. A *negative* forcing generates a cooling effect, and a *positive* value produces heating. Generally speaking, particles produce negative forcing and greenhouse gases produce positive forcing, but secondary phenomena (e.g., particle characteristics and changes in the biosphere) can lead to exceptions to these general effects.

Why Models Are Important

Changes in visibility and climate can be both directly measured and mathematically modeled. Direct measurements provide important current data, but good models are required for predicting future changes. Visibility models are more advanced than climate models, so there is currently more controversy with respect to the effects of air pollutants on climate predictions than for the effects on visibility predictions. As the current climate models evolve, predictions are expected to become more accurate (see Section III).

II. VISIBILITY AND AIR POLLUTION

Vision

Human vision (as shown in simplified form in **Figure 5–3**) is far more complex and interesting than either a "digital" or a "conventional film" camera. Both an eye and a camera rely on a lens to focus an image on a surface that transforms light into another form of energy. During this energy transformation process the informa-

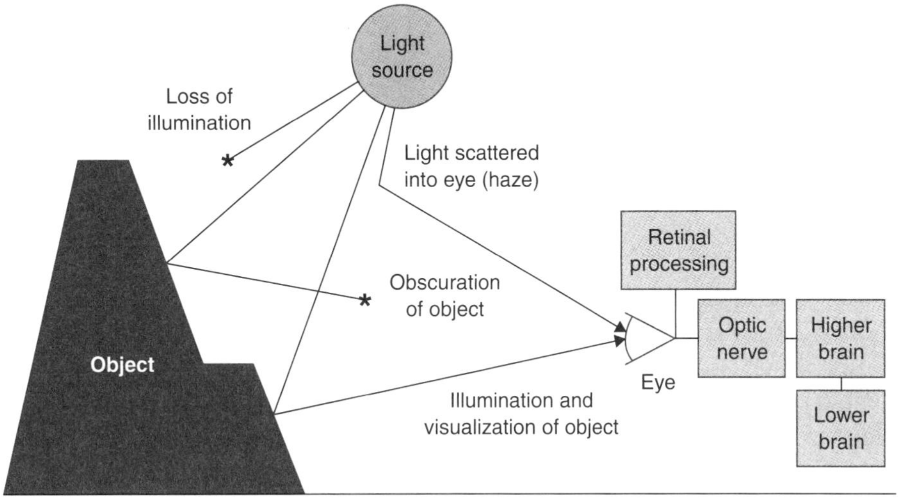

Figure 5–3 Simplified diagram of how light, smog, and the eye interact to determine the visibility of a distant surface feature. Absorption of light by particles and gases (indicated by an asterisk) decrease the illumination, and extraneous light scattered into the eye generates haze. Note that the image on the retina is modified by retinal processing and by several regions of the brain.
Source: The University of California Air Pollution Health Effects Laboratory, with kind permission.

tion content of the original image can be altered. The digital camera converts light directly (via the *photoelectric effect*) into an electrical signal, which then produces a digital image. In contrast, the retina of the eye generates a chemical image, and then modifies the image depending on the image intensity, constancy, movement, contrast, and other factors (e.g., what the brain has directed the eye to look for), before a final image is sent as electrical signals by the optic nerve to the brain. Further processing takes place in multiple regions of the brain before the image is consciously perceived. In short, human vision is highly interactive and subject to *intention, training,* and *experience.* Unlike cameras, humans can "choose" to see the flowers or the weeds, the trees or the haze, and even to totally ignore parts of the visual field. Human vision cannot be replicated by cameras or by other scientific instruments. This complicates the concept of just what visibility means, and how it can be measured in the presence of air pollutants.

Visibility

Visibility is conveniently defined in terms of the ability of the human visual system to discern objects in a background. An object's shape, color, contrast, and its illumination intensity affect its visibility. *Contrast*, which can be defined as a *difference* in light intensity between an object and its background, must be about 2 to 3 percent in order to be perceived by the human visual system. Daytime visibility is also defined as the *distance* at which a prominent dark object on the horizon can be seen. *Nighttime visibility* is based on measuring the loss of intensity of a distant unfocussed light source, such as a beacon or a star. There are other definitions of visibility, such as those used in aviation, but simply the ability to see distant surface objects during daytime is the most applicable to air pollution considerations. Clean technologies and air quality regulations have steadily improved visibility. In our time, the anthropogenic pollutant effects on visibility are modest, so visibility distances are measured in terms of tens or more kilometers, rather than meters. As a result, modern anthropogenic pollutants in developed nations may need improvement, but they do not usually produce a significant visibility-linked hazard. There are some exceptions, such as the effects of glare on aircraft pilots and vehicle drivers. However, modern efforts to improve visibility are based on *psychological* and *economic* considerations, rather than on improvements in safety.

Visibility potentially influences the enjoyment of daily activities, some psychological states (including the sense of well-being), and behavior. Dr. Nicole Hyslop

(2009) briefly reviewed the importance of good tropospheric visibility and expansive natural vistas on phenomena such as stress-reduction, the aesthetic experience, and property values. Good visibility is known to improve the value of real estate. The effects of good visibility on mood and behavior are neither well established, nor well understood. More research in this area is clearly needed. Although Dr. Gary Evans, et al. (1988) found changes in psychological parameters that were associated with photochemical smog in Los Angeles, they used only the ozone concentration, which is only *associated* with poor visibility, as their air pollution measure. Therefore, the actual contribution of visibility to the results of their study, although likely, is unknown. They found no significant influences on depression or hostility, but there was a significant association between smog levels and anxiety. The association remained after they controlled for the effects of socioeconomic status, age, and ambient temperature. Although the psychological effects of poor visibility have not been extensively studied, what is known supports the widespread belief that good visibility is indeed valuable. Thus, it is important to quantify the effects of anthropogenic air pollution on visibility and to estimate the costs associated with improving visibility so that cost–benefit analyses can be performed when improvements are desired. Clearly, the costs of air pollutant abatements to improve visibility are easier to evaluate than are the emotional benefits of improved visibility.

Light Scattering by Air Molecules (and Very Small Particles)

When light of a given frequency is scattered, as by air molecules or by particles, the scattered light may have the same, or a different, frequency. If the frequency is unchanged, Rayleigh (i.e., elastic) scattering has occurred. *Rayleigh scattering* is more effective at short wavelengths (e.g., as for blue light), which contributes to the blue color of the sky. If the incident light is absorbed and reemitted at a different wavelength then *Raman scattering*, a form of inelastic scattering has occurred, and the scattering molecule either gains (i.e., is excited) or loses energy in the process. Raman scattering can emit light photons, along with energy changes in the vibration, rotational, and electronic states of the scattering molecules. As a result, Raman scattering of laser beams (both land-based and satellite-based) can be used to remotely monitor air pollutants. *Remote sensing* is useful for sampling gases in hostile or inaccessible environments such as smokestacks, flames, and planetary atmospheres. Remote sensing is one of the environment-related functions of the U.S. National Aeronautics and Space Administration (NASA).

Light Scattering by Particles

The scattering of light by particles is strongly dependent on the particle size, as well as its index of refraction. If the particle's diameter (or another characteristic dimension) is less than 1/10 of the wavelength of the incident light, then the scattering is weak and predominantly of the Rayleigh type, as was previously described. For particles with diameters that are on the order of the size of the wavelength of light (i.e., particles with diameters between 0.1 and a few μm), *Mie theory* (from the German physicist, Gustav Adolf Feodor Wilhelm Ludwig Mie, 1869–1957) is used to predict the scattering intensity and directional patterns (Figure 5–1). The theory is complex, so scattering calculations are performed on modern computers. Mie scattering by larger particles is more intense in the forward direction (the same direction as the incident light is traveling) than is seen for smaller particles. Because Mie scattering patterns for visible light are not strongly dependent on the wavelength, the scattered light is similar to the incident light in color. Hence, fog appears to be white when illuminated by an automobile headlight.

Mie scattering is extensively exploited in particle sizing instruments because the directional pattern of scattered light is strongly influenced by the particle's diameter.

A *rainbow* is produced by the refraction and reflection of incident light (e.g., sunlight) by water droplets. Refraction is the bending of a light beam by an interface (e.g., between air and a water droplet) at which the index of refraction, and hence the speed of light change. The angle of refractive bending is dependent on the color (i.e., wavelength) of light as is seen when using a glass prism. Therefore, as white light enters a water droplet, it is refracted at the surface, separated into component colors, and then reflected back toward the source by the back of the droplet (which acts as a mirror). The net effect is a circular rainbow if viewed from a large height above the droplets. The rainbow will appear as an arc by a viewer on the surface, because the bottom half of the rainbow is intercepted by the ground.

Light Absorption by Particles

Although all particles scatter light, both elemental carbon and some soil particles also significantly *absorb* light. Such absorption adds energy to the particle in the form of heat and/or molecular excitation. Light absorption is represented by the "imaginary" part, *iy*, of the index of refraction, *M*, of a substance. The "real" part, *x*, represents light scattering, which can be directly observed:

$$M = x + iy. \quad \text{(Eq. 5–1)}$$

The light absorption loss is not readily observed by outdoor instruments. For visibility and climate modeling, both light scattering and light absorption must be considered.

Air Pollutants that Impair Visibility

The absorption and scattering of light by air pollutants produces both light extinction and haze effects and, thus, degrade visibility (Figure 5–3). Light extinction is the loss of intensity of a light beam that is caused by the combined effects of absorption and scattering by gases and particles. Haze is extraneous light scattered into the eye by gases and particles in the air.

Particulate pollutants are almost always more effective than gaseous pollutants with respect to both light absorption and scattering. Although all gas molecules will scatter light elastically (without an energy change, as in Rayleigh scattering), only colored gases absorb light (which involves an energy exchange in the absorber). Since molecular scattering by gases has a relatively small effect on light extinction, the only anthropogenic pollutant gas that significantly affects visibility is the brownish nitrogen dioxide (NO_2). In contrast, all of the particulate air pollutants that are in the fine and coarse size ranges (excluding ultrafine particles) will produce measurable light extinction when they are present in significant concentrations. Environmental monitoring and laboratory research have identified the following gas and particulate classes as important with respect to visibility degradation:

- gaseous nitrogen dioxide;
- particulate sulfates;
- particulate nitrates;
- organic particles;
- elemental carbon particles;
- fine soil particles;
- coarse particle mass (both natural and anthropogenic); and
- sea salt particles.

Light scattering can change for water-soluble particles, because they will grow in high relative humidity conditions. The effect of relative humidity on particle sizes, and thus their light extinction properties, is dependent on the particle composition, the current relative humidity, and even the relative-humidity history. Therefore, two additional factors must be taken into account for the visibility effects of particles: the original size range and the relationship between particle size and relative humidity. With respect to the original size range, the two categories "small" and "large" are used for assessing visibility, mainly because routine monitoring data are available on those ranges. Each class of particles, *i*, that grow in high humidity conditions has a characteristic mathematical water growth function, $f_i(RH)$. This function must be considered for each class of particles that are *hygroscopic* (i.e., they absorb water). The light extinction model used in the U.S. EPA's *Integrated Science Assessment for Particulate Matter* (U.S. EPA, 2009) employs three water growth functions:

- $f_s(RH)$, for small sulfate and nitrate particles;
- $f_L(RH)$, for large sulfate and nitrate particles; and
- $f_{ss}(RH)$, for sea salt.

Modeling Light Extinction

In order to model the effects of air pollution on visibility, the effects of individual pollutants must be identified. Quantifying the effects of specific air pollutants on visibility is not a trivial task. The first problem is to define the visibility parameter of interest. For regulatory considerations that parameter is *light extinction*, which is the loss of intensity of a light beam that is caused by the combined effects of absorption and scattering by gases and particles. The light extinction coefficient, b_{ext}, is used to represent the loss of light intensity per unit of path length:

$$b_{ext} = b_{scat} + b_{abs}, \quad \text{(Eq. 5–2)}$$

where b_{scat} is the scattering coefficient and b_{abs} is the absorption coefficient. When a light beam of intensity I_o passes through the air, its intensity is decreased to *I*. The loss of intensity of light when passing through a path length, *l*, is

$$I/I_o = \exp(-b_{ext} \times l). \quad \text{(Eq. 5–3)}$$

This loss of light intensity is what is used to model visibility reduction. Each air pollutant has its own extinction coefficient. Total extinction is calculated using the sum of individual extinction coefficients. The coefficient will depend on the relative humidity for water-soluble particles because growth by water absorption must be taken into account. One model for the extinction coefficient, called "IMPROVE" (because it uses data from the IMPROVE aerosol monitoring network), uses specific 24-hour pollutant concentrations at over 150 remote area monitoring sites. The concentrations are measured in μg/m^3, and the units for b_{ext} are in inverse megameters (Mm^{-1}). The b_{ext} used for pristine air is the Rayleigh coefficient of 10 Mm^{-1} (which implies an altitude above sea level). A simple IMPROVE model for particles, but not significant light-absorbing gases, approximates the extinction coefficient for each particle type by the product of a numerical weighting factor, a humidity function, and an airborne mass concentration:

$$\begin{aligned} b_{ext} \approx\ & 3 \times f(RH) \times [Sulfate] + 3 \times f(RH) \times [Nitrate] \\ & + 4 \times [Organic\ Mass] + 10 \times [Elemental\ Carbon] \\ & + 1 \times [Fine\ Soil] + 0.6 \times [Coarse\ Mass] \\ & + 10[\text{Rayleigh scattering effect}]. \end{aligned} \quad \text{(Eq. 5–4)}$$

The rather crude IMPROVE model did not work well in urban regions (where NO_2 levels could be important), coastal areas (with sea spray), or when the relative humidity was high. Because particle size can change the light extinction properties, and nitrogen dioxide absorbs light, the extinction coefficient model was revised by Pitchford, et al. (2007):

$$\begin{aligned} b_{ext} \approx\ & 2.2 \times f_s(RH) \times [Small\ Sulfate] + 4.8 \times f_L(RH) \\ & \times [Large\ Sulfate] + 2.4 \times f_s(RH) \\ & \times [Small\ Nitrate] + 5.1 \times f_L(RH) \\ & \times [Large\ Nitrate] + 2.8 \times [Small\ Organic\ Mass] \\ & + 6.1 \times [Large\ Organic\ Mass] \\ & + 10 \times [Elemental\ Carbon] + 1.0 \times [Fine\ Soil] \\ & + 1.7 \times f_{ss}(RH) \times [Sea\ Salt] + 0.6 \\ & \times [Coarse\ Mass] + \text{Rayleigh Scattering} \\ & \quad (\textit{site specific, as a function of elevation}) \\ & + 0.33 \times [NO_2(ppb)]. \end{aligned} \quad \text{(Eq. 5–5)}$$

In the new model the "small" particles have a geometric mean diameter of 0.2 μm (GSD = 2.2), and the "large" particles geometric mean diameter is 0.5 μm (GSD = 1.5) (The GSD, which describes the spread of the size distribution is discussed in Chapter 3). Mathematical models such as that in Equation 5–5 can predict the effect of control strategies on visibility for any location that has adequate 24-hour monitoring data for the relevant pollutants. This attribute is important for regulatory compliance purposes, because when a community is attempting to meet a visibility standard, the local officials need to know which pollutant controls will be the most cost effective at their location. Also, the clean air Rayleigh scattering factor now allows for considering the elevation at each site.

Spatial and Temporal Trends in Visibility

Overview

The U.S. EPA has reviewed studies on the differences in visibility by region, season, and local sources, as well as the temporal trends in visibility in the United States (U.S. EPA, 2009). Because this well-referenced review includes a large number of studies with discussions of the results, it is the basis of what follows in the next two sections.

Regional Differences in Visibility in the United States

The United States of America (U.S.) is geographically large. It has substantial regional differences in weather, population densities, types of fuels used, agricultural practices, and industrial characteristics. The regions also differ with respect to visibility expectations by the public. The U.S. can be roughly divided into geographical regions for visibility considerations. One such division, based on sulfate emissions, is the Western, North Eastern, South Middle, and South Eastern, as shown in **Figure 5–4**. A similar division is Northeast, Southeast, Northwest, Southwest, and Central. Other regional divisions are possible, such as Urban, Industrial, Agricultural, Mountain, Coastal, and Remote. As no single regional division is ideal, the above, and variations, are all used as needed.

Each region has its own characteristic mix of the most important air pollutants that influence visibility. Sulfate particles are the greatest anthropogenic contributors to visibility impairment in the Eastern U.S. due to the use of coal as a fuel. Sulfate particles are secondary, being formed in the atmosphere from the primary pollutant gas, SO_2. Sulfate particles are hygroscopic and produce

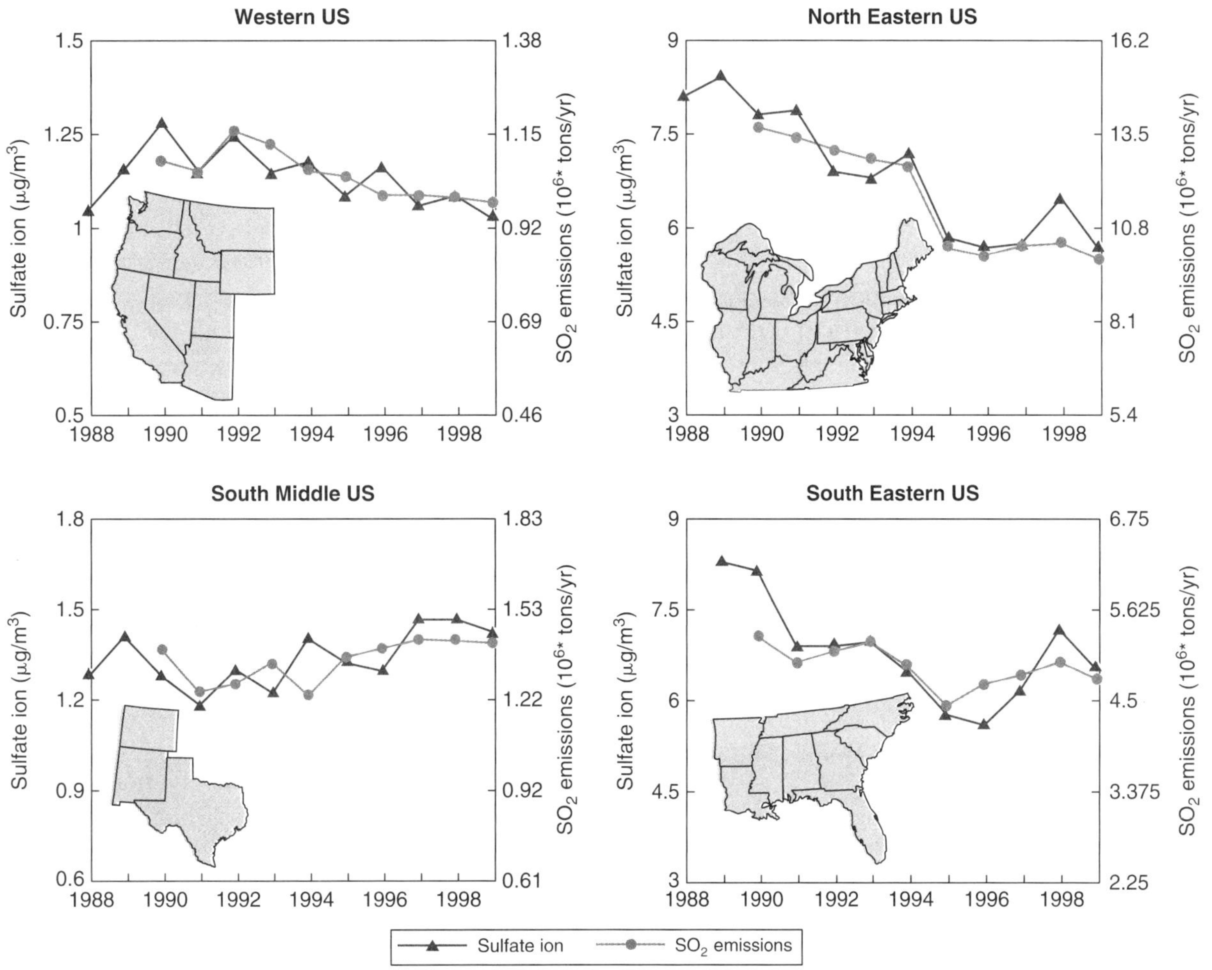

Figure 5–4 One useful regional depiction of the United States, along with ten-year and sulfate ion (SO) trends. The sulfate ion concentrations track sulfur dioxide (SO_2) emissions.
Source: U.S. EPA (2009) and Malm, W. C. et al., Journal of Geophysical Research, 2002, with permission of the American Geophysical Union.

more than 50 percent of the haze in regions such as the Northeast where the relative humidity is generally high. Other regions with lower SO_2 emissions also have visibility degradation by sulfate particles, but to a smaller extent.

In the Northwest, where sulfur emissions are low, elemental carbon (EC) and organic carbon (OC) produce the dominant light extinction under dry conditions. Such carbon particles are dark in color because they strongly absorb visible light. Note that EC has the highest extinction coefficient strength (= 10) in Equation 5–5. Organic aerosol mass and EC aerosols are primarily produced by the combustion of carbonaceous fuels (e.g., coal, diesel fuel, oil, wood, and biomass), and secondarily from atmospheric reactions of gaseous carbon-containing compounds.

Nitrate particles, also secondary pollutants, are similar to sulfate particles with respect to light scattering and hygroscopic growth, so they are perhaps third in importance for affecting visibility in the U.S.; particulate nitrate can dominate visibility in California, and the Midwest during winter. The formation of nitrate particles from vehicle combustion sources of NO_x (oxides of nitrogen) is facilitated by atmospheric ammonia. Ammonia (NH_3) is largely a product of agriculture, industry, and natural organic decay. This increases the importance of nitrate on visibility in large urban centers and major agricultural areas. Long-range transport of ammonia can also impact downwind areas. It is probable that some of the visibility impact of nitrate particles in the Eastern U.S. is due to precursor ammonia transported from the upper Midwest agricultural region. Such long-range transport can complicate the design of local effective abatement strategies to improve visibility.

Crustal material (fine soil) and coarse particulate mass are significant contributors to visibility degradation in dry regions. Such regions include the Southwestern desert areas and the Central Great Plains. The absorption coefficient strengths for fine soil and coarse mass (Equation 5–5) are small compared to those for sulfate, nitrate, organic mass, and elemental carbon. Thus, these latter pollutants can dominate over fine soil and coarse mass in dry regions that have significant combustion pollutant sources.

Haze Events

Haze events produced by "dust" can be characterized by the source of soil and other resuspended particles, including:

- local windblown dust;
- transported regional dust;
- long-range transported dust (e.g., Asian dust clouds that travel thousands of kilometers); and
- miscellaneous undetermined sources (e.g., unpaved roads, construction, mining, etc.).

Using these four source categories, an analysis of 610 "worst dust haze" periods in the Western U.S. was conducted. Regionally-transported dust produced 39 percent of the haze episodes, mainly in Arizona, New Mexico, Colorado, Western Texas, and Southern California. Thirty-three percent of the haze periods were attributed to local windblown dust. Asian dust influenced 9 percent of the periods, almost all in the spring and mainly in the Northwest (where local ground cover is substantial). The undetermined category was responsible for 19 percent of the haze episodes. Clearly, the control of local haze can be difficult in view of long-range transport contributions. Also, dust control can be costly considering that it is generated by winds, and has a significant natural component.

Visibility Temporal Trends

Because the effects of weather on visibility are highly variable, it is necessary to have long-term (e.g., five years or more) data in order to see the effects of anthropogenic emissions. The U.S. EPA (2009) reported trend data between 1995 and 2004 in 47 sites, including U.S. cities, national parks, and other non-urban locations. The haze days were classified as the 20 percent "worst," and the 20 percent "best" days; the remaining 60 percent "average" days were not reported. The "worst" haze days improved at 28 percent of the sites, showed no trend at 64 percent of the sites, and deteriorated at 9 percent of the sites (the total, 101 percent, is due to round-off errors). The "best" haze days had 68 percent of the sites showing an improvement, 21 percent showing no trend, and 2 percent (only the Great Sand Dunes in Colorado) showing a deterioration. These favorable trends were generally consistent with trends in emissions, which have steadily decreased.

III. CLIMATE AND AIR POLLUTION

Introduction

In order to begin to understand climate, the energy (i.e., heat) balance of the Earth must be appreciated. Figure 5–2 depicts the fates of the incoming solar radiation (also called *short-wave radiation*) and the Earth's heat radiation *(also called long-wave).*

The *climate* is influenced by both small and large scale phenomena including solar events, the Earth's orbit, land and ocean masses, and the entire Earth's atmosphere. Diverse interactions, between ocean and air, biota and temperature, and clouds and thermal radiations, plus possible *chaotic* behaviors (i.e., large changes are produced by tiny perturbations in initial conditions) also influence climate. Weather, which is only one of the relevant phenomena, is highly-chaotic, and thus difficult to accurately predict more than a few days into the future. Thus, climate is more difficult to understand than is visibility.

Unlike visibility, where an increase in air pollutant concentrations will produce straightforward degradations, the effects of air pollutants on climate can produce both cooling, and/or warming effects. Also, the magnitudes of the effects of pollutants on climate are not as easy to measure or even estimate as those for visibility. As a result, climate-change predictions tend to be controversial.

Another problem in assessing the effects of anthropogenic air pollutants on climate is caused by the very large effects of natural forces. The natural forces that affect the Earth's climate include the following:

- *the solar constant*—energy flux from the sun measured at the Earth (e.g., at the top of the atmosphere, or on the surface of the Earth).
- *the Earth's orbit*—The *eccentricity* (shape of the orbital ellipse), *obliquity* (axis tilt with respect to the sun), and *precession* (wobbling of the Earth's axis) all change periodically.
- *volcanic eruptions*—emit particles and gases.
- *ocean temperature profile*—which varies in association with various natural parameters (e.g., volcanic activity and weather phenomena).
- *sea ice*—related to salinity and temperature of the ocean.
- *continental drift*—which slowly changes the Earth's albedo.
- *natural greenhouse gases*—e.g., water vapor, methane, and carbon dioxide.

The anthropogenic factors that influence the Earth's climate include the following air pollutants:

- *carbon dioxide*—produced mainly by combustion.
- *methane*—produced by agricultural plant and animal waste and other organic decay.
- *nitrous oxide*—from combustion.
- *chlorofluorocarbons*—associated with manufactured chemicals.
- *particulate material*—from many sources.

Underlying these factors are:

- *the population*—more people means more pollution will be generated.
- *the level of prosperity*—more prosperous economies can afford to incorporate cleaner, greener, more efficient technologies.
- *a commitment to environmental quality*—which includes commitments to perform research and use cleaner technologies.
- *the knowledge base*—improvements in knowledge, instrumentation, and modeling are required to identify and control factors that adversely affect the Earth's climate.

As can be deduced from the previous material, effective climate actions may require considerable international cooperation. The emissions from food production, power generation, manufacturing, and transportation have their impacts on the greater climate regardless of where they are released. However, one of the still unresolved questions is whether or not the anthropogenic contributors to climate changes are significant in comparison with the natural forces. In order to resolve this question valid climate models are required.

The Greenhouse Effect and Greenhouse Gases

The *greenhouse effect,* one of the known climate forcing factors, is itself well-understood scientifically. The term "greenhouse," derived from horticulture, refers to a building with a transparent roof and sometimes one or more transparent sides. Solar energy in the visible spectrum (*short-wave radiation*) passes freely into the greenhouse, where it is largely converted to heat, thereby warming the building's interior. Heat loss from the building is due to two factors, *convection* (classical heat flow) and *radiation* in the infrared region (*long-wave radiation*). The building's insulating roof and walls reduce the convective heat loss, and they absorb infrared radiation. Thus, heat input exceeds heat loss, and the greenhouse becomes significantly warmer than the outside environment. Greenhouses are used to germinate and grow vegetables and ornamental plants during seasons when the outside temperature is low. Given information on the outdoor temperatures, winds, and sunshine, greenhouses can usually be designed to produce a distinct indoor climate. Greenhouses are important contributors to food production, as well as simple models for understanding the Earth's climate.

The Earth and its atmosphere act like a greenhouse in that certain *greenhouse gases,* such as water vapor, carbon dioxide, and methane permit solar short-wave radiation to warm the Earth's surface, but absorb long-wave radiation, which reduces radiant heat loss. Greenhouse gases, and their warming effects, are essential to life as we know it. The primary greenhouse gases are produced by natural sources, but anthropogenic activities are known to increase atmospheric levels of carbon dioxide and methane. Human activities also produce chlorofluorocarbons, nitrous oxide, ammonia, and other infrared absorbing gases. The relative importance of natural vs. anthropogenic sources of greenhouse gases in climate forcing is not well understood. Significant increases in carbon dioxide coincide with the history of increased use of fossil fuels, which has raised concern over an anthropogenic role in climate change.

Predicting the influence of greenhouse gas concentrations on the Earth's climate falls into the realm of *climate modeling.* The models must take into account not only greenhouse gas concentrations, but also their lifetimes in the atmosphere, their sinks, and their short-term and long-term influences on other climate forcing factors. One poorly-understood factor is the effect of greenhouse gases on the Earth's vast biota, which utilizes carbon dioxide and produces oxygen.

Climate Models

Mathematical models both give and take: They are indispensable scientific tools, but they live in an artificial reality (**Exhibits 5–1**, **5–2**, and **5–3**). Models are not only useful for their predictive potential, but they are also important for identifying the relevant factors and data that are needed to understand a system, problem, or question. Models also link the presumed relevant factors in some logical manner in order to represent interactions that influence a system's behavior. Such interactions include up-regulation, positive feedback, inhibition, and negative feedback among the model elements. When a system is as complex as the Earth's climate, model formulation is difficult because each component of the system can have poorly-understood, even undiscovered, impacts on other components. Furthermore, the Earth's climate is influenced by small-scale events that occur in an uncountable number of spatial domains in the Earth's seas, land, atmosphere, and biosphere, as well as events occurring on the sun.

From a physics perspective, climate is a large-scale phenomenon that emerges from the complex behaviors of many smaller-scale sub systems. As a result, modern climate models and their projections require enormous computing power and large computer codes. The codes, which currently include interactions of energy with air and water, are under continual development. Although incomplete, they have had some impressive early successes in modeling the climatic effects of past catastrophic air pollution events. The volcanic eruption of Mount Pinatubo in 1991 ejected more than 5 cubic km (about 1 cubic mile) of material (solids and gases) as high as 35 km (22 miles) into the atmosphere, and some of the ejecta traveled several times around the Earth reducing the amount of solar energy that reached the surface. Over the year following the eruption, the global temperature dropped 4°C in the Northern and Southern hemispheres. The measured drop in temperature closely matched the predictions of the climate code, "GISS Model EGCM."

Exhibit 5–1 Comments on models.

Mathematical models and "real-world" physical systems exist in separate realms. The real world can be defined as the totality of all things (e.g., objects, events, and phenomena), independent of human perceptions, languages, and beliefs. Mathematical models, which include *simulations* of the real world, are human constructions. Such models are limited by human perceptions, languages, and beliefs.

Using the Earth's climate as an example of a real-world system, its characteristics can be compared and contrasted to corresponding mathematical-model simulations. The climate is unlimited in complexity, as all of the atoms (and even sub-atomic structures) in the sun and the Earth theoretically contribute to the state of the climate and its future. Mathematical simulations (e.g., computer codes) are finite and limited in their complexity. These simulations include (1) large-scale components (e.g., known layers of the atmosphere and oceans, winds and ocean currents, continents, desserts, and mountain ranges); (2) the behavior of electromagnetic radiation and heat; and (3) a model structure that describes selected interactions. The climate has a *fractal structure* in which the level of complexity exists at all scales of size. It exhibits *chaotic behavior* that is characterized by sensitivity to small-scale perturbations, and *stability* in the sense that it seeks quasi-stable states (called attractors). Also, complex systems, such as the climate, exhibit short- and long-term *hysteresis* (a type of memory in which the history modifies the response to a given stimulus) and multiple *feedback loops* (both positive and negative), which can either lead to rapid changes in state or a return to the original state after a perturbation. Systems with negative feedback, and therefore stability, are preferred in nature because unstable systems tend to vanish in time. However, even stable systems can become unstable if they encounter a perturbation that is sufficiently large. Model simulations cannot easily represent the natural range of fractal structure, especially at the small scale, nor many of the characteristics of chaotic and complex systems. In sum, computer codes and other mathematical models are created by humans and are therefore limited by perception, language (especially the language of known mathematics), belief, and reason. Real-world systems have none of these limitations. The task faced by climate modelers is incredibly difficult.

Exhibit 5–2 Additional comments on models.

The topics presented in this chapter, visibility, climate, and stratospheric ozone, are model intensive. That is, mathematical models provide much of our understanding about these phenomena. Mathematical models use the precise language of mathematics to describe real-world systems. The success of mathematical modeling is indisputable, as is evidenced by the astounding accomplishments of physics, engineering, chemistry, physiology, and countless other science-based disciplines. One can argue that all physical systems obey (or are constrained by) the rules of mathematics. Philosophers even debate whether humans *discover* or *create* mathematics. Whatever the case, scientists place a considerable amount of confidence in the power of mathematical models as a tool for understanding and predicting the behavior of natural phenomena, including those described in this chapter.

Another way of looking at mathematical models of natural phenomena is that they are often only "educated guesses" about the real world. The degree to which a model describes the real world, including its future behavior, is the primary measure of its success. History shows that the early models are seldom successful, so models generally evolve over time in order to reduce their errors. Climate models are no exception. Simple global heat-balance models treat the Earth as a homogenous ball that intercepts solar energy and radiates some heat back into space. Constants in the simple models include the solar constant, the Earth's albedo, the Earth's radius, the Stefan–Boltzmann constant (the radiant energy per unit time of a blackbody per degree absolute temperature), and the Earth's emissivity (which corrects for its deviation from a perfect blackbody). Such models can predict the Earth's average temperature accurately by adjusting the values selected for the albedo and the emissivity. The simple models can be improved by adding sophisticated sub-models that represent the multilayered atmosphere, cloud effects, a dynamic ocean, ice and snow variations, vegetation coverage, the land-to-ocean heat transfer, and other factors. The construction of complex models requires input from a large number of experts and a computer code that challenges the capabilities of modern computers. Complex climate models are under development by several groups including NASA, NOAA, MIT, NCAR, Princeton's CFDL, the Max Planck Institute, etc. Each model can then be used to make predictions of the future climate of the Earth, given assumptions about future land use, natural and anthropogenic air pollutant emissions, solar emissions, etc. In short, simple models eventually lead to complex models, each of which is unique.

In addition to their utility, models also have their dark side. It is easy to confuse the neat rows of numbers from printed computer outputs with factual information about the modeled systems (e.g., the Earth's climate). Physicists, climatologists, and other qualified scientists may understand the model's limitations and assumptions (and the difference between computer output and truth). However, the general public, legislators, and others may not. Hence, uninformed debates occur over the validity of complex models. To make the situation worse, some modelers refer to their computer runs as "experiments," and the (perhaps erroneous) computer outputs as "experimental data." Also, for computer programs, "validation" can mean that the original mathematical specifications are met, which may not guarantee that the computer results are accurate representations of the behavior of the "real-world" system.

Considerable effort is applied to testing and validating complex models in order to evaluate their applicability to the real world. However, the absolute validation of complex models is still, in most cases, an elusive goal. This goal is an exciting and engrossing ongoing endeavor for the involved scientists.

With respect to climate models, the ultimate goal of climate prediction has not yet been reached. For updates on climate models, the Web sites of the groups that are developing them are informative. The Intergovernmental Panel On Climate Change (IPCC), established by the United Nations Environment Programme (UNEP), reviews the state of climate models, their predictions, and the implications for governmental actions to protect the Earths' climate (http://www.ipcc.ch/).

In order to include land masses and biological processes in the codes, these phenomena are first modeled in sub-codes that can eventually be added to the larger climate codes. Such modular codes have enormous complexity. Dozens of climate codes are now under development by various agencies and scientific groups. Each code differs in their assumptions and selection of phenomena that are modeled. Interestingly, although the models vary in their structures and detailed predictions, they all successfully demonstrate the trend that large increases in greenhouse gases might have on increasing global temperatures. However, it is not clear that the negative feedback mechanisms, such as changes in the biota on greenhouse gases, are sufficiently represented.

How Accurate Are the Climate Models?

Validation of climate models is both important and complex. As a part of this process, a given model is used to address past events and their known effects on

Exhibit 5–3 The NCAR community climate model.

The National Center for Atmospheric Research (NCAR) (in Boulder, Colorado) was established as an institute upon a 1956 recommendation from the National Academy of Sciences. NCAR had initial funding from the National Science Foundation. Included in its mission, NCAR was to work on "fundamental problems of the atmosphere," to provide large-scale facilities, and to institute an interdisciplinary approach. The "institute" later became a "center," but the mission was essentially unchanged.

In 1983 the NCAR created a global atmospheric model called the "Community Climate Model" (CCM) and made it available to the research community. This early model of the atmosphere has evolved considerably since its inception. The CCM is still under development, largely by incorporating sub-models, with additional financial support from the U.S. Department of Energy (DOE) and the National Aeronautics and Space Administration (NASA). Major steps in the development of the CCM involved adding (1) global ocean and sea ice models and (2) the impacts on climate of sulfate aerosols and greenhouse gases, including carbon dioxide. Recognition of the importance of land use and biological factors in modifying climate has led to coupling several other detailed models to the CCM. As a result, the model's name was changed to the Community Climate System Model (CCSM). The CCSM is now under development with several ambitious goals including (1) coupling several component models into a single framework; (2) making the model available to the broad climate research community; (3) engaging the community of climate researchers in development of the (CCSM); (4) addressing important questions, such as the global climate system; and (5) using the model to support policy decisions.

In the coming years the NCAR expects to increase its computing power (which will improve spatial resolution and allow for longer time simulations); provide more-sophisticated physics, chemistry, and biogeochemical phenomena; incorporate improved fluid dynamics simulations; and incorporate advances in satellite observations and other improved measurements. For those interested in following the progress of the CCSM the NCAR Web site (http://www.ncar.ucar.edu/organization/about) is a good informative starting point.

climate. An event as significant as the Pinatubo eruption can usually be effectively modeled, but such events are rare. How well do the climate models predict temperature changes for other, more subtle, forcing events? In order to answer this question, accurate records of the climate plus equally accurate records of the forcing events are required. Also, long-term records are needed to test climate models because of the variability in the data, and the small resultant changes in climate. Adequately reliable long-term climate data comes from ice cores and seabed cores that provide a measure of historical air and ocean temperatures. Such cores can be read like tree rings to obtain temperatures over the past several thousand years. Ice cores also have trapped air bubbles that can provide a record of the historic atmospheric composition. In addition, aerosol particle compositions and concentrations that were produced by major fires, volcanic eruptions, and meteoritic impacts can be estimated by ice core analyses.

The basic climate models show correlations between solar irradiation levels and periods of climate extremes, such as ice ages. Yet, changes in solar heating of the Earth due to orbital eccentricity, axial inclination, and precession alone are not great enough to induce the observed changes. It appears that small changes in the Earth's ocean temperature can trigger the release of large quantities of the greenhouse gases, carbon dioxide, and methane. The higher atmospheric levels of such gases may then amplify the climate changes. Thus, greenhouse gases can produce climate changes and/or result from such changes. The answer to the question (How accurate are climate models?) posed at the beginning of this section is difficult to answer. However, the modern climate models have evolved to fit the Earth's past temperature data fairly well.

Successful climate models must consider many factors, including:

- solar irradiation levels;
- levels of greenhouse gases;
- atmospheric aerosol levels;
- heat exchange between the ocean and atmosphere;
- ocean currents and other oceanic phenomena;
- deforestation and other land changes;
- heat from the Earth's core and undersea vents;
- snow and ice coverage;
- cloud formation and cloud dissipation phenomena; and
- dynamics of plant, animal, and microbial populations.

The above phenomena are not all well understood, so various models and modelers incorporate them in different ways. Also, some phenomena, such as cloud effects, cannot be accurately modeled because their small scale cannot be resolved in global-scale models. Other difficult problems in the climate models include:

- the effects of some phenomena, such as *El Niño;*
- feedback mechanisms that reverse the climate changes;
- "tipping-point" phenomena, for which corrective feedback mechanisms cannot reverse the climate changes;
- the effects of human population growth on fuel usage, greenhouse gases, and land utilization; and
- the future volcanic eruptions, their locations, magnitudes, and timing.

Before projecting climate change with any of the available models, one must first adjust the model's many parameters so that the model accurately predicts past climate periods. This "setup" process will be done differently by different modelers. The choices in setting up the models are numerous, so a given computer code could be set up in perhaps hundreds of different ways. Since all of the versions of computer codes could not be valid, there is a need to evaluate which of their predictions are correct. One method used for dealing with this complex task is to perform a large number of independent computer runs, and then treat their results more-or-less as votes for the various outcomes: warming vs. cooling vs. no change. Thousands of runs using various models on the effect of, say doubling the level of atmospheric carbon dioxide, generates a "large ensemble" of different results. Some results predict no warming, and a larger number predict significant warming by the year 2100. Most of the warming predictions are in the 1 to 5°C range. Experiments such as this raise several questions, including the following:

- Are the climate models advanced enough to be trusted?
- Is the "large ensemble" approach to prediction reliable?
- How serious is the predicted warming over this century?
- Can the observed carbon dioxide increase, and the predicted warming, be prevented?
- What are the economic costs of limiting future anthropogenic greenhouse gas emissions?

What Should be Done to Protect the Climate?

Many would argue that the modeling efforts are sufficiently compelling to prompt instituting immediate restrictions on greenhouse gas emissions. Many such controls, such as promoting greater energy efficiency (e.g., more efficient vehicles, appliances, home heating and cooling systems, etc.) are safe and easy choices to make whether or not they protect the climate. Measures such as replacing oil-fired and coal-fired electric power plants with nuclear, solar, wind, geothermal, and hydroelectric technologies raise issues of increased cost and/or decreased capacity to meet peak demands. Other control measures, such as externally-enforced restrictions on combustion emissions by undeveloped and/or developing nations and heavily taxing carbon dioxide emissions, are highly controversial and contentious. Greater use of nuclear power, which trades waste disposal concerns for freedom from greenhouse gas emissions, is gaining in popularity as a climate-protecting strategy among governments and environmental advocates. As an aside, the issue of climate modeling has entered the political realm, which clouds the scientific development of climate science (see **Exhibit 5–4**).

Climate and Particulate Air Pollution

Natural and anthropogenic phenomena generate a great variety of aerosol particles. As such particles interact with electromagnetic radiation, they may also impact the climate. The U.S. EPA performed an analysis of the effects of aerosol particles on the Earth's climate (U.S. EPA, 2009), addressing both direct and indirect impacts. The "direct" effects of aerosol particles relate to their effects on the Earth's albedo. Sulfate particles, and to a lesser extent, some other particle types, increase the albedo and thus reduce the solar radiation that reaches the Earth's surface. In contrast, black carbon and other colored particles absorb solar radiative energy and thus heat the atmosphere. Thus, the direct effects include both cooling and heating of the Earth. The "indirect" effects of aerosol particles involve their roles in cloud formation and in precipitation from clouds. The major impact of clouds is that they reflect incident solar energy. The predicted net effect of the direct and indirect effects produces a cooling effect on the Earth's climate. An estimate of the net radiative global forcing by aerosol particles is 1.3 negative watts/m^2 (range: −2.2 to −0.5 watts/cm^2). This range of reduction is supported both by satellite measurements and by

Exhibit 5–4 "Climate-Gate."

"Climate-Gate" (also "Climategate") is a derogatory term that refers to an incident that brought international attention on climate-change scientists. The scientists were accused of misrepresenting, deleting, and suppressing data, and not making their computer codes available for examination. The evidence undermined the claim that anthropogenic carbon dioxide emissions were a major cause of catastrophic global warming. The term "Climate-Gate" was derived from the Watergate scandal that destroyed the U.S. presidency of Richard M. Nixon in the early 1970s. The Watergate, a hotel and office complex in Washington, D.C., housed the Democratic National Committee headquarters. On June 17, 1972 five men that were connected to president Nixon broke into the headquarters. The investigation of the break-in led to the discovery of other questionable acts in which the president was involved, eventually leading to his resignation. Since that time, the suffix "gate" has been used to represent a scandal of major significance and ruinous consequences.

As climate models evolved they began to predict that the Earth's future climate could be impacted by human activities. Both land use (e.g., the destruction of forests), and especially the burning of fossil fuels, were projected to produce catastrophic global warming. Because the remedies to prevent this warming also had dire consequences, scrutiny of the climate modeling, and the modelers themselves, became intense. The science behind the models was challenged on many fronts, and skeptics squared-off against the modeling proponents. Most major scientific bodies and individual climatologists defended the predicted warming and its anthropogenic causes. The topic became politicized and both sides sought ways to discredit each other.

The international Intergovernmental Panel on Climate Change (IPCC) was established to objectively evaluate the climate projections and make recommendations to governments. The IPCC supported the idea that anthropogenic activities, especially carbon dioxide emissions, were responsible for global warming, and that immediate action was required.

Although many sources of data on the climate were used by the IPCC, the Climatic Research Unit (CRU) of the University of East Anglia in England supplied some of the key temperature data. In November, 2009 thousands of e-mails and other documents were illegally hacked and distributed worldwide via the Internet. The two sides of the climate warming debate accused each other of misconduct. The contents of some of the e-mails were troublesome. It was clear that the CRU scientists were concerned about attacks on the anthropogenic global warming hypothesis, and they discussed ways of fighting back in their private e-mail messages. Accusers claimed that the emails provided evidence of hiding important data, deleting sensitive files, thwarting requests made under the United Kingdom's Freedom of Information Acts, making plans to interfere with the peer-review process, and not sharing computer codes. The CRU scientists claimed that their private communications had been taken out of context and misrepresented.

Investigations were quickly launched by the University of East Anglia and British police. The CRU Director resigned his position pending the results of the investigation. The IPCC also planned to investigate. The lessons are clear. When a research finding becomes politicized, the normal slow deliberate scientific process that provides for challenge and eventual acceptance (or rejection) is disrupted. Warring sides, interested in winning instead of seeking the truth, can engage in personal attacks, dishonesty, and other destructive and wasteful activities.

---

climate models. It should be noted that the global radiative forcing by greenhouse gases is positive (about 2.9 ± 0.3 watts/m^2), which opposes the apparent cooling effects of particles. Also, although particles tend to reduce the incident solar radiation at the top of the atmosphere, smog can also absorb solar energy and lead to local heating at lower altitudes.

IV. STRATOSPHERIC OZONE

Why is Stratospheric Ozone Important?

Ozone (O_3) has a dual role with respect to its effects on living systems (i.e., microbes, plants, and animals). In the troposphere (the lowest 10 km of the Earth's atmosphere), inhaled ozone in elevated concentrations has adverse effects on the respiratory tracts of mammals. Thus, tropospheric ozone is sometimes called *bad ozone.* Conversely, in the stratosphere ozone absorbs solar ultraviolet (UV) radiation. If all of the solar UV radiation that falls on the Earth reached the surface, terrestrial life as we know it would vanish. Thus, stratospheric ozone is commonly called *good ozone.* Even at relatively low levels (e.g., normal levels) UV radiation can contribute to skin cancer and eye damage in humans. It is important for people to limit their skin exposure to sunlight by using topical sunscreens, and also to wear UV protective glasses in order to control the adverse effects of UV radiation on their eyes. The "suntan," which became popular in the 20th century, is no longer regarded as a sign of good health by medical experts. Both basal cell and squamous cell skin carcinomas are known to be produced by sunlight, especially in light-skinned individuals. Eye damage from UV exposure includes, acutely, an inflammation of the tissues, and chronically, cataracts, independent of skin color.

Fortunately, the Earth has a substantial layer of ozone in the stratosphere (**Figure 5–5**). This layer is very effective in absorbing UV radiation, especially in the shorter wavelengths (**Figure 5–6**). However, a substantial long-term thinning of the stratospheric ozone "blanket" would pose a threat to life and health. As a result, a considerable amount of effort is being directed toward monitoring, modeling, and otherwise understanding ozone chemistry in the stratosphere.

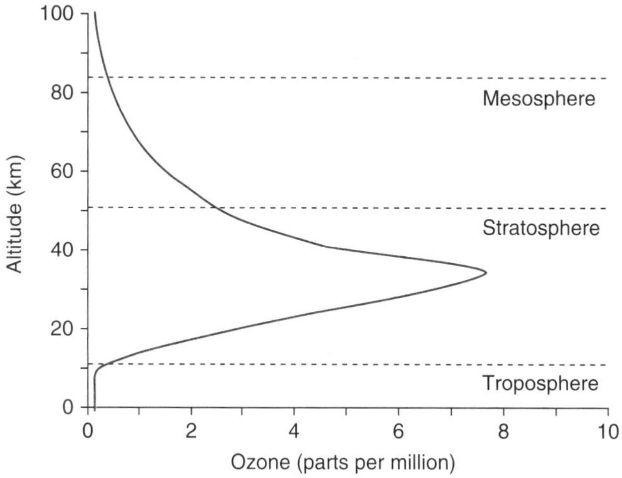

Figure 5–5 Ozone concentration vs altitude. Modified from NASA's "Ozone Facts" (http://ozonewatch.gsfc.nasa.gov/facts/ozone.html; assessed 12/26/2009). *Source:* The University of California Air Pollution Health Effects Laboratory, with kind permission.

Ozone Measurement, Formation, and Destruction

Stratospheric Ozone Measurement

The preferred unit of measurement for ozone depends on the location. In the *troposphere*, where ozone levels are small, ppm (parts per million) or ppb (parts per billion) units are used; such units are ratios of the volume fraction of ozone in relation to the total air. So, if one knows the volume of air inhaled, the dose of ozone to the body can be calculated. In the *stratosphere*, although ppm units of measurement are sometimes used, the *Dobson Unit* (DU) is more common because as a *thickness* unit it is related more directly to the absorption of solar UV energy.

One Dobson Unit (after G. M. B. Dobson, British physicist and meteorologist, 1889–1976) is the number of molecules of ozone that are needed to form a layer of pure ozone 0.01 mm thick (at 0°C and 1 atm. pressure)

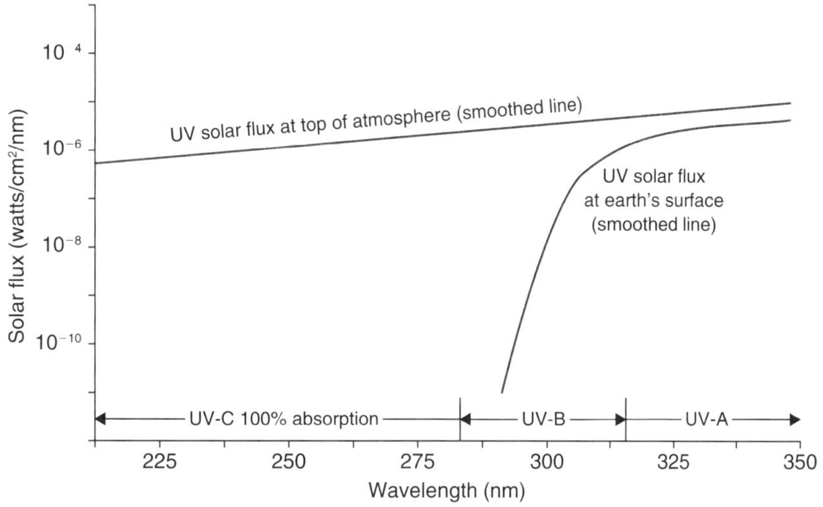

Figure 5–6 The absorption of ultraviolet radiation in the Earth's atmosphere. After NASA's "Ozone Facts" (http://ozonewatch.gsfc.nasa.gov/facts/ozone.html; accessed 12/26/2009).
Source: The University of California Air Pollution Health Effects Laboratory, with kind permission.

that would, like a blanket, cover the Earth's surface (assuming that the Earth was a smooth ball).

The average thickness of the Earth's ozone layer is 300 DU. The Dobson unit effectively describes how thick the ozone blanket is, which is what is needed to understand its UV protective function. A value of less than 220 DU at the Antarctic is taken as the definition of an *ozone loss,* as lower values were not measured before 1979. The area bounded by a 220 DU line defines an *ozone hole,* which represents a thin layer of ozone. The size of the Antarctic ozone hole is measured in tens of millions of square km; roughly centered on the South Pole, the hole covers a significant portion of the Earth's surface.

The thickness of the ozone layer varies both seasonally and spatially over the surface of the Earth. As ozone is created by the interaction of solar UV radiation with oxygen, the stratospheric ozone concentration should be greatest at the equator. However, other factors, especially wind transport from the equator toward the poles, produces greater ozone concentrations at the Earth's north and south polar regions. The northern hemisphere (the Arctic region) has high stratospheric concentrations of ozone in the spring (March and April). But the southern (Antarctic region) has its highest concentrations in its spring (September and October). During the other seasons, stratospheric ozone is depleted in the Arctic and Antarctic regions, producing natural cyclic ozone "holes." Superimposed on this natural cycle is the breakdown of ozone by chemicals such as nitrogen, chlorine, bromine, and hydrogen compounds that reach the stratosphere. The movement of gases from the troposphere is normally slow, so persistent ozone-depleting molecules are of concern. The chlorine, bromine, and some other compounds are mainly anthropogenic, and they are also persistent. Thus, increased attention is directed to them.

Ozone Formation and Depletion

Stratospheric ozone chemistry is complex, but well studied. Ultraviolet radiation (at wavelengths, *hv*, less than 240 nanometers) breaks O_2 molecules into singlet oxygen (O). The singlet oxygen quickly reacts to re-form either O_2 or ozone (O_3):

$$O_2 + hv \rightarrow O + O \qquad \text{(Eq. 5–6)}$$

$$O + O \rightarrow O_2 \qquad \text{(Eq. 5–7)}$$

$$O + O_2 \rightarrow O_3. \qquad \text{(Eq. 5–8)}$$

These recombinations reaction require a third molecule, typically N_2 or O_2, to absorb energy during the reaction, but this third molecule is not shown in the above reactions.

The destruction of ozone can occur due to either direct destruction by ultraviolet light, or by chlorine, bromine, or hydrogen atoms:

$$O_3 + hv \rightarrow O_2 + O \qquad \text{(Eq. 5–9)}$$

$$ClO + O \rightarrow O_2 + Cl \qquad \text{(Eq. 5–10)}$$

$$Cl + O_3 \rightarrow O_2 + ClO. \qquad \text{(Eq. 5–11)}$$

The reactions, Equations 5–10 and 5–11, are *cyclic* in that they utilize and regenerate Cl and ClO. The regenerated chlorine (Cl) can continue to break down ozone for many cycles. Thus, a single molecule of Cl can destroy 1,000 or more ozone molecules before it is converted to a relatively unreactive form, such as hydrochloric acid (HCl). Similar fates apply to the other molecules that deplete stratospheric ozone (e.g., bromine, nitrogen, and hydrogen). Kinetic considerations must be used to model the lifetimes of ozone and the chemical species that deplete it. Such chemical kinetics are complex, so they will not be further discussed here. For more information, atmospheric chemistry texts can be consulted.

Under natural circumstances the ozone levels will reach an equilibrium concentration that varies seasonally. However, anthropogenic compounds such as CFCs (chlorofluorocarbons) can upset the equilibrium and steadily deplete stratospheric ozone. When this happens, it may take several years for the natural ozone levels to be reestablished.

CFCs and Ozone Destruction

CFCs are primarily used as refrigerants (as replacements for the more toxic, highly-irritating, and flammable ammonia) and as propellants for aerosol spray cans. CFCs are normally stable, but they are broken down in the stratosphere by ultraviolet radiation where they participate in ozone destruction. When the ozone depletion problem was recognized, restrictions on the production and use of CFCs were quickly implemented. As a result of the 1987 landmark *Montreal Protocol* and its amendments, tropospheric concentrations of CFCs (and thus their transport to the stratosphere) have decreased. This success story predicts eventual recovery of the natural stratospheric ozone levels. However, there are other potential anthropogenic threats to the ozone layer. Supersonic high altitude transport aircraft exhaust can also potentially alter stratospheric chemistry. Scientists and engineers must collaborate on engine and fuel designs in order to make sure that this promising technology will also be environmentally friendly. Knowledge of the chemistry of stratospheric ozone is now well advanced, and future generations can be protected from exposure to harmful ultraviolet radiation, if new technologies that might threaten ozone are introduced in a responsible manner.

V. SUMMARY OF MAJOR POINTS

This chapter has provided an introduction to three important aspects of air pollution science: (1) visibility degradation; (2) climate change; and (3) stratospheric ozone levels. The importance of these topics results from their potential impacts on the quality of life, health, and perhaps even the existence of life on Earth. Substantial recent scientific progress has greatly improved our understanding of the potential impacts of air pollutants on each of these phenomena. As a result, by careful planning and the deliberate introduction of cleaner technologies, the quality of life for future generations can be protected.

Visibility has both physical and psychological components. Under pristine (pollution-free) conditions, high-contrast objects such as mountain ranges can be seen at distances of hundreds of kilometers. Natural phenomena such as fogs, precipitation, fires, and volcanic eruptions can greatly impair visibility. Anthropogenic pollutants such as particulate material and the colored gas nitrogen dioxide (NO_2) scatter and absorb visible light. Such interactions both diminish light transmission and scatter stray light into the visual field, producing *haze*. Thus, objects lose their *contrast* and are obscured by haze. In order to make improvements in visibility, *direct measurements* and *computational models* are used to identify problematic pollutants. Regulatory actions to improve visibility have economic consequences that must be weighed against the psychological and economic (e.g., property values) benefits of improved visibility.

The Earth's climate is primarily controlled by natural phenomena, but the potential impacts of anthropogenic phenomena must also be considered in order to prevent unacceptable changes. The *solar constant* (which varies), the Earth's orbital variation, inclination, and precession are beyond our control. However, anthropogenic air pollutants, especially those related to fuel combustion, produce aerosol particles and gases that can influence the average temperature of the Earth. The net effect of particulate pollutants appears to be cooling, primarily by increasing the Earth's *albedo*, which reduces the solar radiant energy that reaches the surface. *Greenhouse gases*, such as water vapor, carbon dioxide, and methane, have large natural sources. Such gases trap *long-wave radiation*, which results in climate warming. Recent concern over the anthropogenic production of carbon dioxide has led to debates over the role of human activities on the climate. Sophisticated and rapidly-evolving computational models are being used in an attempt to ascertain the various impacts of natural and anthropogenic phenomena on the future climate. Such models have numerous input assumptions that can make the predicted role of human activities significant or negligible. When the models are run multiple times with different assumptions, they produce a distribution (also called an ensemble) of predictions about climate changes. Since a large majority of the predictions indicate a possible significant global warming, there is pressure to curtail anthropogenic carbon dioxide emissions. The issue is a difficult one, as fuel combustion is essential to prosperity and the provision of lifesaving essential goods, services, and electrical power. Aside from the obvious benefits of increased energy efficiency, with an ever-growing human population, the substitution of fossil-fuel technologies with cleaner, more sustainable technologies will be beneficial for generations to come.

Ozone in the stratosphere is often called *good ozone* because it absorbs harmful solar ultraviolet radiation. Without this protective *ozone layer,* life as we know it could not exist on the Earth. Ozone is created in the stratosphere by solar ultraviolet radiation. Stratospheric ozone is measured by the Dobson Unit (DU), which describes the average thickness of the stratospheric ozone. The normal thickness, 300 DU (i.e., 3 millimeters), varies with the solar radiation levels and stratospheric meteorology. When the DU is 220 or less, it is called an *ozone hole* (which is actually the area covered by a thinning of the protective layer). Compounds (notably CFCs) used in refrigeration and aerosol spray cans that are released into the troposphere can reach the stratosphere and produce cyclic reactions that destroy ozone molecules. International actions to reduce CFC emissions have greatly reduced tropospheric concentrations to low levels. Current models predict that such reductions will prevent the long-term thinning of the stratospheric ozone layer.

Important future technologies must be carefully designed with their potential impacts on stratospheric ozone considered. Scientists and engineers must collaborate to insure that future technologies are effective and responsible.

VI. QUIZ AND PROBLEMS

Quiz Questions

(select the best answer)

1. Under "pristine" (pollution-free) conditions, scenic visibility is:
   a. unlimited.
   b. less than 20 km.
   c. between 20 and 100 km.
   d. greater than 100 km.
2. Haze:
   a. is produced by pollutant gases but not by pollutant particles.
   b. is produced by the absorption of light by air contaminants.
   c. is produced by the scattering of light by air contaminants.
   d. is a psychological phenomenon with no real physical basis.
3. Which statement about light extinction is *untrue*?
   a. It is a result of both scattering and absorption.
   b. It depends on the extinction coefficient and the path length through which light travels.
   c. It is produced by both gases and particles.
   d. It is produced by particles but not gases.
4. Factors that significantly influence the Earth's climate include:
   a. the solar constant.
   b. the Earth's albedo.
   c. volcanic eruptions.
   d. All of the above are true.
   e. None of the above is true.
5. With respect to changing the Earth's climate:
   a. greenhouse gases have a negligible effect.
   b. greenhouse gases produce a cooling trend.
   c. greenhouse gases produce a warming trend.
   d. chaotic effects are negligible.
6. Anthropogenic air pollutants that might influence climate:
   a. are the exclusive forcing factors.
   b. are among the forcing factors included in climate models.
   c. include air pollutant particles but not gases.
   d. include power plant fuel combustion emissions but not car and truck emissions.
7. Which statement is true about atmospheric ozone?
   a. Stratospheric ozone is considered to be "bad ozone."
   b. Tropospheric ozone is considered to be "good ozone."
   c. All ozone is considered to be "bad ozone."
   d. None of the above is true.
8. Ultraviolet solar radiation:
   a. has a net beneficial effect on mammals.
   b. is considered to be harmful to life on Earth.
   c. is completely absorbed in the stratosphere.
   d. is not absorbed in the stratosphere.
9. CFCs (chlorofluorocarbons):
   a. are natural greenhouse gases.
   b. build stratospheric ozone concentrations.
   c. are inert and thus do not influence ozone concentrations.
   d. lead to stratospheric ozone depletion.
10. Why are computational models important?
    a. They are used to identify factors that influence visibility, climate, and the ozone layer.
    b. They can show how various factors interact to produce effects on visibility, climate, and the ozone layer.
    c. They evolve over time to include more complex phenomena.
    d. All of the above are true.

Problems

1. The following table shows the albedo (in the visible region) of several planets:

Planet	Albedo
Mercury	0.076
Venus	~0.67
Earth	0.36
Mars	0.15
Jupiter	0.54
Saturn	0.57

   a. When viewed at the same distance from an observer, which planet would appear to be the brightest?
   b. How much brighter is the Earth than Mars when viewed from the same distance?

   c. List three factors that might make the albedo of the Earth larger than that of Mars.
2. Calculate the ratio of transmitted light intensity (I) to incident light intensity (I_o) for a beam of light passing through a path length of 100 km given a value of b_{ext} of 10^{-5} m^{-1}.
3. Define "haze" and discuss how it impacts the visibility of high-contrast distant objects.
4. The CFCs used in refrigeration and as dispersants in "spray cans" are generally considered to be chemically stable in the troposphere. Explain how the chlorine in CFCs can deplete the stratospheric ozone layer.
5. Why is the *Dobson Unit* (DU) used to quantify stratospheric ozone instead of the unit, ppm, which is used to quantify tropospheric ozone levels?

VII. DISCUSSION TOPICS

1. Why might the permissible visibility in an urban area differ from that in a national park such as the Grand Canyon?
2. What pollution controls could be used to improve visibility in a large city, and what are the potential effects on the economy?
3. Why is the "climate-change" issue so hotly debated? How might stringent control of anthropogenic greenhouse gases differentially affect developed nations and developing nations?
4. What sources of electrical power are the most "climate friendly?" Can only "climate friendly" sources of power be used in the near-term (next 20 years) to satisfy the world wide need for domestic and industrial electricity?

References and Recommended Reading

Cravens, G., *Power to Save the World: The Truth About Nuclear Energy,* Vintage (Random House), Lincolnshire, IL, 2008 (paperback edition).

Evans, G. W., Colome, S. D., and Schearer, D. F., Psychological reactions to air pollution. *Environ. Res.,* 45: 1–15, 1988.

Evans, G. W. and Jacobs, S. V., Air pollution and human behavior, *J. Social Issues,* 37:95–125, 1981.

Evans, G. W., Jacobs, S. V., and Frager, N. B., Human adaptation to smog, *J. Air. Pollut. Ctl. Assoc.,* 32: 1054–1057, 1982.

Finlayson–Pitts, B. J. and Pitts, J. N. Jr., *Chemistry of the Upper and Lower Atmosphere: Theory, Experiments, and Applications,* Academic Press, London, UK, 2000.

Hinds, W. C., *Aerosol Technology: Properties, Behavior, and Measurement of Airborne Particles, 2nd Edition,* John Wiley & Sons, Inc., New York, 1999.

Hyslop, N. P., Impaired visibility: The air pollution people see, *Atmos. Environ.,* 43:182–195, 2009.

Jacobs, S. V., Evans, G. W., Catalano, R., and Dooley, D., Air pollution and depressive symptomology—exploratory analyses of intervening psychosocial factors, *Population Environ.,* 7:260–272, 1984.

Malm, W. C., Trijonis, J., Sisler, J., Pitchford, M., and Dennis, R. L., Assessing the effects of SO2 emission changes on visibility, *Atmos. Environ.,* 28:1023–1034, 1994.

Malm, W. C., Schichtel, B. A., Ames, R. B., and Gebhart, K. A., A ten-year spatial and temporal trend in sulfate across the United States, *J. Geophys. Res.,* 107(D22):4627, doi: 10.129/2002JD002107, 2002.

NASA (National Acronautics and Space Administration), Ozone Facts, http://ozonewatch.gsfc.nasa.gov/facts/ozone.html (accessed December 12, 2009).

Norton, T., Sun, D. W., Grant, J., Fallon, R., and Dodd, V., Applications of computational fluid dynamics (CFD) in the modeling and design of ventilation systems in the agricultural industry: A review, *Bioresource Technol.,* 98:2386–2414, 2007.

Pitchford, M., Malm, W., Schichtel, B., Kumar, N., Lowenthal, D., and Hand, J., Revised algorithm for estimating light extinction from IMPROVE particle speciation data, *J. Air Waste Manag. Assoc.,* 57:1326–1336, 2007.

Schmidt, G. A., The physics of climate modeling, *Physics Today,* 60:72–73, 2007.

Seinfeld, J. H. and Pandis, S. N., *Atmospheric Chemistry and Physics: From Air Pollution to Climate Change,* John Wiley & Sons, New York, 2006.

U.S. EPA (U.S. Environmental Protection Agency), *Integrated Science Assessment for Particulate Matter,* EPA 600/R–08/139F, U.S. Environmental Protection Agency, Research Triangle Park, NC, December 2009.

Chapter 6

Regulation and Abatement of Air Pollutants

LEARNING OBJECTIVES

By the end of this chapter the reader will be able to:

- describe the motivation behind the regulation of air pollutants
- discuss the positive and negative impacts of environmental regulations
- define "abatement" and discuss its relationship to regulations
- describe the four components of an air standard
- plan an abatement strategy for fine particulate air pollutants

CHAPTER OUTLINE

I. INTRODUCTION AND SCOPE

Introduction

The Justification for Regulation and Abatement of Air Pollutants

Are air quality standards really necessary? More to the point is the question: "Are controls on anthropogenic (human-made) air contaminant emissions necessary to protect public health and the environment?" The answer to both questions is "yes," considering: (1) the great air pollution disasters (Chapter 1); (2) current data on human health (Chapters 8 and 10); (3) air pollutant toxicology studies (Chapter 9); and (4) modeling predictions for climate change (Chapter 5). The scope of the major potential adverse effects of air pollutants is depicted in **Figure 6–1**. As can be seen, air pollutants can have adverse effects on human health, animals, plants, ecosystems, building materials, artwork, visibility, and climate. Because the dose makes the poison, very high levels of air pollution can adversely affect all of the systems shown in the figure. It is difficult, if not impossible, to define air pollution levels that have no absolutely adverse effects. Thus, the regulation, monitoring, and control of air pollutants are justified. Avoiding unacceptable health, welfare, and environmental effects is the primary motivation for the regulations. Given the necessity for air quality regulations, control strategies (sometimes called *abatements*) must also be devised. Regulations and abatements are tightly intertwined, which is why they are introduced together in this large chapter.

The Complications

From an idealistic viewpoint, the clean air problem is easy to solve; just eliminate all of the pollutant sources. But a bit of reflection on this solution is sobering. It means we must close the factories, shut down the electric power plants, ban all vehicles, suspend new construction, halt agricultural operations, stop cooking our food, abolish wood burning, and even go as far as eliminating most living things. Yes, plants, animals and other living organisms generate significant amounts of air pollutants. Therefore, the regulation of air pollutants

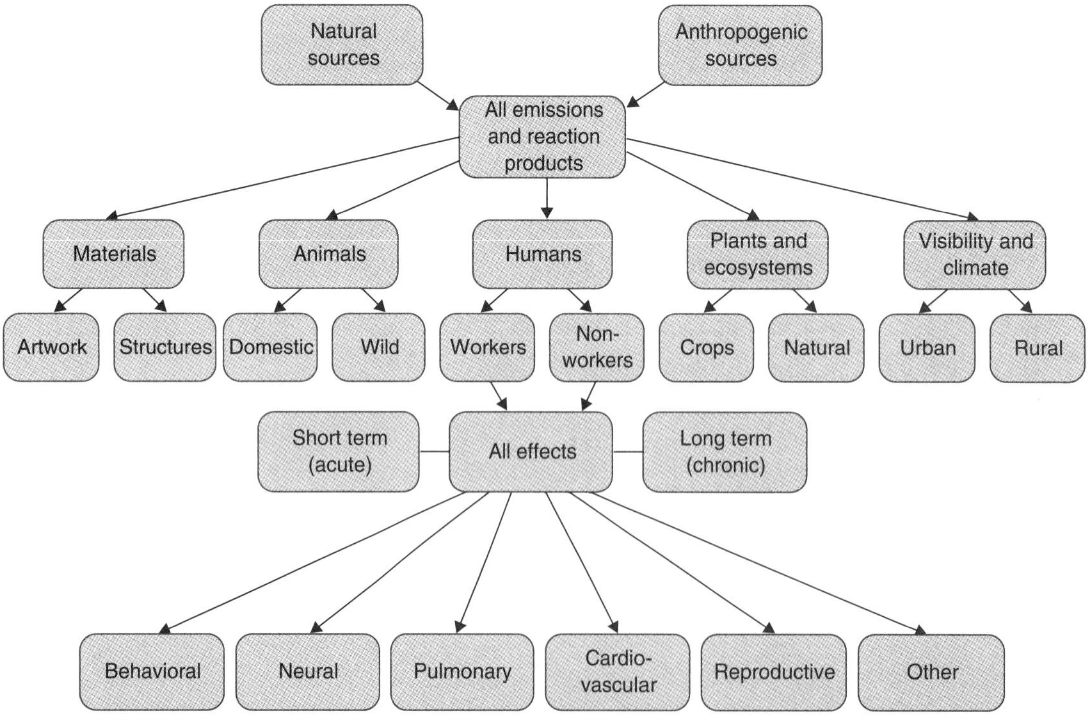

Figure 6–1 Simplified depiction of the sources and potential effects of excess levels of air pollutants. Effects on humans have been expanded. Regulators look for points of intervention in order to reduce the adverse outcomes.
Source: The University of California Air Pollution Health Effects Laboratory, with kind permission.

must be done carefully, and in a manner that is practical, affordable, and consistent with continued survival, happiness, and prosperity. The very sources that pollute the air also make life possible and worthwhile.

A holistic approach to air pollution regulation is needed; one that considers and balances significant competing factors, including human health, prosperity, and freedom vs. having a pristine natural environment. Unfortunately, that ideal has not been reached, nor is it likely to be achieved in the foreseeable future. Several factors, including economic and political realities, the pressures of various stakeholders, and the lack of an accepted integrated public health model, all stand in the way. Yet, regulation and abatement activities have progressed and are evolving at a rapid pace.

Today, several relatively-independent governmental agencies pursue their specific *mandates* (i.e., legal responsibilities that are driven by laws, court decisions, and policy choices) and promulgate air pollution regulations. These regulations must be met by states, communities, businesses, and individual citizens. At times, the regulations are difficult to meet, because some regulatory agencies, including the U.S. Environmental Protection Agency (U.S. EPA), are prevented by statute from considering practical factors, such as the cost, trade-offs, or feasibility associated with meeting a standard. No one can argue with the goal of protecting public health, and it is often unpopular to question new air quality regulations. However, there must be compromise and balance in the system. As an example, economic stability and equitable living standards (e.g., elevated prices have a greater effect on the poor) are two key trade-offs worthy of consideration for any nation. Thus, a holistic approach to air pollution regulation and abatement is necessary to prevent serious unintended consequences.

Strategic vs. Tactical Air Quality Regulations

Regulations generally fall into two categories: (1) long-term (to achieve future goals) and (2) short-term (to deal with current problems). Long-term goals are also called "strategic" and short-term goals "tactical." For example, the phased-in control of ozone-friendly chemicals is a strategy to protect the ozone layer, but temporary subsidies for fuel-efficient hybrid vehicles is a tactic used to reduce the potential harmful effects particles and gases on human health and the environment. Other tactical measures include the restrictions on traffic imposed during Olympic events in the United States and China.

Compliance Strategies

Compliance strategies for meeting air regulations and standards take several forms, including (1) introducing cleaner technologies; (2) cleaning existing fuels; (3) improving industrial processes; (4) encouraging restraints on energy use; and (5) banning some activities. Some regulatory pressures have forced needed engineering advances, including cleaner technologies and emission controls, which might not have otherwise occurred. Thus, new regulations are frequently described as *technology-forcing,* as they require select industries to use new technologies.

Scope of this Chapter

This chapter covers several topics related to the regulation and abatement of air pollutants. Some of the governmental agencies and their responsibilities for protecting air quality will be introduced. The benefits and some of the *trade-offs* (e.g., direct abatement costs, effects on jobs, and indirect health effects) of regulatory actions will also be presented. Later, this chapter will cover some of the methods used to meet air pollution standards. Two case studies will be presented; one for coal-fired electric power plants and one for automobiles and trucks. This chapter demonstrates the intimate relationship between air pollution regulations and abatement techniques used to comply with the regulations.

II. REGULATORY AGENCIES

As was addressed in Chapter 1, the introduction of modern air quality regulations were responses to specific incidents or problems. In the mid-twentieth century, air quality was regulated in the United States at the state, county, city, or even sub-city level; there was essentially no national coordination or standardization. In a review of the status of air pollution regulation as of 1960, Samuel Rogers of the U.S. Public Health Service pointed out the need for better coordination among federal and state agencies. About that time, the U.S. Departments of Defense, Agriculture, Health, Commerce, and Interior, along with the Atomic Energy Commission and the National Science Foundation formed the *ad hoc Interdepartmental Committee on Community Air Pollution.* This committee recognized the need for better coordination and recommended a "broad federal program of research and technical assistance" related to the regulation and control of air pollutants. This recommendation was timely, as the state of regulation was not only poorly coordinated, but it was also based on a lack of adequate science. Much of the problem was due to a lack of both effective national financial support and clear guidance. The Committee's recommendations had a positive impact, as air pollution science has advanced and national standards are in place. **Table 6–1** shows a small sample of agencies that currently promulgate air pollution regulations, recommendations, and standards. Numerous provinces, states, districts, municipalities, and other entities still have their own local standards and enforcement agencies. The resulting differences in local regulations can usually be traced to differences in air pollutant sources, economic structures, and prevailing meteorological conditions.

When various nations are considered, the diversity in air pollution standards becomes even more apparent. For example, **Table 6–2** compares standards for specific air pollutants in the United States and in China. The dissimilarities are partly due to differences in economic development. Some less-developed or developing nations have no effective environmental standards at all. One consequence of this diversity among nations is that some businesses relocate operations to countries with lax air quality standards. In addition, the cost and availability of labor are factors in the relocation of industrial operations. The impacts of such relocations on local loss of jobs and revenue can be painful. In some cases, the loss of local businesses has been offset by the eventual replacement with lower-pollutant-generating businesses.

The World Health Organization (WHO, http://www.who.int), comprised of nearly 200 member states (representative nations), considers air quality "a major environmental health problem." The WHO recommends a set of air quality "guidelines" for application to all nations

Table 6–1 A small sample of governmental agencies that issue air pollution regulations, recommendations, and standards.

Agency	*Air-Related Responsibility*
U.S. Environmental Protection Agency	Sets and enforces primary and secondary standards for criteria air pollutants (NAAQS) and Hazardous Air Pollutants (HAPs) and New Source Performance Standards (NSPS), also guides and performs research and educates the public
U.S. State Air Quality Agencies	Set and enforce statewide air quality regulations, perform public education, and sometimes supports research
European Environment Agency	Establishes both standards and objectives for air pollutants, analyzes research, and provides educational material
Japan: Ministry of the Environment	Establishes and enforces environmental standards and provides educational material for the public, businesses, and the government
U.K.: Department for Environment, Food and Rural Affairs	Defines the air quality (and other) standards and compliance strategies for England, Scotland, Wales, and Northern Ireland
Canada: Council of Ministers of the Environment	Sets national goals for environmental improvement and emissions of a variety of air pollutants, which are to be achieved by various other levels of government
Germany: Federal Environment Agency	Prepares and helps enforce air quality (and other) standards, provides services and support for federal environmental research, maintains informational material, and publishes reports
India: Ministry of Environment and Forests	Plans, promotes, and oversees implementation of environmental programs and establishes National Ambient Air Quality Standards

Table 6–2 A comparison of air quality standards in the United States and China. The Chinese standard classifications are Class I, for no harmful effects in selected areas (resorts, tourist areas, etc.); Class II, to protect plants and humans in urban, commercial, residential, and rural areas; and Class III for protecting animals and resistant plants in industrial and heavy-traffic areas.

Pollutant	*U.S. Standards (μg/m³ [ppm])*	*Chinese Standards (μg/m³)*		
		Class I	*Class II*	*Class III*
Particles less than 10 μm in diam.	24-hour: 150	daily: 50	150	250
		maximum: 150	500	700
Sulfur dioxide	1-hour: 198 [0.075]	daily: 50	150	250
	24-hour: 367 [0.14]	annual: 20	60	100
	annual: 79 [0.03]	maximum: 150	500	700
Nitrogen dioxide	annual: 100 [0.053]	daily: 50	100	150
		maximum: 100	150	300
Carbon monoxide	1-hour: 1,000 [9]	daily: 4,000	4,000	6,000
	8-hour: 4,000 [35]	maximum: 10,000	10,000	20,000
Photochemical oxidant (ozone)	1-hour: 240 [0.12]	hourly: 120	160	200
	8-hour: 147 [0.075]			

Source of data for China: Niu (1987).

(**Table 6–3**). The guidelines cover particulate matter (PM), ozone (O_3), nitrogen dioxide (NO_2), and sulfur dioxide (SO_2). They were designed as interim targets to significantly reduce the health risks associated with air pollution. However, due to their high implementation costs and potential impacts on national productivity and security, many nations, including the United States do not meet many of the WHO recommended guidelines.

Table 6–3 Current World Health Organization's worldwide Air Quality Guidelines. Issued in 2005.

Air Pollutant	*Guideline Values*
Particulate matter	
$PM_{2.5}$ (under 2.5 μm in diameter)	10 μg/m³, annual mean
	25 μg/m³, 24-hour mean
PM_{10} (under 10 μm in diameter)	20 μg/m³, annual mean
	50 μg/m³, 24-hour mean
Ozone (O_3)	100 μg/m³, 8-hour daily max.
Nitrogen dioxide (NO_2)	40 μg/m³, annual mean
	200 μg/m³, 1-hour mean
Sulfur dioxide (SO_2)	20 μg/m³, 24-hour mean
	500 μg/m³, 10-minute mean

Source: http://www.who.int/en.

III. REGULATIONS AND STANDARDS

Air Pollution Regulations and Air Quality Standards

Air pollution regulations are *rules* that must be followed. Air pollution regulations are of three basic types: (1) emission, (2) air quality, and (3) process. *Emission standards* define release limits, which are usually measured directly at their sources. The sources include *tailpipe emissions* from cars and trucks, *stack emissions* from factories and power plants, and *general air emissions* from businesses and the public (e.g., vapors from fueling operations, smoke from food preparation, and solvents from dry-cleaning establishments). *Air quality standards* define acceptable air concentrations of pollutants without regard to their source. *Process standards* relate to how an activity is performed. For example, many local governments have banned or restricted the use of volatile organic cleaning solvents (degreasers) for cleaning parts within the automotive service industry. Instead, less-volatile, water-based degreasers were required to reduce air pollution due to solvent evaporation and incineration (a common waste disposal method for organic solvents). Regulators have also encouraged dry cleaners to replace volatile solvent processes ("dry" processes) with water-based technologies. *Fuel-quality standards,* such as the allowable sulfur in coal, oil, and transport fuels are also examples of process standards.

Components of an Air Quality Standard

The air quality standards issued by the U.S. EPA have four major components, as follows:

1. The *indicator*—which defines what is to be measured (e.g., $PM_{2.5}$, SO_2, O_3, etc.) and how it is to be measured.
2. An *averaging time*—e.g., 1-hour, 8-hours, 24-hours (i.e., daily), or 1-year (i.e., annually).
3. The *form*—a statistical method for defining *compliance* (i.e., conformity with the standard), such as the number of exceedances permitted in a year.
4. The *level*—the numerical value of an indicator, such as 15 μg/m^3 (as for the $PM_{2.5}$ annual average) or 0.075 ppm (as for the ozone 8-hour average).

The Indicator

The *indicator* is typically a chemical or a chemical class (e.g., O_3, SO_2, NO_x, polycyclic aromatic hydrocarbons, or oxidants). An exception is particulate matter (PM), which is mass-based (e.g., particle size fractions, $PM_{2.5}$, or PM_{10}) without regard for composition. Mass-based standards are more controversial than chemical-specific standards because of the potential contribution of natural sources and the substantial variations seen in the adverse effects of specific components (e.g., bioaerosols, quartz, vanadium, soil, etc.). Mass-based PM standards are likely to be eventually replaced by more specific (e.g., composition-specific) standards.

The Averaging Time

The *averaging time* selected for an air quality standard relates to the adverse effects the standard is directed toward averting. Peak or short-term concentrations of gases such as carbon monoxide and ozone can trigger acute responses in the lung, so the averaging times typically include 1-hour or 8-hours. Sulfur dioxide (SO_2) at environmental levels can act rapidly or slowly on the respiratory tract, so the averaging times used in the United States include 1-hour, 24-hours, and 1-year. Recognizing the potential acute effects of SO_2 on asthmatics, the WHO has a 10-minute standard. Particulate material can have both short-term and long-term effects, hence 24-hour and annual averaging times are typically used. The chronic effects (e.g., that occur over decades) of exposures associated with many pollutants are covered by annual averaging times. In summary, the averaging times vary extensively, can appear to be somewhat arbitrary, and a given air pollutant can have more than one averaging time. The ultimate goal of the averaging time is to aid in the recognition and control of harmful air pollution levels.

The Level

The *level* of an air pollutant is the numerical value that separates *compliance* from *non-compliance* with a regulation. The exact level that is safe is difficult to determine, so it is often the most contentious component of the standard. The level also determines how expensive and/or difficult it will be to meet the air standard in a given community. Regulators, their advisors, and other interested parties (e.g., those industries impacted by the regulation) can be expected to debate the appropriate level, and they often fail to agree on a best value. A stringent air quality level may mean (1) some businesses will close or relocate; (2) automobiles, fuels, and other goods will be more expensive; and (3) lifestyles will change (e.g., via restrictions or prohibitions on outdoor cooking and vehicular off-road recreational activities). To quote the first U.S. EPA Administrator:

> "In the case of CO, one of the most important automotive pollutants, we have set a standard to protect against effects reported by investigations that prompt arguments even among our own scientists." (William Ruckelshaus, U.S. EPA Administrator, April 1971, http://www.epa.gov/air/caa/Part 2.pdf)

Primary vs. Secondary Standards

Table 6–4 shows the components of the original primary U.S. National Ambient Air Quality Standards (NAAQS, pronounced "knacs") of 1971. Note that *primary standards* are those issued for the protection of human health and *secondary standards* are those issued for the protection of human welfare. *Welfare effects,* which are not considered to be direct health effects, include adverse impacts on:

- soil
- water
- crops
- non-crop vegetation
- animals (domestic and wild)
- materials

Table 6–4 The four components (indicator, level, averaging time, and form) of the original 1971 primary U.S. National Ambient Air Quality Standards issued by the U.S. EPA.

Pollutant (Indicator)	*Level*	*Averaging Time*	*Form*
Particulate Matter (TSP)	75 μg/m³	Annual	Geometric Mean
	260 μg/m³	24-hour	Not to be exceeded more than once per year
Sulfur Oxides	0.03 ppm	Annual	Arithmetic mean
(as SO_2)	0.14 ppm	24-hour	Not to be exceeded more than once per year
Carbon Monoxide (CO)	10 mg/m³ (9 ppm)	8-hour	Not to be exceeded more than once per year
	40 mg/m³ (35 ppm)	1-hour	Not to be exceeded more than once per year
Photochemical Oxidant (as O_3)	160 μg/m³ (0.08 ppm)	1-hour	Not to be exceeded more than once per year
Hydrocarbons (HC)	160 μg/m³ (0.24 ppm)	3-hour, 6–9 am	Not to be exceeded more than once per year
Nitrogen Oxide (as NO_2)	100 μg/m³ (0.053 ppm)	Annual	Arithmetic mean

- property (including their economic values)
- transportation
- personal comfort and general well-being
- weather
- visibility
- climate

As a general rule, primary standards are more restrictive than secondary standards, in part, because there is more research information on human health effects than on welfare effects. The U.S. EPA National Ambient Air Quality Standards (NAAQS), as of 2010, are presented in **Exhibit 6–1**. The NAAQS periodically change; updates can be found at http://www.epa.gov/air/criteria.html.

Sampling for Achieving Compliance with Air Quality Standards

Acceptable sampling procedures, including the samplers that must be used for achieving *monitoring compliance* (i.e., conformity with regulations), are specified by regulations. For *area samplers* (i.e., those that are stationary), their number and placement within a community are important considerations. Population density, meteorology, geography, and locations of sources should all be considered. For compliance purposes, sampling arrays must gather information that is keyed to specific regulated pollutants (e.g., $PM_{2.5}$, ozone, sulfur dioxide, oxides of nitrogen, lead, etc.). In addition, as most regulatory agencies are also involved in research efforts, area samplers may also make measurements that do not directly relate to specific air standards. For example, additional sampling can be performed for peak pollutant concentrations, ultrafine particle counts, aerosol particle size distributions, and specific chemical substances not currently regulated. Chapter 4 covers air sampling in more detail.

Evolution of Air Quality Standards

Air quality standards continue to evolve over time, taking into account relevant new research findings. The U.S. EPA uses a 5-year review cycle to update standards for the NAAQS. **Table 6–5** shows the evolution of the U.S. NAAQS for ozone (O_3) and particulate matter (PM). In general, air quality standards are set at levels that are difficult (but hopefully not impossible) to achieve. William Ruckelshaus summarized the situation in 1971 when the first NAAQS were issued:

> "These are tough standards. They are based on investigations conducted at the outer limits of our capability to measure connections between levels of pollution and effects on man. . . . our standards approach levels that occur fairly commonly in nature." (William Ruckelshaus, U.S. EPA Administrator, April 1971, http://www.epa.gov/air/caa/Part2.pdf)

Exhibit 6–1 The U.S. EPA's National Ambient Air Quality Standards (NAAQS) as updated April 16, 2010 (http://www.epa.gov/air/criteria.html).

The *Clean Air Act* requires the U.S. EPA to set *National Ambient Air Quality Standards* (40 CFR part 50) for pollutants considered harmful to public health and the environment. The Clean Air Act established two types of national air quality standards. *Primary standards* set limits to protect public health, including the health of "sensitive" populations such as asthmatics, children, and the elderly. *Secondary standards* set limits to protect public welfare, including protection against decreased visibility, damage to animals, crops, vegetation, and buildings.

The U.S. EPA has set National Ambient Air Quality Standards for several principal pollutants, which are called "criteria" pollutants. They are listed below. Units of measure for the standards are parts per million (ppm) by volume, parts per billion by volume, milligrams per cubic meter of air (mg/m^3), and micrograms per cubic meter of air ($\mu g/m^3$).

National Ambient Air Quality Standards

	Primary Standards		*Secondary Standards*	
Pollutant	*Level*	*Averaging Time*	*Level*	*Averaging Time*
Carbon Monoxide	9 ppm (10 mg/m^3)	8-hour(1)	None	
	35 ppm (40 mg/m^3)	1-hour(1)		
Lead	0.15 $\mu g/m^3$ (2)	Rolling 3-Month Average	Same as Primary	
	1.5 $\mu g/m^3$	Quarterly Average	Same as Primary	
Nitrogen Dioxide	53 ppb(3)	Annual (Arithmetic Average)	Same as Primary	
	100 ppb	1-hour(4)	None	
Particulate Matter (PM_{10})	150 $\mu g/m^3$	24-hour(5)	Same as Primary	
Particulate Matter ($PM_{2.5}$)	15.0 $\mu g/m^3$	Annual(6) (Arithmetic Average)	Same as Primary	
	35 $\mu g/m^3$	24-hour(7)	Same as Primary	
Ozone	0.075 ppm (2008 std)	8-hour(8)	Same as Primary	
	0.08 ppm (1997 std)	8-hour(9)	Same as Primary	
	0.12 ppm	1-hour(10)	Same as Primary	
Sulfur Dioxide	0.03 ppm	Annual (Arithmetic Average)	0.5 ppm	3-hour(1)
	0.14 ppm	24-hour(1)		
	75 ppb	1-hour(11)	None	

(1) Not to be exceeded more than once per year.

(2) Final rule signed October 15, 2008.

(3) The official level of the annual NO_2 standard is 0.053 ppm, equal to 53 ppb, which is shown here for the purpose of clearer comparison to the 1-hour standard.

(4) To attain this standard, the 3-year average of the 98th percentile of the daily maximum 1-hour average at each monitor within an area must not exceed 0.100 ppm (effective January 22, 2010).

(5) Not to be exceeded more than once per year on average over 3 years.

(6) To attain this standard, the 3-year average of the weighted annual mean $PM_{2.5}$ concentrations from single or multiple community-oriented monitors must not exceed 15.0 $\mu g/m^3$.

(7) To attain this standard, the 3-year average of the 98th percentile of 24-hour concentrations at each population-oriented monitor within an area must not exceed 35 $\mu g/m^3$ (effective December 17, 2006).

(8) To attain this standard, the 3-year average of the fourth-highest daily maximum 8-hour average ozone concentrations measured at each monitor within an area over each year must not exceed 0.075 ppm (effective May 27, 2008).

(9) (a) To attain this standard, the 3-year average of the fourth-highest daily maximum 8-hour average ozone concentrations measured at each monitor within an area over each year must not exceed 0.08 ppm.

(b) The 1997 standard, and the implementation rules for that standard, will remain in place for implementation purposes as the U.S. EPA undertakes rulemaking to address the transition from the 1997 ozone standard to the 2008 ozone standard.

(c) The U.S. EPA is in the process of reconsidering these standards (set in March 2008).

(10) (a) The U.S. EPA revoked the 1-hour ozone standard in all areas, although some areas have continuing obligations under that standard ("anti-backsliding").

(b) The standard is attained when the expected number of days per calendar year with maximum hourly average concentrations above 0.12 ppm is ≤ 1.

(11) (a) Final rule signed June 2, 2010. To attain this standard, the 3-year average of the 99th percentile of the daily maximum 1-hour average at each monitor within an area must not exceed 75 ppb.

Table 6–5 History of U.S. Primary National Ambient Air Quality Standards for ozone and particulate matter.

Year	*Standards (averaging time)*
Ozone	
1971	0.08 ppm (1-hour)
1979	0.12 ppm (8-hour)
1997	0.12 ppm (1-hour)
	0.08 ppm (8-hour)
2008	0.12 ppm (1-hour)
	0.075 ppm (8-hour)
Particulate Material	
1971	TSP: 75 μg/m^3 (annual)
	TSP: 260 μg/m^3 (24-hour)
1987	PM_{10}: 50 μg/m^3 (annual)
	150 μg/m^3 (24-hour)
1997	$PM_{2.5}$: 15 μg/m^3 (annual)
	65 μg/m^3 (24-hour)
	PM_{10}: 50 μg/m^3 (annual)
	150 μg/m^3 (24-hour)
2006	$PM_{2.5}$: 15 μg/m^3 (annual)
	35 μg/m^3 (24-hour)
	PM_{10}: 150 μg/m^3 (24-hour)

TSP = Total Suspended Particulates; PM_{10} = Particle Mass under 10 μm in aerodynamic diameter; $PM_{2.5}$ = Particle Mass under 2.5 μm in aerodynamic diameter.

Occupational Particle Standards

Unlike mass-based environmental particle standards, composition-specific regulations and standards are common practice in the occupational health and safety arena. For example, the U.S. Occupational Safety and Health Administration (OSHA) lists over 150 separate composition-specific occupational standards for particles in the air (see Title 29 of the Code of Federal Regulations, Section 1910.1000). Examples of these include aluminum, coal dust, gypsum, kaolin clay, limestone, nickel, pesticides (various), Portland cement, silica (various forms), silver, vanadium, and zinc. Many of these occur naturally and may enter the air by wind erosion as well as from industrial processes. The American Conference of Governmental Industrial Hygienists (ACGIH®) has established and published *recommended* "maximum allowable concentrations" and "threshold limit values" for airborne particles since the 1940s. Each standard is designed to minimize the harm associated with a specific disease(s), outcome, or condition. When OSHA was created in 1970 it adopted the 1968 ACGIH® values and established them as regulatory *permissible exposure levels.* It is likely that future *environmental* regulatory activities will also establish composition-specific standards.

The U.S. Clean Air Act

The original *Clean Air Act* (CAA) of 1963 passed by the U.S. Congress established funding for two major purposes: (1) the study of air pollution monitoring and control and (2) the cleanup of air at several significantly-polluted sites. This act was followed by a stronger, revised CAA in 1970, and the creation of the U.S. EPA in the same year. The 1970 CAA firmly established the federal and state roles in limiting emissions from both stationary and mobile sources, and it established four major regulatory programs:

- The National Ambient Air Quality Standards (NAAQS)
- State Implementation Plans (SIPs) to meet air quality regulations and standards
- New Source Performance Standards (NSPS)
- National Emission Standards for Hazardous Air Pollutants (NESHAPs)

These regulatory programs and enforcement authority were delegated to the U.S. EPA in the 1970 CAA.

Major amendments to the CAA were made in 1977 and 1990, which established (1) permit reviews; (2) acid deposition (also called *acid rain;* see Exhibit 2–1) control; (3) stratospheric ozone protection; (4) expansion of the NESHAPs (for Hazardous Air Pollutants, HAPs); (5) increased enforcement authority; and (6) expansion of research programs. Legislation passed since 1990 has only made minor changes. As of 1990, the major sections of the CAA, called "Titles," include the following:

- Title I—Air Pollution Prevention and Control
- Title II—Emission Standards for Moving Sources
- Title III—Air Toxics
- Title IV—Acid Deposition Control
- Title V—Permits
- Title VI—Stratospheric Ozone Protection
- Title VII—Provisions Relating to Enforcement (see **Exhibit 6–2** for more on enforcement)

Exhibit 6–2 The U.S. EPA's enforcement activity.

The U.S. EPA's goal of achieving cleaner air is supported by both its regulations and enforcement powers. Enforcement falls under three "programs": civil, clean up, and criminal.

Civil enforcement can be either "informal" or "formal." An informal action involves a communication that notifies a regulated "entity" that there is a "problem." The expectation is that those responsible for the problem will take steps to resolve it. The next level of civil enforcement is a "Formal Administrative Action" or an "Administrative Order." At this level, penalties, including stiff fines, may be imposed.

"Civil judicial actions" are used against persons or entities who do not comply with statutory or regulatory requirements, or with an Administrative Order. In these cases, formal lawsuits are filed by the U.S. Department of Justice on behalf of the U.S. EPA, or by State's Attorneys General (for a state). The judicial action can be resolved by "Consent Decrease" signed by all parties, monetary settlements, remediation activities, or other actions.

In the most serious violations, "Criminal Enforcements" are used. Such enforcements may include arrests, jail sentences, and other consequences. The U.S. EPA claims that the conviction rate is 90 percent in its criminal enforcement action. These enforcement actions are published on the Web site, http://www.epa.gov/oecaerth/criminal/index.html.

Although the CAA per se has expired, its major provisions have been incorporated into the U.S. Code of Federal Regulations (i.e., permanent law).

Other Clean Air Acts

In addition to the U.S. CAA, the United Kingdom, Canada, and other nations had similar federal acts. The United Kingdom responded to the London air pollution disaster of 1952 with its 1956 and 1968 Clean Air Acts, which initially specified the fuels that could be used in certain legislated zones. Subsequent revisions expanded the scope; defined the specific pollutants that must be measured; specified required industrial chimney heights; defined acceptable fuels; identified the sources of information derived from research; addressed special sources (e.g., railroads, vessels, and mines); and established regulatory and enforcement powers. Canada's Clean Air Act was passed in 1970 in order to regulate specific air pollutants (asbestos, lead, mercury, and vinyl chloride). Subsequent Canadian legislation expanded the scope of the act to include additional measures for control of both ground-level pollution and greenhouse emissions. Many other nations have air pollution regulations and standards that are periodically revised.

Tobacco-Use Controls

Health Effects of Smoking

The adverse health effects of tobacco smoking have been recognized for more than 150 years. In a paper published in the *London Journal of Medicine,* John Webster, M.D. a Fellow of the Royal College of Physicians wrote:

> "But although intemperance in spirituous liquors be almost unknown in certain ranks, a vice appears to have taken its place, which I think as great an abomination, and equally destructive to health, namely, tobacco-smoking, especially amongst the younger portion of the male population." (Webster, 1850)

Over 100 years later, the U.S. Surgeon General issued several reports detailing the adverse health effects associated with cigarette smoking. In 1964, the U.S. Surgeon General's report concluded that:

> "The risk of developing lung cancer increases with duration of smoking and the number of cigarettes smoked per day, and is diminished by discontinuing smoking." (U.S. DHHS, 1964)

The 1964 report also linked cigarette smoking with heart disease. Since that time, additional health risks associated with cigarette smoking have been solidified. In 1966, the U.S. *Federal Cigarette Labeling and Advertising Act* took effect. As a result, clear warning statements on packages of cigarettes, other tobacco products, and on tobacco advertisements are required in order to inform the public of their risks.

Environmental Tobacco Smoke

The possible risks of *environmental tobacco smoke* (ETS), also called *second-hand tobacco smoke,* were reviewed by the National Research Council (NRC).

The NRC concluded that the relative risks (*RRs*, see Chapter 10) of lung cancer in non-smokers exposed to ETS were elevated: ranging from 1.24 to 1.61 for non-smokers exposed in homes, workplaces, and elsewhere (NIOSH, 1991). As a result, the U.S. National Institute for Occupational Safety and Health (NIOSH) recommended:

> "Employers should therefore assess conditions that may result in worker exposure to ETS and take steps to reduce exposures to the lowest feasible concentrations." (NIOSH, 1991)

Although the claimed adverse health effects of ETS have been supported by research, they have also been challenged. Dr. James Enstrom studied 35,561 never-smokers of smoking spouses and found no significant associations with coronary heart disease, lung cancer, or chronic obstructive pulmonary disease; participants were followed from 1960 to 1998 (Enstrom and Kabat, 2003).

Tobacco Regulation

Currently, most U.S. states have bans on smoking in public areas. *The Family Smoking Prevention and Tobacco Control Act* (Public Law 111–31, H.R. 1256, Enacted June 22, 2009), geared primarily to protect children and adolescents, gives the U.S. Food and Drug Administration (FDA) the authority to regulate tobacco. The Act is based on nearly 50 findings, including the following:

- Nicotine is addictive.
- A scientific consensus affirms that tobacco products cause cancer, heart disease, and other adverse health effects.
- Tobacco use is the foremost preventable cause of death in America.
- Reducing minors' use of tobacco by 50 percent will prevent over 10 million regular daily users.
- Interventions should target all smokers, not just minors.
- Advertising contributes to attracting new users under the legal age to purchase tobacco products.
- The sale, distribution, and advertising of tobacco products affect interstate commerce (hence, federal control is justified).
- Because of its scientific expertise, the FDA can effectively protect public health with respect to the use of tobacco products.

The FDA has the authority to regulate nearly all aspects of tobacco in the United States, including its manufacture, labeling, advertising, distribution, and sales. Oversight and enforcement powers are also delegated to the FDA. To assist in the evaluation of new products and tobacco-related research, the FDA Secretary has established a *Tobacco Products Scientific Advisory Committee* (the "Advisory Committee"). The Advisory Committee has 12 members: seven from the medical, scientific, and health-care professions, and one each from or representing (1) a state or local government; (2) the general public; (3) the tobacco manufacturing industry; (4) the small business tobacco industry; and (5) tobacco growers. Those members representing the tobacco industry are non-voting members.

Because the U.S. Tobacco Control Act is new, it is too early to evaluate its effectiveness in preventing youth from becoming habitual smokers, or to measure its effectiveness in reducing tobacco-related diseases.

IV. TRENDS, BENEFITS, AND TRADE-OFFS

Trends, Benefits, and New Questions

Trends and Benefits

Air quality regulations and standards have had dramatic effects on the levels and components of smog (**Figure 6–2**), and on the emission of ozone-depleting substances. London no longer has the choking acidic air that produced thousands of deaths in December 1952. The phase-out of lead in automobile fuels has dramatically lowered the concentrations of this pollutant in the air (**Figure 6–2B**). Controls on the use of chlorofluorocarbons that are capable of destroying stratospheric ozone have reversed their previous rapidly increasing emission rates. Levels of ozone, sulfur dioxide, oxides of nitrogen, and particulate material have steadily declined in urban areas in many parts of the world. Today, many fuels, such has coal, fuel oil, diesel oil, and gasoline contain lower levels of sulfur, and burn cleaner than was the case 20 years ago. Modern automobile and truck engines are also designed to be more fuel efficient and to burn much cleaner than in the recent past. Workplace air has similarly become more healthful, and respiratory diseases in workers are declining. Tobacco use is also declining in many nations, and substantial reductions in new cases of lung cancer are expected to occur in the coming decades. In spite of these important steps, lung diseases still persist, and new cases of asthma are actually on the rise in many

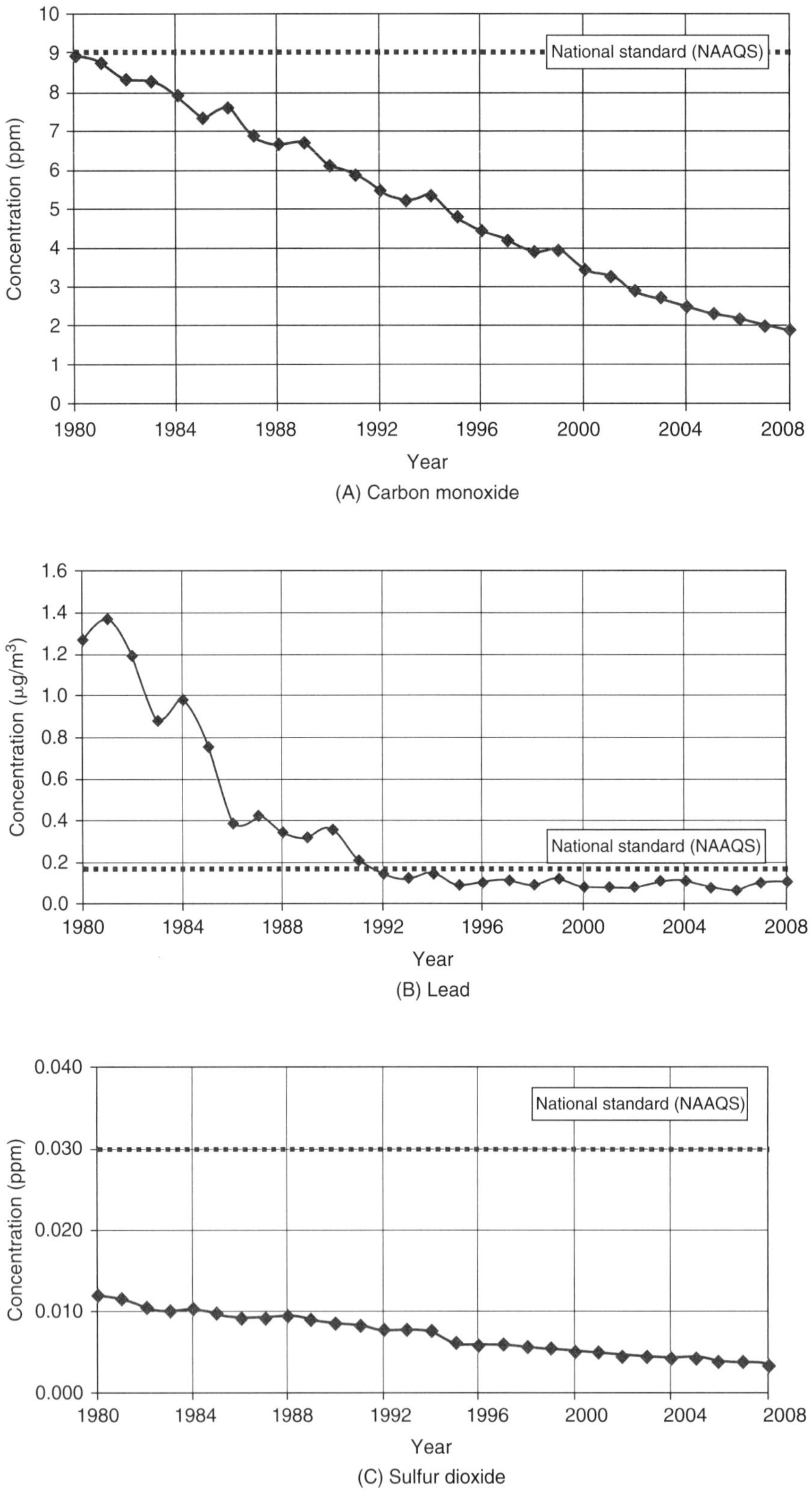

Figure 6–2 Declines in tropospheric levels of selected components of smog in the United States: A. Carbon monoxide, B. Lead, C. Sulfur dioxide, D. PM_{10}, and E. Nitrogen dioxide.
Source: The University of California Air Pollution Health Effects Laboratory, with kind permission.

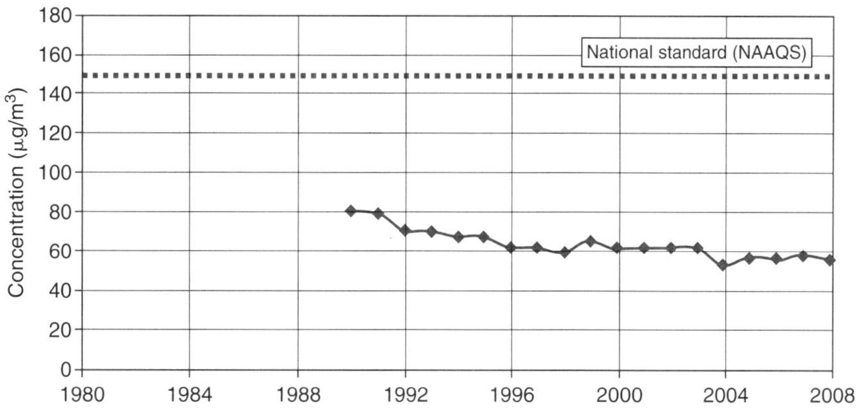

(D) PM_{10}

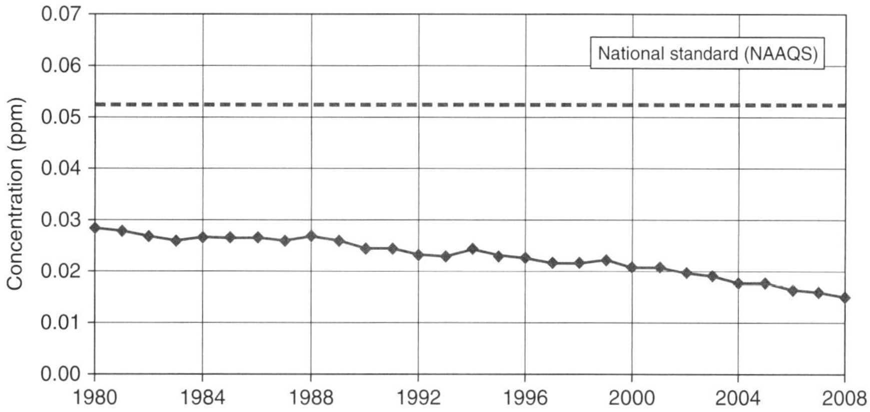

(E) Nitrogen dioxide

Figure 6–2 *(Continued)*

areas of the world. It appears that new approaches, including improvements in basic individual health and general prosperity, may be more effective in controlling lung diseases than simply more stringent air quality regulations.

New Concerns

Although much has been achieved by air quality regulations, the case is strong for shifting to a more targeted and holistic control strategy for air pollutants. For example, the control of particle mass without regard to composition is a problem, as is controlling each air pollutant separately (instead of considering mixtures). A public health perspective instead of a pollutant perspective should be considered. Also, all of the benefits and consequences of a regulation, pros and cons, should be evaluated and presented to the public for input before a final rule is promulgated.

It is clear that the level of scientific knowledge needed to significantly improve air standards is still inadequate. As examples, it appears that exposure to some air pollutants in young children can protect them from eventually developing asthma, and that exposure to air pollutants in adults can maintain essential lung defenses. This effect, in both children and adults, sometimes called "the hygiene hypothesis," is described in **Exhibit 6–3**. In addition, low socioeconomic status (SES) is a factor in health, lifespan, and increased susceptibility to the harmful effects of air pollutants (see **Exhibit 6–4**).

The major anthropogenic sources of air pollutants have largely been controlled in prosperous nations. Important current questions include (1) Do economic

Exhibit 6–3 The "Hygiene Hypothesis."

In a broad sense, *the hygiene hypothesis* asserts that the human environment can actually be too clean. The basis for this claim is that if *physiological systems* are not continuously challenged, then they will decline to the point that they cannot protect from the adverse effects of future challenges. Microbiologists caution that living in ever-cleaner microbial environments make populations more susceptible to acquiring future infections. When astronauts spend extended time in zero gravity, their bones lose significant amounts of calcium, and they are susceptible to fractures when they return to Earth. Scientists who spend months in clean air environments, such as polar outposts, have difficulty adapting to the normal air quality when they return home. The time scales for loss of physiological functions can be short, as many cell types, including those with defensive functions, have turnover times on the order of a few days. The caveat that applies in these cases is, "use it or lose it."

In a narrower, air pollution sense, the hygiene hypothesis has been proposed as a mechanism to explain the lower rates of asthma in children who grew up in "dirty" environments, such as those on U.S. farms and ranches, or in rural China, where they were exposed to animal dander, high microbial counts, open fires, and soot from cooking and heating fuels. Supporting evidence includes the realization that as air quality has improved, asthma rates have increased.

Other explanations for the asthma epidemic that have been supported by evidence include fewer childhood infections, frequent use of antibiotics, and isolation from pets. Such factors alter the childhood development of immune competency, and increase the chance of becoming asthmatics. Lack of adequate exercise and obesity are also associated with childhood asthma. The potential roles of environmental air pollution in the asthma epidemic are unclear.

The intriguing, unanswered question is, "Can the air really be too clean?"

costs, including unemployment, associated with progressively tighter air pollution regulations have significant adverse effects on public health and well-being? and (2) Do we need to understand more about the multiple factors that affect human health before we continue to further reduce air pollutant concentrations?

The reduction of air pollutant levels has benefited human health and the environment, as well as improved the efficiency of many industrial and non-industrial processes. It is also clear that these benefits have been associated with significant costs. Debatably, future controls are likely to be more expensive and to have smaller pay-offs in terms of benefits. More research, broader analyses, and better cooperation among a more diverse assembly of stakeholders (e.g., the unemployed, economists, businesses, regulators, and health advocates) are clearly needed.

Regulatory Trade-offs of Air Pollution Regulations

Air pollution regulations have their associated regulatory trade-offs. *Regulatory trade-offs* are adverse conse-

Exhibit 6–4 Socioeconomic status (SES) and susceptibility to air pollutants.

One of the unintended consequences of ever-tightening air quality standards is the economic impacts of increased costs of goods and services and the loss of employment. Low socioeconomic status (SES), e.g., as seen in the poorly educated and the unemployed, is associated with a number of adverse health consequences. Substandard living conditions, exposure to violence and other crime, lack of access to high-quality health care, low educational attainment, depression, drug and alcohol abuse, and unfavorable birth outcomes, are only a few of the problems of the poor. In addition, low SES populations have been identified as being more susceptible to the direct harmful effects of air pollution. Several factors have been postulated to explain this increased susceptibility. First, those with low SES tend to live in areas with higher levels of air pollution, such as near industrial operations and heavy automobile and truck traffic. Second, their indoor air quality tends to be poor; deteriorating paint and wood, dampness, lack of air conditioning, and indoor infestations of rodents and insects are common. Third, their many social stressors (as mentioned above) amplify the effects of air pollutants. The bottom line with regard to the effects of air quality regulations on low SES populations and health is uncertain. Do stricter air quality standards convey significant direct benefits to the health of low SES populations? Or alternatively, do the associated economic impacts (job loss and increased costs of goods and services) outweigh the benefits of better outdoor air quality?

quences (also called *unintended consequences*) that offset the desired benefits of regulations. Essentially all well-intended decisions that affect what people do will also have some adverse consequences; air pollution regulations are no exception. Unless the significant trade-offs are considered, a given regulation could do more harm than good. For example, the current practice by the U.S. EPA in establishing the NAAQS is to set the primary standards for criteria air pollutants one-at-a-time at levels that protect the general population and sensitive sub-populations from the *direct* health effects. By law and by policy, no consideration may be given to the economic costs of compliance, nor to a variety of *secondary* health effects (including adverse effects). Thus, the broad-scope of positive and negative health effects of these regulations is not considered. An enlightening monograph, *Risk vs. Risk: Trade-offs in Protecting Health and the Environment* (Graham and Wiener, 1995), argues that risks should be treated in a holistic manner, such that efforts to minimize risks are made in the context of the whole picture. Graham and Wiener caution those who control risk-reduction decisions to be more aware of the danger that the regulation of one risk can lead to significant increases in other risks. With these points in mind, **Figure 6–3** lists some of the potential adverse consequences of air pollution regulations. For any given case, some of these consequences may be significant, and others may be trivial. However, it is both useful and ethical to consider the elements in Figure 6–3 as part of the process of setting mandatory regulations. The important ethical principles that support such a holistic approach to regulation include:

- the principle of *minimizing the harms and maximizing the benefits* that proceed from a decision;
- the principle of *informed-consent* that requires those affected by a decision be told of the *known and suspected risks* they face, and that they consent to accept those risks; and
- the principle of *equity* that requires some groups or individuals not bear excessive burdens in order to provide benefits to others.

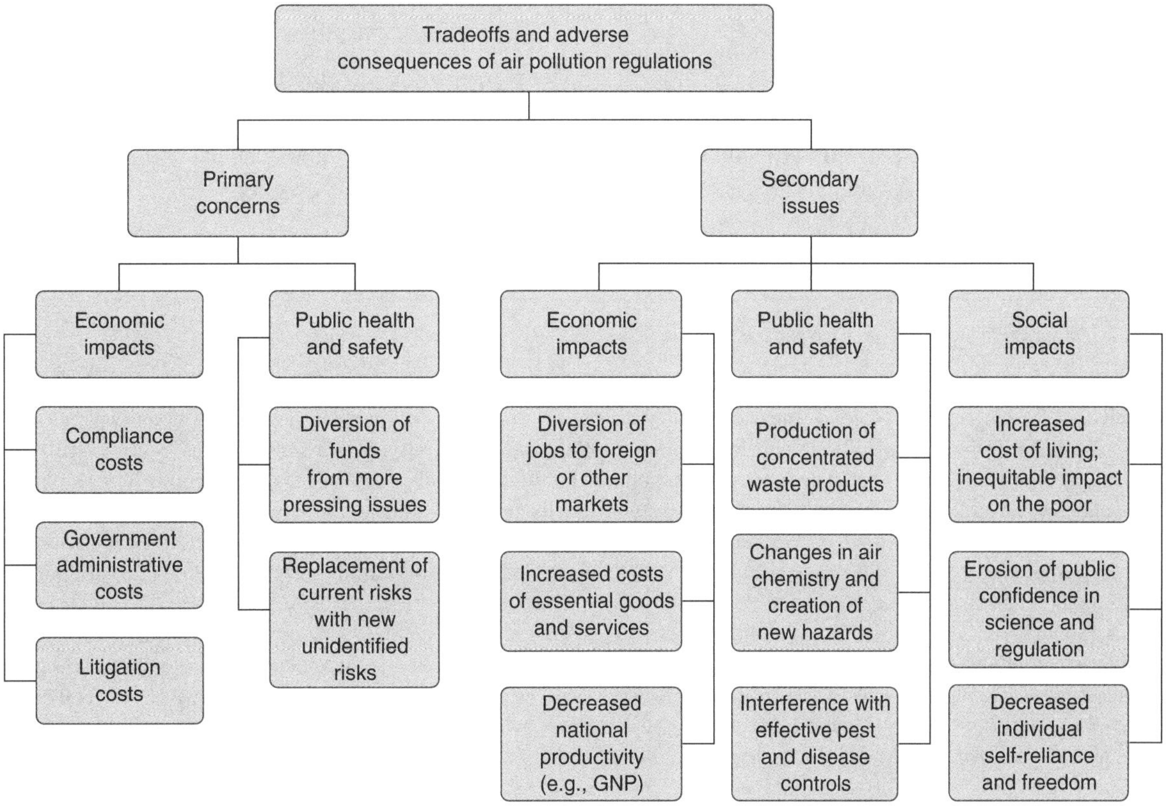

Figure 6–3 Some potential tradeoffs and adverse consequences associated with air pollution regulations.
Source: The University of California Air Pollution Health Effects Laboratory, with kind permission.

It is clear that it may not be possible to meet all of these ethical principles when setting air quality standards, but they should be given serious consideration. Chapter 12 covers ethical principles in more detail.

V. ABATEMENT AND COMPLIANCE STRATEGIES

Introduction: Definitions and Scope

The word "*abatement*" is a noun that relates to the act or process of abating. The verb *abate* means to put an end to, or reduce in intensity, especially of a nuisance. Thus, "abatement" and "abate" are frequently used in legal contexts. From an air pollution perspective, a successful abatement is one that achieves *compliance* (i.e., conformity) with a regulation, be it an emission standard, an air quality standard, or a process standard. Compliance is established by measurements using approved sampling and analysis techniques, such as those described in Chapter 4. In many instances the same pollutant collection method that is used for sampling and analysis is also used for abatement purposes. In both sampling and abatement, a central objective is to achieve near 100 percent collection efficiency of air pollutants within the volume of air that is processed. The primary difference between sampling and abatement is in the amount of material collected. Sampling involves the collection of small quantities (e.g., μg and mg amounts), whereas abatement must collect much larger amounts (e.g., kg and higher) in order to prevent releases into the environment.

Abatement and compliance strategies call on expertise from several disciplines including (1) the sciences (e.g., chemistry, physics, meteorology, economics, etc.); (2) engineering (e.g., chemical, civil, mechanical, industrial, electrical, etc.); (3) psychology and sociology (as related to culture, institutions, crime, law, politics, and government); and (4) the health professions (e.g., industrial hygiene, public health, medicine, safety, etc.). Each of these involved disciplines may overlap with each other; abatement is a large endeavor. In this chapter, the primary emphasis is on engineering controls, as the reduction or removal of pollutants before they enter the environment is preferred. Engineering approaches to air pollution abatement fall into several major categories, as follows:

- developing new cleaner technologies;
- developing cleaner fuels;
- developing cleaner industrial processes;
- improving energy-use efficiencies; and
- removing or scrubbing gases and particles from emission sources (i.e., from effluent streams) prior to release into the environment.

Exhibit 6–5 provides some useful references related to engineering methods of abatement, including the designs of particle and gas collection equipment that are described in this chapter.

Exhibit 6–5 References related to air pollution abatement.

Several specialized references cover the wealth of information available on air pollution control. *Fundamentals of Air Pollution Engineering* by Richard Flagan and John Seinfeld (1988) covers the basic physics and chemistry associated with air pollutants and their transformations in the atmosphere; combustion fundamentals; fuels and emission controls; and removal of particles and gases from effluent streams. This book is appropriate for upper division and graduate engineering courses. *The Air Pollution Engineering Manual* edited by Wayne Davis (2000) is geared toward governmental, university, and industrial professionals, instead of students. This 900-page book provides useful and detailed information on the major sources of air pollutants. It has 20 main chapters and over 60 subchapters, each written by knowledgeable experts. Topics covered include control of gaseous and particulate pollutants; fugitive dust; odors; ventilation exhausts; combustion sources (coal, fuel oil, natural gas, wood, internal combustion sources, and stationary gas turbines); waste incineration; semiconductor manufacturing; evaporative emissions sources; graphic arts; chemical process industries; food and agriculture; metallurgical industry; mineral products industry; pharmaceutical industry; petroleum industry; wood processing industry; soil, waste, and water; and accidental and catastrophic releases. *Air Quality Management in the United States*, published by the National Academies Press (NRC, 2004), provides an overview of air quality management in the United States and important recommendations for its improvement. This book and related books can be accessed free online at http://www.nap.edu. Perusing the topics in these books provides an overview of the vast scope of air pollution abatement.

VI. CONTROL OF PARTICULATE AND GASEOUS EMISSIONS

Basic Principles for the Collection of Particles

Particle Physics

Physics, the science of mass, energy, and their interactions, provides essential insight into the design of particle collection devices. Aerosol particle collection (i.e., deposition on or in a collector) occurs when the particle leaves the entraining air and contacts a collecting surface. The *aerosol deposition mechanisms* that cause aerosol particles to depart from their airstreams and contact a collection surface include:

- *diffusion*—The diffusional velocity is inversely related to the particle diameter.
- *sedimentation*—The rate of which is proportional to the particle mass and the gravitational acceleration, or other artificially-imposed acceleration (e.g., in a cyclone aerosol collector).
- *impaction*—The probability of impaction is approximately proportional to the particle density, velocity, and diameter squared. It is also inversely proportional to the entraining air viscosity and a characteristic dimension (e.g., an impactor jet width) of the device in which impaction occurs.
- *electrical* and/or *magnetic*—The force depends on the charge sign (+ or −), the field strength, and the level of particle charge, and/or its magnetic properties.

Treatments of the basic physics of particle motion are found in Chapter 3. An assumption is that upon touching a surface (e.g., in a collector) the particle sticks to it; i.e., the *sticking coefficient* is 1 (100 percent collection efficiency). Deviations from this assumption are rare, but particles can bounce upon contacting a surface when (1) the particle and collection surfaces are not deformable (e.g., steel or glass); (2) the collision velocity is high, and/or the angle of contact is small (i.e., the particle velocity vector is nearly parallel to the collection surface); or (3) a strong repulsive electrical force exists.

Another form of particle removal from the air is *particle destruction,* which may occur by thermal degradation, dissolution in a collection fluid, or chemical conversion. The physical and chemical properties of the particle, along with the environmental characteristics determine its elimination by such means.

Particle Adhesion

Once deposited, particles attach to surfaces via contact and attraction forces including: van der Waals (electric dipole); electrostatic (for particles with net charges); and surface tension (e.g., from thin liquid films, which are usually present on particles and sometimes collection surfaces). These three adhesive forces are approximately proportional to the first power of the particle diameter (d). So the adhesion force (F_{adh}) is

$$F_{adh} \propto d. \qquad \text{(Eq. 6–1)}$$

Particle Resuspension

Once attached to a collection surface, a particle faces the possibility of being dislodged. If particle resuspension occurs in a collection device following deposition, the collection attempt has failed. Resuspension can result from mechanical and electrical forces. Mechanical forces include those produced by air currents, liquid sprays (or flows), gravity, vibration, scraping, impaction of other particles, and abrasion. The removal efficiency of such forces acting on a deposited particle depends on (1) the strength of the applied force; (2) the particle composition, conductivity, and size; (3) the characteristics of the collection surface; (4) the depth of the collected particle deposit; and (5) the environmental humidity and/or liquid content of the deposit. Resuspension is a complex process, but when all of the mechanisms are considered the probability of resuspension is roughly proportional to the square of the particle diameter or d^2:

$$F_{res} \propto d^2. \qquad \text{(Eq. 6–2)}$$

So the ratio of adhesion to resuspension forces is approximately

$$F_{adh}/F_{res} \propto 1/d. \qquad \text{(Eq. 6–3)}$$

This relationship implies that larger particles will be easier to dislodge from collection surfaces than smaller particles. This result, although approximate, is consistent with both intuition and experience. For example, coarse sand is much easier to dislodge from a surface than is fine soot.

Particle Collection Devices

Particle collection device designs vary considerably, but they fall into a few major types. Each type has its advantages, disadvantages, and application to specific collection challenges. Collection devices for particles (and sometimes gases) fall into several categories:

- elutriators and settling chambers;
- cyclones and centrifuges;
- electrostatic precipitators;
- filters and sieves (e.g., baghouses and other filter configurations);
- wet scrubbers;
- diffusion batteries and diffusion scrubbers;
- supersaturators (to increase particle size for more efficient subsequent collection);
- condensers (e.g., cold traps for water vapor); and
- acoustic agglomerators (to increase particle size and thus improve collection efficiency).

Elutriators for Particle Collection

Elutriators used for aerosol particle collection can have either vertical or horizontal orientations. A simplified *horizontal elutriator* design is depicted in **Figure 6–4**. Aerosol is introduced into air flowing along the top of the elutriator channel at a height, H, above the bottom collecting surface. Clean winnowing air is introduced below the aerosol stream at a velocity, V_x, which matches the aerosol inlet velocity. In passing down the length of the elutriator L, the particles follow a trajectory that is the vector sum of V_x and V_{ts} (the particle's terminal settling velocity). If the trajectory causes the particle to settle to the collecting surface before exiting the elutriator, it is captured. Otherwise, the particle passes through and exits the device, avoiding capture. The relationship that defines the V_{ts} of the smallest particle that will be collected is

$$V_{ts} = \frac{H \cdot V_x}{L}. \qquad \text{(Eq. 6–4)}$$

The critical diameter, above which the particle will be collected, can be found in tables relating the particle diameter to its terminal settling velocity (e.g., Table 3–4). Since V_{ts} is proportional to d^2, the duct height to length ratio must be small and/or V_x small in order to collect small particles.

Given a reasonable airflow rate (e.g., $V_x = 1$ m/sec) and an elutriator channel height of $H = 1$ m, in order to collect all particles with diameters larger than 1 μm ($V_{ts} = 3.5 \times 10^{-5}$ m/sec), the length (L) of the elutriator must be at least about 30,000 m. Normally, a 30 km (18.6 miles) elutriator length is unreasonable. If the elutriator length is restricted to 10 m (and the height is 1 m), the aerodynamic diameter of the smallest particle that will be collected (with a flow velocity of 1 m/sec) is about 60 μm. Therefore, one must conclude that this type of horizontal elutriator is only useful for collecting very large particles. The particles of interest from the point of view of urban air standards are usually below 10 μm in diameter, which limits the usefulness of horizontal elutriators for cleaning the air.

Cyclones for Particle Collection

Cyclones (and aerosol centrifuges) impart a rotational velocity to the particle-containing air stream that produces an increased artificial g-force (gravitational acceleration), sometimes called a "centrifugal" force, to the particles. The increased g-force increases the collection efficiency, especially for small particles that normally settle slowly. A common cyclone design used for collecting particles is shown in **Figure 6–5**. Normal gravity produces a force, F_g, that is the product of the particle's mass, m, and the acceleration due to gravity, g:

$$F_g = mg. \qquad \text{(Eq. 6–5)}$$

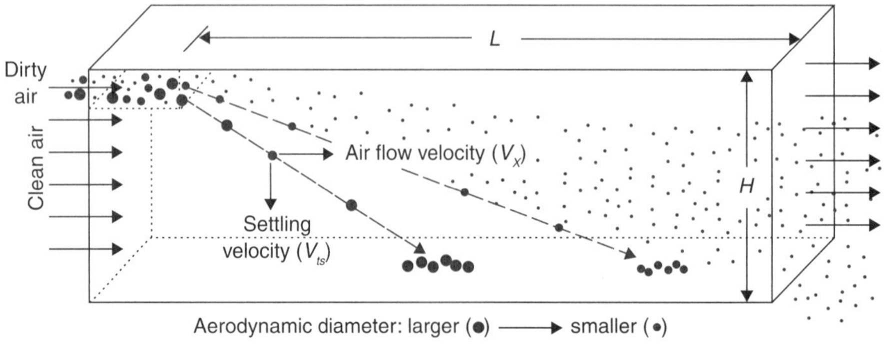

Figure 6–4 Cross sectional view of a simplified horizontal elutriator used to collect aerosol particles.
Source: The University of California Air Pollution Health Effects Laboratory, with kind permission.

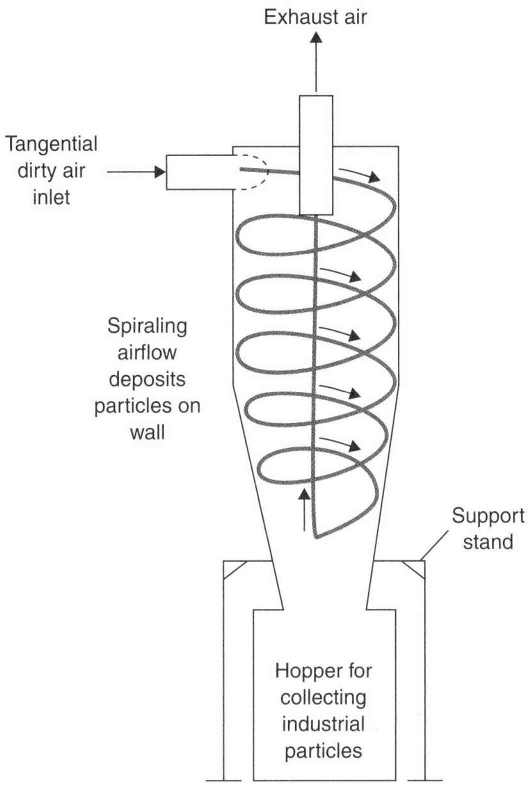

Figure 6–5 A simplified aerosol cyclone for collecting particles.
Source: The University of California Air Pollution Health Effects Laboratory, with kind permission.

In a cyclone, the rotational motion of the fluid air produces a force, F_c, that depends on the tangential velocity, V_t, and the radius of the circular path, R:

$$F_c = mV_t^2/R. \quad \text{(Eq. 6–6)}$$

The ratio of the induced force to the gravitational force is

$$F_c/F_g = V_t^2/R \cdot g. \quad \text{(Eq. 6–7)}$$

The purpose of a cyclone is to enhance the particle collection efficiency by increasing the force that generates its settling rate. A high rotational speed and a small circular-path radius can greatly increase the force on a particle. For example, a V_t of 10 m/s would require an R of 1 m to produce an F_c 10 times greater than F_g:

$$F_c/F_g = V_t^2/R \cdot g = (10\,\text{m/s})^2/((1\,\text{m})(10\,\text{m/s}^2)) = 10. \quad \text{(Eq. 6–8)}$$

Aerosol centrifuges are similar to cyclones, but in a centrifuge the rotational motion is produced by spinning a duct or channel with the particle-containing air inside. The mechanical simplicity of the cyclone favors it over the centrifuge for removing particles from their entraining fluid.

Cyclones and centrifuges can be used to collect smaller particles than can a simple horizontal elutriator. However, there is a major drawback associated with cyclones and centrifuges. The pressure drop, as well as the energy to move air (e.g., by using air pumps) through the device, increases as V_t^2 increases. As the initial investment and maintenance costs for powerful pumps are substantial, this additional drawback can be severe. Also, there are better ways to collect large quantities of small particles, such as electrostatic precipitators, wet scrubbers, and filters.

Electrostatic Precipitators

Charged particles in an electrical field can encounter electrostatic forces that greatly exceed resistive drag and gravitational forces. Charged particles are attracted to oppositely-charged and grounded bodies. The velocity with which a charged particle moves toward a collecting surface can be more than 1,000 times greater than the particle's terminal settling velocity. The movement of an electrically-charged particle toward a collector is the mechanism by which an *electrostatic precipitator* (ESP) operates. The particles that are to be collected need not carry an initial electrical charge. Practical industrial ESPs perform three sequential operations: (1) *charging* the entering aerosol particles; (2) *depositing* the charged particles on a collector; and (3) *cleaning* the collector by removing the particle deposit to a reservoir such as a "hopper." Cleaning the ESP is periodically required in order to prevent reduced collection efficiency. Both solid and liquid particles can be collected with an ESP.

In an ESP, applying a charge to the entering particles is usually accomplished by air ions (positive or negative) generated by applying a high voltage (e.g., thousands of volts) to a metal charging device, such as a wire or wires (**Figure 6–6**). As the voltage is increased, the electrical field potential at the surface of the wire increases to a point where it both ionizes the nearby gas (usually air) molecules and ejects free ions (e.g., electrons from a negatively-charged wire). Orbital electron transitions in the surrounding ionized gas generate a faint blue visible glow called a *corona.* During operation, the ions generated by the corona discharge, move toward the oppositely-charged collector. Particles in the

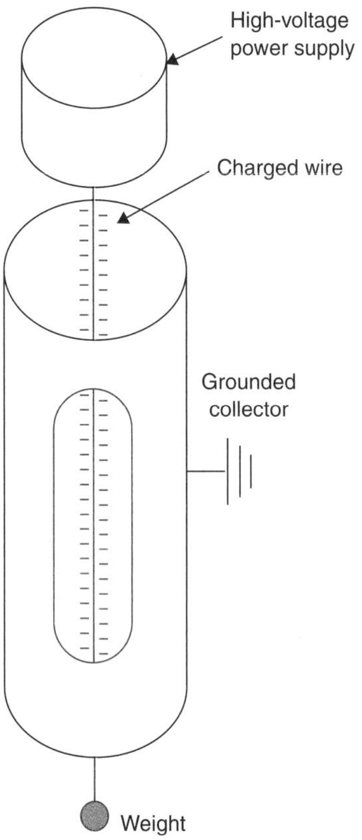

Figure 6–6 Schematic diagram of an electrostatic precipitator (ESP) with a cylindrical collector and a central charging wire.
Source: The University of California Air Pollution Health Effects Laboratory, with kind permission.

input air intercept the moving ions, become electrically charged, and then move to the collector where they are deposited. The air that exits the ESP thus contains fewer particles. If the voltage is further increased, eventually a spark discharge between the charging wire and the collecting surface occurs. The spark briefly interrupts the operation of the ESP. In most ESP designs, intermittent sparking is permitted, as the higher voltage improves the overall collection efficiency. If the ESP is properly designed and operated, 99 percent or more of the particles can be removed from the input air. The removal efficiency, E_{esp}, varies with the charging and collecting voltage gradients, V_a and V_b, the collection surface area, A_c, the particle diameter, d, and the airflow rate, Q:

$$E_{esp} \propto V_a V_b A_c \, d/Q. \qquad \text{(Eq. 6–9)}$$

As one might guess, the collection surface in the ESP eventually becomes coated with a *particle cake* that must be removed. If the deposit is thick enough, it can reduce the electrical potential difference between the charging wire and the collecting surface, thereby decreasing the collection efficiency. Also, the deposited particles can lose their charge (or acquire an opposite charge) and be ejected back into the outlet air stream. To prevent this loss of collection efficiency, practical ESPs have sonic horns, water washers, or *rappers* (i.e., hammers) that clean the collectors by moving the collected particles into a *hopper.* The dislodging process (e.g., vibrating, washing, or rapping) can also resuspend the particles. These resuspended particles may be removed by additional downstream ESP collector sections, or bag filtration systems at the end of the collection process.

ESP instillations can be large enough to process more than a million cubic feet of air per minute (1 m^3 = 35.3 cubic feet), or small enough to be used as home air cleaners. Operating voltages typically range from 2 kV to greater than 100 kV. Designs vary, and new design variations are continually emerging. Some types of particles cannot be collected in ESPs because of their extreme high or low intrinsic electrical conductivities. In addition, the presence of flammable or combustible conditions may restrict use of an ESP.

Wet Scrubbers for the Collection of Particles and Gases

In a *wet scrubber* the collection medium is a liquid, which is usually water or an aqueous solution. The collection liquid can be in the form of droplets, a bulk volume, or a flowing sheet (as on a wetted collection surface). Designs vary considerably, but wet scrubbers are used for the efficient collection of both particles and gases. Among the advantages of wet scrubbers is their ability to collect substances that, if collected in a dry state, would be combustible or otherwise hazardous. A disadvantage is that considerable volumes of pollutant-contaminated liquids may be generated, which may cause disposal problems. Common designs include the following:

- spray chambers, or spray towers
- venturi scrubbers
- wet cyclone scrubbers
- wet filters
- wet packed towers

The mechanisms for particle and gas collection include:

- Brownian diffusion (for gases and small particles);
- impaction on wet surfaces or on aerosolized liquid collector droplets;
- electrostatic attraction (for charged particles); and
- condensation growth of particles followed by a secondary collection method (e.g., impaction).

Spray Towers

Spray towers (**Figure 6–7**) typically introduce a water spray that falls downward through an upward-flowing "dirty" air stream. Contaminant particles are collected on the falling water droplets by impaction and diffusion. Gases are collected by diffusion to the water droplets where they are absorbed. A *demister* (e.g., wire screens) above the sprayer section collects contaminated water droplets before they exit at the top of the tower. Cleaned air exits above the demister at the top of the tower. Water droplets and condensed water (from the demister) fall into a sump or drain at the bottom of the tower. In order to prevent gases from diffusing back out of the droplets after collection, they can be chemically fixed by additives in the collection droplets. The collection efficiency of the spray tower is a function of several parameters, including the following:

- the collector droplet size distribution;
- the number of collection droplets per unit volume of air;
- the upward flow velocity of the contaminated air stream;
- the height and cross-sectional area of the tower;
- the efficiency of the demister for condensing contaminated water droplets;
- the contaminant solubility in the sprayed collection liquid (for a gas); and
- the chemical fixation efficiency of a gas in the collection liquid.

Venturi Scrubbers

Venturi scrubbers (**Figure 6–8**) operate on principles similar to spray towers. They differ in that the contaminated air usually flows downward toward a

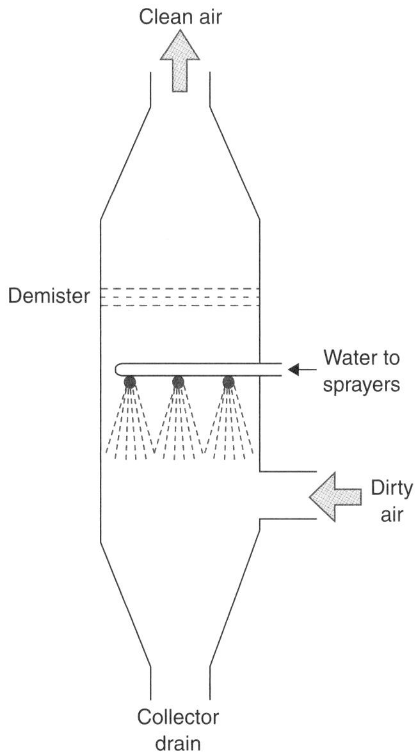

Figure 6–7 A simplified spray tower for the collection of particles and gases.
Source: The University of California Air Pollution Health Effects Laboratory, with kind permission.

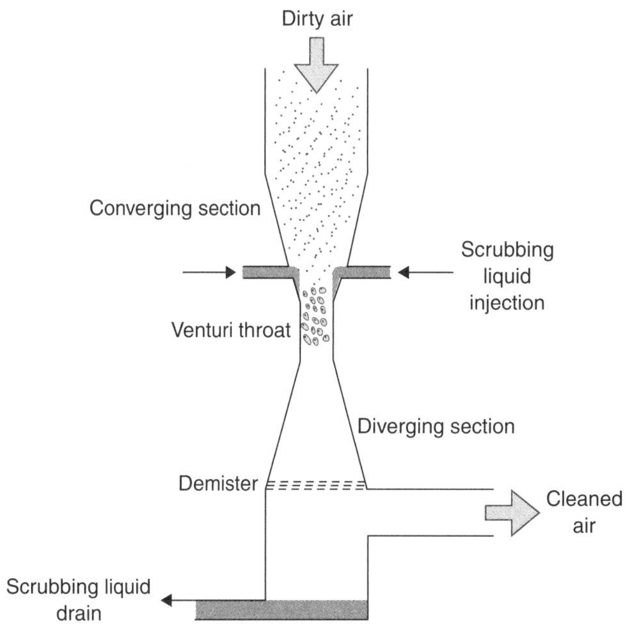

Figure 6–8 Schematic diagram of a venturi scrubber.
Source: The University of California Air Pollution Health Effects Laboratory, with kind permission.

converging "throat" where it is rapidly accelerated. Liquid is introduced along the wall above the converging section, where it is aerosolized upon encountering the reduced pressure in the throat. The aerosolized and falling collection liquid contacts and collects the downward flowing particles and gases in a diverging lower section of the scrubber. Contaminated liquid droplets are typically collected by a demister before exiting the scrubber. The collection efficiency of contaminated wet particles in the demister by impaction will be enhanced if they grow in the low pressure humid environment of the scrubber.

Demisters and Water Vapor Removal Techniques

Demisters (also called "entrainment separators," "mist eliminators," or "moisture separators") are used to remove water droplets, steam, and water vapor from an air stream. Wire-mesh, packed-filters, flat or bent plates, and perforated plates are used to collect water droplets (e.g., by the impaction mechanism). In order to collect small droplets (e.g., under 10 μm in diameter), these devices must have significant pressure drops, which requires larger, more costly pumps to force air through them.

Water vapor can be removed efficiently by cold surfaces that act as condensers, or by sorbents, such as physical adsorbents and chemical absorbents. Many sorbents for water can be easily regenerated by heating them to drive off the water. Common water-vapor collection materials include:

- calcium sulfate
- silica gel
- activated alumina
- anhydrous magnesium perchlorate
- sulfuric acid

The removal of water vapor by cooling the air is relatively simple and effective. The performance of a cooling device can be predicted using the reference data provided in **Table 6–6**, which shows the maximum (saturation) water content of air at a pressure of 1 atmosphere and at various temperatures. To illustrate the removal of water vapor by cooling, consider an airstream that is saturated with water vapor heated to a temperature of 71.1°C. The air will contain about 207 grams of water vapor per m^3 of air. By cooling the air to −17.8°C and collecting the water by condensation, the maximum water content is then reduced to 1.23 grams/m^3, equivalent to a 99.4 percent reduction in water vapor in the air. Upon re-warming the air to 21.1°C the new relative humidity (*R.H.*) will be

$$R.H. = 100(1.23\ \text{grams}/18.45\ \text{grams}) = 6.7\ \text{percent}. \quad \text{(Eq. 6–10)}$$

This represents a significant reduction of water vapor from the air.

Table 6–6 Amount of water vapor in clean air at saturation (i.e., 100% relative humidity) at a pressure of 1 atmosphere.

Air Temperature (T)		*Maximum*
°F	*°C*	*Grams of Water per m^3 of Air (at temp. T)*
0	−17.8	1.23
10	−12.2	1.86
20	−6.7	2.94
30	−1.1	4.49
40	4.4	6.55
50	10.0	9.40
60	15.6	13.27
70	21.1	18.45
80	26.7	25.31
90	32.2	34.23
100	37.8	45.74
110	43.3	60.39
120	48.9	78.86
130	54.4	101.8
140	60.0	130.2
150	65.6	165.1
160	71.1	207.3
170	76.7	258.1
180	82.2	319.0
190	87.8	391.2
200	93.3	476.2

Data from Nelson (1971).

Bag Filters

Bag filters are specialized filters that are preferable to simple flat filters because they can be repeatedly cleaned and reused. Thus, they do not require frequent replacement. Bag filters are widely used for the collection of particles in high-throughput applications, such as coal-fired power plants, or for other industrial processes that require efficient dust emission controls. In a *baghouse* (**Figure 6–9**), dirty air enters below a parallel array of

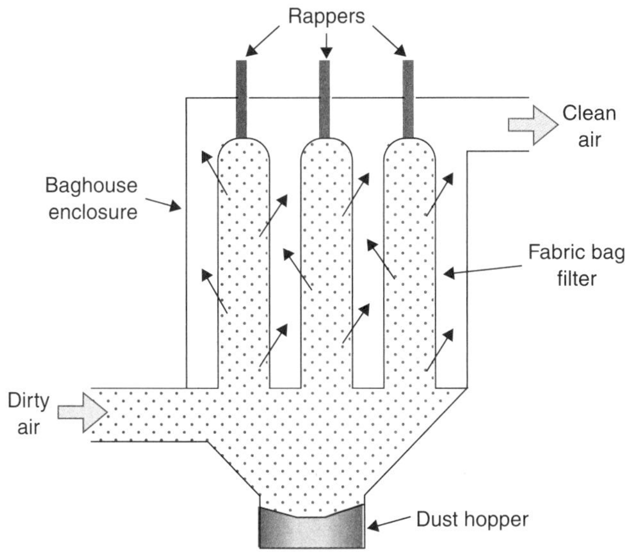

Figure 6–9 Example of a baghouse dust collector design. *Source:* The University of California Air Pollution Health Effects Laboratory, with kind permission.

tubular fabric bag filters. (e.g., shaped like socks, sleeves, or elongated bags). The airflow is initially upward, inflating the tubular filters. The air then flows radially outward through the filter material where it exits the baghouse containment. The dust collection efficiency (E_{bag}) can be very high (e.g., 99.9 percent or more for all sizes of particles), but it varies during operation, as the collected dust bridges gaps in the fabric. The collection efficiency actually increases as dust collects on the filters. However, this increased efficiency is accompanied by an increase in pressure drop, which decreases airflow and/or increases the cost of forcing air through the filters. Eventually the dust forms a "cake" that occludes the filters. Periodic cleaning via shaking, acoustically vibrating, or pulsing the filters with air jets, dislodges the cake, which is then collected in a hopper for bulk removal.

The filter bag can be made of varied fibrous materials, including cotton, wool, minerals, metals, and plastics.

Weave type, fiber diameter, fiber density (e.g., number of fibers per cm^2), fiber coating (e.g., with oil), and fiber electrification (via "permanent" static charges, or externally-applied voltages) can all influence the collection efficiency. The overall efficiency, E_{bag}, is

$$E_{bag} = f(E_{fiber} + E_{weave} + E_{cake} + E_{elect}), \qquad \text{(Eq. 6–11)}$$

where E_{fiber} is the single fiber collection efficiency, E_{weave} is the enhanced filter efficiency due to the woven fabric, E_{cake} is the filter efficiency of the collected dust, and E_{elect} is the electrostatic charge contribution to the collection efficiency. The latter two terms are time-dependent because they change during particle collection. The collection mechanisms of a bag filter include the following:

- impaction
- diffusion
- interception
- sedimentation
- electrostatic attraction

As a result of mechanical damage during cleaning and potential damage produced by chemical degradation, the filters will eventually require replacement. It is therefore important to select bag materials that are appropriate for the specific application. For example, caustic dusts and gases require corrosion-resistant fabrics. Wet inlet air is another problem that can degrade filter performance due to water clogging the fabric's pores. If the air entering the system contains excessive water, a drier may be required upstream of the filter to prevent clogging.

Packed Beds

Packed beds, such as those containing activated charcoal or collections of glass, metal, or plastic beads, are used for efficiently collecting particles and gases, including caustic substances. The beads can be coated with various substances to improve their performance. The disadvantages of packed beds include the difficulty in removing the collected materials from the bed substrate, plus the tendency to develop uneven airflow patterns (e.g., channeling) through the filter bed. In order to avoid *channeling* (i.e., preferential flow through loosely-packed regions), the bed may be vibrated and/or uniformly packed with monodisperse spheres. With packed spheres the uniform air flow may disrupted if (1) contaminants build up within the bed; (2) spheres degrade with use; or (3) the spheres were not adequately uniform. If gases are to be collected, the bed may contain high-surface-area packing materials, such as activated charcoal, molecular-sieve spheres (e.g., porous glass or ceramic spheres) or other stable adsorbent materials (e.g., clay or zeolites). If the collected pollutants are hazardous, they may require treatment in the bed prior to disposal. The destruction of gases on adsorbent beds can be accomplished by catalytic metals, oxidants such as $KMnO_4$ (potassium permanganate), and by heating to a high temperature. Safety can be a *major* concern, especially if the bed is packed with a flammable substance such as activated charcoal; charcoal will adsorb flammable organics, which increases the fire hazard.

Acoustic Agglomerators

In general, large particles can usually be collected more efficiently than smaller ones, so methods for increasing particle size are used in some industrial scrubbing applications. In the presence of an intense acoustic field, aerosol particles will undergo rapid motion that produces facilitated agglomeration (i.e., they collide and stick together to form larger particles). The use of acoustic agglomerators can make subsequent capture of the particles by inertial and other collection mechanisms more efficient. Although the power required to produce agglomeration can be significant, the method can be applied in situations where high collection efficiencies of small particles are required. One such application is found in mines where high concentrations of fine particles are produced by blasting. Acoustic agglomeration may also be used to control smokes, as may occur in fires, and dense clouds of radioactive particles, as may occur in nuclear accidents.

Figure 6–10 shows applicable size ranges for aerosol collectors.

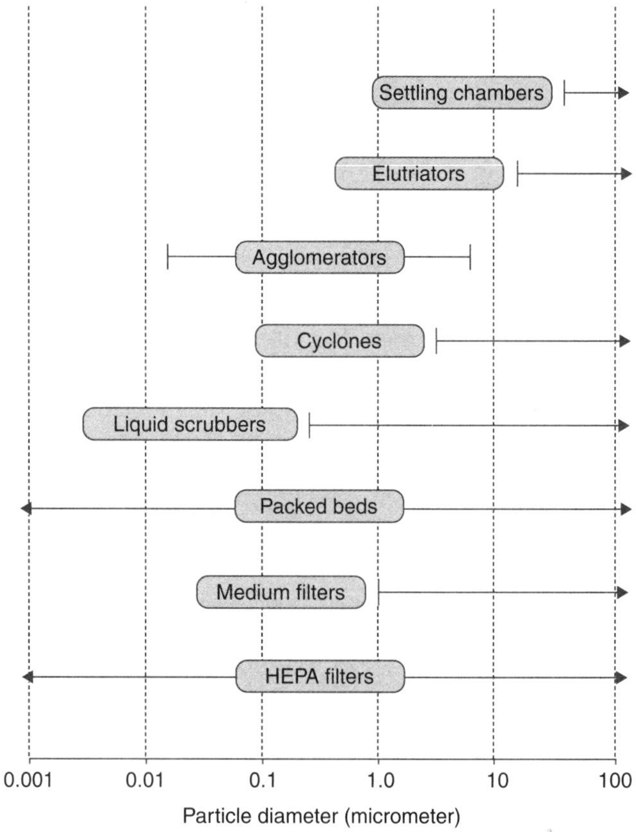

Figure 6–10 Approximate size ranges for particle collection for various air pollution control devices. *Source:* The University of California Air Pollution Health Effects Laboratory, with kind permission.

Additional Methods for Controlling Gas Emissions

Thermal and Catalytic Oxidation

Thermal oxidizers (also called "afterburners") are used to destroy combustible gases and vapors. In a typical application, the contaminated airstream is introduced into an open flame or a high-temperature heater. Temperatures in thermal oxidizers are typically between 1,000 and 1,500°F (540–815°C). Combustion may be followed by entry into a *residence chamber* where oxidation proceeds and less-toxic combustion products (i.e., carbon dioxide and water vapor) are formed. A variation, the *direct combustor,* introduces a fuel (e.g., natural gas or oil) that is mixed with the pollutant particles and gases in order to facilitate their subsequent combustion. Thermal oxidizer systems are relatively simple in design and, if properly engineered, require minimal maintenance. As with any combustion-related process, *safety* and air pollution concerns regarding the formation of CO_2, CO, and nitrogen oxides exist. In addition, hot effluents and containment surfaces can produce burn and fire hazards, so cooling sections and/or thermal insulation may be required.

Catalytic oxidizers are catalyst-containing filter beds operated at high temperature or high pressure, or both, in order to facilitate the oxidation of particles and gases. Platinum is commonly used as a catalyst; oxidation temperatures range between 700 and 900°F (370–480°C). *Extreme caution must be used when the treated airstream contains gases and/or particles at concentrations near or above their explosive limits.* Explosive limits of particles and gases in the airstream must be known; and appropriate safety precautions, such as prior dilution, must be implemented. Explosive concentration limits are found in various handbooks including the *CRC Handbook of Chemistry and Physics* (Lide, 2009). Fires and explosions in air cleaning systems must be anticipated and prevented at all times. Fine particulate material, which is not usually considered flammable, can be explosive at high concentrations.

The presence of non-combustible or combustible particles can clog catalyst beds and even poison catalysts. Therefore, the particulate material may require removal from the air prior to the combustion of gases.

Sequestration of Anthropogenic CO_2

Sequestration, which refers to the stabilization and storage of a substance, is a technique that is being actively explored for preventing the greenhouse gas, carbon dioxide (CO_2), from accumulating in the atmosphere. The major advantage of sequestration is the vast available storage capacities of natural geologic, oceanic, and biotic reservoirs. In theory, all of the current, and future, anthropogenic CO_2 production could be placed in these reservoirs. The oceans alone could probably sequester about 10,000 years of the current anthropogenic CO_2 production. The disadvantages of sequestration include (1) potential environmental effects (including substrate acidification); (2) excessive costs, e.g., when using current technologies; and (3) safety issues associated with the available and proposed technologies.

Abiotic sequestration is based on the engineering application of physical and chemical mechanisms that do *not* involve living organisms. *Ocean sequestration* of CO_2 is an example of abiotic sequestration. Deep-sea injection of liquefied CO_2 derived from industrial operations is a promising technique, especially given the enormous available capacity of the Earth's oceans for storage, as mentioned earlier. The potential storage capacity of the Earth's oceans exceeds the total future CO_2 production that would result from burning all of the known reserves of coal, oil, and natural gas. In order to prevent CO_2 from escaping, it can first be liquefied and then injected at significant ocean depths (e.g., 500–3,000 m), depending on the injection technology used. However, the production and safe handling of liquid CO_2 can be costly. Ocean sequestration can therefore have significant impacts on the costs of electrical power production and other important industrial processes. In addition, large undersea lakes of CO_2 could form and have negative impacts on sea life, and/or the potential for leakage into the atmosphere. Before deep-sea sequestration can be implemented on a large scale, additional research on the technologies and potential long-term adverse effects must be performed.

Geologic sequestration also has great potential for the disposal of all current and projected anthropogenic CO_2 production. In addition, the injection of liquid CO_2 into subsurface saline aquifers below impermeable geologic strata provides CO_2 sequestration by (1) dissolution in water, and (2) chemical reactions with minerals to form solid carbonates. Injection of liquid CO_2 into low-producing oil deposits has already been used successfully to increase the recovery rates of oil and gas; the technique can be used in both onshore and offshore operations. Carbon dioxide leakage into the air and subsurface environments are potential concerns, but the technique is more likely to be used with current oil recovery efforts than deep-sea disposal. *Mineral sequestration,* which is another form of geologic storage, involves either (1) the formation of solid minerals, such as calcium carbonate and magnesium carbonate, or (2) the injection of CO_2 dissolved in water into porous geological formations, such as sandstone. In the latter case, the CO_2 is either trapped in subsurface caverns or taken up within porous rock. As with other abiotic methods, the development of cost-effective and environmentally-acceptable techniques requires additional research and development.

Biological sequestration involves the ecological management of microorganisms or complex plants to increase their storage capacities of CO_2. Biological sequestration (or substitution) is based on either (1) increasing the biotic mass (e.g., by increasing the areas of wetlands, forests, crops, and natural plants), or (2) using natural soil enhancers and fertilizers (e.g., manure, wood, and charcoal) to replace fossil-fuel based products. The carbon in such natural fertilizers does not enrich CO_2 in the atmosphere. These substitution techniques are described as "win-win" methods, as they have environmental, economic, and food production benefits. However, they require changing the ways land is currently used, and they must compete with current food production methods.

It is clear that some significant technical challenges must be overcome before many of the biological CO_2 sequestration techniques can be implemented on a large scale. It is also clear that such techniques hold great promise for future generations. The review by Lal (2008) provides an excellent introduction to several basic concepts, as well as the pros and cons associated with biological CO_2 sequestration.

Selection of Aerosol and Gas Collectors

As there are several options available, it is not always easy to select the most appropriate aerosol and gas collectors. Performance, initial cost, space requirements,

Table 6–7 Comparison of selected air pollution control devices.

Device Types	*Applications*	*Advantages*	*Disadvantages*
Settling Chambers, Impingers, Cyclones	large to medium size particles of all types	low cost simplicity durability	low efficiency large space required erosion of components
Filter Systems	dusts, fumes, all sizes of particles that are not sticky	high collection efficiency for all particle sizes modest power requirements dry disposal product	low initial cost costly bag replacements R.H. and temperature limitations can have fire hazards
Spray Chambers, Venturi Scrubbers, Wet Cyclones	high temperature moisture-laden air, and all but the smallest particles	compact constant pressure drop not dusty	cost of disposal of waste water
Electrostatic Precipitators	all sizes of particles and fine mists, including corrosive and high temperature air streams	high efficiency low pressure drop dry disposal product low maintenance costs	large space required high initial cost can't collect some materials
Catalytic Converters	gases and vapors	low maintenance low pressure drop efficient	inefficient when cold clogged by particles catalyst expense
Gas Adsorbers (in liquids)	toxic, radioactive, and odorous gases	collection fluid may be recycled	high costs corrosion problems contamination problems

Data from NIOSH (1973).

maintenance costs, serviceable lifetime of the collection medium, power requirements, generation of hazardous waste, and safety are all important considerations. **Table 6–7** provides a comparative overview of some air pollution control devices. Figure 6–10 shows the typical particle collection ranges for various devices. Although it is difficult to concisely describe industry-specific air pollution control methods, some general information is provided in **Table 6–8**. It must be noted that both regulatory air standards and accepted industrial practices are evolving. Therefore, air pollution control equipment must also change in order to meet accepted safety practices and new air quality regulations.

VII. CASE STUDY: COAL-FIRED POWER PLANT

Overview

Because of its availability, low cost, and high-energy output, coal is the primary fuel used for electric power generation worldwide. The availability of affordable electrical power is essential for public health and economic prosperity. Although approximately 40 percent of the world's electricity is generated by coal-fired power plants, 20 percent comes from other combustion fuels (i.e., oil, gas, biomass, and burnable waste), 18 percent from hydroenergy (e.g., hydroelectric dams), 14 percent from nuclear energy (77 percent in France), and 1 percent from *renewable energy* technologies (e.g., geothermal, wind, solar, and tidal). The simplified chemical energy balance equation for hydrocarbon fuel combustion is

$$C_aH_b + \left(a + \frac{b}{4}\right)O_2 \rightarrow \text{Heat} + aCO_2 + \frac{b}{2}H_2O \quad \text{(Eq. 6–12)}$$

where a and b are the chemical-balance (stoichiometric) coefficients. Equation 6–12 applies to complete combustion. In reality, most combustion is incomplete, which produces CO, CO_2, soot, and other products of incomplete combustion (e.g., unburned fuel, partially-burned fuel and oxides of nitrogen).

The heat energy from fuel combustion in a power plant is transformed into mechanical energy that spins large electric generators. In most cases, mechanical

Table 6–8 Sample industrial operations and typical particulate air pollution control equipment used.

Operation	*Typical Particle Diameter (μm)*	*Cyclones*	*High Efficiency Centrifugal*	*Wet Collectors*	*Fabric Filters*	*Electrostatic*
Pulverized coal burning	under 5	seldom	frequently	no	no	frequently
Abrasive cleaning	5–15	no	sometimes	frequently	frequently	no
Grain handling	5–15	yes	sometimes	seldom	frequently	no
Metal grinding & machining	15–100	frequently	frequently	frequently	frequently	no
Woodworking	5–100	frequently	sometimes	seldom	frequently	no
Chemicals handling	5–15	sometimes	frequently	frequently	frequently	rarely
Steel blast furnace	varied	frequently	rarely	frequently	no	frequently
Cement kiln	5–15	rarely	frequently	rarely	yes	yes
Steel open hearth	<5 to >15	no	no	seldom	sometimes	yes

Data from ACGIH® (1976).

steam pressure is used to turn steam turbines attached to electric generators. Approximately 33 percent of the fuel's heat energy is converted to electrical energy in a typical coal-fired electric power plant. Airborne combustion by-products, non-combustible contaminants in coal (e.g., minerals and metals), products of incomplete coal combustion, along with carbon dioxide (CO_2) and sulfur dioxide (SO_2) are present in the outlet emissions from a coal-combustion chamber. These products are the major potential environmental air pollutants from coal combustion. The major air pollutants of environmental concern include sulfur oxides (SO_x), nitrogen oxides (NO_x), particulate material, carbon dioxide (CO_2), and trace elements (e.g., vanadium, iron, mercury, etc.). At the present time, viable technologies exist for collecting SO_x, NO_x, and particles from the effluent airstream. However, carbon dioxide control is more difficult, and new technologies are currently under development, as were previously described. Capture and storage, or sequestration in geologic formations, biota, and oceans, are all under consideration as CO_2 control technologies. Research efforts are also underway to convert CO_2 into a practical fuel (e.g., methanol).

Pulverized Fuel Coal-Fired Power Plants

Operations

Modern coal-fired power plants are technically complex, and their designs are focused on increasing the efficiency of conversion of heat to electrical energy, as well as minimizing the emission of air pollutants. In a simplified form, the basic operation of a *pulverized fuel* (PF) coal-fired power plant involves the following sequence:

1. Coal is delivered to a *bunker.*
2. The bunker coal enters a *pulverizing mill.*
3. Pulverized coal, along with heated air is blown into a furnace (the combustion chamber or *boiler*) where it rapidly burns.
4. Purified recirculating water flows through closed tubes (a heat exchanger) in the boiler where it is turned into superheated steam.
5. The steam passes through a series of turbines that turn a shaft that spins one or more large electric generators.
6. Steam exiting the turbines is condensed (via cooling) for return to the boiler's heat exchanger.
7. Hot combustion-chamber air, along with by-products (called *flue gas*), passes through air pollutant removal equipment and is blown by a fan (or fans) into a tall *stack* where it enters the ambient air.
8. Solid wastes from the boiler and air pollutant collectors are removed and transported for disposal or sale.

In order to appreciate the scale of the process, a typical coal-fired power plant that supplies 700,000 homes with electricity heats water to about 1,000°F (540°C) in order to produce steam at nearly 2,000 pounds per square inch (1 psi = 704 kilograms per square meter) to spin one or more 20,000 volt electric generators at

3,600 revolutions per minute. The yearly output of a single plant can be 10 billion kilowatt-hours, and the daily coal consumption about 14,000 tons. Each day, about 140 railroad cars of coal is burned. The daily production of fly ash is about 1 ton, and about 50 tons of NO_x and SO_2 are also produced. In a well-controlled and modern power plant, the daily emissions of fly ash are only about 25 pounds (11.3 kg), with about 500 pounds (227 kg) of NO_x and SO_2. Thus, about 99.5 percent of the potential air pollutants are captured at the more modern power plants.

Pollutant Controls

Focusing on the pollutant controls, contaminant reduction ideally begins prior to combustion by removing sulfur and sometimes other non-fuel materials from the coal. During combustion, the conditions are managed to limit formation of some potential air pollutants. Post-combustion, SO_2, particles, and other pollutants are removed from the flue gas before they enter the environment through an emission stack. The stack height is optimized to ensure effective dilution within the troposphere and minimize regional ground-level exposures to pollutants (see Chapter 2).

The flue gas contains particulate fly ash, which consists of the non-combustible solids (such as metals, minerals, and dirt) found in the coal. Fly ash consists of airborne polydisperse (variable in diameter range, e.g., from < 1 μm to > 50 μm) glass or ceramic spherical particles. In the formation of fly ash, minerals with high melting points (such as silicon dioxide, aluminum oxide, and iron oxide) initially condense to form the fly-ash spheres, after which, the substances with lower melting points condense on the surfaces; thus the fly ash has surface coatings of metals and other substances. Common metals that coat fly-ash include vanadium, nickel, iron, arsenic, barium, mercury, beryllium, lead, copper, and zinc. Many of the surface coatings on fly ash have significant intrinsic toxicities and are leachable when in contact with water, as found in living organisms, soils, lakes, and watersheds. Therefore, storage of high-metal-content fly ash must be done in a manner to protect people and the environment. Smaller particles of fly ash are inhalable, which potentially exposes downwind populations to minerals and toxic metals. For these reasons, fly ash is a primary target of emission control equipment in modern coal-fired power plants. It should be noted that older plants, especially in developing nations, may have little to no emission controls. If released into the upper regions of the troposphere, the small-particle fly ash, with low terminal settling velocities, can travel long distances and cross national and international borders.

Particle Controls

Removal of fly ash from the flue-gas stream is typically a multi-stage process. A post-combustion settling chamber (similar to a horizontal elutriator) can remove the larger particles, and baghouse filters, electrostatic precipitators, and/or wet scrubbers can collect the smaller particles. Collected fly ash, which resembles fine sand, can sometimes be sold, as it has some commercial uses, including:

- as an additive to increase the strength of cement and grout;
- as a mineral filler in asphalt;
- for structural fills (e.g., embankments);
- as road base material; and
- as a stabilizer when mixed with other waste materials prior to disposal.

However, due to the variable quality of power-plant fly ash, and the cost of its transportation, about 70 to 75 percent is not commercially viable. Most coal fly ash must be disposed of as waste in landfills and storage lagoons.

Sulfur Dioxide Controls

Sulfur dioxide is removed from the flue gas (i.e., *flue gas desulfurization,* FGD) because of its potential for acidifying rain and for generating fine-particulate sulfate. Sulfate along with its precursor, SO_2, is associated with adverse health effects when inhaled in sufficient concentrations. One benefit of FGD is the production of commercially-useful sulfur, sulfuric acid, and relatively-pure SO_2 gas. Several processes are available for SO_2 removal from flue gas including adsorption on charcoal, absorption in sulfuric acid, catalytic oxidation, and wet scrubbing (e.g., with a limestone ($CaCO_3$) slurry). The wet scrubbing process produces calcium sulfite ($CaSO_3$), which is not a directly commercially-useful product, so it must often be disposed of as waste, or converted to a useful product:

$$SO_2 + CaCO_3 \rightarrow CaSO_3 + CO_2. \qquad \text{(Eq. 6–13)}$$

Further oxidation and treatment of calcium sulfite can be used to produce calcium sulfate ($CaSO_4$), which can be used in plasters and drywall. However, contaminants (e.g., unfixed sulfur dioxide) from the flue gas can sometimes make these products unsafe for use in homes or buildings.

NO_x Controls

Nitrogen oxides (NO_x) also form acid rain, and they contribute to the chemistry and the adverse health effects of urban smog. The most convenient method for decreasing the NO_x concentration in flue gas is by modifying the combustion process (e.g., by lowering the combustion temperature, or by further combustion, or "reburning," of the flue gas to convert NO_x to nitrogen and oxygen). If further reductions of flue-gas NO_x are required, a more expensive *selective catalytic reduction* (SCR) process can be used. The SCR process involves injecting ammonia (as a reducing agent) into the flue gas stream and passing the mixture over an oxidation catalyst (e.g., vanadium) coating a plate or honeycomb surface.

Mercury and Radioactivity Controls

Coal-fired power plants also emit mercury and radioactive materials that are present in soil and ore deposits. To date, major efforts have not been implemented to remove these and other trace emissions. An excellent resource for more information on flue gas treatment is the International Energy Agency (IEA) Clean Coal Centre (http:/www.iea-coal.org.uk).

VIII. CASE STUDY: AUTOMOBILES AND TRUCKS

Regulatory Pressure and Overview of Controls

The U.S. 1990 Clean Air Act strengthened air pollution regulations related to automobiles and trucks on several fronts, namely:

- use of cleaner burning fuels;
- stricter tailpipe emission standards; and
- inclusion of non-road engines (i.e., in boats, farm equipment, bulldozers, construction machinery, and lawn and garden devices).

Fuel standards, emission standards, and evaporative fuel controls were devised. *Reformulated gasoline,* which was required for use in targeted areas with high environmental ozone levels, was intended to reduce emissions of hydrocarbons, nitrogen oxides, and other pollutants. A major goal of the Clean Air Act was to limit emissions of ozone-forming substances in the air of impacted regions. In order to achieve this goal, *fuel standards* for gasoline were developed that defined a minimum oxygen content and maximum benzene content. Tightened *tailpipe emissions* were established for total hydrocarbons, carbon monoxide, and nitrogen oxides in order to control local smog levels. *Evaporative fuel emissions* were also regulated, as fuel vapors contribute to tropospheric ozone formation. Evaporative emissions have three major sources: (1) permeation of fuel through fuel lines and some non-metal fuel tanks; (2) the engine heat that increases the evaporation and permeation of fuel; and (3) spillage and evaporation of fuel during refueling (**Figure 6–11**).

Emission Controls

Gasoline and diesel fueled engines differ somewhat in their emission control technologies. Currently, emission control techniques are rapidly evolving in the transportation industry. The Manufacturers of Emissions Controls Association (MECA, http://www.meca.org) is a useful source of updated information on emission control technologies. MECA was created in order to (1) support

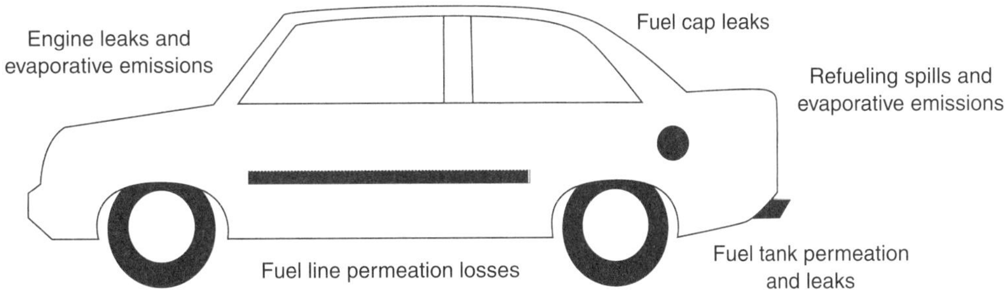

Figure 6–11 Spillage and evaporation of fuel during refueling and other non-driving conditions.
Source: The University of California Air Pollution Health Effects Laboratory, with kind permission.

compliance with air regulations and (2) *promote* the use of effective air quality programs. Thus, MECA exists to protect human health and the environment and to facilitate industrial progress. This enlightened holistic viewpoint (i.e., considering both "health" and "progress") should be noted and appreciated. Much of what follows is abstracted from the MECA website.

Evaporative Emission Controls

The goal of evaporative emission controls is to reduce, or prevent, the release of hydrocarbon vapors from vehicles into the atmosphere. There are two main motivations for this goal. A primary motivation is to prevent the formation of the secondary pollutant, ozone, which results from reactions of volatile organic compounds with oxides of nitrogen. Another motivation is to reduce the direct contribution of hydrocarbons that contribute to the formation of smog.

Evaporative emissions have several sources, which require multiple controls. Evaporation of fuel from the engine compartment has been greatly reduced by the adoption of *fuel injection systems* as replacements for conventional fuel delivery technologies, such as the carburetor. Carburetors allow fuel vapor to escape during operation, tend to suffer from fuel overflow, and have large gasket-sealed surfaces that can leak when worn or damaged. Another technology that reduces fuel evaporative emissions from the engine is the use of carbon-containing canisters that scavenge unburned (and partially-burned) fuel vapor from the lubricating oil crankcase. The scavenged fuel is then passed back to the engine's combustion chambers, which prevents its evaporation into the ambient air. One common emission control device is the *positive crankcase ventilation* (PCV) *valve* that is a one-way valve allowing unburned crankcase gases to re-enter the combustion chambers of the engine. The PCV valve must be replaced periodically as oils accumulate on the inner surfaces, but the cost is minimal (e.g., a few U.S. dollars) compared with other technologies.

Permeation of fuel from fuel tanks and fuel distribution hoses is controlled by the use of impermeable materials, effective seals, ventilated fuel tank caps, and when needed, layered polymer coatings on fuel tanks and fuel hoses.

Evaporative emissions during fueling have been reduced by adding flexible vapor-barrier seals on fueling nozzles and using vapor-recovery systems that reduce the atmospheric pressure within hoses to collect vapors at the nozzle exit. These advancements have essentially done away with the smell of fuel at fueling stations, in parking lots, and on urban streets, which is an indication that these efforts are largely effective.

Catalytic Converters

Modern gasoline engines are equipped with catalytic *three-way converters* (TWCs), which are visible as metal canisters placed in the exhaust line. A catalyst is a substance that modifies or increases the rate of a chemical reaction without being directly consumed in the reaction.

Most TWCs for gasoline engines contain an internal coating of a precious metal catalyst (e.g., platinum, palladium, or rhodium) on high surface-area materials arranged in parallel flow channels. This general design increases the contact area for the catalyst with passing exhaust gases. The TWCs convert carbon monoxide, nitrogen oxides, and unburned hydrocarbons (hence "three-way") to carbon dioxide (CO_2), water, elemental nitrogen (N_2), and oxygen (O_2).

Thermal Management of Catalytic Converters

In order to work properly, catalytic converters must be sufficiently hot. As a result, a majority of the emissions from automobiles and truck engines are formed shortly after a "cold startup." Two methods are used to deal with this problem. First, a small upstream catalytic converter placed near the hot engine exhaust manifold rapidly generates exothermic heat that warms the larger downstream catalytic converter. Second, heat can be preserved in the exhaust line upstream of the converter using double-walled and insulated pipes; this is a common strategy for dealing with the cold catalyst problem.

Particle Traps

Gasoline-fueled engines do not require particle traps, but diesel engines unavoidably generate significant quantities of fine carbonaceous particles that must be removed from the exhaust emissions. These particles can have significant hydrocarbon surface coatings, which may pose an additional health hazard. Thus, diesel engines require high collection-efficiency particle filters. These filters typically have a porous honeycomb structure arranged in parallel channels with plugged alternate channels. This design enhances the filtration efficiency, as it forces the exhaust to flow into half of the porous honeycomb channels, pass through the channel walls, and then flow out of the adjacent channels. The honeycomb traps collect particles by diffusion, interception, and impaction mechanisms, providing greater than 90 percent particle capture rates. Gases are allowed to pass freely through the trap. Over time, deposited particles build up and require removal. *Self-regenerating* traps have a catalyst that produces high-temperature exothermic oxidation reactions that literally burn away the soot accumulations. In some cases, self-regeneration fails, and the trap must be periodically removed and heated to a high temperature in order to remove the deposits. In other designs, an electrical heater within the particle-trap canister or an internal fuel burning system is used to destroy the accumulated soot. Overall, both catalytic converters and particle traps are highly effective in controlling vehicular emissions.

Engine Management

Effective minimization of exhaust emissions requires a *systems approach.* In this approach, the interactions among engine performance and emission-control components are optimized and managed. The previously described cold-start problem demonstrated the importance of upstream thermal management on catalytic converter performance. Some other systems approaches that are designed to decrease pollutant emissions and/or formation during the combustion process include the following:

- Retarding the ignition timing in cold conditions is used to allow the resulting increased hydrocarbons to raise the temperature of catalytic converters.
- *Variable valve timing* (VVT) is used to passively re-introduce some of the exhaust gas into the combustion chambers, which further decreases hydrocarbon and NO_x emissions.
- A technique similar to VVT is *active exhaust gas recirculation* (EGR), in which a fraction of the exhaust gas is forced back into the intake air. EGR reduces combustion chamber temperatures, which reduces NO_x formation.
- *Direct injection* of fuel into the combustion chamber (rather than injection into an upstream zone) is used to improve the fuel-to-air ratio, which improves fuel efficiency.
- *Turbulent mixing* of fuel and air prior to entering the combustion chamber is also used to produce significant improvements in fuel efficiency.

Modern "clean" diesel engine designs have significantly lowered NO_x emissions by the addition of sophisticated controls for the fuel injection system. *Electronic control* of the fuel injection and ignition processes allows for precise shaping of the fuel injection rate and periodic enrichment of the fuel-to-air ratio. These techniques improve fuel efficiency, decrease the initial formation of NO_x, and reduce the burden on downstream catalytic converters and particle traps.

There Are Positive Results, but Some Persistent Problems

As a result of these and other emission control strategies, modern vehicles emit only a small percentage of the pollutants emitted by cars and trucks that were on the roads in the 1970s. However, the ultimate dream of producing internal combustion engines that emit only water and carbon dioxide will be difficult to achieve. One unresolved issue is whether or not to control carbon dioxide, which is not toxic, but has been implicated as a climate change factor (see Chapter 5). Another stubborn problem is the contamination of the combustion zone with lubricating oil from the crankcase. Piston rings, which are designed to reduce friction between the engine pistons and cylinder walls, wet the combustion chambers with engine oil as they travel up and down. Engine oil contains an incredibly large number of additives such as:

- viscosity adjusters for maintaining proper viscosity at various engine temperatures;
- corrosion and rust inhibitors and detergents to keep engine parts clean;
- anti-wear agents to coat and reduce the friction on moving metal surfaces;
- dispersants to reduce sludge formation;
- anti-foaming agents;
- buffers for neutralizing acidity; and
- antioxidants to prevent oil breakdown.

These are just a sampling of some of the more important lubricating oil additives. Since the oil and additives enter the combustion chambers, they contribute to the pollutants in the exhaust. This lubricating oil problem means that even combustion engines that run on clean fuels, such as pure hydrogen or natural gas, will not be completely free of polluting emissions. As electric motor-driven vehicle engines do not require engine lubricating oil sumps, they can achieve much lower emission rates. However, efficient battery designs and other advances must be made before gasoline and diesel engines can be completely replaced. Diesel engines are critically important for construction, shipping, and power generation purposes. Therefore, increases in their costs can have serious public health and economic consequences.

IX. SUMMARY OF MAJOR POINTS

Air quality *regulations* and *abatements* are tightly coupled, as abatements are the methods used to achieve *compliance* with regulations. Regulations are motivated by the desire to mitigate the adverse effects of air pollutants on human health, human welfare, plants and animals, ecosystems, and materials (e.g., buildings and artwork). Regulations are either *strategic* (to achieve long-term goals) or *tactical* (to relieve existing problems). Preserving the stratospheric *ozone layer* is an example of a strategic objective, but preventing asthmatic attacks in the population is a tactical goal. Air quality regulation has evolved in the last 60 years, with the creation of national governmental agencies, such as the U.S. Environmental Protection Agency (in 1970), and similar agencies in Canada, Japan, the United Kingdom, and other countries. An unsolved issue is how to deal with the effects of national disparities in environmental regulations on the relocation of polluting industries to less-regulated countries.

Air pollution regulations (i.e., rules that must be followed) include *fuel standards, process standards,* and *air quality standards.* Air quality standards are numerical values that must not be exceeded. They typically have four major components: (1) an *indicator,* (2) an *averaging time,* (3) a *form,* and (4) a *level.* In addition to regulations and standards for industry, transportation, and electric power plants, tobacco products have also received considerable attention due to their direct effects on users, as well as the potential secondary effects of environmental tobacco smoke (ETS) on susceptible populations.

New regulations often drive the development of abatement technologies necessary to achieve compliance. Engineering approaches to achieve compliance fall into several categories including those that (1) introduce new cleaner technologies; (2) develop cleaner fuels; (3) improve existing industrial processes; (4) improve energy-use efficiencies; and (5) scrub gaseous and particulate emissions at their sources. The effective removal of particulate and gaseous air pollutants at their sources is based on knowledge of physics and chemistry. Aerosol particle collection devices exploit mechanisms that move particles out of their airstreams and deposit them on, or in, collection media. Such mechanisms include *sedimentation, impaction, interception, electrostatic attraction,* and *diffusion.* Gases are removed from the air by diffusion to *adsorbers* or *absorbers,* by sequestration or by destruction. Devices for collecting air pollutants include *elutriators, cyclones, electrostatic precipitators, wet scrubbers,* and *bag filters.* In contrast, *thermal oxidizers* (including *catalytic converters*) are designed to *destroy* air pollutants before they are released into the environment. Air pollution control devices must be carefully designed in order to both meet specific abatement needs, and also to prevent local hazards such as explosions and fires.

Two regulated technologies that demonstrate the use of air pollution controls are (1) coal-fired electric power-generating plants and (2) vehicles powered by gasoline or diesel fuels. Coal-fired power plants use several methods to limit the formation of, and to capture, air contaminants. Coal is often pre-treated to remove sulfur, and crushed to improve its combustibility. *Flue gas* from the burner (i.e., the coal combustion unit) is treated to remove particles and gases before being discharged into the air through a tall emission stack. Particles, called *fly-ash,* are collected in *settling chambers, electrostatic precipitators,* and sometimes *wet scrubbers* and *cyclones.* Gases, such as sulfur dioxide are collected and/or destroyed in *charcoal beds, sulfuric acid absorbers, catalytic oxidizers,* and *wet scrubbers.* In some cases, the cost of controlling pollutants can be offset by simultaneously producing commercial-grade products, such as SO_2, sulfuric acid, and fly-ash products.

Automobiles and trucks have evolved to burn cleaner fuels and to incorporate pollutant control devices in their designs. The minimum oxygen, maximum benzene, and maximum sulfur content of fuels have also been regulated to reduce exhaust emissions. The combustion process has been modified in order to increase fuel efficiency and minimize the formation of pollutants such as nitrogen oxides, hydrocarbons, carbon monoxide and particles. Collection devices and catalytic oxidizers used on modern automobiles and trucks include particle traps and precious-metal catalyst beds (*catalytic converters*) for destroying gaseous emissions. Current challenges include the control of carbon dioxide (a greenhouse gas), and combustion products from lubricating oil and oil additives. Evaporating fuel from (1) fueling operations, (2) the hot engine environment, and (3) permeable gas tanks and fuel lines have been controlled by *vapor recovery systems,* improved seals, the use of impermeable materials and coatings, and by the replacement of carburetors by *fuel-injection systems.* As a result, modern vehicles emit a very small percentage of the air pollutants emitted by 1970s vintage vehicles.

Because internal combustion engines perform numerous tasks that are essential to human health and well-being, the cost and availability of gasoline-fueled and diesel-fueled engines must be carefully considered by regulators. Inevitably, new and cleaner technologies will replace the current internal combustion engines. These changes occur with continued support for research and development, and when the benefits of their use outweigh their potential adverse economic consequences.

X. QUIZ AND PROBLEMS

Quiz Questions

(select the best answer)

1. Which factors motivate the U.S. EPA to issue National Ambient Air Quality Standards?
   a. Human health effects, compliance costs, and materials damage.
   b. Human health effects, compliance costs, and ecological effects.
   c. Human health effects, materials damage, and ecological effects.
   d. None of the above are true.
2. Air quality standards typically have which of the following components?
   a. Indicator, averaging time, form, and level.
   b. Ambient temperature, season, size-distribution, and relative humidity.
   c. Abatement strategies, non-attainment penalties, and enforcement policies.
   d. Maximum and minimum pollutant levels, number of exceedances permitted, and adjustments for co-pollutants.
3. The U.S. EPA's National Ambient Air Quality Standards (NAAQS):
   a. are on a 5-year review cycle.
   b. consider "primary" and "secondary" effects.
   c. consider both particles and gases.
   d. All of the above are true.
4. A regulatory standard for the control of CO_2 emissions in order to protect the Earth's climate would be an example of:
   a. a "tactical" standard.
   b. a "strategic" standard.
   c. an "intermediate" standard.
   d. a "voluntary" standard.
5. Air quality regulations and standards:
   a. have not had a significant impact on urban air quality in developed nations.
   b. have actually led to increases in air concentrations of secondary air pollutants such as lead.
   c. have often produced dramatic improvements in air quality in areas where they have been enforced.
   d. have affected emissions from industry but not emissions from automobiles.
6. Abatement strategies are designed to:
   a. comply with regulations.
   b. challenge regulations via legal actions.
   c. shift the blame for poor air quality from industry to truck and automobile traffic.
   d. relax standards in those cities with economic problems.
7. Engineering approaches to air pollution control include:
   a. development of cleaner technologies.
   b. development of cleaner fuels.
   c. scrubbing pollutants at their sources.
   d. All of the above are true.
8. Electrostatic precipitators:
   a. are no longer used due to their electric shock hazard.
   b. use electrically-charged charcoal to collect toxic gases.
   c. cannot be used effectively to collect liquid particles.
   d. are useful for collecting nearly all types of particles.
9. Baghouse filters:
   a. are used on diesel trucks, but not automobiles.
   b. are not good for collecting particles with diameters smaller than 10 μm.
   c. are not used in industry due to their tendency to catch fire.
   d. None of the above are true.
10. Automotive catalytic converters:
   a. operate most effectively when they are cold.
   b. operate most effectively when they are hot.
   c. have been phased out of service due to the cost of precious-metal catalysts.
   d. have had only minor impacts on automotive emissions.

Problems

1. Calculate the length of a horizontal elutriator that will collect particles with diameters larger than

10 μm, given an elutriator height of 0.1 m, and elutriator airflow of 0.1 m/sec. The V_{ts} of a 10 μm particle is 3×10^{-3} m/sec.

2. If one particle collector is 50 percent efficient for capturing all particles, how many collectors placed in series (such that the output of each flows into the input of the next one) must be used to capture at least 90 percent of the particles?
3. Referring to Figure 6–10, which particle collection devices would benefit by installing an upstream agglomerator?
4. It is necessary to double the induced centrifugal force in a cyclone particle collector. The collector has a tangential velocity of 10 m/sec and a radius of 1 m. If the size of the collector cannot be changed, what should the new tangential velocity be?
5. Design a system for collecting CO_2 from diesel engine exhaust, and (1) explain how it works and (2) what problems might be encountered in maintaining its effectiveness.
6. Search the scientific literature on converting CO_2 into a practical fuel and describe how this might be accomplished.
7. Search the scientific literature on "dust explosions" and propose two methods for their prevention in an industrial operation.
8. How long would it take for a 10 μm aerodynamic diameter particle to settle to the ground when emitted from a smoke stack 100 m tall?
9. How far would the particle in problem 8 travel before settling to the ground if the environmental wind speed was 10 m per second?

XI. DISCUSSION TOPICS

1. Should all nations, rich and poor, be required to adhere to the same ambient air quality standards? List the reasons, both "pros" and "cons," and apply them to a country such as France or Japan, and then to a country such as the Sudan or Ethiopia.
2. Should health-based ambient air quality standards consider the monetary costs to communities trying to achieve compliance?
3. Consider an essential laboratory process (e.g., for vaccine development) that creates exceptionally-hazardous particulate air pollutants, such as viable pathogens for which there are no existing vaccines or cures to protect humans. What practices could be used to insure the safety of the laboratory workers and the general public?
4. It has been suggested that air quality regulations are becoming so strict that some of the unintended consequences now outweigh the direct health benefits. Is there merit in this suggestion? What unintended consequences of air quality rules, if any, do you think might harm your community?
5. What research should be performed on a new fuel for cars and trucks before it is made available for use by the public?

References and Recommended Reading

ACGIH® (American Conference of Governmental Industrial Hygienists), *Industrial Ventilation, 14th Edition,* ACGIH®, Cincinnati, OH, 1976, Fig. 11–17.

ACGIH® (American Conference of Governmental Industrial Hygienists), *Industrial Ventilation: A Manual for Recommended Practice for Design, 27th Edition,* ACGIH®, Cincinnati, OH, 2010.

Davis, W. T., ed., *Air Pollution Engineering Manual, 2nd Edition,* Air and Waste Management Association, John Wiley & Sons, Inc., New York, 2000.

Enstrom, J. E. and Kabat, G. C., Environmental tobacco smoke and tobacco related mortality in a prospective study of Californians, 1960–98, *Brit. Med. J.,* 326: 1057–1061, 2003.

Flagan, R. C. and Seinfeld, J. H., *Fundamentals of Air Pollution Engineering,* Prentice–Hall, Englewood Cliffs, NJ, 1988.

Graham, J. D. and Wiener, J. B., eds., *Risk vs. Risk: Tradeoffs in Protecting Health and the Environment,* Harvard University Press, Cambridge, MA, 1995.

Hesketh, H. D. and Schifftner, K. C., *Wet Scrubbers, 2nd Edition,* CRC Press, Boca Raton, FL, 1996.

Hinds, W. C., *Aerosol Technology: Properties, Behavior, and Measurement of Airborne Particles, 2nd Edition,* John Wiley & Sons, Inc., New York, 1999.

Hoffmann, T. L., Environmental implications of acoustic aerosol agglomeration, *Ultrasonics,* 38:353–357, 2000.

IEA (International Energy Agency), Control and Minimisation of Coal-fired Power Plant Emissions, http://www.iea.org/papers/2003/Coal_Fired_Fossil_Fuels.pdf (accessed May 12, 2010).

Lal, R., Carbon sequestration, *Phil. Trans. R. Soc. B,* 363:815–830, 2008.

Lide, D. R., Ed., *CRC Handbook of Chemistry and Physics, 90th Edition,* CRC Press (an imprint of Taylor & Francis Group), Boca Raton, FL, 2009.

List, J. A., McHone, W. W., and Millimet, D. L., Effects of air quality regulation on the destination choice of relocating plants, *Oxford Econ. Papers,* 55:657–678, 2003.

Nelson, G. O., *Controlled Test Atmospheres: Principles and Techniques,* Ann Arbor Science, Ann Arbor, MI, 1971.

NIOSH (National Institute for Occupational Safety and Health). *The Industrial Environment: Its Evaluation and Control,* NIOSH, Cincinnati, OH, 1973, Chapter 43.

NIOSH (National Institute for Occupational Safety and Health). *Environmental Tobacco Smoke in the Workplace: Lung Cancer and other Health Effects,* Current Intelligence Bulletin 54,

NIOSH, Cincinnati, OH, 1991.

Niu, S., Outdoor and indoor air pollution in China, *Asia-Pacific J. Public Health,* 1:46–50, 1987.

NRC (National Research Council), *Air Quality Management in the United States,* The National Academies Press, Washington, D.C., 2004.

Phalen, R. F., Cleaning the air, in *Methods in Inhalation Toxicology,* Phalen, R. F. ed., CRC Press, Boca Raton, FL, 1997, pp. 25–38.

Rogers, S. M., Air pollution legislation: A review of current developments, *Am. J. Public Health,* 50:642–648, 1960.

Thomas, W. F. and Ong, P., Location adjustments to pollution regulations: The South Coast Air Quality Management District and the Furniture Industry, *Econ. Devel. Quarterly,* 18:220–235, 2004.

U.S. Congress, *The Family Smoking Prevention and Tobacco Control Act,* Public Law 111–31, U.S. House of Representatives, H.R. 1256, June 22, 2009.

U.S. DHHS (U.S. Department of Health and Human Services), *The Health Consequences of Smoking: A Report of the Advisory Committee to the Surgeon General of the Public Health Service,* U.S. Department of Health and Human Services, Atlanta, GA, 1964.

U.S. DHHS (U.S. Department of Health and Human Services), *The Health Consequences of Smoking: Cancer A Report of the Surgeon General, U.S. Department of Health and Human Services,* Atlanta, GA, 1973.

U.S. DHHS (U.S. Department of Health and Human Services), *The Health Consequences of Smoking: Cancer; A Report of the Surgeon General, U.S. Department of Health and Human Services,* Atlanta, GA, 1982.

Vincent, J. H., *Aerosol Science for Industrial Hygienists,* Elsevier Science, Inc., Tarrytown, NY, 1995.

Webster, J., On the health of London during the six months terminating September 28, 1850, *London J. Med.,* Number XXIV, December 1850.

Chapter 7

Human Exposures to Air Pollutants

LEARNING OBJECTIVES

By the end of this chapter the reader will be able to:

- describe why inhalation is a major exposure route for environmental contaminants
- describe the major regions of the respiratory tract and the deposition of inhaled air pollutant particles of different sizes in each region
- discuss the factors that cause variability in population doses to air pollutants
- contrast the general population's exposures and workers' exposures to air contaminants

CHAPTER OUTLINE

I. Introduction: Breathing—An Old Habit
II. Respiratory Tract Compartments for Inhalation Considerations
III. Pollutant Deposition in the Body
IV. Fates of Air Pollutants in the Body
V. Population Variability
VI. Exposure in the Workplace
VII. Summary of Major Points
VIII. Quiz and Problems
IX. Discussion Topics
References and Recommended Reading

I. INTRODUCTION: BREATHING—AN OLD HABIT

Gas Exchange

The internal exposure to air pollutants is a consequence of an old habit, breathing. From the first breath to the last, and the approximately 600 million breaths over an 80-year lifespan, breathing is a necessary physiological function for humans. Breathing is so critical, that along with the heart beat, a lack of respiration for more than a few minutes usually results in death. The gas exchange functions of breathing are (1) to bring large quantities of oxygen into continuous contact with the blood, and (2) to carry carbon dioxide out of the body. In order to accomplish these functions the adult human respiratory tract has an elaborate structure that moves air to and from hundreds of millions of alveoli (tiny air sacs) where gas exchange takes place.

Other Critical Functions

There are other critical functions of the respiratory tract, including: (1) adjusting the temperature and humidity of inspired air; (2) trapping air pollutants in the upper airways, which protects the delicate alveoli; (3) eliminating volatile pollutants from the body; (4) adjusting the chemistry and viscosity of the blood (e.g., via removal of tiny blood clots); (5) eliminating excess body heat by exhaling warm air; (6) providing for the immunologic sampling of allergens; (7) detecting irritants in the air and triggering defenses; (8) generating speech and other phonation; (9) providing for olfaction (the sense of smell); and (10) dealing with inhaled threats (e.g., killing infectious agents). The respiratory system accomplishes these functions with about 40 different types of cells that are organized into dozens of specialized tissues. A failure of any of the functions of the respiratory system can impair the quality of life and also can be life threatening.

Inhaled Air Volumes

Breathing also introduces an enormous variety of inhaled pollutants to the body. **Figure 7–1** and **Table 7–1** depict the daily intakes of air, water, and food for a *reference man.* The reference man (a term used in radiation protection) is an average size worker who inhales about 23 m^3 of air daily (21 m^3 for a reference woman),

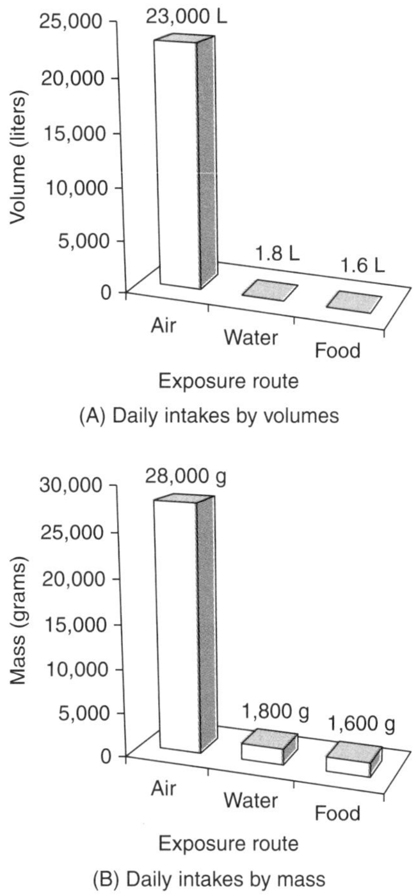

Figure 7–1 Daily intakes of air water and food for an average adult man. A: by volume, and B: by mass. *Source:* The University of California Air Pollution Health Effects Laboratory, with kind permission.

and takes in about 1.8 liters of fluid and about 1.6 liters of food in the same period. Inhaled air is the largest intake both by mass and by volume. Thus, if air, water, and food are all contaminated with 1 μg per liter of a contaminant, the daily exposure would be about 23 mg via breathing, 1.8 μg via drinking, and 1.6 μg via eating. So the daily inhaled dose could be more than 10,000 times that delivered from water or food! This dose ratio is calculated on the basis of *volumes* of air, water, and food taken in. Because air is less dense (density = 1.2 g/L) than water or food (density = approximately 1,000 g/L), if the contaminant is present at a concentration of 1 μg/g of air, water, and food, the air exposure (on the basis of *mass* intakes) would be over 10 times that for water or food. The main point is that breathing is a more efficient

Table 7–1 Average intakes for air, water, and food for an adult male (70 kg) and the dose for pollutant exposure by these routes assuming 100% uptake.

		Normal Rates of Intake	
Exposure Media/Vehicle	*Notes*	*Mass/day (grams)*	*Volume/day (Liters)*
Air	Resting	20,000	17,000
Air	8 hr light work per day	28,000	23,000
Drinking water	Normal temperatures	1,800	1.8
Food	Including water content	1,610	1.6

Data from Schleien et al. (1998).

mechanism for introducing pollutants into the body than most other mechanisms. Of course, some contaminants, especially those that do not normally become airborne, can have a greater impact on health via oral or dermal exposure routes; examples include many poisons (natural and man-made), numerous irritants (including solvents, strong acids, and strong bases), and several types of infectious bacteria and viruses. Thus, safety or toxicity testing of air pollutants must include all significant routes of exposure, not just inhalation.

Breathing volumes and capacities for an adult man are depicted in **Figure 7–2** and **Table 7–2**. The tracings are obtained using a spirometer. Figure 7–2 shows a spirometer tracing in which normal resting breathing for 24 seconds is followed by a maximum inhalation, a maximum exhalation, and then resting breathing. This type of tracing is one of the tools used by physicians to diagnose respiratory problems and by researchers to study the health effects of air pollutants. The *tidal volume* (volume of air inhaled in each normal breath) of an adult is about 500 mL. The adult's maximum lung volume is about 6 L (the *total lung capacity*), and the air volume in the lung after a maximum exhalation is about 1.2 L (the *residual volume*). The lung cannot expel all of its air, unless the organ is removed from the body and forcefully collapsed. This fact is important because the residual volume dilutes air pollutants that enter the deep lung. Such dilution serves as a protective mechanism.

The spirometric values in Figure 7–2 and Table 7–2 will vary from person to person; body size, state of health, and level of physical exertion are all major modifiers of respiratory tract volumes and capacities. Values for women, children, and various ethnic groups can be found in several references including U.S. EPA (1985),

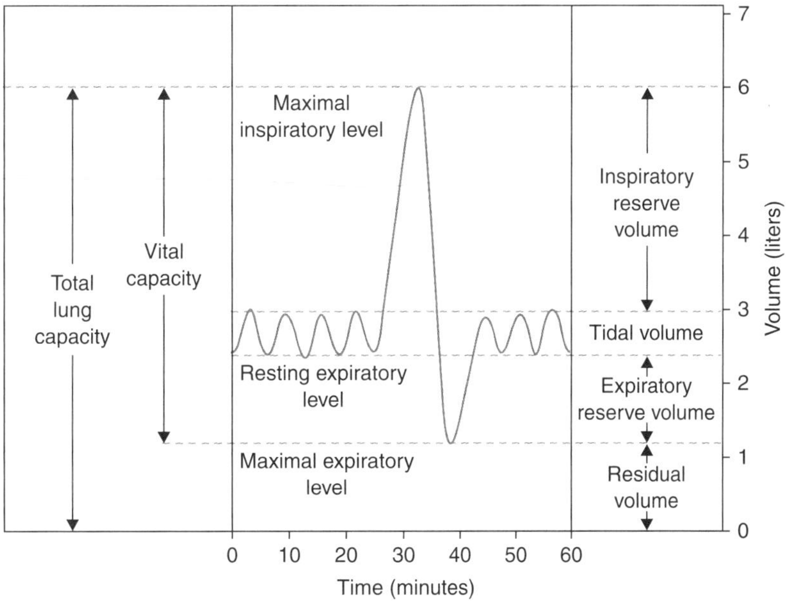

Figure 7–2 Capacities and volumes of the human lung as measured by spirometry.
Source: The University of California Air Pollution Health Effects Laboratory, with kind permission.

Table 7–2 Lung volumes and capacities for mammals, including humans.

1. *Tidal Volume:* the volume of gas inspired or expired during each respiratory cycle. Also the depth of breathing.
2. *Inspiratory Reserve Volume:* the maximum volume of gas that can be inspired from the end-inspiratory level.
3. *Expiratory Reserve Volume:* the maximum volume of gas that can be expired from the end-expiratory level.
4. *Total Lung Capacity:* the volume of gas in the lungs at the end of a maximum inspiration (the maximal inspiratory level).
5. *Vital Capacity:* the maximum volume of gas that can be expelled from the lungs by forceful effort following a maximum inspiration.
6. *Functional Residual Capacity:* the volume of gas remaining in the lungs at the resting expiratory level.

Schleien, et al. (1998), ICRP (1994, Annexe B), and NCRP (1997, Chap. 3).

Such values for respiratory tract volumes are used in risk assessments for calculating inhaled doses from air pollutants; they are also used by researchers and physicians for various purposes (e.g., measuring the effects of air pollutants, assessing lung diseases, and dispensing inhaled medicines). In the following sections we will explore the fates of inhaled air pollutants.

II. RESPIRATORY TRACT COMPARTMENTS FOR INHALATION CONSIDERATIONS

Compartmental Models

The respiratory tract can be classified as a *complex system.* Complex systems are often described mathematically by *compartmental models.* Each compartment represents a part of the whole system, and *transfer coefficients* (transfer rates) are used to describe the movements of matter (e.g., air or air pollutant particles), value (e.g., dollars or euros), or information (e.g., data) among the compartments. If the compartments and transfer coefficients are well chosen, the compartment model can be used to describe and to predict the behavior of the original complex system. Also, compartmental models isolate the important *parts* and *functions* of living and other systems, making them easier to understand. The number and types of structures selected for a given compartmental model will depend on the intended purpose of the model. If the purpose is to understand the deposition and/or health effects of inhaled air pollutants, 4 to 20 or more compartments are normally used. A key consideration is the deposition and the removal of air pollutants that are inhaled.

Pollutant Deposition and Clearance Models

For the purpose of establishing air standards, the deposition and subsequent fates of inhaled pollutants must be known or at least be estimated. Sophisticated compartmental models for the initial deposition and subsequent movement of deposited air pollutants out of the respiratory tract have been developed and used for many decades. **Table 7–3** shows four respiratory tract compartmental models used for modeling inhaled particle deposition and clearance. The models are also used for inhaled gases and vapors and for predicting the health effects of inhaled substances. **Figures 7–3** and **7–4** show a respiratory tract compartmental model used by the National Council on Radiation Protection and Measurements (NCRP, 1997) for modeling the deposition and clearance of inhaled particles. The International Commission on Radiological Protection (ICRP, 1994) compartmental model is similar but more complete in that it has more compartments. Both models require large and sophisticated computer programs to calculate the doses received from inhaled air pollutants. Both the ICRP and NCRP models were initially developed for application to inhaled radioactive air contaminants, but they are also used for non-radioactive substances in the air.

III. POLLUTANT DEPOSITION IN THE BODY

Inhaled Particle Deposition

When one takes a pill, for example a vitamin tablet, the *dose* is defined as the mass (or weight) of active ingredient (e.g., 100 mg). When inhaled air pollutants are considered, the concept of the actual *delivered dose*

Table 7–3 Compartments of the human respiratory tract used for analyzing particle inhalation.

Region	*Anatomic Structures*	*ACGIH®*	*ICRP*	*NCRP*	*TGLD*
Head airways	• Nose • Mouth • Nasopharynx • Oropharynx • Larynx	Head airways region (HAR)	Extrathoracic region (ET)	Naso-oro-pharyngo-laryngeal region (NOPL)	Nasopharynx (NP)
Tracheo-bronchial tree	• Trachea • Bronchi • Bronchioles (to terminal bronchioles)	Tracheobronchial region (TBR)	Bronchial region (BB) and bronchiolar region (bb)	Tracheobronchial region (TB)	Tracheo-bronchial region (TB)
Gas exchange	• Respiratory bronchioles • Alveolar ducts • Alveolar sacs • Alveoli	Gas exchange region (GER)	Alveolar-interstitial region (AI)	Pulmonary region (P)	Pulmonary region (P)

Data from (ACGIH®, 1985; ICRP, 1994; NCRP, 1997; TGLD, 1966).

is far more important and interesting. The simple product of air concentration *(C)*, ventilation rate *(V)*, and exposure time *(T)* is a useful, but crude estimate of the total inhaled dose (D_{TOT}), but it is not specific with respect to where the material deposits in each respiratory tract compartment:

$$D_{TOT} = C \cdot V \cdot T. \qquad \text{(Eq. 7–1)}$$

The dose, D_{TOT}, is in mass units (e.g., mg), *C* is in mass per unit volume of air (e.g., mg/L), *V* is in volume breathed per unit time (e.g., L/min), and *T* is in units of time (duration of exposure). The equation is useful, but several facts make Equation 7–1 an inadequate measure of the inhaled doses from air pollutants.

First, not all of the particulate pollutants originally in the inhaled air volume actually *enter* the nose or mouth.

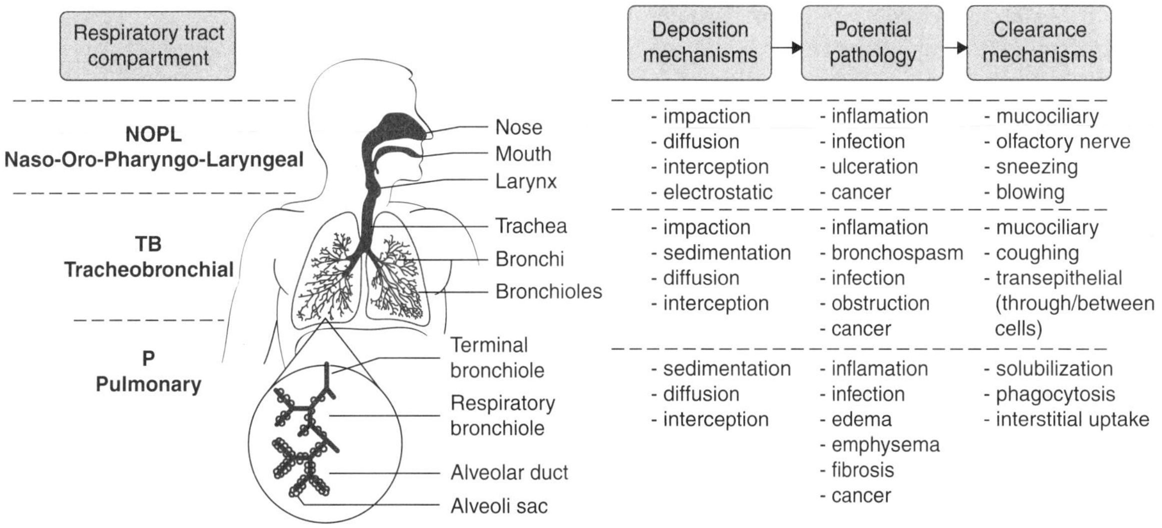

Figure 7–3 Compartmental model of the NCRP (1997).
Source: The University of California Air Pollution Health Effects Laboratory, with kind permission.

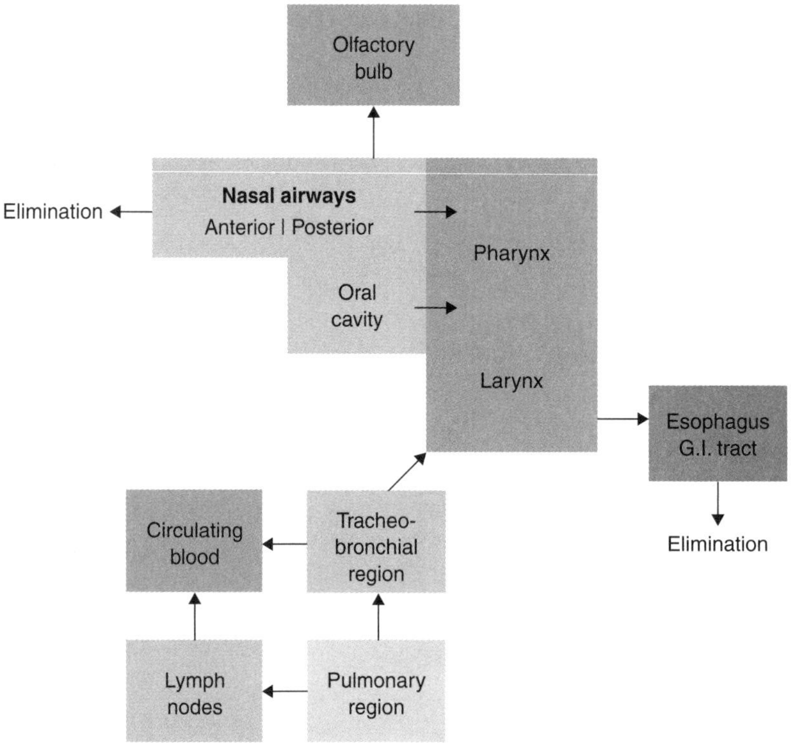

Figure 7–4 Compartmental model of the NCRP (1997) used for describing the clearance of inhaled, deposited particles. *Source:* The University of California Air Pollution Health Effects Laboratory, with kind permission.

This is mainly because very large particles are falling through the air (due to gravity) so fast that some of them cannot be drawn upward into the mouth or the nostrils. The fraction of particles actually inhaled is called the *inhalability (I),* which varies from 1.0 (100 percent) for small particles to 0.5 (50 percent) for larger particles up to 100 μm in diameter (actually, the *aerodynamic equivalent diameter;* see Chapter 3). The inhalability is known from tests in wind tunnels using adult manikins. These tests are conducted in low wind speed conditions, with the wind direction averaged over a complete 360° rotation of the manikin. This concept of inhalability was originally developed for adult male workers, but it is probably valid for most members of the general population. The inhalability changes if there is no relative motion of the person to the wind, wind speeds are high, or the person is lying down.

Second, not all inhaled particles and gases deposit in the respiratory tract, as some fraction may be exhaled. Thus, for calculating inhaled doses, the efficiency *(E)* of deposition in each compartment of the respiratory tract should be taken into account: E_{NOPL} (for the nasal, oral, pharyngeal, and laryngeal regions), E_{TB} (for the tracheobronchial region), and E_P (for the pulmonary region). There are now three particle deposition doses to calculate; one for each respiratory tract compartment:

$$D_{NOPL} = C \cdot V \cdot T \cdot I \cdot E_{NOPL} \quad \text{(Eq. 7–2)}$$

$$D_{TB} = C \cdot V \cdot T \cdot I \cdot E_{TB} \quad \text{(Eq. 7–3)}$$

$$D_P = C \cdot V \cdot T \cdot I \cdot E_P. \quad \text{(Eq. 7–4)}$$

The deposition efficiencies of particles in each compartment, where *I* and *E* are unitless ratios or percents, depend on particle size, as shown in **Figure 7–5** for the reference man. Values for women, children, and some laboratory animals are also obtainable from laboratory measurements, or by using computer modeling software. Figure 7–5 is corrected for inhalability *(I).* It is clear that some particles are exhaled, and thus do not deposit in the body.

Inhaled Dose per Unit Body Mass or Airway Surface

Equations (7–2) to (7–4) provide the total particulate pollutant deposition in three respiratory tract compart-

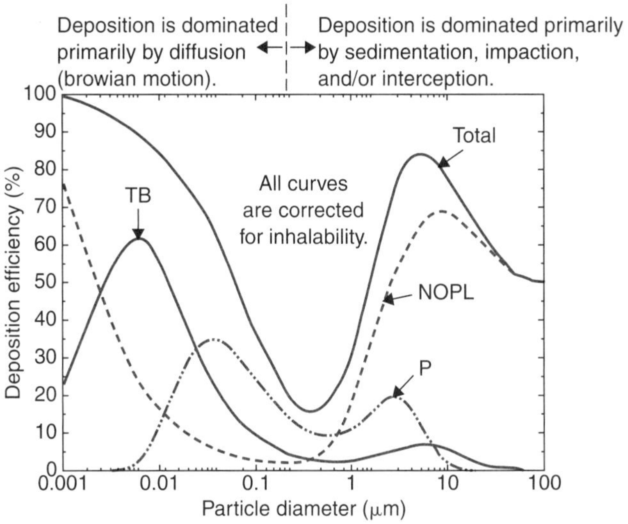

Figure 7–5 Particle deposition in the major regions of the human respiratory tract during normal respiration corrected for the size-dependent inhalability. *Source:* The University of California Air Pollution Health Effects Laboratory, with kind permission.

ments. However, the *biological effects* may depend on the deposition dose per unit of body mass (M_B). For example, the same amount of air pollutant deposition may have more impact on a tiny child than an adult. So, for an individual the effective dose (D_{EF}) might be

$$D_{EF} = D/M_B, \qquad \text{(Eq. 7–5)}$$

where M_B is in units of mass (or weight, if gravity is present). Alternatively, the effective dose in a specific respiratory tract compartment may depend on the pollutant deposition per unit surface area (S) of that compartment. The deposition per unit surface area tells us what the individual cells are exposed to. For the tracheobronchial region the effective dose may depend on the total deposited dose (D_{TB}) divided by the surface area of the tracheobronchial airways (S_{TB}):

$$D_{EF,TB} = D_{TB}/S_{TB}. \qquad \text{(Eq. 7–6)}$$

As an example, consider a nose-breathing resting adult (V=10 L/min), inhaling some standard density (1g/cm^3) particles with an average diameter of 5 μm. The inhalability is 0.87 (87 percent), the deposition efficiency in the pulmonary compartment (E_P) is about 0.14 (14 percent), and the surface area of the pulmonary compartment is enormous, about 70 m^2. If the air concentration of the particles is 1 μg/m^3 and the duration of exposure is 10 hours, the deposited effective dose $(D_{EF,P})$ in μg of particles per m^2 of pulmonary tissue surface is

$$\begin{aligned} D_{EF,P} &= 1\,\mu g/m^3 \cdot 0.01\,m^3/min \cdot 600\,min \\ &\quad 0.87 \cdot 0.14/70m^2 \\ &= 0.01\,\mu g/m^2 \qquad \text{(Eq. 7–7)} \end{aligned}$$

Although this surface dose is small, if the particles are very toxic or consist of infectious bacteria the adverse health impact could be serious. On the other hand, this single dose will be inconsequential for healthy adults breathing most types of particles found in the air. Whether or not an individual is adversely affected by a given pollutant will depend on his/her *susceptibility* to harm in relation to the doses received. Note that the above dose computation ignores the effects of particle clearance, which reduces the dose over time. Also, the lung defenses can deal with many toxic chemical substances, and it can inactivate most microorganisms. So the deposition of inhaled particles is not necessarily harmful to human health. An average breath will contain about a million particles, of which we are unaware, and which does not harm the health of most people.

Inhaled Gases

Gas molecules rapidly diffuse in random directions due to the continual bombardment by other molecules. This motion (i.e., diffusion) causes gas molecules to contact respiratory tract surfaces, where they might be absorbed. Much is known about the deposition doses of inhaled air pollutant gases and vapors. The major biological factors are (1) airway anatomy, (2) mode of breathing (nasal, oral, or a combination of nasal and oral), and (3) breathing air flow rate. In addition, the solubility of the gas in water is a factor that determines how much will deposit in each compartment of the respiratory tract. Gases that are highly water-soluble, such as ammonia, sulfur dioxide, and formaldehyde, will have high deposition efficiencies in the nose, mouth, and large airways. Gases that are poorly-soluble in water, such as oxygen, ozone, nitrogen, and carbon monoxide, will saturate the upper airways, and thus have a greater relative deposition deep in the respiratory tract. Thus, water-soluble gases will tend to exert their effects, if any, on the upper airways (and the eyes), while poorly water-soluble gases may impact the deep lung. The body in general may also be affected if the gases are taken up and distributed by the blood. Dose modeling for inhaled gases is discussed in some detail in the ICRP (ICRP, 1994) model, and in Chapter 3. As a note, *vapors,* which are produced by volatile liquids (e.g., gasoline, water, and paint-thinners) and solids (e.g., naphthalene and phenol) are indistinguishable from true gases for deposition modeling purposes.

Complicating Factors in Inhaled Particle and Gas Deposition

Much more could be presented on the *dosimetry* (the measurement and calculation of dose) of inhaled air pollutants, including: (1) individual subject differences (which are significant); (2) the effects of mixtures of particles and gases (which is quite complicated); (3) the effects of respiratory tract diseases; and (4) the effects of physical exercise. Such factors must be taken into account by health and safety professionals. However, for the purposes of this textbook, the basic principles have been presented.

IV. FATES OF AIR POLLUTANTS IN THE BODY

Introductory Comments

Once deposited, the initial fates of air pollutants in the body depend on several factors, including: (1) where they have deposited; (2) their toxicity; (3) their solubility (actually, for particles, their rates of dissolution) in respiratory tract and body fluids and tissues (e.g., mucus, blood, fat, protein, and pulmonary surfactant); (4) the health status of the individual; (5) their metabolism (chemical transformations) in the body; and (6) several other characteristics. Accordingly, what follows is simplified in order to demonstrate the basic principles that apply to public health.

In order to illustrate the wide range of differences in toxicity of some sample air contaminants it is useful to examine the *exposure guidelines* (that determine permissible inhaled doses) established for workers. Workers usually breathe specific air pollutants at higher concentrations than the general population. **Table 7–4** shows some of the air concentrations recommended for workers (e.g., Threshold Limit Values, TLVs®) by the American Conference of Governmental Industrial Hygienists that are well-tolerated by most workers for 8 hours/day, 5 days/week, 50 weeks/year for several years. The table also shows other workplace air recommendations, such as short-term emergency concentrations, and the type of air sample that should be obtained. Note that there are no simple rules for defining acceptable exposures, these concentrations may not protect all workers, and that they are definitely not applicable to the general population. Permissible air concentrations for the general population, which includes children, the elderly, and the ill, will generally be lower (i.e., more strict) than those for healthy workers. Although the permissible concentrations may be lower, it is still anticipated that the relative toxicities will be similar for the general population. The overall variation in relative toxicity can be seen in Table 7–4, with the permissible concentrations ranging from 0.025 mg/m³ for silica to 1,000 ppm (1,890 mg/m³) for ethanol.

Table 7–4 Examples of Threshold Limit Values (TLVs®) for air quality in workplaces.

Substances	*TWA*	*STEL*	*Sample Type*
Aluminum	1 mg/m³	—	R
Asphalt fume	0.5 mg/m³	—	I
Calcium sulfate	10 mg/m³	—	I
Flour dust	0.5 mg/m³	—	I
PVC	1 mg/m³	—	R
Silica	0.025 mg/m³	—	R
Sulfuric Acid	0.2 mg/m³	—	T
Zinc Oxide	2 mg/m³	10 mg/m³	R
Acetone	500 ppm	750 ppm	—
Ammonia	25 ppm	35 ppm	—
Chlorine	0.5 ppm	1 ppm	—
Ethanol	1,000 ppm	—	—
Phosgene	0.1 ppm	—	—
Vinyl chloride	1 ppm	—	—
Xylene	100 ppm	—	—

TWA = Time Weighted Average; STEL = Short term Exposure Limit; I = Inhalable Particulate Matter sample; T = Thoracic Particulate Matter sample; R = Respirable Matter sample; ppm = parts per million.
Data from ACGIH® (2010).

Fates of Deposited Particles

Particles that Dissolve Rapidly in the Body

Particles that are rapidly dissolved in respiratory tract fluids (which are solutions and suspensions in water) have several possible fates. They can enter the respiratory tract tissues or the circulating blood and lymphatic fluid, where they enter the metabolic environment of the body. Such particles are said to be *soluble, unstable,* or *rapidly dissolving.* Particles that are very tiny have large *specific surfaces* (their surface area divided by their mass), which increases the *rate* at which they dissolve in the body, even if they are made of substances that are normally not very water-soluble. *The equilibrium solubility* in pure water (as found in chemistry books) will generally underestimate the rate of dissolution in the body. Even metals, such as lead and silver, that do not dissolve when large pieces are placed in water, can quickly dissolve in the lung if they are present as tiny particles (see Figure 3–5). Some dissolved substances will attach to blood-borne proteins and be taken up by the liver and other organs and tissues. They may reside there for long periods, or be transformed and/or eliminated, for example by the kidneys, liver, or intestines. Some inhaled substances, lead or mercury, for example, may be retained in the nervous system. If the amounts retained are large enough, then they may produce debilitating diseases. The "Mad Hatter" in Lewis Carroll's *Alice in Wonderland* represents the perils of significant mercury poisoning (mercury was used to prevent mold growth on wet felt). If the amounts of the retained pollutants are sufficiently low, no adverse effects will be seen. Remember, in toxicology, Paracelsus taught that "The dose makes the poison."

Particles that Dissolve Slowly in the Body

Particles with slow dissolution rates (also called *persistent, poorly-soluble, stable,* or *insoluble*) in lung fluids are subjected to several clearance mechanisms (Figure 7–3). The specific particle clearance mechanisms will depend on: (1) the particle sizes; (2) regions in which they deposit; (3) the exposed animal species (e.g., rat, mouse, dog, human, etc.); (4) the particle's composition and toxicity; and (5) the state of health of the exposed subject. In addition, if excessively large amounts of persistent particles deposit in (and effectively clog) the alveoli, clearance stasis (lack of clearance) may occur. Without clearance, the lung disease *pneumoconiosis* (meaning "*lung dust disease*") may be initiated and possibly lead to debilitation and death. One form of pneumoconiosis is *black lung* (i.e., coal miner's disease), which, along with the disease *silicosis,* still claims many workers lives every year.

In the *nose,* persistent particles that deposit at or near the nasal entrances (nares) are generally cleared by so-called "extrinsic" means, such as sneezing, blowing, dripping, and wiping. The reason these mechanical mechanisms of clearance are necessary is that the first third of the nasal cavity in humans is lined by a type of epithelium (lining cells) similar to the facial skin, and thus not covered by a rearward moving blanket of mucus. Deeper in the nose, the epithelial lining has mucus-secreting cells and glands (**Table 7–5**). The ciliated cells (cells that have tiny hair like projections) propel mucus toward the throat, where it is swallowed.

Table 7–5 Simplified cellular and other components of human airways.

Region	*Surface Cells*	*Secretory Cells & Glands*	*Smooth Muscle*	*Innervation*
Nose				
a. anterior	squamous	numerous	absent	present
b. posterior	mucociliary	numerous	absent	present
Larynx	mucociliary	present	absent	present
Trachea	mucociliary	present	present	present
Bronchi				
a. large	mucociliary	present	present	present
b. small	mucociliary	few	present	present
Bronchioles	mucociliary & some alveolar	few	present	present
Alveolar Ducts	mucociliary & alveolar	surfactant cells	present	present
Alveoli	alveolar & macrophages	surfactant cells	present	present

Therefore, most particles that land on the mucus are transported and swallowed. Very tiny particles (e.g., a few nm in diameter) deposited in the olfactory (smell-receptor) region in the rear upper portion of the nose can travel up the olfactory nerve directly to the brain. Although the importance of this transport to the brain on health is not yet clearly understood, it is of current interest to air pollution toxicologists and regulators.

The post-nasal and post-oral passages, called the *pharynxes,* are lined by mucociliary epithelium. Poorly-soluble particles that deposit in this region are moved by mucus to the throat where they are swallowed and enter the gastrointestinal (GI) tract. Once in the GI tract, the particles are generally expelled from the body, but some may be absorbed in the intestines and taken up in the circulating blood.

The *trachea* (windpipe) and its branches *(bronchi and bronchioles)* are lined by mucociliary epithelium. In this region, mucus is also moved toward the throat, where deposited poorly-soluble particles are swallowed. In healthy mammals, including humans, it is generally assumed that poorly-soluble particles deposited in this region are completely cleared by swallowing and passage into the GI tract within about 24 hours. However, a portion of the deposited particles appear to be retained for longer periods. This effect, called "slow tracheobronchial clearance," increases as particle size decreases. So, half of the insoluble particles less than 4 μm in diameter that deposit in this region might be retained for several days to several weeks. Recently, the magnitude of the slow bronchial clearance effect has been challenged, so it should be applied to risk assessments with great caution, if at all. In addition, respiratory tract infections (including the common cold) and some toxic exposures can damage ciliated cells and inhibit normal mucus flow for several weeks post-illness. In this case, mucus build-up stimulates a residual lifesaving *cough.* Several "coughing fits" may be necessary to clear the tracheobronchial airways of excess mucus. For deep-lying mucus, 10 or more rapid coughs may be needed to clear the mucus. If cough is suppressed, mucus may plug small airways and lead to deep-lung infections (potentially life-threatening conditions). Although coughing is annoying, it is an important mechanism for preserving health when the mucociliary clearance system fails.

In the deep *alveolar region* of the respiratory tract, particle clearance is less well understood. The known mechanisms include: (1) dissolution in surface fluid (i.e., alveolar surfactant); (2) engulfment and removal by macrophages (large mobile cells that engulf particles and liquids); and (3) uptake by non-mobile alveolar cells, with possible transport to the surrounding tissues, blood, and lymphatic fluids. *Macrophages* (the name means "big eater") are not only effective in maintaining the sterility of the deep lung by killing microorganisms, but they are also capable of *harming* the delicate alveolar structure. It is the strong chemicals (including hydrogen peroxide) in macrophages that both kill bacteria and sometimes damage valuable lung tissue. In prolonged heavy cigarette smoke exposure, the induced increased number of alveolar macrophages is believed to participate in producing emphysema (a disease characterized by permanent loss of delicate deep lung tissue). Certain types of particles, such as some forms of quartz and asbestos, are particularly toxic to macrophages in high doses, possibly leading to lung diseases including fibrosis (scarring), emphysema, and pneumoconiosis. Persons with such deep-lung diseases may require supplemental oxygen in order to survive and function.

In summary, the healthy respiratory tract has many effective mechanisms for removing deposited particles. However, large doses and particle toxicity can overwhelm the protective mechanisms and lead to poor clearance, and even produce serious disease states. The particle doses required to produce diseases can be expected to vary in humans due to individual susceptibility factors (including health status and genetic make-up).

Species Differences

There are known species differences in the efficiency of particle clearance from the respiratory tract. In general, rats and mice have more rapid particle clearance than do humans, dogs, guinea pigs, and hamsters. On the other hand, rats are very susceptible to developing deep lung clearance failure (also called "clearance overload") when particle deposits are heavy. Experts are required to interpret inhalation studies conducted in laboratory animals, especially studies with high concentrations of particles in rats. Although laboratory animal studies are essential in air pollution toxicology, species differences must be understood and taken into account.

More on Fates of Inhaled Gases

The fates of inhaled gases (including vapors) mainly depend on several factors, including: (1) their chemical reactivity with tissue components; (2) their solubility in respiratory-tract fluids; (3) their relative solubility in air, water, and lipids (e.g., fats in the body); and (4) the presence of particles that can carry adsorbed or absorbed gases. Reactive gases (such as ozone and sulfur dioxide) will react chemically after contact with surfaces in the respiratory tract. As a result of these reactions, the gas will not only be transformed but potentially induce harmful chemical changes in the local tissue (see Chapters 8, 9, and 10).

The deposition and distribution of inhaled gases was covered in Chapter 3. Here, we will briefly review that material. Inhaled gas molecules rapidly diffuse to airway walls. For reacting and non-reacting gases, their water solubility determines where they will initially deposit in the respiratory tract. Highly *water-soluble gases* (e.g., sulfur dioxide, carbon dioxide, and ammonia) will be effectively taken up by tissues in the nose, mouth, throat, larynx, and major bronchial tubes (such as the trachea and first few generations of the bronchi). It is in these tissues that the gases may trigger reflexes and thus exert their effects, such as initiating changes in breathing patterns, and producing cough, excess mucus secretion, and bronchial constriction (e.g., as in an asthma attack). Again, the dose delivered and sensitivity of the respiratory tract will determine the severity of the response. Persons with asthma and/or bronchitis (a state of chronic inflammation) may respond to lower concentrations of air pollutants than do healthy individuals.

In contrast, gases *poorly-soluble* in water will *not* be taken up avidly in the upper airways, and will therefore flow into to the small bronchial airways and the alveoli. Ozone, for example, will produce its damage largely in the terminal bronchioles and alveoli, and in sufficient concentrations kill cells and even produce deep-lung edema (leakage of fluid into the airspaces of the alveoli). Low concentrations of ozone (e.g., less than 0.1 ppm) will produce minor effects that are repairable. High concentrations (e.g., approaching 1.0 ppm or more) produce serious acute effects throughout the respiratory tract due to the killing of cells, damaging cilia, destroying surfactant, and disrupting barriers to fluid leakage.

Gases of *intermediate* water solubility are capable of depositing throughout the respiratory tract. Such gases must be considered as being capable of producing a variety of adverse effects if they are inhaled in sufficient concentrations. Also, mouth breathing, as occurs during heavy exercise or nasal congestion, bypasses the nose (which is especially efficient in scrubbing water soluble gases) and allows for deeper penetration of many types of gases.

As covered in Chapter 3, the relative solubility of non-reacting gases in air, water, and lipids determines their post-deposition distribution and clearance from the various organs of the body. For example, a gas (such as the anesthetic ether) will rapidly enter the aqueous portion of the blood and then accumulate in the fatty tissues (such as are found in the brain and body fat). In the brain, the anesthetic gas in sufficient concentrations can produce unconsciousness (and death if the dose is too high). When the external air concentration is reduced to zero, the anesthetic gas is depleted rapidly from the brain but only slowly removed from the fatty deposits in the rest of the body. This is but one of many examples of how the relative solubility of a gas in air, water, and lipids determines its uptake, distribution, and clearance.

ADMSE and PBPK Models

For inhaled deposited materials (including particles and dissolved gases) that leave the respiratory tract to enter other organs and tissues, sophisticated mathematical models are used to describe their absorption, distribution, metabolism, storage, and elimination (ADMSE, also see Chapter 9). Such models, which are useful for extrapolating results from one species to another, include *PBPK* (physiologically-based pharmacokinetic) models. These and other models are used by the U.S. EPA for extrapolating animal study results to humans, and for establishing inhalation RfCs (reference concentrations) to adjust for species differences.

From the foregoing material, one can appreciate the large scientific effort on the part of regulators and other health professions that goes into the research and analysis used in estimating exposures, and in setting air quality and exposure standards.

V. POPULATION VARIABILITY

How Variable is Exposure?

When considering internal exposures to air pollutants, the observed human population variability must be considered. Individual differences in anatomy, physiology, health status, and exposure locations lead to a very wide range of exposures and internal doses of air pollutants.

When Dr. Richard Cuddihy and colleagues in Albuquerque, NM measured the titanium and aluminum content of the internal organs of humans, they found a wide range of levels. These soil-derived metals

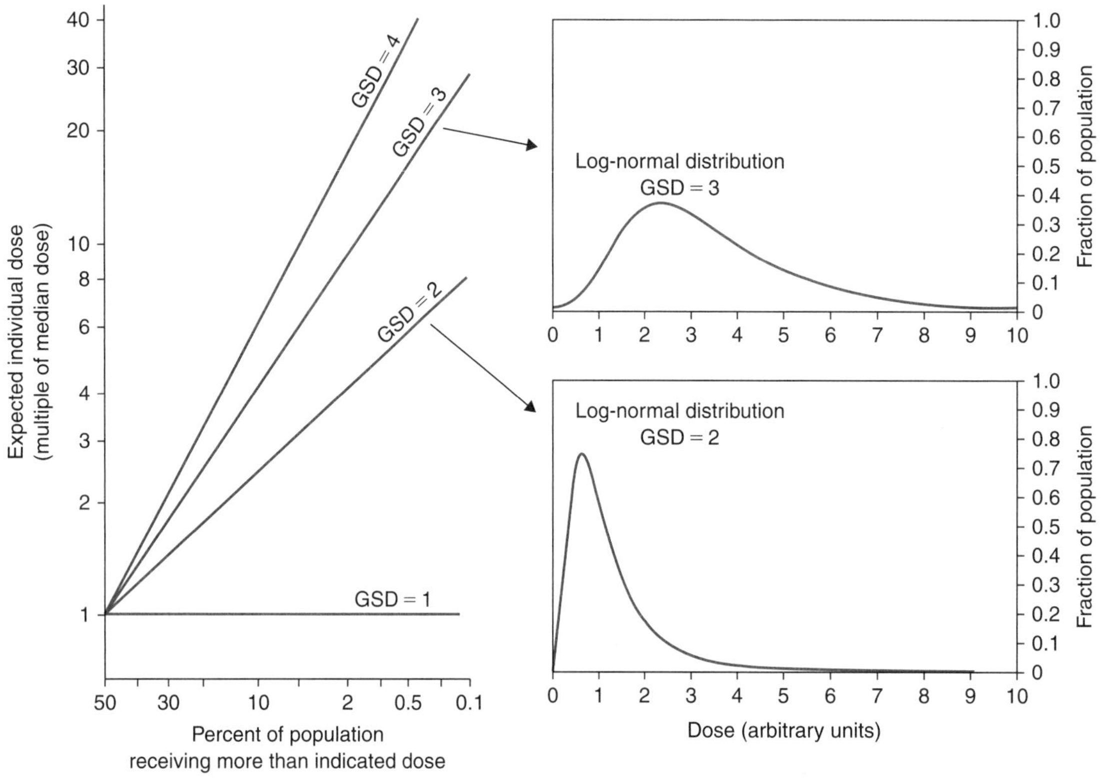

Figure 7–6 Distribution of individual organ doses (as a function of geometric standard deviation, GSD) in a population exposed by inhalation or ingestion to a material that accumulates in tissue. Lognormal distributions are assumed to be valid. See Chapter 3 for information on lognormal distribution and GSD.
Source: The University of California Air Pollution Health Effects Laboratory, with kind permission.

were assumed to have been inhaled in approximately equal air concentrations by everyone, as they lived in the same area. However, the variations in lung deposition and post-deposition kinetics were great enough to indicate that ⅙ of the population had an internal exposure 340 percent above the median level (half of the population is above the median and half is below), and 1 percent of the population had a predicted exposure over 10 times (1,000 percent) the median value! The range of varied population organ doses is shown in **Figure 7–6**. The implication is very clear; some individuals in a population will receive comparatively large internal doses (e.g., doses to various organs) of air pollutants. Pharmacokinetic variations in the absorption, distribution, metabolism, storage, and elimination of pollutants only partially explain the variability of population doses. Exposure studies have clearly demonstrated significant variations in doses are also based on proximity to sources (e.g., living near busy streets and roads). The effects of exercise include increased oral breathing and ventilation, which also greatly increase the internal exposure to particles and gases.

There are several implications of such wide population variability in internal doses. First, accepted and enforceable air quality standards may not protect everyone, so individual protective actions, such as reducing exposure by avoiding exercise and/or staying indoors on some days, may be necessary. Second, extremely-sensitive people may even require masks or medications during a "smog alert." Finally, it is important to define susceptible individuals through research, so that they can be warned, or otherwise protected.

Where the Exposure Occurs and the Personal Cloud

Air concentrations of contaminant particles and gases are highly variable with respect to both location and time. Proximity to contaminant sources were previously mentioned as an important factor that affects personal exposures. The term "*personal cloud*" refers to the generation of contaminants in the immediate vicinity of a person as a result of their own activities. Such activities include vacuuming, smoking, cooking, gardening, and working in a dusty trade, to name a few. Figure 2–10 depicts some personal-cloud exposures in homes.

Personal-cloud exposures are usually difficult to measure, as they are not captured by *stationary* indoor or outdoor air samplers, and they depend on just how and where activities are performed. For example, a person digging into a backyard compost pile will have greater exposure (to soil, plant debris, microorganisms, spores, etc.) if they are working downwind, the compost is dry, and the digging activity is vigorous. Conversely, working upwind with damp compost may produce almost no exposure at all. Similarly, painting in a small enclosed space leads to a larger exposure than painting outdoors with a wind that blows the volatile paint emissions away from the user. As an aside, significant personal exposures can affect the embryo or fetus during pregnancy. Women who are pregnant should avoid painting without copious ventilation; even "non-toxic" paint may contain potentially toxic volatile substances before it is completely dry. Local meteorologic factors (wind, humidity, etc.) are clearly important exposure modifiers. Dust and vapor masks and/or substantial ventilation are needed when performing activities that generate substantial exposure to air contaminants.

Proximity to Significant Pollutant Sources

Proximity to non-personal sources of air contaminants deserves emphasis as an exposure modifier. If one is downwind and near heavy traffic, a significant combustion source (such as waste burning), or an active construction site, the exposure can be much greater than that measured by a distant air monitoring station. Interestingly, the exposure is generally less if the wind velocity blowing *from* the source to the exposed subject is high. This is because the wind not only disperses, but also dilutes the air contaminants. Outdoors, near-stagnant or stagnant air is usually associated with greater local air concentrations for a given source strength. When large *area sources,* such as cities, farms, or industrial regions are considered, local winds may produce significant downwind air concentrations, especially if a local *air inversion* traps the air contaminants near the ground. As was presented in Chapter 1, the historic major "air pollution disasters" occurred in heavily industrialized regions during air inversions, high humidity, and stagnant wind conditions.

Geographical Factors

As is discussed in Chapter 5, different geographical locations are also associated with different air contaminant exposures. For example, the Southwestern United States, which is generally dry and often windy, has elevated levels of soil-derived dust in relation to the Northeastern United States, where coal combustion, high humidity, and lower wind speeds are common factors. In the Northwestern United States, wood burning in homes is an additional contributor to local indoor and outdoor air quality. Such regional differences are sometimes cited in arguments against the use of uniform national particulate mass standards.

Comments

Many other examples could be presented, but the principles are clear. Exposures are highly variable, in large part due to individual characteristics, personal activities, local source strengths, proximity to sources, and meteorological conditions.

More on How Biological and Physiological Factors Influence Exposures

Individual Variations

Individuals differ greatly in their deposition efficiencies of inhaled particles and gases in the compartments of the respiratory tract. If the population variability is examined for the deposition of 3 μm diameter particles in the NOPL compartment, the range would be approximately 0.01 to 0.90 (1 percent to 90 percent deposition). In simpler terms, some people (as well as some races) have noses that trap particles efficiently, while others do not.

Small Children

Small children, like all small mammals, rapidly lose body heat to the environment, and thus must breathe more air per unit body mass in order to maintain a normal body temperature (**Table 7–6**), so they are expected to have greater tissue concentrations of inhaled substances than do adults. As an example, a newborn child at rest must breathe more than three times the volume of air per unit of body mass (or unit of body weight) than an adult. This greater exposure of children is enhanced by their tendency to exercise more than adults. Thus, air

Table 7–6 The effect of age on body size and ventilation (L/min).

Personal Characteristics			Minute Ventilation (liters)*		
Age (years)	*Body Mass (kg)*	*Height (cm)*	*Low Activity*	*Light Exertion*	*Heavy Exertion*
0	3.3	50	1.52	3.00	8.92
2	13.0	88	2.75	5.48	16.4
4	16.4	104	3.18	6.34	19.0
6	22.0	115	3.89	7.77	23.2
8	27.0	127	4.53	9.05	27.1
10	34.0	138	5.42	10.8	32.4
12	43.0	150	6.56	13.1	39.3
14	54.0	162	7.96	15.9	47.8
16	63.0	170	9.10	18.2	54.6
18	70.0	175	10.0	20.0	60.0

*Terminology used for describing ventilation is arbitrary as no standard accepted definitions exist.
Data from NCRP (1997).

standards for protecting adults may not be acceptable for very small children, such as newborn infants.

Potentially Susceptible Populations

Given the large individual differences in deposited doses of inhaled air pollutants, it is not surprising that some people have great difficulty during periods of high air pollution, while others are not affected at all. The U.S. EPA considers *potentially susceptible subpopulations* in setting National Ambient Air Quality Standards (NAAQS). Such populations include the very young, the elderly, and persons with chronic heart and/or lung disease (see Chapter 10). For the most vulnerable asthmatics, a single inhaled pollen grain may produce discomfort, so even strict air standards will not be able to protect them from harm. It is also important to consider the need for maintaining good general health, as well as healthful indoor air quality for vulnerable individuals.

VI. EXPOSURE IN THE WORKPLACE

Introduction

Because many chemical and industrial workers traditionally work in elevated levels of air contaminants, considerable research and effort have been directed toward worker protection. In addition, occupational exposures are usually to a limited number of known substances, so defining the exposure is simplified in relation to the general population. Although this section describes exposures that apply to workers, many of the same principles apply to other exposure scenarios, especially community exposures to airborne contaminants following a catastrophic release.

Exposure Characteristics

Workplace exposures to air pollutants are generally assumed to be 8 hours/day, 5 days/week, 50 weeks/year, for 20 to 40 years. **Table 7–7** provides a comparison of typical workplace exposures with several other exposure scenarios. Workplace co-stressors can also modify the effects of air pollutants. Such co-stressors include noise, vibration, temperature extremes, exertion (possibly extreme), repetitive motion, ergonomic stress (relating the potentially stressful interaction of the human body with objects and activities), and safety-related factors. As an example, consider construction workers. They can be exposed simultaneously to high levels of dust, gases, vapors, significant vibration (from equipment), excessive noise, extreme heat and cold, heavy and repeated lifting, and physically awkward operations. Such workers also face accidents that include falls, lacerations, burns, and crushing injuries. Although construction work is known for its risks, several other occupations have similar co-stressors. Such factors can both amplify and mask the adverse health effects produced by inhaled particles and gases. Maintaining a healthy

Table 7–7 Exposure characteristics for various scenarios.

Exposure	*Population*	*Duration*	*Concentrations*	*Co-stressors*
Workplace	Healthy adults	8 h/d, 5 d/wk, 50 wk/yr, 20–40 yrs	High, limited to a few substances, relatively high concentrations	Heat, cold, vibration, exertion, noise
Environmental	All ages, all states of health	24 h/d, lifetime	Low, a large number of substances, variable concentrations	Heat, cold, brief periods of exertion, disease
Vehicular	All ages, all states of health	1–3 h/d, 6 d/wk, 52 wk/year, lifetime	Mixed: CO and other exhaust components	Fatigue, low RH if air conditioned
Schools (public)	Ages 4–18 yrs, usually healthy, some chronic and acute illnesses	6 h/d, 5d/wk, 40 wk/year 12–13 years	Low to modest, mostly indoor pollutants	Periodic exercise, crowding
Organized Athletics	Healthy youth and young adults	1–4 hr/d, 1–5 d/wk, 20–50 wk/yr, 5–30 years	Low indoor & outdoor levels	High levels of exertion, temperature extremes

h = hour; d = day; wk = week; yrs = years.

workforce has spawned several certified specialties including industrial hygienists, safety professionals, occupational physicians, and occupational health nurses. In addition, researchers specializing in inhalation toxicology, stress-physiology, exercise-physiology, and respiratory protection, study air contaminant exposures and their effects on worker health.

Inhaled Dose vs. Exposure Dose

Instead of using compartmental inhaled particle *deposition modeling* (as depicted in Figure 7–5), the ACGIH® uses *exposure modeling* when recommending permissible air concentrations in workplaces. Although the recommended concentrations do not have the force of law, they are respected and widely-used by industrial hygienists for worker protection. Similar permissible air concentrations are established by the Occupational Safety and Health Administration (OSHA), the National Institute for Occupational Safety and Health (NIOSH), the U.S. EPA, and by the Federal Republic of Germany (DFG), to name a few. For airborne substances, the ACGIH® recommends TLVs® (Threshold Limit Values) of three types: (1) TWAs (time-weighted averages, averaged over 8 hr/d, 40 hr/week, working lifetime exposures) that prevent adverse effects in "nearly all workers"; (2) STELs (short-term exposure limits) that prevent irritation, chronic or irreversible tissue damage, dose-rate dependent toxic effects, or narcosis (that could impair rescue efforts); and (3) C (ceiling) values that should not be exceeded, even briefly.

For particles, the TLVs® consider three particle aerodynamic size ranges: (1) *inhalable particulate matter* (IPM) that is hazardous when deposited anywhere in the respiratory tract; (2) *thoracic particulate matter* (TPM) that is hazardous only when deposited in tracheobronchial or alveolar airways; and (3) *respirable particulate matter* (RPM) that is hazardous only when deposited in the alveoli. As shown in **Figure 7–7,** each of these TLVs® has particle size ranges that can be measured using specific air samplers that have particle collection efficiency curves that mimic human respiratory-tract exposures, based on the IPM, TPM, and RPM criteria. In contrast, the U.S. EPA uses respiratory tract regional deposition considerations in considering particle mass fraction standards for environmental aerosols. The size

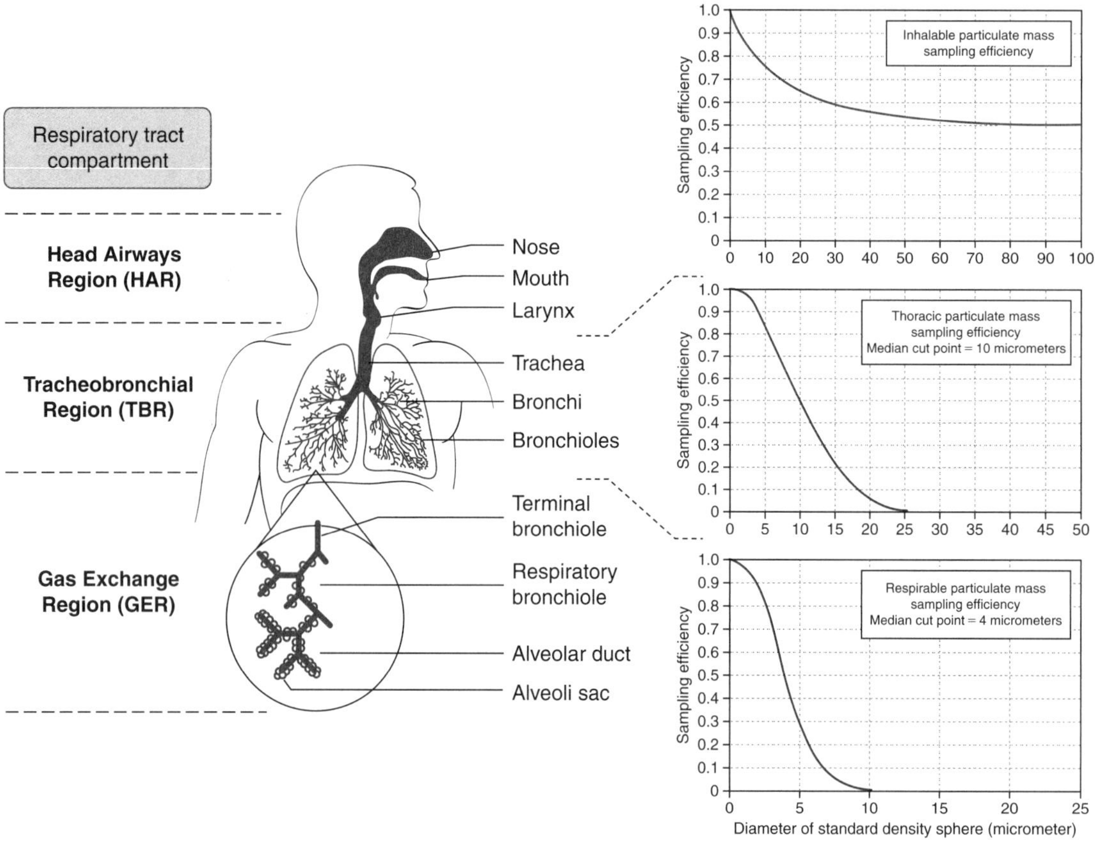

Figure 7–7 The three aerosol mass fractions recommended for particle size-selective sampling by the ACGIH®. Note that the respiratory tract compartments used here for worker protection differ in terminology from those in Figure 7-3. *Source:* The University of California Air Pollution Health Effects Laboratory, with kind permission.

ranges are PM_{10} (particles under 10 μm in diameter); $PM_{2.5}$ (fine particles, under 2.5 μm in diameter); and PM_{10} to $PM_{2.5}$ (coarse particles). Environmental samplers are used to determine whether or not the air quality is acceptable for each of the particle size ranges. However, the EPA's air samplers are based on particle *deposition,* while the workplace samplers are based on inhaled particle *exposure* for regions of the respiratory tract.

Exposure Control Methods Used to Protect Workers

In order to maintain acceptable air concentrations, three basic methods are used in the workplace: (1) *engineering controls* such as the substitution of less hazardous materials or enclosing processes that produce air contaminants; (2) *administrative controls* such as restricting worker access to hazardous processes; and (3) *personal respiratory protection* (use of respiratory masks, full-face respirators, air supplied hoods, and air supplied total-body suits). The purpose of these control measures is to ensure that concentrations of air contaminants in worker's breathing air are harmless for the majority of workers. Some enlightened employers recognize and promote good general health as an important additional protection against workplace injuries and illnesses. When considering the hierarchy of controls it must be noted that personal respiratory protection is used *only* when the other controls are inadequate, as a secondary measure, or to protect a susceptible individual. Also, some susceptible individual workers may need reassignment to other tasks, or to take individual actions to protect their health. In the workplace, engineering-type controls are preferred, followed by administrative and then personal protective measures. By analogy, environmental air standards for the general population are primarily maintained by *engineering-type controls* in the form of *emission controls* for vehicles, factories, construction sites, and agricultural operations (see Chapter 6).

VII. SUMMARY OF MAJOR POINTS

The human respiratory tract is a living air sampler that brings large volumes of oxygen, and any air contaminants that are present, into the three compartmental regions of the respiratory tract (e.g., the NOPL, TB, and P airways). Some or all of the inhaled pollutants will deposit in the body, depending primarily on the particle size for aerosols and the water-solubility for gases. The deposition (and exposure) probabilities are estimated by sophisticated computer models for the primary purpose of establishing acceptable health-based standards for air contaminants. Health-based air standards are described in more detail in Chapters 6 and 11.

Once in the body, air pollutants have several possible fates, including incorporation into tissues (storage), biotransformation (i.e., metabolism) to other substances, and elimination from the body (either rapidly or slowly). The fate will depend on the region of deposition in the respiratory tract, the physical and chemical properties of the pollutant, and the exposed individual's biological characteristics. Accordingly, some people will receive significantly higher internal exposures than an average member of the population.

Workplace exposures and the associated techniques for exposure control provide a good case study for air pollutant exposures, because occupational exposures can be high and to a small number of well-known substances. Also, significant workplace co-stressors can modify exposures to and effects of inhaled particles and gases. In considering workplace exposures to particles, the effects of aerosol size must be taken into account. Thus, workplace air standards are set for various particle size ranges (IPM, TPM, and RPM) that will *expose* all or some portions of the respiratory tract. In order to ensure that workers do not breathe harmful concentrations of air contaminants, *engineering* and *administrative* controls are preferred, but *respiratory protective* devices are also used.

Because members of the general population are quite variable, their exposures are more complex than those for the working population. Therefore, environmental air standards are usually more stringent than workplace standards. Yet, the most sensitive individuals in the population may require personal protection, and some may need medical assistance, even when the air quality is within acceptable levels for the majority of the population.

VIII. QUIZ AND PROBLEMS

Quiz Questions

(select the best answer)

1. In addition to gas exchange, the respiratory system also:
   a. delivers air pollutants to the rest of the body.
   b. adjusts the temperature and humidity of inspired air.
   c. eliminates some volatile air pollutants from the body.
   d. All of the above are true.
2. Compared to environmental contaminant exposures from eating and drinking, inhalation:
   a. brings larger daily volumes of the environment into the body.
   b. brings smaller daily volumes of the environment into the body.
   c. brings equal daily volumes of the environment into the body.
   d. brings gases but not particles into the body.
3. The tidal volume is:
   a. the volumc of air in the lungs.
   b. the volume of air inhaled or exhaled in a single breath.
   c. the ratio of resting lung volume to the maximum lung volume.
   d. None of the above is true.
4. Inhaled particles will:
   a. deposit in the body with efficiencies that depend on the particle sizes.
   b. always deposit with 100 percent efficiency in the respiratory tract.
   c. deposit in the nose but not the alveoli.
   d. None of the above is true.
5. Inhaled gases will:
   a. not deposit in the body.
   b. will deposit in the body depending on their solubility in respiratory tract fluids (such as mucus).
   c. will always deposit in the body with 100 percent efficiency.
   d. will deposit in the body only if they are accompanied by airborne particles.
6. Insoluble particles:
   a. cannot be cleared from the body once they are inhaled and deposit in the respiratory tract.
   b. will rapidly dissolve in the respiratory tract after they deposit.

   c. can produce lung diseases if they deposit in sufficient concentrations in the respiratory tract.
   d. are not capable of producing lung diseases in healthy people.
7. The term "personal cloud" refers to:
   a. air pollutants derived from atmospheric clouds that travel from industrial sources.
   b. air pollutants that are produced by a person's own activities.
   c. air pollutants produced by nearby people if the wind is blowing from their direction.
   d. air pollutants that are produced by motor vehicles.
8. Potentially susceptible subpopulations for adverse effects of air pollutants include:
   a. young children.
   b. persons with chronic heart and/or lung diseases.
   c. both a and b.
   d. Neither a nor b are true.

Problems

1. Calculate the volume of air that an average adult male engaged in light work 8 hours per day inhales at work in a 5-day work week.
2. If a 2-year-old child and an adult are breathing during low activity, how much air will be inhaled by each one in a 1-hour period? Which one will inhale the most air per unit body mass (kg) during that hour? (see Table 7–6).
3. Using Figure 7–5, what is the expected percent deposition of an inhaled 1 μm diameter particle in the pulmonary region? Is this greater than or less than the expected deposition of the same size particle in the tracheobronchial region?
4. List some diseases that might be produced by breathing high concentrations of (a) water soluble gases and (b) gases that are insoluble in water.
5. Explain why small quartz particles produce lung fibrosis, but very large quartz particles do not.
6. Using Figure 7–6, if the GSD is 3, what fraction of the population will have a predicted dose that is 10 times the "average" dose?

IX. DISCUSSION TOPICS

1. How do children differ from adults in their exposures to air pollutants?
2. How do particle sizes influence their deposition in the various parts of the respiratory tract? In general, would you expect either large or small particles to be more hazardous if the composition and air mass-concentrations are equal?
3. Should workers be allowed to have greater risks from airborne contaminants than the general public? Why or why not?
4. Why are "administrative" controls preferred in the workplace over "personal protection" controls when there are toxic airborne particles present?

References and Recommended Reading

ACGIH®, Technical Committee on Air Sampling Procedures, *Particle Size-Selective Sampling in the Workplace,* ACGIH®, Cincinnati, OH, 1985.

ACGIH®, *Guide to Occupational Exposure Values—1999,* ACGIH®, Cincinnati, OH, 1999.

ACGIH®, *2010 TLVs® and BEIs®*, ACGIH®, Cincinnati, OH, 2010.

Asgharian, B., Kelly, J. T., and Tewksbury, E. W., Respiratory deposition and inhalability of monodisperse aerosols in Long-Evans rats, *Toxicol. Sci.,* 71:104–111, 2003.

Asgharian, B., Ménache, M. G., and Miller, F. J., Modeling age-related particle deposition in humans, *J. Aerosol Med.*, 17:213–224, 2004.

Brown, J. S., Wilson, W. E., and Grant, L. D., Dosimetric comparisons of particle deposition and retention in rats and humans, *Inhal. Toxicol.,* 17:355–385, 2005.

Costa, D. L., Air pollution, in *Casarett and Doull's Essentials of Toxicology,* Klaassen, C. D. and Watkins, J. B. III, eds., McGraw–Hill, New York, 2003, pp. 407–418.

Cuddihy, R. G., McClellan, R. O., and Griffith, W. C., Variability in target organ deposition among individuals exposed to toxic substances, *Toxicol. Appl. Pharmacol.,* 49:179–187, 1979.

Dorman, D. C., Brenneman, K. A., McElveen, A. M., Lynch, S. E., Roberts, K. C., and Wong, B. A., Olfactory transport: A direct route of delivery of inhaled manganese phosphate to the rat brain, *J. Toxicol. Environ. Health Part A,* 65:1493–1511, 2002.

Ferro, A. R., Kopperud, R. J., and Hildemann, L. M., Elevated personal exposure to particulate matter from human activities in a residence, *J. Exposure Analysis Envir. Epidemiol.,* 14:S34–S40, 2004.

ICRP (International Commission on Radiological Protection, Task Group of Committee 2), *Human Respiratory Tract Model for Radiological Protection,* Publication 66, Pergamon Press, New York, 1994.

Jarabek, A. M., The application of dosimetry models to identify key processes and parameters for default dose-response assessment approaches, *Toxicol. Lett.,* 79:171–184, 1995.

Jarabek, A. M., Asgharian, B., and Miller, F. J., Dosimetric adjustments for interspecies extrapolation of inhaled poorly soluble particles (PSP), *Inhal. Toxicol.,* 17: 317–334, 2005.

Kreyling, W. G. and Scheuch, G., Clearance of particles deposited in the lungs, in *Particle-Lung Interactions,* Gehr, P. and Heyder, J., eds., Marcel Dekker, New York, 2000, pp. 323–376.

Lidén, G. and Harper, M., The need for an international sampling convention for inhalable dust in calm air, *J. Occup. Envir. Hygiene,* 3:D94–D101, 2006.

Mauderly, J. L. and McCunney, R. J., eds., *Particle Overload in the Rat Lung and Lung Cancer: Implications for Human Risk Assessment,* Taylor & Francis, Washington, DC, 1996.

NCRP (National Council on Radiation Protection and Measurements), *Deposition Retention and Dosimetry of Inhaled Radioactive Substances,* NCRP SC 57–2 Report, National Council on Radiation Protection and Measurements, Bethesda, MD, 1997.

Oberdörster, G., Sharp, Z., Atudorei, V., et al., Translocation of inhaled ultrafine particles to the brain, *Inhal. Toxicol.,* 16:437–445, 2004.

Pavia, D., Acute respiratory infections and mucociliary clearance, *Eur. J. Respir. Dis.,* 71:219–226, 1987.

Phalen, R. F., Stuart, B. O., and Lioy, P. J., Rationale for and implications of particle size-selective sampling, in *Advances in Air Sampling,* ACGIH®, Lewis Publishers, Inc., Chelsea, MI., 1988, pp. 3–15.

Schleien, B., Slaback, L. A. Jr., and Birky, B. K., Eds., *Handbook of Health Physics and Radiological Health, 3rd Edition,* Williams & Wilkins, Baltimore, MD, 1998, pp. 12-23 to 12-38.

Soderholm, S. C., Proposed international conventions for particle size-selective sampling, *Ann. Occup. Hygiene,* 33:301–320, 1989.

Stahlhofen, W., Scheuch, G., and Bailey, M. R., Investigations of retention of inhaled particles in the human bronchial tree, *Radiat. Prot. Dosim.,* 60: 311–319, 1995.

TGLD (Task Group on Lung Dynamics), ICRP Committee II, Deposition and retention models for internal dosimetry of the human respiratory tract, *Health Phys.,* 12:173–207, 1966.

U.S. EPA (U.S. Environmental Protection Agency), *Development of Statistical Distributions or Ranges of Standard Factors Used in Exposure Assessments,* EPA/600/8–85/010, Office of Health and Environmental Assessment, Washington, DC, 1985.

Wolff, R. K., Experimental investigation of deposition and fate of particles: Animal models and interspecies differences, in *Aerosol Inhalation: Recent Research Frontiers,* Marijnissen, J. C. M. and Gradón, L., eds., Kluwer Academic Publishers, Norwell, MA, 1996, pp. 247–263.

Chapter 8

Effects on Human Health

LEARNING OBJECTIVES

By the end of this chapter the reader will be able to:

- discuss the measures of human health that relate to air pollutants
- describe the environmental and other factors that correlate with human health
- describe the direct and indirect health effects of air pollutants
- name three human population groups that are especially susceptible to air pollutants

CHAPTER OUTLINE

I. Introduction to Air Pollution and Health
II. Sources of Health Data
III. Health Effects of Selected Air Pollutants
IV. Susceptible Populations
V. Sources of Information on Health Effects of Air Pollutants
VI. Summary of Major Points
VII. Quiz and Problems
VIII. Discussion Topics
References and Recommended Reading

I. INTRODUCTION TO AIR POLLUTION AND HEALTH

Key Concepts

Definition, Measures, and Determinants of Human Health

> Health is "... a state of optimal physical, mental, and social well-being...." (Anderson, 2003)

Health and well-being are easier to define than to achieve or to measure. For populations, average or median lifespan, age-specific mortality rates, prevalence of diseases, and even perceptions of happiness, are among the many criteria used to quantitatively describe health status. Substantial variations in health are observed in the nations of the world. **Table 8–1** depicts some indicators of national health along with income and literacy data. Air pollution is not explicitly included in this table. Except for airborne infections, major air pollution episodes (e.g., from wild fires), and industrial accidents, it is a relatively minor determinant of health. From a national perspective, family income, availability of medical services, food and water quality, and sanitation practices are major public health factors. The health effects of environmental air pollution usually become an important public health consideration only after the major killers, such as infections (including parasites), starvation, and poor nutrition are under control. For many less developed nations, such killers still destroy hundreds of thousands of lives each year, so they have priority in the competition for meager public health expenditures.

For individuals, the direct effects of air pollutants on health can be measured by premature death, and several specific disease states, such as asthma, bronchitis, and cardiovascular diseases. Related factors, such as hospital admissions, medication usage (e.g., bronchodilators), absenteeism from work and school, and pulmonary function measures are commonly used to measure air pollutant effects in clinical and epidemiological investigations. In a broad sense, essentially any physical, mental, or social parameter that can be reliably measured has the potential to be used as an indicator of a health effect, either adverse or beneficial. Within the definition of human health, *welfare* effects, such as good visibility and climate (see Chapter 5), the health of domestic and wild plants and animals, and damage to buildings and artwork, could be included. Both health and welfare effects are usually considered by regulators when they establish air quality standards.

The many factors that contribute to public health are intimately interrelated. A strong economy, a stable family income level, well-organized and active public health agencies, accessible medical services, public literacy, and environmental quality are all important and are all competing for limited monetary and human resources. Improvement in one area, such as better air quality through taxes on transportation fuels, can adversely affect local economies and the availability of affordable food. Wage earners need jobs, such as those

Table 8–1 World health as measured by under age 5 mortality per 1,000 live births, lifespan in relation to per capita income in U.S. $, and literacy.

Region/Level of Development	*Under Age 5 Mortality in 2006*	*Adult Lifespan, Years*	*Income per Capita, U.S. $*	*Percent Adult Literacy, 2005*
Sub-Saharan Africa	160	50	851	58
Middle East and North Africa	46	69	2,104	73
South Asia	83	64	777	58
East Asia and Pacific	29	72	2,371	91
Latin America and Caribbean	27	73	4,847	90
Industrialized Countries	6	79	37,217	Not available
Developing Countries	79	66	1,967	76
Least Developed Countries	142	55	438	55
World	72	68	7,406	78

Data from UNICEF (2007).

in the transportation industry, to support their families and keep the economy strong. Likewise, increased transportation expenses can elevate the costs of nutritious foods and essential goods, which can have a disproportionate impact on lower-income families. Therefore, industry contributes to public health as well as to air pollution. The way to optimize the distribution of talent and other resources among these competing priorities has not yet been perfected.

Even in developed nations, there is intense competition for resources, and too much attention to one area impacting public health means that other areas may not be adequately covered. The overall prosperity of a nation, state, city, or region will usually determine the level of expenditures available for improving air quality. Perhaps at some future date, sufficient resources will be available to provide all people with the necessities that provide for a high-quality life plus acceptable environmental quality. The main point is that clean air cannot be considered in isolation from other factors that impact public health. Human health is impacted by a multitude of environmental, behavioral, social, and economic factors, each potentially influencing the other. Thus, the practice of public health is a multidisciplinary endeavor.

Dose-Response Concepts and Complications

Living organisms are *complex systems* with multifaceted properties and behaviors. The relationship between the internal dose of a substance and the biological response is anything but simple. Each organ, tissue, and cell in the body can have a different exposure, and exhibit different responses. Environmental scientists and toxicologists work to uncover these various responses and determine which ones may lead to adverse health effects in humans. In some cases the responses may have a beneficial health effect, especially if the immune system is enhanced in the process.

The *route of exposure,* inhalation, ingestion, dermal, etc., and the dose (both the level and rate of delivery) are among the important modifiers of a subject's responses. The *responses* to a foreign substance will become new "internal exposures" that also further affect the body's responses. It is the body's response, or lack thereof, to a pollutant that determines the health outcome. Some of the body's responses are *adaptive,* in that future exposures will produce either *diminished* or *exaggerated* responses, in comparison to the initial responses. The entire suite of consequences of an exposed subject to the initial and subsequent dose, or doses, of air pollutants is important. The responses are not always harmful. In many cases, they may be beneficial or even harmful initially but beneficial later, such as in the acquisition of immunity, or up-regulation of a defensive function. This latter point is pertinent to both individuals and populations, in that some air pollutant exposures may harm some individuals (e.g., trigger bronchial constriction in sensitive asthmatics) and benefit others (e.g., up-regulate immunological defenses in healthy young children). The co-existing harmful and beneficial effects of air pollutant exposure complicate the process of establishing acceptable air standards. It should also be apparent that the response to a given air pollutant dose can even vary from time to time in an individual. In summary, dose-response relationships are only simplified surrogates for exceedingly-complicated phenomena. Dose-response relationships do have value as one of several inputs that guide regulatory, scientific, and personal decisions about exposure to air pollutants. More detailed information on dose-response relationships can be found in Chapters 9 and 10.

Acute vs. Chronic Exposures and Effects

In toxicology "acute" means "brief," as opposed to the common meanings, such as "sharp," "intense," or "critical." *Acute* actually means *brief in relation to biological time scales,* such as cell turnover rates, physiological adjustments, and the time required for a disease to manifest, or adaptation to occur. *Acute exposures* and acute responses typically range from those with durations of minutes to several days.

Chronic exposures and responses usually have durations of months to years or longer. During prolonged or repeated exposures many internal biological events can occur. Environmental air pollutant exposures are combinations of repeated acute exposures in a chronic exposure background, with short-term and long-term variations in concentration (e.g., weekly and seasonal). Some of the variation is due to source cycles (e.g., high-traffic times, weekday industrial operations, and seasonal demands), and some by people moving about (i.e., time-activity patterns). Air pollutant exposures also have long-term trends (over years), as the technology associated with sources and emission controls changes. Similarly, biological responses have mixed characteristics including both rapid and slow processes, such as injury and repair phenomena, and changes due to diseases and aging. Biological phenomena interact with the complex exposure patterns, making it difficult to accurately predict the health effects of air pollutants. As a result, many regulatory agencies rely heavily on observational epidemiologic studies, especially those in which small changes in health status of a large group can be detected and statistically associated with air quality. Epidemiology reflects human reality, but it is not always strong with respect to uncovering causal agents (see Chapter 10). Other types of studies (e.g., *in vitro*) and controlled animal studies are supplemental and used to identify causal agents, mechanisms of action, etc. Toxicologists sometimes view epidemiology as an aid in defining hypotheses for their studies.

As previously noted, acute and chronic exposures and responses to various individual air pollutants can interact and influence each other. An acute inhalation exposure to ozone can cause initial injury but induce protection against future exposures to ozone and other oxidants. In contrast, some acute exposures (e.g., to allergens and some irritants) can sensitize subjects and lead to exaggerated (even fatal) responses upon future exposures. There is current uncertainty over whether co-exposure to both particles and ozone decreases or increases the toxicity of ozone; both possibilities may be true depending on the effects that are studied. Scientists are a long way from understanding the effects of real-world air pollutant exposures. Some liken the current understanding of air pollution's health effects to that of the understanding of chemistry before the *periodic table of the elements* was devised. There are many challenges and opportunities for future air pollution researchers.

More About Responses

Biological systems are, in general, acutely sensitive to *changes* in their environments; even the sense organs of complex organisms are finely-tuned to detect change. Most such changes (even small ones) will produce a biological response (or responses), whether or not the responses are measurable. Some responses to air pollutants are clearly adverse with respect to health status, and some are beneficial (e.g., those that indicate the presence of danger, or that induce protective defenses). However, many responses (such as avoidance behavior), although measurable, may have no significant impact on the health of the subject.

The principle of *parsimony* (stinginess) of biological systems, which is captured in the phrase "use it or lose it," should be taken into account when evaluating the health effects of air pollutants. Biological systems, being adaptive, will attempt to maintain necessary capabilities, but not necessarily those that are not needed for their near-term success. So *challenge* (e.g., biological stress) seems to be essential for maintaining defensive capabilities and other useful functions. Examples of the lack of challenge leading to loss of capabilities are well-known. For example, astronauts begin losing bone mass within hours of encountering microgravity conditions (in which forces on bones are greatly reduced). Unless heroic efforts are directed toward preserving the "useless" portions of the skeleton, bone strength may be compromised to the point that the astronaut cannot safely return to the Earth with its normal gravity. Also, microbiologists warn that living in a "sterile" microbial environment (e.g., by overuse of antibiotics) will predispose people (and other complex life forms) to future infections as a result of loss of microbial defenses. It is sobering to contemplate how labile the anatomy and physiology of complex lifeforms can be.

Mammalian defenses against the harmful effects of inhaled substances can be modulated by previous exposures. Inhalation toxicologists are aware that laboratory animals that have not had recent prior exposures to air pollutants show greater responses to low-level inhalation challenges. Thus, in the laboratory, a few days of living in clean-air housing can be used to amplify experimental responses. Also, upon repetition of low-level air pollutant exposures, acute responses can diminish, even while chronic injury may be accumulating. Outside of the laboratory, many people may have difficulty in moving from a clean air environment to a polluted one. As developed nations have improved their air quality it is increasingly uncomfortable for their citizens to travel to nations with poor air quality. Hence, *significant changes* in air pollutant levels, or composition, can have adverse effects.

Multi-Causation and Sensitive Populations

Massive inhalation exposures, as occur in natural disasters (such as volcanic eruptions and fires), in military operations, and severe industrial accidents, can surely injure and kill healthy individuals. Also, some air pollutants, such as virulent infectious agents and (a few) super-toxic substances, can harm healthy people when inhaled in modest and even small doses. In contrast, normal levels of modern air pollutants alone are not acutely lethal to healthy people. Yet, even modest air pollution episodes are *associated* with detectable adverse effects, including deaths, in epidemiologic studies (see Chapter 10). Thus, many realistic air pollutant exposures appear to be merely *contributors,* along with other factors, to adverse health outcomes. In other words, air pollution episodes may add an additional stress that harms individuals who are already coping with existing health problems. The observed health effects of ordinary levels of air pollutants appear to be multi-causal.

Underlying serious diseases are perhaps the major contributing factors to air pollutant related injuries and deaths. Persons with advanced heart, vascular, and respiratory tract diseases, along with existing hypersensitivity conditions (e.g., bronchial asthma), may only require small additional triggers to produce serious effects. The triggers include: (1) changes in ambient temperature, humidity, pressure, and dust levels; and (2) physical exercise, fear, anxiety, and other emotional states. Hence, published warnings such as, "Persons with preexisting cardiopulmonary conditions should avoid outdoor exercise," are issued when air pollution levels are significantly elevated. Such warnings should be heeded.

Other factors can modify susceptibility to the adverse effects of air pollutants. Such factors include low socioeconomic status (SES), age, gender, genetics, and race. A more complete description of sensitive populations is presented in Chapter 10. For the purposes of this chapter it is important to understand that modern air pollution is mainly an additional contributor to other factors, external and internal, that can cause or intensify adverse health responses. The fact that many adverse responses have multiple contributing causes complicates establishing clear cause-and-effect relationships for specific air pollutants. Air quality regulations are just one of the means for protecting public health. Improvements in general health will also reduce the adverse impacts of air pollution.

Direct vs. Indirect Health Effects

The *direct health effects* of air pollutants are those produced in a subject as a result of *actual contact* with the pollutants. The contact can be due to deposition in the respiratory tract or deposition on external surfaces such as the skin, mucous membranes, or the eyes. Direct effects are relatively easy to study in the laboratory, and thus they are the main target of air pollution regulations.

Indirect health effects (also called *secondary health effects*) are those that do *not* result from the physical and chemical reactions in the body as a result of pollutant deposition. Poor visibility, especially in areas such as national parks, can adversely affect a sense of well-being; thus, it is an adverse secondary or indirect effect. Stratospheric ozone, and to a much smaller extent, tropospheric ozone, absorbs solar ultraviolet rays, which reduces cancer risks, and can also be thought of as a positive indirect health effect. Other indirect health effects include the effects on other persons as a result of the injury to some. Living below the poverty level is associated with decreases in well-being and significant loss of life span. A disabled or unemployed wage-earner can potentially cause adverse health effects in their dependents for many reasons, including stress and the effects of loss of income. Air pollution regulations can be costly, and may result in job loss, decreased access to goods (e.g., food), and access to services (e.g., medical care), which can produce indirect adverse health effects.

II. SOURCES OF HEALTH DATA

Much is currently known about the health effects of thousands of commonly encountered air pollutants. The main sources of this knowledge are:

- laboratory animal studies (primarily using mice and rats);
- human clinical studies (using both healthy and compromised subjects);
- *in vitro* studies (using cell cultures, genetic material, biochemicals, etc.);
- environmental and other accidental exposures;
- military actions;
- intentional and accidental poisonings; and
- computer simulations (of exposures, fates in the body, and effects).

Each of the above sources provides unique but limited information (**Table 8–2**). The full range of such information is needed to understand the impacts of, and to devise effective regulations for controlling, air pollutants. Intact animals enrolled in inhalation studies are essential sources of key information on lethality, birth defects, fertility, chemical sensitization, cancer production, dose-response relationships, and other phenomena that require whole animals to manifest but cannot be ethically conducted in human subjects (see

Table 8–2 Sources of information on the health effects of air pollutants.

Source	*Types of data*
Animal Models (inhalation studies are preferred over other types of exposure)	Information is obtained on short- and long-term exposures to controlled doses of single pollutants and simple mixtures on intact subjects: fertility, birth defects, cancer, sensitization, adaptation, dose-response information, behavior, etc., are evaluated.
Human Clinical Studies	Information is obtained on clinical effects using "safe" doses of pollutants on normal, diseased subjects, and potentially-susceptible volunteers; exposures are acute or repeated over a few days or weeks.
In vitro Research (mainly cell and organ cultures and biochemical preparations)	Mechanisms of injury, metabolism, comparative toxicity of pollutants, mutagenicity, etc., are studied.
Computer Simulations	Detailed data is obtained on exposures, uptake, internal distribution, metabolism, storage, and elimination; the effects of species, gender, age, etc., provide information on extrapolation, study design and data interpretation.
Accidents, Military Operations, and Poisonings	High-dose effects, including lethality, acute and chronic effects, and information on medical treatments for effects are obtained.

Chapter 12 for more on research ethics). Lifetime, even multi-generational exposures, can be conducted in animal models. Such animal studies have led in discovering and controlling the effects of numerous air pollutants on domestic and wild animals as well as human populations.

One important goal of animal studies is to initially identify doses below which serious harm does not occur (i.e., NOELs, *no observed effect levels*). Such levels are the initial estimates of acceptable exposures in humans. After enough data has been acquired in animal and *in vitro* studies to ensure the safety of intentional human exposures, follow-up clinical research can be ethically conducted. Clinical studies (typically supervised by physicians) are needed to refine the knowledge obtained in animal and *in vitro* investigations. Clinical studies use both healthy and health-compromised subjects. Clinical studies can examine pollutant levels that exceed environmental concentrations to establish cause-and-effect relationships. The study of actual environmental exposures (including normal day-to-day exposures and accidental pollutant releases) by clinicians and epidemiologists provide key data that are used to establish air quality criteria for both the general population and workers. Because the environmental exposures are not usually controlled, the causal factors may not be accurately characterized.

Military operations and *poisonings* (accidental and intentional) are mainly used to define both lethal doses and the medical interventions that can be used to preserve life and health as a result of massive exposures.

In recent decades, sophisticated *computer programs* have played increasingly important roles in air pollution health-effects research. The uptake of inhaled air pollutants into the body can be modeled for humans and animals from a knowledge of air concentrations, breathing rates, airway sizes, and physicochemical properties of the pollutants. Once in the body, the pollutants can be followed in PBPK (*physiologically-based pharmacokinetic models*) simulations. Such modeling is useful for identifying target organs and tissues and performing extrapolations of laboratory animal data to predict effects in humans.

By assembling and integrating the varied knowledge base on air pollutants, the health effects associated with various pollutant doses can be determined. If the data base is complete enough, acceptable exposure concentrations can be identified. Yet, the knowledge base on realistic air pollutant combinations and individual differences in responses is far from complete.

III. HEALTH EFFECTS OF SELECTED AIR POLLUTANTS

Introduction

As essentially any particulate or gaseous substance can be found in the air, the range of human exposures and potential health effects is large. Therefore, only a few representative air contaminants will be presented here in order to demonstrate (1) their health effects and (2) air concentrations that are currently considered to be acceptable. Acceptable air quality standards differ between the general population and workers. Not included here are agents involved in serious accidents, military operations, and criminal attacks. The emphasis is on the U.S. Environmental Protection Agency's (U.S. EPA) criteria air pollutants, but some other common air contaminants are covered in order to introduce basic principles.

U.S. EPA's Criteria Pollutants

Carbon Monoxide

Carbon monoxide (CO) is an odorless and tasteless gas that is a product of fuel combustion, more often incomplete combustion. Because it is odorless, it is said to have poor *warning properties.* Thus, people can be exposed to lethal concentrations without perceiving the danger. Cars, trucks, furnaces, and numerous industrial operations are common sources of CO. It has been responsible for large numbers of human deaths in homes, vehicles, and workplaces. Although humans are exposed to CO via inhalation, the *critical effects* do not occur directly in the respiratory tract. The lungs are merely delivery organs that supply CO to the body. Carbon monoxide binds to hemoglobin in the blood with 200 to 250 times greater affinity than does oxygen, and thus decreases the oxygen delivery to *all* tissues of the body. The primary toxicant class of CO is an *asphyxiant.* The main critical target organs for acute adverse health effects are the *brain* and the *heart,* because of their essential immediate functions and their high rates of oxygen utilization. Thus, a diminished oxygen supply can quickly produce striking cardiac and nerve-related symptoms. However, all organs may be affected by sufficient CO doses. There are also non-asphyxiant effects of CO, but they are less well-recognized, and are not controlling factors with respect to acute effects or air quality criteria.

When CO binds to hemoglobin in red blood cells, carboxyhemoglobin (COHb) is formed, and the toxicity is presumed to be correlated with the percent of COHb in the blood. The air concentration of CO, the lung ventilation rate, pulmonary diffusing capacity, and exposure duration all affect the air-to-lung uptake and the blood COHb concentration. During prolonged exposures, CO builds up over time to reach equilibrium in the body. Therefore, both concentration and exposure time are key internal exposure factors. When the environmental CO concentration drops, CO washes out of the body with a half-time of about four hours. *Half-time* is the time required for a concentration to decrease by 50 percent. Therefore, the relationships between the air concentration of CO and COHb blood levels are dynamic and mathematically complex.

The adverse health effects of CO have been extensively studied and recently reviewed by the U.S. EPA (U.S. EPA, 2010a). The normal COHb levels in the blood of non-smokers, which is believed to be produced in the body, is about 0.5 percent of the total hemoglobin. Cigarette smokers average about 5 percent. Fresh tobacco smoke is estimated to contain up to 100,000 components, with CO being in the top three by mass. Levels of COHb less than 10 percent are generally considered to be safe, at least in healthy adults. However, people with cardiovascular disease, anemia, and diabetes, along with the developing fetus, and visitors to high-altitudes (including tourists and aircraft occupants) already have compromised tissue oxygenation capabilities, which is expected to place them at greater risk from given COHb levels than other individuals.

At high COHb levels, the effects, even in healthy adults, can include headache, poor coordination, visual and speech disturbances, coma and even death. Intermediate levels of CO have progressively debilitating effects that can occur without warning. As CO is usually a component of combustion, the smell of vehicle exhaust (or any organic fuel combustion) indicates the need for immediate escape and ventilation with fresh air. This potentially lifesaving response is urgent if accompanied by a headache or other of the symptoms listed in **Table 8–3**. As previously noted, the lack of odor of CO makes it an exceptionally dangerous air pollutant, even to rescue personnel.

Exposure standards are based on CO concentrations in the air, rather than COHb levels, as such blood levels are more difficult to measure. The 2010 U.S. EPA's National Ambient Air Quality Standards (NAAQS) for the general population are 9 ppm averaged over 8 hours and 35 ppm averaged over 1 hour. The NAAQS are intended to protect nearly everyone, including susceptible groups. In the workplace, the American Conference of Governmental Hygienists (ACGIH®) recommends a maximum TWA (time-weighted, 8-hour average, 5 days/week) CO concentration of 25 ppm. The TWA is believed to be the level that "the typical worker can experience

Table 8–3 Effects of percent COHb in blood.

% COHb	*Acute Effects*	*Approx. ppm CO in Air for a Few Hours Exposure*
0–10	Usually none in healthy individuals; angina and reduced performance in patients with lung or coronary heart diseases	100
10–20	Headache, dyspnea (anxiety about breathing) on exertion, angina in patients with coronary heart disease	150
20–30	Headache, nausea, fatigue, irritability, decreased concentration	200
30–40	Severe headache, dizziness, fatigue, weakness, poor concentration	400
40–50	Fainting, heart-rate disturbances, confusion	500
50–60	Collapse, respiratory failure, convulsions, seizures, coma	700
60–70	Loss of blood pressure, respiratory failure, often fatal	1,000
over 70	Coma, death	

Data from Ilano and Raffin, (1990); Varon et al. (1999).

without adverse health effects." Note that workers are generally healthier than the larger population and thus more tolerant to the adverse effects of CO.

Lead

Lead (Pb) is a workable corrosion-resistant low melting point (327°C) metal with many important applications. It has been used in alloys, solder, plumbing, pottery, medicines, candies, eating utensils, bullets, weights, cans, and seals for thousands of years. It also is used as a pigment and rust inhibitor in paints and ceramics. Pb is found in soil, water, and air. Pb is directly toxic to the body's cells, including neurons (nerve cells). Both inhalation and ingestion are associated with lead uptake in the body, where it accumulates in bone and other target organs. Even after the decreased use of Pb in gasoline, oral exposures in children (e.g., by eating paint chips in old buildings or consuming Pb-containing foods, medicines, and candies) continue to be a major health concern. This is mostly due to lead's heightened neurotoxicity in children.

Occupational and environmental exposures must be controlled to prevent a variety of adverse effects, including neurotoxicity, anemia, cardiovascular and kidney damage, learning problems, and other potential problems (including possibly cancer). Workers can transport Pb dust into their homes on contaminated clothing, shoes, hair, and skin. Pb is excreted from the body with a primary half-time of about 30 days. Pb is relatively easy to control in the environment, primarily by introducing substitutes, both in industrial operations and in Pb-containing products.

Health protection from airborne exposure is obtained for the general public by a NAAQS of 0.15 $\mu g/m^3$ (as a 3-month average air concentration) and for workers by a time-weighted daily 8-hour average air concentration of 50 $\mu g/m^3$. The continuing substitution of other materials for Pb in consumer products, and monitoring its presence in blood, are the major means for controlling toxicities. The exceptional vulnerability of children to its neurotoxicity and other serious effects must be emphasized, because few children (and their parents) are aware of their exposures and the dangers such exposures may pose.

Ozone

Due to its chemical reactivity, and its photochemical formation in the troposphere from a variety of natural and anthropogenic precursors, ozone (O_3) is a major daytime outdoor air pollutant. It is also produced around electrical arcs and ultraviolet light producing devices. It is an intense oxidizing irritant affecting the eyes, mucus membranes, and lungs. Upon inhalation, O_3 reacts with and damages cell membranes on all respiratory tract surfaces from the nose down to the deep-lying alveoli. Thus, the respiratory tract is the most critical organ for toxicity. Ozone's characteristic odor (a pungent smell) around strong ultraviolet light, some so-called "air cleaners," and electrical discharges is a warning that levels are unhealthy to breathe for more than a few minutes. The odor detection threshold of 0.5 ppm or less decreases over time after initial exposure due to "*odor fatigue*." Therefore, exposed people may think that the O_3 has dissipated, when it is still present. Even at levels in the air below the odor threshold, O_3 can have adverse health effects, so the warning cues cannot be relied on. Levels in air above 1 ppm are acutely harmful. Very low levels, even at or near the natural background concentrations, can harm sensitive individuals, such as those with lung disease. Exercise is an additional risk factor, as increased ventilation delivers greater doses to the lungs.

Much of what is known about the adverse effects of O_3 comes from laboratory animal studies. Ciliated cells (that propel mucus) and thin alveolar cells in the respiratory tract are primary targets of O_3. Air concentrations as low as 0.2 ppm cause transient, repairable cell damage, but it is unknown whether some irreparable damage accumulates on repeated exposures. In this regard, O_3 damage has been compared to sunburn, which has a cumulative effect, even after the initial burn injury resolves. As previously mentioned, exercise increases the inhaled dose rate, and changes in pulmonary function have been seen in exercising laboratory animals and exercising adult humans exposed to concentrations as low as 0.1 ppm for 4 to 6 hours. Other effects of low levels of O_3 include increased number of respiratory infections, increased tissue permeability, exacerbation of asthma, and pulmonary function impairments. O_3 is controlled by controlling its photochemical precursors (i.e., NO_2 and hydrocarbons), securely enclosing O_3 generating sources, and passing O_3 containing air through filters, such as activated charcoal.

The 2008 NAAQS for O_3 are 0.075 ppm (8-hour average) and 0.12 ppm (1-hour average). The ACGIH® occupational air standard depends on the level of physical exercise for exposed workers. For an 8-hour exposure during heavy work the TWA is 0.05 ppm and for light work 0.1 ppm. If the exposure is two hours or more, a maximum of 0.2 ppm is recommended for all workloads, including those at resting ventilation. Epidemiological studies continue to identify children as a susceptible population.

Nitrogen Dioxide

Nitrogen dioxide (NO_2) gas is similar to O_3 both chemically and with respect to its effects when inhaled. However, higher concentrations, as occur in occupational settings, are required to produce acute injury. As normal outdoor air concentrations are low (about 15 ppb), elevated levels in enclosed silos and in homes with poorly-vented gas appliances such as stoves and heaters are of concern. NO and NO_2 are often grouped together as NO_x because they are simultaneously emitted from combustion sources, and they can react in the atmosphere to produce each other (along with several other reactive products). Therefore, health effects studies often have the additional complication of the presence of variable mixtures of NO, NO_2, nitrogen-containing acids, and other potentially toxic substances. Accordingly, the health effects as seen in clinical, epidemiological, and toxicological investigations are variable. Effects at modest concentrations may include impaired lung defenses against infections, increased airway responsiveness in asthmatics, and airway inflammation.

The NAAQS for NO_2 are 53 ppb (annual average) and 100 ppb (1-hour average). In the workplace the recommended time-weighted, 8-hour average is 3 ppm, and the short-term emergency concentration (permitted for exposures less than 15 minutes) is 5 ppm. Again, workers are presumed to be more resistant to the adverse effects of NO_2 than are sensitive population groups in the general population.

Particulate Matter

Particulate matter (PM) is the only criteria air pollutant that is not chemically defined; it is defined gravimetrically, hence "PM" also means *particle mass*. Therefore, PM in different locations has variable chemical compositions. PM toxicities range from relatively high in those containing substantial amounts of very toxic metals to low in those containing relatively chemically-benign substances such as fine soil or dust. Because particle size has been shown to alter the deposition pattern and the translocation of poorly-soluble particles post deposition in the respiratory tract, the EPA currently regulates PM in two size ranges, PM_{10} and $PM_{2.5}$. PM_{10} is the total mass of particles less than 10 micrometers (μm) in aerodynamic diameter. $PM_{2.5}$ refers to the mass of particles less than 2.5 μm in diameter. $PM_{2.5}$ is also called *fine particle mass,* and PM_{10} minus $PM_{2.5}$ is called *coarse particle mass* (i.e., the mass of particles in the diameter range of 2.5 to 10 μm). Because the available PM air samplers do not have sharp particle size cut-offs, each PM sample may have particles outside of the intended size range. For more information on the deposition of inhaled PM see Chapter 7 on human exposures.

A currently unregulated size range, *ultrafine particles* (UFP) is being considered by the U.S. EPA for future regulation. Ultrafine particles are defined as the number of particles less than 0.1 μm in physical diameter.

Because the individual particles in UFP have very little mass, particle count, or even surface area they are logical targets for toxicity testing and regulation. Poorly-soluble ultrafine particles have some unique behaviors in the body. When they deposit by diffusion in the olfactory region of the nose they can travel along the olfactory nerve into the brain, and when they deposit throughout the respiratory tract, they can pass into epithelial tissues due to their tiny dimensions (see Chapter 3). The health implications of UFP that leaves the respiratory tract to expose other organs and tissues is a topic of current interest on the part of researchers.

The observed health effects of PM are quite variable, as would be expected given its diverse composition. Associations with adverse acute and chronic cardiovascular and pulmonary outcomes are seen in numerous studies as detailed in Chapters 9 and 10.

As of 2010, the NAAQS are:

- PM_{10}, 150 µg/m^3 (24-hour average);
- $PM_{2.5}$, 15 µg/m^3 (annual average); and
- $PM_{2.5}$, 35 µg/m^3 (24-hour average).

The NAAQS are likely to change as more research is performed. The toxicity of PM is modified by gaseous co-pollutants, which vary significantly at different exposure locations. As composition, solubility, and other particle properties are known to influence toxicity, the application of one set of PM NAAQS to all of the United States is controversial. In contrast, hundreds of existing workplace-recommended standards take particle composition and size into account.

Sulfur Dioxide

Sulfur dioxide (SO_2), a major air pollutant with a strong disagreeable sulfurous odor, is both water-soluble and a strong respiratory irritant. In the atmosphere, SO_2 is a precursor to acid sulfates (including sulfuric acid), which are neutralized by ammonia, which is usually present. SO_2 and its reaction products have been steadily dropping in concentration in the atmosphere, mainly due to decreasing sulfur levels in fuels.

When inhaled, SO_2 and some of its sulfate products contribute acidity to the respiratory tract, which can produce inflammation. Interestingly, ammonia levels in the human nasal and oral airways are often very high (up to 350 ppm), which is capable of neutralizing considerable acidity. SO_2 and acid sulfates have been intensively studied in humans and animals, so the major health effects are relatively well understood. In sufficient concentrations, SO_2 related irritancy causes cough, bronchial constriction, and alterations in the rate of bronchial mucociliary clearance. More specifically, low doses may accelerate mucus clearance, while higher doses produce a slowing of mucus flow. Chronic exposure to high levels may produce bronchitis, an inflammation of the major airways of the lungs.

In order to protect general populations, including asthmatics, the NAAQS for SO_2 are:

- 0.03 ppm (annual average);
- 0.14 ppm (24-hour average); and
- 0.075 ppm (3-year average of daily 1-hour maximum).

The workplace recommended SO_2 short-term exposure limit of 0.25 ppm is based on changes in pulmonary function and respiratory tract irritation (ACGIH®, 2010). As an aside, it is reasonable to ask whether aggressive control of ammonia emissions (e.g., from farms, dairies, poultry houses, and ranches) could have an adverse effect on human health by the mechanism of increasing the acidity of downwind air pollution. Recall that ammonia neutralizes airborne acidity.

Other Air Pollutants

Formaldehyde

Formaldehyde (HCHO) is a water-soluble gas with highly irritant properties. In solution it can kill cells and microorganisms, and thus preserve tissues; it is used as a fixative in mortuaries and medical laboratories. As a gas, it is used to sterilize buildings, rooms, and equipment. It is the most common aldehyde in outdoor air, and indoor concentrations may also be significant. Indoor sources include tobacco smoke, cosmetics, fresh paint, particle board, new carpeting, and some furniture. Formaldehyde is used as a potent microbiocide in consumer products, including paint and cosmetics. It is widely used in industry as a chemical intermediate. Therefore, exposure is essentially unavoidable for most of the population. Note that so-called "non-toxic" paint can be preserved with HCHO, so it must be applied and allowed to completely dry under conditions of adequate ventilation.

In sufficient air concentrations formaldehyde is a potent irritant of mucous-membranes, the eyes, and upper airways. It can cause skin reactions from contact, and it may be a carcinogen in prolonged heavy exposures. It has an irritating odor detectable at concentrations in the range of 0.5 to 1 ppm. Odor detection rapidly declines due to olfactory fatigue. Therefore, it does not have good odor warning properties. HCHO produces tissue irritation at airborne concentrations around 2 to 3 ppm. Levels above about 4 ppm can be intolerable due to its immediate irritant effects. It is similar to SO_2 with respect to toxicity, and is capable of producing nasal cancer in rats (exposed to 6 ppm for 2 years), and nasal cell proliferation in mice (exposed to 14 ppm for 5 days). Such rodent data generate concern over its potential carcinogenicity in humans. The recommended workplace short-term emergency concentration is 0.3 ppm, based on its irritation of upper respiratory tract airways and the eyes.

Bioaerosols

Aerosol particles that are either living or the products of living things make up the largely unseen world of *bioaerosols.* The human eye can resolve individual objects larger than about 70 μm in diameter. The particle diameter range of bioaerosols is less than 0.01 μm to 100 μm or greater. Bioaerosols include: (1) living and dead whole microorganisms (e.g., viruses, bacteria, and fungi); (2) cells and fragments of plants and animals; (3) spores; (4) fragments of excreta; (5) animal dander; and (6) wet or dry airborne animal secretions. Many bioaerosols have little health significance, but some are infectious, allergenic, irritant, or poisonous. They are found in nearly all exposure-related media, including indoor and outdoor air, water, soil, food, etc. Plants, insects, and animals are significant sources of bioaerosols; humans, for example, shed millions of dead-skin cells per minute (the shed cells also carry bacteria and viruses). Airborne transmission of bioaerosols is a mode of infection of some of the most common, and also some of the most deadly diseases.

Aside from being essential components of a healthy natural environment, bioaerosols probably produce more human death, debilitation, and discomfort than any other air contaminant. Because they are natural, their control is difficult. Bioaerosols can travel in air currents for great distances, exceeding thousands of miles or kilometers. Key references covering characteristics, sampling, health effects, and control of bioaerosols are found in Exhibit 3–3. The references describe dozens of health-related conditions, the responsible bioaerosol agents, and methods of control.

Hazardous Air Pollutants

The U.S. Clean Air Act specifies a group of anthropogenic air pollutants called *Hazardous Air Pollutants* (HAPs), or *Air Toxics* (see Chapter 3 and Table 3–2). HAPs are not included in the NAAQS, but they are considered to be important enough to be regulated at their sources. HAPs require risk assessments and emission controls to prevent adverse health and environmental effects; some are known or suspected carcinogens. Any person can make a request to the U.S. EPA for addition of a substance to the list of HAPs. As a result, the number of HAPs, which is currently about 200, is not fixed. The HAPs list is varied, including benzene, asbestos, mercury, pesticides, polycyclic hydrocarbon particles, etc., so their health and environmental effects must be examined individually. A good starting point for information on specific HAPs is the U.S. EPA Technology Transfer Network (http://www.epa.gov/ttn).

Acrolein

Acrolein is an irritant aldehyde that is more chemically reactive than formaldehyde, but it is found in lower concentrations in the air. Its effects include eye and mucous membrane irritation. Studies in guinea pigs indicated reversible (i.e., transient) increases in airflow resistance at 0.6 ppm and above. Guinea pigs are an exquisitely-sensitive animal model with respect to bronchial constriction, so they probably represent the most sensitive human asthmatics in toxicology studies. In prolonged inhalation exposures (to 0.4, 1.4, or 4.0 ppm, 6 hours/day, 5 days/week, for 13 weeks), rats exhibited a complex dose response. At the low dose there was lung hyperinflation (attributed to a decrease in airway resistance). At 1.4 ppm the effects were similar to those seen in controls (i.e., exposed to clean air). At 4.0 ppm there was inflammation and deep-lung fibrosis (i.e., scar formation, which can be life-threatening). For acrolein, the lower-airway effects at the highest dose, contrasted with the predominantly upper airway effects of inhaled formaldehyde. Some of this disparity is likely due to the lower solubility of acrolein in the mucous membranes of the upper airways. However, acrolein is still moderately soluble in water. The short-term exposure limit recommended for workers is 0.1 ppm on the basis of eye and upper respiratory tract irritation, lung edema (i.e., engorgement with blood or tissue liquids), and emphysema. These effects indicate that acrolein has both acute effects on upper airways and chronic effects in the deep lung.

Metals

Metals are common components of both outdoor air pollution and of the air in many occupational settings. Due to their varied intrinsic toxicities and vast number of chemical compounds, it is not possible to generalize their health effects. The effects of inhaled metals vary from benign to highly toxic, and even carcinogenic. Catalytic metals can intensify the effects of other inhaled air contaminants. The greatest concern for the adverse effects of metal-related aerosols is in the workplace, where air concentrations can be orders of magnitude greater than those in urban or rural air. Metal-fume fever, lung fibrosis, and cancer still plague metal workers. As expected, recommended workplace air standards for specific metals vary considerably, from as high as 5 mg/m^3 (Tungsten, Iron Oxide) to as low as 0.00005 mg/m^3 (Beryllium).

The problem with respect to the general population primarily derives from exposure to metals in emissions from the combustion of fuels. Coal and fuel oil are contaminated with several metals, depending on the local mineralogy at their sources. More recently, epidemiologists have associated adverse effects in elderly populations of fuel-related airborne metals such as vanadium, carbon, and nickel in $PM_{2.5}$. This, and similar findings, indicate the need for additional research on the effects of metal components of PM, as all PM is not equally harmful. Metal air pollutants are generally controlled in the environment by the use of clean fuels and filtration of combustion emissions (see Chapter 6).

Paradoxical Effects of Low Dose Exposures

The effects of administered chemical substances on living subjects can be complex. One of the non-intuitive interactions that is quite common is the triggering of *paradoxical effects* of toxicants at very *low doses,* also called *beneficial effects, biphasic dose-response phenomena,* or *hormesis* (Calabrese, 2008). See Figure 9-3 in Chapter 9 for a biphasic dose-response curve. There are hundreds of documented biphasic dose response observations in plants, animals, and microbes after exposure to chemical and physical agents (such as ionizing radiation). Therefore, the phenomenon is firmly established in toxicology and medicine. However, hormesis is still controversial, and thus not considered by regulators in establishing environmental regulations. Perhaps the lack of recognition of paradoxical effects of low doses of air pollutants in regulatory circles relates to the great individual variations in dose-response curves. More specifically, a low dose of an agent that may benefit the majority might also harm sensitive individuals or groups. Hormesis is in fact a familiar phenomenon. It is well known that (1) low doses of some sedatives (e.g., barbiturates and alcohol) can also produce a strong excitement phase; (2) the repeated administration of very small quantities of allergens can sometimes desensitize allergic patients; and (3) immunizations using small doses can prevent future infections. The later is a common practice in public health. In addition, essential minerals, vitamins, and nutrients have biphasic dose-response curves, as too little or too much of them produces adverse health effects. Currently, it is not feasible to integrate hormesis into air pollution regulatory decisions. However, as more is learned, this may change.

Sick-Building Syndrome vs. Building-Related Illness

Sick-building syndrome (SBS) (also known as *building-related symptoms)* is actually a collection of symptoms that persist for at least two weeks in 20 percent of the building's occupants, and *there is no specific known cause.* The symptoms include the following:

- eye and upper airway irritation;
- fatigue;
- headache and nausea;
- nasal congestion and/or bleeding;
- psychological irritation;
- reduced attention span;
- breathing problems; and
- skin dryness.

In SBS the symptoms diminish when people leave the building (e.g., on weekends or during vacations). Suspected causes are:

- inadequate ventilation;
- bioaerosols;
- combustion products;
- volatile organic compounds; and
- psychological stress.

Building-related illnesses (BRI), unlike SBS, have *well-documented causes* for the health problems associated with building occupancy. The causes include chemical air contaminants (CO, formaldehyde, NO_2, etc.), and bioaerosols (e.g., dust-mite and other insect-related allergens, infectious viruses, bacteria, mold-related products, etc.). The symptoms are specific to the causal agent(s), and not general, as in SBS. As an example, several deaths following a 1976 Legionnaires' conference in Philadelphia, Pennsylvania were found to be caused by a bacterial infection subsequently named *Legionnaires' disease.* The source of the pathogenic airborne agent, *Legionella pneumophila,* was contaminated cooling tower water. In general, industrial hygienists are responsible for the identification and control of such causal agents. When the causes are controlled the adverse effects vanish.

IV. SUSCEPTIBLE POPULATIONS

Susceptible populations include those who might require a more restrictive protective air standard than the general population. Chapter 10 covers this topic from the viewpoint of the epidemiologist, so only the basic principles will be presented here. Some groups are more *sensitive* because of poor nutrition or health, unfavorable genetics, age (very young or very old), or temporary conditions (e.g., acute infections) that intensify the response to a given dose. Others, such as school children, those who are exercising, or those exposed nearby significant local air pollution sources are more susceptible because they receive greater doses than the average population. Those with low socioeconomic status (e.g., those below the poverty level) are consistently seen in epidemiological studies as having greater adverse health effects (in comparison to the general population) from air pollutant exposures. Such sensitive populations may or may not be protected from harm by air pollution regulations. The regulations and normal control actions cannot adjust the air quality to levels that protect absolutely everyone. In many instances, extremely sensitive individuals must act to protect themselves by avoiding exposures, avoiding exercise during air pollution episodes, or even by use of doctor-prescribed medications. An interesting finding is that low educational attainment is associated with increased susceptibility to air pollutants; the reason is not well understood.

V. SOURCES OF INFORMATION ON HEALTH EFFECTS OF AIR POLLUTANTS

It is clear that brief summaries of the potential health effects of air pollutants are inadequate at best, and misleading at worst. So how can one obtain more complete information on specific air pollutants? Other than exhaustive searches of the peer-reviewed scientific journal article databases, a good place to start is TOXNET® (http://www.toxnet.nlm.nih.gov). TOXNET® is a collection of toxicology and environmental health databases, created and maintained by the Division of Specialized Information Services (SIS) of the National Library of Medicine (NLM). The searchable TOXNET® website opens over a dozen important databases that cover hazardous chemicals, environmental health, and environmental releases. The databases include:

- ChemIDplus®—structures, names, synonyms, and links for nearly, 400,000 chemicals;
- HSDB®—Hazardous Substances Data Bank, peer-reviewed toxicity data on about 5,000 chemicals;
- TOXLINE®—extensive toxicology literature online;
- CCRIS—Chemical Carcinogenesis Research Information Systems, covers carcinogenicity and mutagenicity information on over 8,000 chemicals;
- DART—Developmental and Reproductive Toxicology Database;
- GENETOX—Genetic Toxicology Database, with peer-reviewed information on over 500 chemicals;
- IRIS—Integrated Risk Information System, with hazard identification and dose-response information on over 500 chemicals;
- ITER—International Toxicity Estimates for Risk, that covers over 600 chemicals;
- LactMed—containing drugs and lactation (peer-reviewed) information that may be used for breast-feeding mothers;
- Multi-Database—containing searches on several chemical databases;
- TRI—Toxics Release Inventory, with annual U.S. environmental releases for over 600 chemicals;
- Haz-Map—links occupational diseases and their symptoms to hazardous jobs and tasks;
- Household Products—a household products database that covers the potential health effects of chemicals in over 8,500 household products; and
- TOXMAP®—with e-maps for geographic information on TRI data and links to other TOXNET® resources.

Other toxicity databases (e.g., from the U.S. EPA, WHO, and other governmental websites) are also available.

VI. SUMMARY OF MAJOR POINTS

Optimal health can be defined as a state of physical, mental, and social well-being. For many of the world's populations, the health effects of modern air pollutants are minor compared to the effects of infections, parasites, poor nutrition, and lack of adequate medical care. Yet, the adverse health effects of air pollutants are measurable, and even given priority in the competition for public-health monetary resources in many nations.

An unsolved problem is that when air pollutants are considered and controlled one at a time, new *unintended* adverse consequences may arise. Such consequences may be economic, or due to changes in air chemistry that increase the toxicity of other air pollutants. Economic effects have known public health implications, especially on those of lower socioeconomic status. Suppressing one problem (e.g., an individual air pollutant) can allow other public health threats to manifest themselves.

Health effects of, and exposures to, air pollutants may be *acute* or *chronic*. *Acute exposures* are measured in minutes to days, times short with respect to many biological processes, such as the times for cell turnover, and the development and repair of injuries. *Chronic exposures* are usually measured in months to years, which allow for the development of chronic diseases, sensitization, and long-term adaptation. In some cases, adverse acute responses may be protective in that they diminish the adverse consequences of future exposures. In other cases, repeated acute injury can eventually lead to chronic diseases.

In air pollution studies, many types of health-related information are required to support air standards. Controlled inhalation studies with human and laboratory animal subjects, *in vitro* mechanistic data, and real-world human exposures, including accidents, all contribute unique information. Models and computer simulations are used to predict inhalation doses, fates of deposited pollutants in the body, and for extrapolations of laboratory animal and *in vitro* data to humans.

Environmental and workplace air standards are based on critical adverse effects that are usually specific to each air contaminant. For example, workplace air standards are designed to protect humans from the asphyxiant property of carbon monoxide (CO). Environmental lead (Pb) standards are designed to eliminate the adverse effects on learning in children. Ozone (O_3) standards are based on respiratory tract cell damage. Also, air standards can potentially reduce the infections and allergic responses from exposure to bioaerosols.

Sick-building syndrome (SBS) is defined by a set of lasting symptoms in 20 percent of the occupants, which cannot be connected to any specific air pollutants. Lack of adequate ventilation is commonly associated with SBS. *Building-related illnesses* are different in that an identifiable cause, or causes, is known. Common agents that cause building-related illnesses include carbon monoxide, formaldehyde, and bioaerosols.

Susceptible populations include those who require additional protection than the general population from air pollutant exposures. Some people have greater responses than the average person to a given concentration of an air pollutant, while others receive much greater than average doses from a given air concentration of pollutants (e.g., children and exercising adults). Low socioeconomic status is associated with increased susceptibility to air pollution. In some cases, very sensitive people cannot be protected by air quality regulations, so they must take individual actions to protect their health.

An excellent source of health and environmental information on individual air pollutants is the TOXNET® website, which links to databases on chemistry, toxicity, carcinogenicity, accidents, and other useful data.

VII. QUIZ AND PROBLEMS

Quiz Questions

(select the best answer)

1. Human health includes:
   a. physical well-being.
   b. mental well-being.
   c. social well-being.
   d. All of the above are true.
2. Which factor is most important for protecting the public health in *developing* nations?
   a. air quality regulations.
   b. access to medical services.
   c. an outdoor and indoor air quality monitoring network.
   d. "tight" buildings that prevent smog from coming indoors.
3. "Direct," as opposed to "indirect," health effects of air pollutants include:
   a. only those effects that result from their contact with human or animal tissue.

   b. the costs of purchasing and using emission control equipment.
   c. the effect of lost work days on family income.
   d. All of the above are true.
4. Computer simulations:
   a. are irrelevant when studying the health effects of air pollutants.
   b. are useful for devising air quality regulations.
   c. are valid for evaluating the effects of particulate air pollutants but not gases.
   d. can currently replace *in vitro* laboratory and animal studies.
5. The health effects of gaseous air pollutants:
   a. are negligible.
   b. are independent of the specific gas.
   c. are all based on asphyxiation properties.
   d. None of the above are true.
6. The health effects of particulate air pollutants are:
   a. independent of particle size.
   b. independent of particle composition.
   c. negligible.
   d. None of the above are true.
7. Bioaerosols:
   a. are not a health problem because they are natural.
   b. include living microorganisms and dead organisms.
   c. include chemical pesticides.
   d. are outdoor but not indoor air pollutants.
8. Sick-building syndrome:
   a. is the same as building-related illnesses.
   b. is caused by known air contaminants.
   c. must affect at least 20 percent of the occupants.
   d. persists indefinitely after leaving the "sick" building.
9. TOXNET® searches:
   a. provide carcinogenicity information.
   b. provide toxicity information.
   c. provide chemical properties information.
   d. All of the above are true.
10. Hormesis:
   a. refers to paradoxical effects of low-dose exposures.
   b. refers to effects of particles but not gases.
   c. does not occur in air pollutant exposures.
   d. applies to laboratory animals but not humans.

Problems

1. Select two air pollutants and compare their odor threshold, symptoms of excess exposure, and lethal exposure levels. Hint: Use the U.S. EPA Technology Transfer Network, or TOXNET®.
2. Using the information in Table 8–3 determine whether or not the COHb level is a linear function of ppm CO in the air. Hint: Prepare a graph of the data.
3. Using the data in Table 8–1 make three graphs: Per Capita Income vs. Under 5 Infant Mortality; Literacy vs. Under 5 Infant Mortality; and Adult Lifespan vs. Under 5 Infant Mortality. Which factor correlates best with Infant Mortality? Which factor correlates least with Infant Mortality? Briefly discuss each of the three graphs in order to explain possible reasons for their correlations with infant mortality.
4. Using the TOXNET® website, find: (1) formaldehyde, (2) ozone, and (3) Pb (lead). Write a 1-page report on the public health effects of each air pollutant. Which one do you believe is the greatest hazard in your community at the present time?
5. Identify two infectious bioaerosols, one bacterial and one viral. Describe for each (1) their effects on human health and (2) measures for controlling or preventing infections.
6. Compare and contrast the data obtained from human clinical studies versus that from military casualties. Why are both types of data important for protecting public health?
7. Can chronic toxicology studies in mice be used to accurately predict chronic effects in humans? Explain your answer.

VIII. DISCUSSION TOPICS

1. Is it practical to devise particulate air pollution regulations that protect the health of everyone in the population? Explain your answer.
2. How might low educational achievement cause an increased susceptibility to air pollutants?
3. It has been suggested that intentionally inhaling metal aerosols could supply trace metals that are needed for maintaining good health. Is the use of inhalers that contain metals a good idea?

4. Pregnant women sometimes do additional cleaning and painting of rooms in which babies will be housed. Is this safe thing to do if "non-toxic" paints are used?
5. Is the current regulatory philosophy of regulating air pollutants one-by-one without regard to economic impacts in need of replacement?

References and Recommended Reading

ACGIH®, 2010 TLVs® and BEIs®, ACGIH®, Cincinnati, OH, 2010.

Anderson, D. M. (Chief Lexicographer), *Dorland's Illustrated Medical Dictionary, 30th Edition,* Saunders, Philadelphia, PA, 2003.

Bell, M. L., Ebisu, K., Peng, R., Samet, J., and Dominici, F., Hospital admissions and chemical composition of fine particle air pollution, *Am. J. Respir. Crit. Care Med.,* 179:1115–1120, 2009.

Burge, H. A., *Bioaersols,* Lewis Publishers, Boca Raton, FL, 1995.

Calabrese, E. J., Hormesis: Why it is important to toxicology and toxicologists, *Environ. Toxicol. Chem.,* 27:1451–1474, 2008.

Calabrese, E. J., Hormesis is central to toxicology, pharmacology and risk assessment, *Human Exptl. Pharmacol.,* 29:249–261, 2010.

Cohen, B. L., Catalog of risks extended and updated, *Health Phys.,* 61: 317–335, 1991.

Costa, D. L., Air pollution, in *Casarett and Doull's Toxicology: The Basic Science of Poisons, 7th Edition,* Klassen, C. D., Ed., McGraw–Hill, New York, 2008, pp. 1119–1156.

Dennison, J. E., Andersen, M. E., and Yang, R. S. H., Pitfalls and related improvements of *in vivo* gas uptake pharmacokinetic experimental systems, *Inhal. Toxicol.,* 17:539–548, 2005.

Graham, J. D., and Wiener, J. B., eds., *Risk vs. Risk: Tradeoffs in Protecting Health and the Environment,* Harvard University Press, Cambridge, MA, 1995.

HHMI, *The Race Against Lethal Microbes,* a report from the Howard Hughes Medical Institute, Chevy Chase, MD, 1996.

ICRP (International Commission on Radiological Protection, Task Group of Committee 2), *Human Respiratory Tract Model for Radiological Protection,* Publication 66, Pergamon Press, New York, 1994.

Ilano, A. L. and Raffin, T. A., Management of carbon monoxide poisoning, *Chest,* 97:165–169, 1990.

Jenkins, R. A., Guerin, M. K., and Tomkins, B. A., *The Chemistry of Environmental Tobacco Smoke: Composition and Measurement, 2nd Edition,* Lewis Publishers, Boca Raton, FL, 2000, pp. 49–60.

Macher, J., Ammann, H. A., Burge, H. A., Milton, D. K., and Money, P. R., Eds., *Bioaerosols: Assessment and Control,* ACGIH®, Cincinnati, OH, 1999.

Muilenberg, M. and Burge, H., Eds. *Aerobiology,* CRC Press, Boca Raton, FL, 1996.

NCRP, *Deposition Retention and Dosimetry of Inhaled Radioactive Substances,* NCRP SC 57–2 Report, National Council on Radiation Protection and Measurements, Bethesda, MD, 1997.

NRC, *Science and Decisions: Advancing Risk Assessment,* National Academies Press, Washington, DC, 2008.

Phalen, R. F., *The Particulate Air Pollution Controversy: A Case Study and Lessons Learned,* Kluwer Academic Publishers, Boston, MA, 2002, pp. 55–68.

UNICEF, *The State of the World's Children 2008,* United Nations Children's Fund, New York, 2007, (http://www.unicef.org, accessed August 4, 2010).

U.S. EPA, *Air Quality Criteria for Lead, Volume I of II,* EPA/600/R–5/144 aF, U.S. Environmental Protection Agency, Research Triangle Park, NC, 2006.

U.S. EPA, *Integrated Science Assessment for Oxides of Nitrogen-Health Criteria,* EPA/600/R–08/071, U.S. Environmental Protection Agency, Research Triangle Park, NC, 2008.

U.S. EPA, *Integrated Science Assessment for Carbon Monoxide,* EPA/600/R–09/019F, U.S. Environmental Protection Agency, Research Triangle Park, NC, 2010a.

U.S. EPA, *Integrated Science Assessment for Particulate Matter,* EPA 600/R–08/139F, U.S. Environmental Protection Agency, Research Triangle Park, NC, 2010b.

Varon, J., Marik, P. E., Fromm, R. E. Jr., and Gueler, A., Carbon monoxide poisoning: A review for clinicians, *J. Emerg. Med.,* 17:87–93, 1999.

Warheit, D. B., Borm, P. J. A., Hennes, C., and Lademann, J., Testing strategies to establish the safety of nanomaterials: Conclusions of an ECETOC workshop, *Inhal. Toxicol.,* 19:631–643, 2007.

Chapter 9

Toxicology Studies

LEARNING OBJECTIVES

By the end of this chapter the reader will be able to:

- define "toxicology" and describe its application to air pollution issues
- describe *in vitro* studies and what they can contribute to air pollution toxicology
- discuss the role of laboratory animal research in understanding the health effects of inhaled particles and gasses
- discuss the future research needs for understanding the toxicology of new air pollutants

CHAPTER OUTLINE

I. Introduction
II. Air Pollution Toxicology
III. *In Vitro* Studies and Mechanisms of Toxicity
IV. Animal Studies
V. Human Clinical Studies
VI. Exposure Methods
VII. Unsolved Problems in Air Pollution Toxicology
VIII. Summary of Major Points
IX. Quiz and Problems
X. Discussion Topics
References and Recommended Reading

I. INTRODUCTION

Definition, Scope, and Tools

What is Toxicology?

Toxicology is a scientific discipline devoted to the study of the harmful effects of chemical agents. The chemicals of interest in toxicology are those that originate from outside of the body. Such chemicals, also said to be "xenobiotic," "exogenous," or "foreign," can be natural (e.g., venoms, allergens, and vegetable or microbial toxins), or of anthropogenic (human) origin. Most potentially toxic chemicals can also be beneficial to health:

> "All substances are poisons; there is none which is not a poison. The right dose differentiates a poison from a remedy." Paracelsus (1493–1541) (as quoted by Michael Gallo, 2003).

In fact, even at a given dose, a substance can be beneficial to some people and harmful to others, as is true for many medications. Some toxicologists also study agents that are not chemicals, such as vibration, heat, noise, pressure, ionizing radiation, ultraviolet light, and zero-gravity. For the purposes of this text, only chemical air pollutants will be addressed.

The scope of toxicology is also seen in its various specialties, which include:

- *basic toxicology*—involves the mechanisms, both biochemical and physiological, that produce adverse effects;
- *developmental toxicology*—involves effects on developing organisms;
- *regulatory toxicology*—includes assessments of food, medicines, appliances (e.g., implants), industrial chemicals, pesticides, environmental pollutants, and other products regulated by governments;
- *forensic toxicology*—aimed at establishing a cause of death or processing crime-scene evidence; and
- *environmental toxicology*—involves studying environmental contaminants that potentially adversely affect animals (including humans), plants, and microbes.

What Tools are Used by Toxicologists?

The tools of toxicologists are varied. Some of the sciences and disciplines that underpin toxicology include chemistry (inorganic, organic, physical, and biochemistry); physics (especially for defining exposure and dose); mathematics and computer science (for modeling); biology; physiology; anatomy; genetics; medicine (including veterinary); and behavioral science. Toxicologists study chemical structures, chemical reactions, cell cultures, cell components (e.g., cell walls, mitochondria, DNA, etc.), microbes, insects, plants, and animals. The studies can be conducted in the "field" or in laboratories. Air pollution toxicologists often collaborate with other environmental scientists in order to understand the relevant chemical forms and exposure characteristics of the pollutants they are studying. As a result, most toxicologists work in a research group such as a university department, an institute, or a private or governmental laboratory.

Concepts in Toxicology

Dose

The concept of *dose* is of fundamental importance in toxicology, it could even be deemed the most important concept. To the public, the "dose" usually relates to how much medicine is administered in a pill, an injection, etc. To a toxicologist, dose can be a much more complex concept, as it has several general meanings (**Table 9–1**). For air pollutants the *deposition dose* is relevant.

The deposition dose is the amount of inhaled material that deposits in the respiratory tract. The total mass of material that deposits in the respiratory tract can be calculated by

$$Dose = C \cdot T \cdot V \cdot E, \qquad \text{(Eq. 9–1)}$$

where C is the air concentration of aerosol particles or gas (e.g., in milligrams/liter of air), T is the exposure duration (e.g., in minutes), V is the volume of air inhaled per unit time (e.g., liters/minute), and E is the deposition efficiency (the fraction of inhaled material that deposits in the respiratory tract instead of being exhaled). For example, consider an air concentration (of particles 1.0 micrometers in diameter) of 0.001 milligrams per liter of air, which is breathed for 480 minutes (8 hours), at an average inhalation rate of 10 liters per minute, with a deposition efficiency of 0.5. The calculation indicates the total deposition dose is 2.4 milligrams of particles:

$$Dose = 0.001\ \text{mg/L} \times 480\ \text{min} \times 10\ \text{L/min} \times 0.5$$

$$Dose = 2.4\ \text{mg}.$$

Table 9–1 Selected descriptions of the many doses used by toxicologists.

Terminology	*Description*
Dose	The "amount" of a substance in units relevant to the expected effects, e.g., milligrams, number of infective (or active) units, pH, particulate surface area; the dose may be *normalized,* e.g., per pound of body weight, per unit of alveolar surface area, etc.
Exposure dose (for airborne exposure)	An external exposure concentration, e.g., mg/cubic meter of air; or an internal exposure, e.g., air concentration × inhalation volume × time of exposure
Lethal dose	$LD_n(t)$, the lethal dose to *n* percent of the exposed population within time of exposure *t,* e.g., LD_{50} *(30),* dose lethal to 50% of exposed population within 30 days of exposure
Lethal concentration	$LC_n(t)$ the lethal air concentration for *n* percent of the population within time *t*
Lethal concentration low	LC_{LO} the lowest concentration in air that produces lethal toxicity when inhaled
Biologically effective dose	Amount required to produce a response of any type
Target tissue dose	Concentration in a specified tissue such as the lung's bronchial tree; units may be normalized, e.g., milligrams per unit surface area (or mass) of the tissue in question
Toxic dose low	TD_{LO} the lowest dose by any route other than inhalation to produce toxicity
No effect dose or concentration	The dose, or air concentration, below which effects have not been observed

This deposition dose can be made more relevant to the potential biological effect (say triggering an asthma attack) by normalizing it to the total body mass of the exposed individual. A normal average adult has a body mass of 70 kilograms (154 lbs). In this case the total dose is:

$$\textit{Dose per kg} = 2.4 \text{ mg} / 70 \text{ kg}$$
$$= 0.034 \text{ mg/kg}.$$

To make a further point, consider a newborn child breathing the same air concentration of particles for the same time period. The child has a body mass of 3.3 kilograms (7.3 lbs), an inhalation rate of 1.5 liters per minute, and a deposition efficiency of 0.6 (because of the increased collection efficiency of the smaller airways). In this case the normalized dose is:

$$\textit{Dose per kg} = 0.001 \text{ mg/L} \times 480 \text{ min} \times 1.5 \text{ L/min}$$
$$\times 0.6 / 3.3 \text{ kg}$$
$$= 0.13 \text{ mg/kg}.$$

This newborn's dose per kilogram of body mass is 3.8 times greater than that for an adult breathing from the same air concentration! The importance of normalizing the dose to body mass is clear when one is considering the potential biological effects. Body size normalization yields a *toxicologically effective dose.* As a general rule, smaller subjects will receive greater inhalation doses under the same exposure concentrations than are received by larger subjects. Table 7–6 shows the average ventilation rates and average body masses for people of various ages.

The increased ventilation rate per unit body mass holds for smaller mammals in general (**Figure 9–1**). Because mammals are warm-blooded, they must generate sufficient metabolic internal heat to maintain body temperatures (e.g., 37°C for humans). As small mammals have a larger surface-to-volume ratio, they breathe more oxygen per unit body mass to compensate for heat loss. This means that smaller mammals also receive larger *body-mass-specific* doses of air pollutants. This effect (along with others) should be taken into account when designing and interpreting toxicology studies that

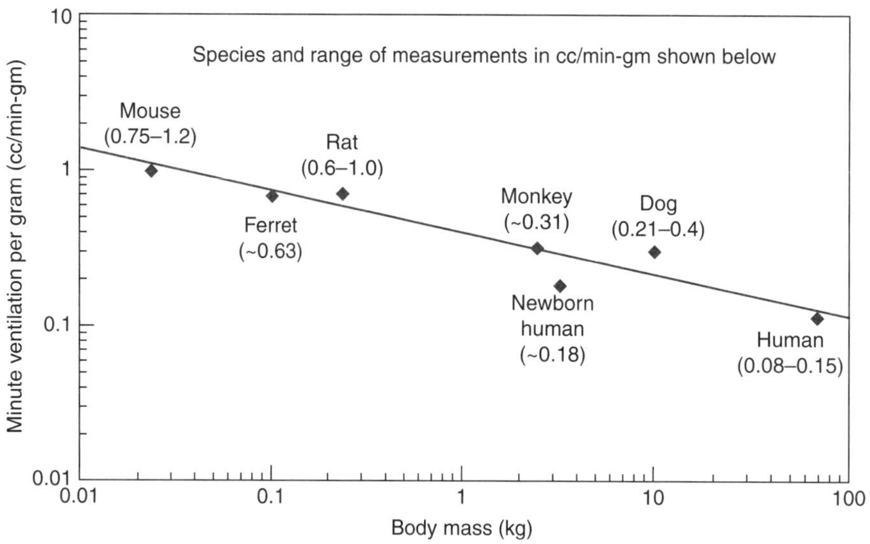

Figure 9–1 Observed log-log relationship between minute ventilation and body mass for several mammals. *Source:* The University of California Air Pollution Health Effects Laboratory, with kind permission.

use laboratory animals, and when calculating air pollution doses to domestic and wild animals.

The eminent American physiologist Arthur C. Guyton described the mathematical relationships between ventilation and body mass (or weight) for mammals in 1947. His relationship for tidal volume (V_t), breathing frequency (f), and body weight (w) are

$$V_t(\text{cm}^3) = 0.0074\, w\ (\text{grams}) \qquad \text{(Eq. 9–2)}$$

$$f(\text{min}^{-1}) = 295/w^{1/4}\ (\text{grams}). \qquad \text{(Eq. 9–3)}$$

Multiplying V_t and f to get minute ventilation V_m and dividing by body weight *(w)* one obtains

$$V_m/w\ (\text{cm}^3/\text{min/g}) = 2.18/w^{1/4}\ (\text{grams}). \qquad \text{(Eq. 9–4)}$$

Thus, as one example, one can use body weight to estimate the amount of air inspired by a subject in one minute, which is the minute ventilation (V_m). For example, for a 20 g mouse the estimated minute ventilation is

$$V_m = 2.18/w^{1/4} \times w = 2.18\ \text{cm}^3/\text{min}/(20\ \text{g})^{1/4} \times 20\ \text{g}$$

$$= 21\ \text{cm}^3/\text{min}.$$

In Figure 9–1, the observed V_m/w for mice ranges from about 0.75 to 1.2 cm^3/min/g, which for a 20 g mouse would provide a comparable V_m between 15 and 24 cm^3/min.

As is seen from the foregoing, the concept of dose in air pollution toxicology is more complicated than the familiar doses used for medicines.

Dose-Response Relationships

Dose-response relationships are fundamental to quantitative toxicology. Such relationships are used to compare the toxicity of various substances, and they are also used by regulators to establish criteria for air quality. There are two basic types of dose-response relationships: those for an *individual,* and those for *groups* of individuals (**Figure 9–2**). Dose-response relationships are also categorized according to their shapes when drawn as *dose-response curves* (**Figure 9–3**). The shaded region in Figure 9–3 is a region of beneficial effect. This effect is also called *hormesis.* For more information on hormesis see Chapter 6 and newsletters from *Biological Effects of Low Exposures* online (http://www.belleonline.com).

Classification of Toxic Agents

It has long been recognized that substances vary considerably in their toxicities. **Table 9–2** shows the lethal oral doses (LD_{50}) of some representative substances. Note the most toxic substances known are usually of

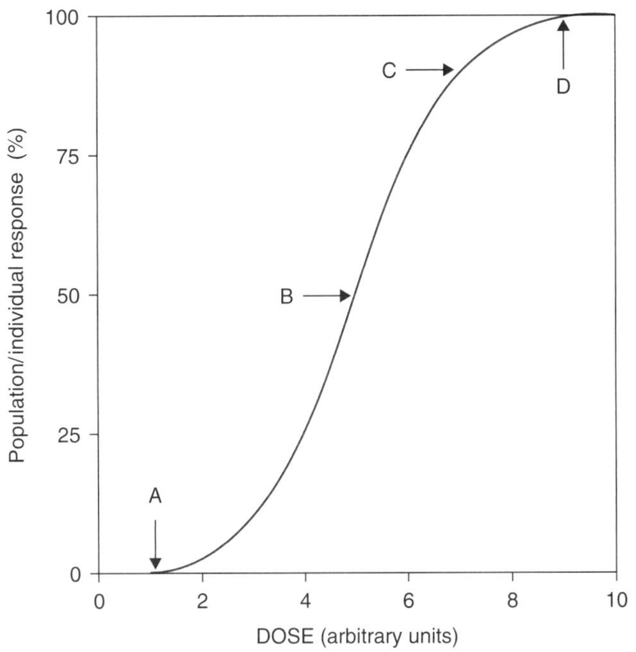

Figure 9–2 A typical sigmoidal dose-response relationship showing: A the threshold of response; B the point of maximum sensitivity; C the region of near saturation of response; and D the maximum response.
Source: The University of California Air Pollution Health Effects Laboratory, with kind permission.

Table 9–2 Approximate acute lethal oral doses (LD_{50}) of some representative substances.

Toxicity Class	*Substances*	*LD_{50} in mg/kg body mass*
Relatively Harmless	sucrose (beet sugar)	29,700
Practically Non-toxic	ethyl alcohol	10,000
Slightly Toxic	sodium chloride,	4,000
	ferrous sulfate,	1,500
	morphine sulfate,	900
Moderately Toxic	phenobarbital sodium	150
Highly Toxic	strychnine sulfate,	2
	nicotine	1
Extremely Toxic	tetrodotoxin (from fish)	0.1
	botulinum toxin (from bacteria)	0.00001

Data from Eaton and Klaassen (2003); and Hodge and Sterner (1949).

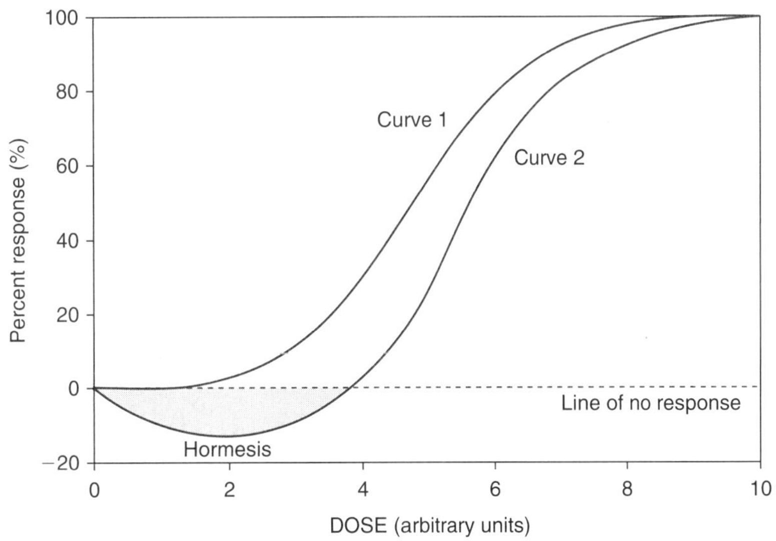

Figure 9–3 Dose-response curves: Curve 1 is a traditional sigmoidal curve; Curve 2 shows a shaded region of beneficial effect at low doses (also called a region of hormesis).
Source: The University of California Air Pollution Health Effects Laboratory, with kind permission.

natural origin. The classification and ranking of toxicities of substances depends on several factors, including the following:

- exposed subject's health, state of exertion, etc.;
- gender and species (e.g., human, rat, mouse);
- route of exposure (e.g., oral, inhalation, skin, injection, etc.);
- rate of exposure (e.g., acute vs. chronic);
- physical form of substances (e.g., solid vs. liquid, large vs. small particles, etc.);
- co-exposures (e.g., mixtures); and
- environmental factors (temperature, pressure, etc.)

Many classifications of toxic substances are possible, such as *carcinogens, teratogens* (produce birth defects), *irritants, allergens,* etc.

For air pollutants, the lowest acute lethal air concentration (LC_{LO}) is one useful way of ranking their toxicities. Another method, which is applicable to workers, is to use the permissible workplace air concentrations (i.e., TLV®). **Table 9–3** lists some gases and aerosol particles along with their acute LC_{LO} values, permissible workplace air concentrations, and the U.S. EPA's National Ambient Air Quality Standards. As is seen in the table, permissible exposure air standards are well below any known lethal concentrations in humans or other mammals.

Routes of Exposure

The specific route of exposure (**Figure 9–4**) to any substance will modify its effects on exposed subjects. Most commonly, the greatest effects are seen at the point of exposure. However, any route of exposure can produce effects in distant organs and tissues by the following mechanisms:

- The substance may enter the blood circulation and be distributed throughout the body.

Table 9–3 Data on air pollutant gases and particles including the lowest observed acute lethal concentrations (LC_{LO}), permissible occupational 8-hour average air levels (TLV®), U.S. EPA National Ambient Air Quality Standards (NAAQS), and key toxic effects.

Substance	*Acute LC_{LO}*	*TLV®*	*NAAQS*	*Key Effects*
Ammonia	3,000 ppm, for 5 min, man	25 ppm	None	eye and lung irritation
Carbon monoxide	5,000 ppm, for 5 min, man	25 ppm	35 ppm, 1-hr. avg. 9 ppm, 8-hr. avg.	asphyxiant
Nitrogen dioxide	200 ppm, 1 min, man	3 ppm	0.053 ppm, annual avg. 0.1 ppm, 1-hr avg.	pulmonary injury
Ozone	100 ppm, 1 min, man	0.05–0.2 ppm, depends on level of exertion	0.075 ppm, 1 hr. avg.	pulmonary injury
Sulfur dioxide	1,000 ppm, 10 min, man	no TWA, short term limit 0.25 ppm	0.03 ppm, annual avg. 0.14 ppm, 8-hr avg.	irritant
Lead particles	unknown	0.05 mg/m³	0.15 µg/m³, 3 mo. avg.	nerve & blood effects
Natural latex particles	unknown	0.1 µg/m³	None	pulmonary sensitizer
Cadmium particles	39 mg/m³, 20 min, man	0.01 mg/m³	None	kidney damage
Urban particles	unknown	not applicable	PM_{10}, 150 µg/m³, 24 hr. avg. $PM_{2.5}$, 35 µg/m³, 24 hr. avg. $PM_{2.5}$, 15 µg/m³, annual avg.	pulmonary, cardiovascular injury

Data from ACGIH® (2010); Lewis (2000); and U.S. EPA (2010).

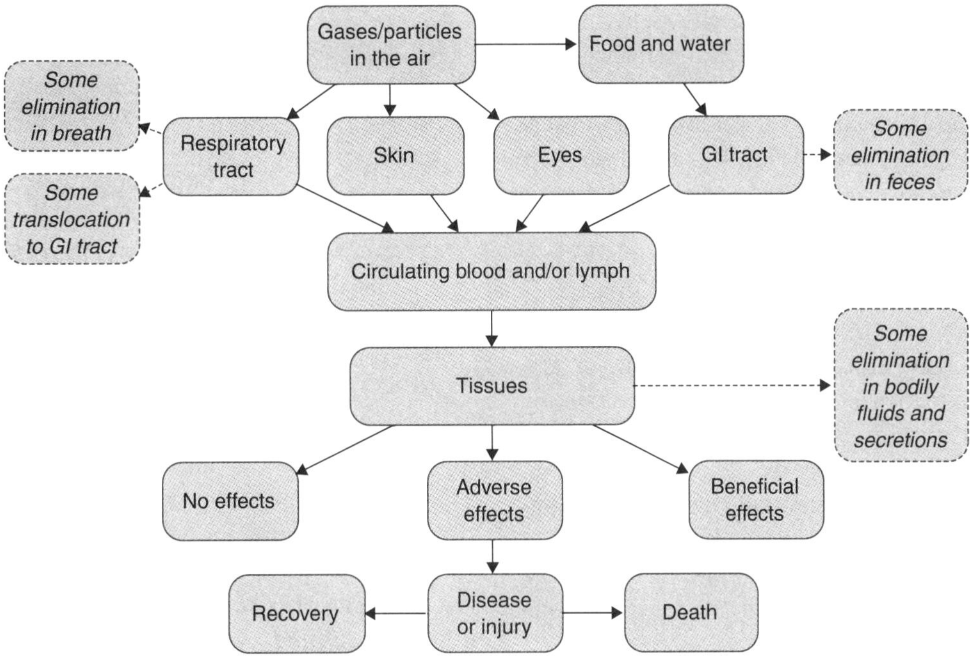

Figure 9–4 Routes of exposure to chemical substances.
Source: The University of California Air Pollution Health Effects Laboratory, with kind permission.

- The substance may generate a local cellular response, which releases biochemicals that can travel to other tissues and elicit responses.
- The substance may trigger central nervous system responses that affect distant organs, glands, and tissues.
- The substance may damage an organ or tissue, which leads to compensatory responses in other organs.

There are many examples of distant effects produced by chemical substances. Inhaled carbon monoxide travels to the blood and binds tightly to hemoglobin (the protein carries and delivers oxygen to all of the body's cells), which then deprives all tissues of oxygen. The brain, eyes, heart, and muscles are the first to be affected if the carbon monoxide exposure is moderate. At high concentrations, loss of consciousness, convulsions, and death occur if the exposure is not terminated, as by providing fresh air or administering oxygen. As another example, dermal or inhalation exposure to solvents also leads to general distribution via the blood circulation. As most solvents are significantly soluble in oil, fatty tissues in the brain extract the solvent from the blood. Symptoms range from light-headedness to unconsciousness and death from respiratory failure (the central nervous system controls the breathing muscles). As a final example, inhaled irritant particles in sufficient concentrations can produce inflammation in the lungs that leads to edema (swelling of the tissues) and thus compression of the pulmonary blood vessels. This effect imposes a heavier than normal workload on the heart, which must continuously pump large quantities of blood through the lungs. If the heart has been compromised, as by a poor cardiac muscle blood supply or a previously damaged heart muscle (e.g., due to a prior heart attack), the heart may fail. These examples underscore the point that the body is highly integrated in its essential functions. Toxicologists must consider *all* adverse effects, not just those at the site of contact with environmental contaminants.

The route of exposure also influences the chemical transformations (biotransformations) that a xenobiotic (foreign to the body) substance undergoes in the body. *Biotransformations* usually decrease the toxicity of xenobiotic chemicals and converts them to forms (e.g., that have greater hydrophilicity or water solubility) that favor excretion. Although such *detoxification* is a common result of biotransformations, *activation* to a more toxic or carcinogenic form can also occur. Xenobiotic biotransformation is achieved by any of numerous enzymes, such as the versatile cytochrome P450s, alcohol dehydrogenase (ADH), esterases, reductases, and many more. The liver, lung, skin, kidney, and other organs and tissues have variable levels of these enzymes depending on genetics, exposure histories, and disease processes.

Toxicokinetics

Toxicokinetics comprises absorption, distribution, metabolism, storage, and excretion of xenobiotic substances. Toxicokinetics is the study of the processes by which toxic substances, and their metabolites, interact within and pass through the body.

In order to simplify the potentially bewildering complexity involved in the fate of a xenobiotic chemical in the body, a modified ADMSE (absorption, distribution, metabolism, storage, and excretion) model is shown in **Figure 9–5**.

Absorption

Except for direct injections (as in intravenous administration) xenobiotic substances must cross lipid-rich cell membranes in order to be absorbed and thus enter the body. There are five main processes by which chemicals cross cell membranes:

1. *passive diffusion*—Lipid-soluble substances can enter cell walls and diffuse through them. The rate of diffusion will depend on the thickness and surface area of the barrier membrane, on the concentration gradient across the membrane, and on the molecular weight and equilibrium partition coefficient (e.g., lipid: water or lipid: air) of the xenobiotic.
2. *diffusion through open ion channels*—Water-soluble compounds and ions cannot dissolve in the cell membrane lipids. Ion channels (pores) in cell membranes allow non-lipid soluble xenobiotics to bypass the lipid barrier and diffuse across cell membranes. As the pores are small, large molecules are excluded.
3. *facilitated diffusion*—Cell-membrane proteins that physically penetrate the lipid barrier can permit free diffusion of selected substances as a result of concentration gradients. Energy is not required for such transport.
4. *active transport*—Transporter proteins that require energy, supplied by ATP (adenosine triphosphate, a high-energy, unstable compound), can transport selected substances across membranes. Such transport can proceed against a concentration gradient.
5. *endocytosis*—Endocytosis, or physical engulfment of extracellular material, can transport substances into a cell. Endocytosis involves the formation of a pocket in the cell membrane followed by closure of the membrane fully inside the cell. There are two types of endocytosis: (1) *pinocytosis* ("cell-drinking") that mainly engulfs liquids and (2) *phagocytosis* ("cell eating") by macrophage cells in the respiratory tract that are active in engulfing microbes and other particles that deposit on respiratory-tract surfaces. Powerful chemicals in macrophages can kill microbes. Macrophages can destroy and/or transport particles out of the lungs (as macrophages are mobile).

The absorption of xenobiotic substances is a relatively well-understood process. However, the route of exposure, ingestion, inhalation, or percutaneous (through the skin) will modify the process. Skin, with its layer of dead cells (the stratum corneum), is a more efficient barrier than the thin alveolar membranes of the lungs.

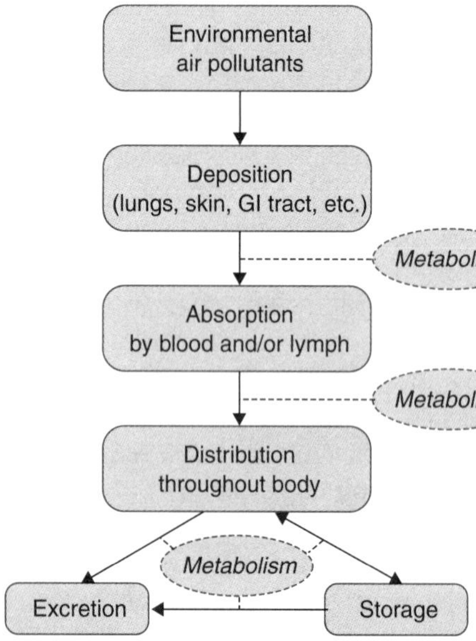

Figure 9–5 The toxicokinetics model.
Source: The University of California Air Pollution Health Effects Laboratory, with kind permission.

In the alimentary tract (mouth, esophagus, stomach, intestines, and rectum) thick mucus layers, digestive enzymes, and large changes in pH are all factors that influence the absorption of xenobiotic substances.

Distribution

Once it is absorbed, a substance is available for distribution throughout the body, although local binding (sequestration) can also occur. Primarily the blood, and secondarily the lymph, are efficient in facilitating such distribution throughout the body. The blood circulation is designed to efficiently transport oxygen and nutrients to all of the tissues of the body and to remove waste products. This system is also capable of transporting toxicants (including air pollutants) throughout the body.

The blood circulation is simplistically depicted in **Figure 9–6**. The blood flow to selected organs and their masses is shown in **Table 9–4**. The blood flow to various organs varies considerably, with body fat deposits being the most poorly perfused. The size of the body fat reservoir is large, which makes it capable of storing significant amounts of fat-soluble toxicants. When considering the distribution of xenobiotics to various tissues and organs the following factors should be taken into account:

- the equilibrium distribution coefficients of the substance in water and lipid;
- the amount of fat (i.e., lipids) and water in the perfused tissues and organs;
- the blood flows (i.e., perfusion) to various organs and tissues at the level of exertion (e.g., rest vs. exercise) during exposure;
- the duration of exposure;
- the rate of elimination of the substances from the blood; and
- other factors, such as metabolism of the substances, and individual differences in body size, health status, genetic variations, etc.

The *lymph* flow returns fluid from the tissues back to the bloodstream. Lymph in lymphatic channels also flows through *lymph nodes,* where some insoluble particles are collected and stored.

Toxicologists who use laboratory animal models to study the internal distribution of xenobiotics, must take species differences into account. For this and other reasons, several species are studied in toxicology laboratories before the results can be confidently extrapolated to humans.

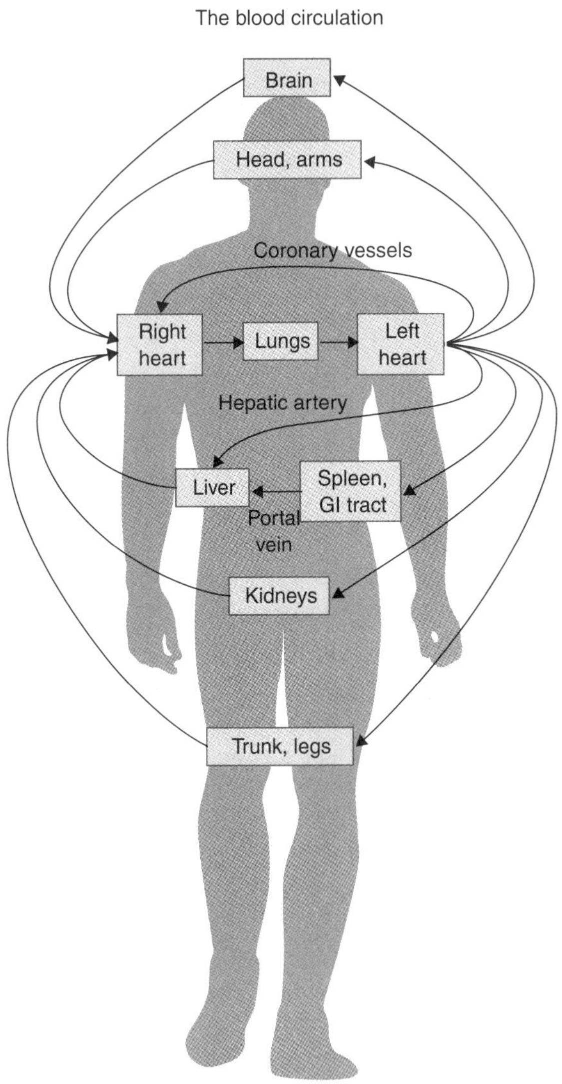

Figure 9–6 The human blood circulation. *Source:* The University of California Air Pollution Health Effects Laboratory, with kind permission.

Metabolism of Xenobiotic Substances

The body is very active in modifying the chemical and physical properties of xenobiotic substances to which it is exposed. *Metabolism* is the process of chemical transformation in living things. Without this capability, toxicants would eventually accumulate and produce death. After all, *xenobiotic substances are by definition materials that do not belong to the body.* Some substances, such as unreactive gases, liquids, and solids, do not undergo significant chemical changes in

Table 9–4 Blood flow in organs and fat tissue of an average 70 kg resting adult male. The lung has two blood supplies: a) the pulmonary artery (pa) supplies blood for oxygenation and b) the bronchial arteries (ba) supply a smaller flow of oxygenated blood to nourish the lung itself.

Organ or Tissue	*Mass (kg)*	*Blood Flow (ml/(kg × min)*
Lungs (2)	1.00	5,400
(a) pa		5,300
(b) ba		100
Kidneys (2)	0.31	950
Liver	1.8	750
Brain	1.4	500
Heart muscle	0.33	210
Skin	3.5	86
Muscles	28.0	30
Skeleton	10.0	29
Fat tissue	13.5	23

Note: Organ mass and weight are used interchangeably, but they differ in altered gravity conditions. In zero gravity, the mass is unchanged, but the weight is zero.

Data from ICRP (1975); Guyton (1991); and Lindstedt and Schaeffer (2002).

the body. For substances that are metabolized, or otherwise altered after absorption, the complex chemistry of the body must be considered by toxicologists. Given there are thousands of potentially significant air pollutants, understanding the metabolism of each one is a formidable task. For the sake of illumination of general principles, the basic types of metabolic reactions are briefly examined here.

Those xenobiotic pollutants that gain access to the body by passive diffusion are usually lipid soluble, and consequently not easily eliminated from the body. The two main excretory organs are the kidneys (which pass substances into the urine) and the liver (which passes substances into the gallbladder that empties into the duodenum of the small intestine). These excretory pathways favor water-soluble substances, so a major function of xenobiotic metabolism is to convert lipid-soluble substances to water-soluble products. This biotransformation decreases the half-life of retention of converted substances (i.e., hastens their elimination from the body).

Biotransformation is divided into two successive steps: *Phase 1* reactions and *Phase 2* reactions. Phase 1 reactions alter the original molecule by the addition of a *functional group* (such as a hydroxyl, OH group). The functional group then facilitates either excretion from the body (e.g., in urine or sweat) or further processing, such as *conjugation.* Conjugation is the covalent linking of a foreign (xenobiotic) compound with an endogenous (found normally in the body) enzymatic compound to yield a product called a conjugate.

Table 9–5 describes some typical conjugation reactions. The conjugate is then ready for further conversion by a Phase 2 reaction.

Phase 2 reactions (also called conjugation reactions) involve the addition of a polar (characterized by an uneven distribution of electrical charge) endogenous moiety to the xenobiotic substrate. The separation of charge in a compound allows it to interact with polar water molecules, which causes it to dissolve in the water solvent. In contrast, non-polar compounds (e.g., lipid-soluble compounds) do not dissolve in water (i.e., they are immiscible, like the oil that forms droplets in a salad dressing). As previously mentioned, water solubility is a key property that facilitates elimination from the body in urine and bile. **Table 9–6** shows some of the important Phase 2 reactions.

Table 9–5 Some typical Phase 1 conjugation reactions of xenobiotics.

Type of Reaction	*Typical Enzymes*	*Sample Reactants*
Oxidation	cytochrome P450	aromatics aliphatics alicyclic rings alkyl group cpds. nitrogen and sulfur cpds.
Reduction	reductases (various)	nitro, azo group cpds. aldehydes ketones disulfides sulfoxides quinones metals etc.
Hydrolysis	esterases; amidases	esters, which include some insecticides, warfare agents, and drugs; amides
Hydration	epoxide hydrolase	alkene epoxides arene oxides

Table 9–6 Some Phase 2 conjugation biotransformation reactions.

Type of Reaction	*Typical Enzymes*	*Sample Reactants*
Sulfation (addition of sulfate)	sulfatransferases	aromatics aliphatics N-hydroxy cpds. amino cpds.
Glucuronidation (addition of glucuronic acid)	glucuronosyl transferases	pesticides, e.g., tecnazene and fenhexamid
Glutathione conjugation (addition of glutathione)	glutathione transferases	epoxides halogen cpds. nitro cpds. alkyl halides unsaturated aliphatics
Acetylation	acetyltransferases, ligases	aromatic amino cpds. hydrazines hydrazides sulfonamides organic acids
Methylation	methyltransferases	metals hydroxyl cpds. amino cpds. thiol cpds.

As can be appreciated by the foregoing brief introduction to Phase 1 and Phase 2 biotransformation reactions, the subject is both vast and complex. A review of this topic by Timbrell and Marrs (2009) covers the basics along with factors such as species, strain, sex, genetic factors, and environmental factors that affect the metabolism of xenobiotic chemicals. Although such biotransformations generally serve to detoxify environmental pollutants, there are also some exceptions that involve the formation of more-toxic compounds.

Storage in the Body

The storage of xenobiotics, which is more accurately called *long-term retention,* or *slow elimination,* can also occur. Mechanisms for long-term retention of contaminants include the following:

- *physical trapping*—Poorly-soluble fine particles (e.g., asbestos, gold, or carbon) can be retained in the lungs for decades. The locations of such stored particles can be in the alveoli, within the bronchi (especially at bifurcations), or in the lymph nodes and channels found in lung tissue and in the lymph nodes external to the lungs (e.g., at the bifurcation of the trachea).
- *incorporation*—A tissue in the body with slow structural turnover, such as bone, can accumulate foreign materials by incorporating them into its structure. An example is the retention of lead (Pb) in bone because Pb is chemically similar to calcium. The clearance half-time of Pb in bone is about 30 years and plutonium (Pu) in bone is about 100 years. Other tissues, such as those in the liver, kidney, and nervous system can also retain xenobiotic substances for decades.
- *dissolution*—Fatty tissue is a long-term storage site for substances soluble in lipids. Such substances are described as lipid-soluble, fat-soluble, or oil-soluble. The long-term retention in fat deposits is related to their slow washout due to poor blood perfusion. On the other hand, the brain is richly perfused by blood, so washout rates are more rapid. A practical health issue arises when the accumulation of lipid-soluble xenobiotic compounds (e.g., several pesticides and solvents) are followed by rapid weight loss. The rapid loss of fatty tissue in such circumstances can lead to high blood levels of toxicants and subsequent toxic effects.

To summarize, the long-term retention of substances is facilitated by their incorporation or dissolution in tissues with slow-turnover rates. Such tissues include bone, fat, neurons, kidney, and liver. Also, particulate material that has low solubility or reactivity can persist throughout life in lung tissue, bronchial airway tissue, lymph nodes, and other tissues. When substances are retained for long periods, chronic internal exposure and toxic effects, including fibrosis and cancer can occur.

Excretion

Pollutants that deposit in the body are eliminated by several routes. Elimination mechanisms include the following:

- Chemical conversion can eliminate substances by destroying them. However, the conversion products may produce toxic effects if present in significant concentrations.

- Urinary excretion is a mechanism for water-soluble xenobiotics that undergo Phase 1 and Phase 2 reactions.
- Fecal excretion occurs for materials that enter the gastrointestinal tract and are not absorbed. Oral doses and swallowed particles cleared from the respiratory tract by the mucociliary escalator can also undergo fecal elimination.
- Exhalation is an efficient route of elimination of volatile substances, including non-reacting gases and volatile metabolic products. Volatility at body temperature, and equilibrium partition coefficients, determine the effectiveness of this route.
- Sweat and saliva are potential, though minor, elimination routes. Significantly toxic compounds may cause dermatitis if excreted in sweat. Substances that pass into the saliva are typically swallowed where they enter the gastrointestinal tract.
- Milk is an important route of elimination because potentially toxic substances may be passed to nursing offspring. Also, the consumption of milk from contaminated cows and goats can lead to the oral dosing of humans, pets, and livestock.
- Expectoration (spitting) and coughing are routes of elimination of irritating substances.
- Cerebrospinal (CS) fluid, which cushions the spinal cord and suspends the brain (the specific gravity of the brain is about four percent less than CS fluid), is renewed at a rate of about 500 milliliters per day, producing a turnover of about three times daily. The CS fluid also functions to transport proteins and particulate material (such as dead cell debris) and other substances into the venous blood. The importance of CS fluid for the clearance of xenobiotics is poorly understood but likely to be important for some contaminants.

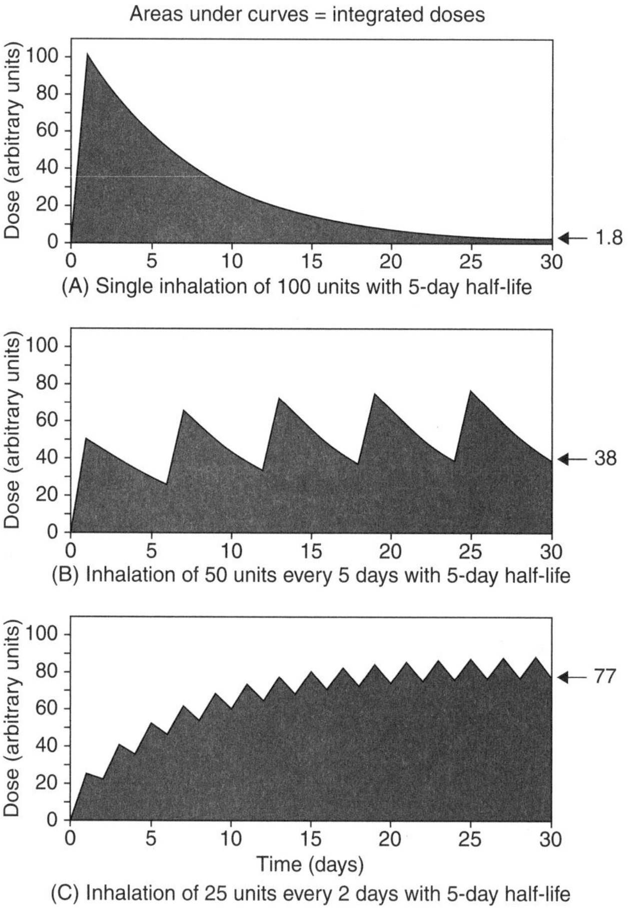

Figure 9–7 Three dosing regimens with 5-day clearance half-life: (A) single dose of 100 units; (B) repeated doses of 50 units every 5 days; (C) repeated doses of 25 units every 2 days. Dose remaining at day 30 is shown at the right for each.
Source: The University of California Air Pollution Health Effects Laboratory, with kind permission.

The elimination of xenobiotics through various routes is addressed (along with their absorption, distribution, metabolism, and storage) by *toxicokinetic modeling,* which is also used to describe the movement and disposition of substances in the whole organism. Toxicokinetic modeling provides *instantaneous* and *integrated* doses of modeled substances in various organs and tissues. **Figure 9–7** provides examples of the uptake and elimination of a material using three different dosing regimens. The combination of exposure timing and clearance rates as seen in the figure has a significant impact on the final body burden. The *area* under each graph is the *integrated dose,* sometimes referred to as the "area under the curve." Toxicokinetic modeling is also essential for applying laboratory animal data to humans.

II. AIR POLLUTION TOXICOLOGY

Introduction

Air pollution toxicologists use several methodologies including *in vitro, whole animal,* and *human clinical* techniques. Each of these methodologies uses mathematical modeling in the design, interpretation,

Exhibit 9–1 Sources of air pollution toxicology information.

Air pollution toxicology is a vast and interesting topic. What is presented here is only a brief and incomplete summary of information sources. Detailed up-to-date information can be found in the scientific journals *Inhalation Toxicology; Critical Reviews in Toxicology; Aerosol Science and Technology; Experimental Lung Research; Environmental Health Perspectives; Environmental Science and Technology; The Journal of Occupational and Environmental Hygiene; The Annals of Occupational Hygiene; Toxicological Sciences; Nonlinearity in Biology-Toxicology-Medicine;* and *Health Physics* (for radioactive substances), to name a few.

Several monographs (books that treat one subject in-depth) and textbooks also provide information on air pollutant toxicology including *Casarett and Doull's Toxicology: The Basic Science of Poisons* (Klaassen, 2008); *General and Applied Toxicology* (Ballantyne et al., 2009); *Air Pollution and Health* (Holgate et al., 1999); *Air Pollution and Community Health* (Lipfert, 1994); *Particle-lung Interactions* (Gehr and Heyder, 2000); *Toxicology of the Lung* (Gardner et al., 1999); *Concepts in Inhalation Toxicology* (McClellan and Henderson, 1995); *Environmental Health Science* (Lippmann et al., 2003); *Particle Toxicology* (Donaldson and Borm, 2007); and others. There is essentially no limit on the resources available for research on air pollution toxicology.

and extrapolation (extending results to unstudied problems) of their studies. It is only through the combined results of the above methods that the toxicology of specific air pollutants can be understood. **Exhibit 9–1** lists some of the sources of detailed information on the toxicology of air pollutants.

Because the relative intake of air on both a volume and mass basis far exceeds those of water, food (see Figure 7–1), or through the skin, the focus of air pollution toxicology is on inhalation exposures. Note that *inhalation toxicology* involves the study of all of the adverse effects of inhaled substances. The respiratory tract is both a portal for entry and a potential target of air pollutants. For many air contaminants, the respiratory tract is not the critical tissue from the toxicology perspective. The heart, central nervous system, peripheral nervous system, the blood, eyes, skeleton, and other organs and tissues have all been shown to be harmed by one or more specific air pollutants.

The following air pollution toxicology concepts are discussed in greater detail:

- *In vitro* studies are essential for screening purposes and for discovering the biological mechanisms of injury and disease.
- Animal studies are critical for understanding whole-animal effects of pollutants, including cancer, heart disease, asthma, birth defects, and others.
- Human clinical studies are needed, especially for developing information to set safe levels of air pollutants and to protect potentially sensitive subgroups.

III. *IN VITRO* STUDIES AND MECHANISMS OF TOXICITY

Overview

In vitro literally translates as "in glass." This translation does not capture the true scope of *in vitro* studies, which include the use of cell cultures; organ cultures; organelles (e.g., mitochondria, cell nuclei, or other components of cells); and biochemical preparations.

The major drawbacks of *in vitro* testing are mostly associated with the isolation of cells, as occurs in cell cultures. This isolation deprives the cells of their normal interactions with (1) other cell types, as only 2 or 3 cell types can be successfully co-cultured; (2) the many substances made in the body and distributed in the blood; and (3) realistic contact with the studied pollutant (and its metabolic products). Thus, *in vitro* testing is primarily used as an important screening tool for hazardous or toxic properties of potential air pollutants.

Macrophages and Other Cells Used to Study Air Pollutants

Although there are a large number of cell types, ranging from those in mammals to yeast and bacterial cells, only a few are well-established in air pollutant investigations. One cell type, human alveolar macrophages, are indispensable as they directly contact the inspired air. They are critical to lung defenses (e.g., by killing infectious microorganisms and engulfing particulate air pollutants). As previously-mentioned, they also excrete highly-toxic chemicals (e.g., hydrogen peroxide) that can initiate lung diseases, such as fibrosis. Importantly, macrophages can be obtained for study by lung lavage (e.g., the washing out of physiological saline that has been instilled into the lungs) and studied outside of the body for several days.

Epithelial cells line the surface of the lungs, so they are also indispensable tools for air pollutant studies *in vitro.* Epithelial cells come in nearly a dozen varieties (e.g., ciliated, secretory, squamous (flat), microvillar, etc.) that all play important roles throughout the body. Epithelial cell cultures are used to study (1) oxidant stress, (2) cell killing, (3) inflammation, (4) genetic damage, and (5) biochemical responses.

Methods for directly exposing macrophages and epithelial cells in culture have been developed for both particles and gases, which more or less simulates their natural direct contact with the atmosphere.

Another cell type, *Salmonella* (rod-shaped bacteria capable of producing severe gastrointestinal infections when they contaminate food), are easy to culture, and they are a mainstay in the *Ames test,* which is used for studying the mutagenicity of xenobiotics. In the Ames test, *Salmonella typhimurium* mutations can be easily detected in cell cultures. Mutagens may also be carcinogens. A difficulty in using salmonella relates to their formidable cell walls that exclude whole particles from entry. Still, soluble products released from particles, and non-particulate pollutants, can be tested in the *Ames assay.*

In Vitro Toxicity Testing Used for Air Pollutants

Cell Viability

Sometimes it is important to test for cell killing in air pollutant studies. *Cell viability* (being alive) can be tested *in vitro* using a variety of methods. *Dye exclusion* using Trypan Blue can detect dead cells; living cells exclude the dye and remain unstained, but dead cells turn blue, as they are penetrated by the dye. Similarly, a fluorescent dye is also excluded by living cells, which makes their detection and quantification less labor-intensive than with Trypan Blue (which requires counting individual cells with a microscope). There are also viability tests based on measuring specific biochemical substances that leak out of killed cells.

Reactive Oxygen Species

Oxygen molecules and many of their derivatives (products of oxidation reactions) are highly active chemicals. In fact, diatomic oxygen (O_2), which is essential for the survival of humans and other complex organisms, is toxic in high concentrations. *Reactive oxygen species* (ROS) are molecules that contain an unpaired electron. ROS are capable of damaging DNA, proteins, and lipids; such damage is currently considered to be a major mechanism of air pollutant toxicity. ROS are generated naturally in the body, where their damaging potential is diminished by antioxidants, including enzymes, vitamin E, glutathione, and ascorbic acid. When unchecked, ROS can produce or exacerbate a variety of diseases including atherosclerosis, neurodegeneration (which is associated with Parkinson's, Alzheimer's, and Huntington's diseases), diabetes, blindness (especially in newborn infants), lung fibrosis, and cancer. The production of ROS is also one of the concerns associated with the inhalation of fine particulate quartz and ultrafine poorly-soluble particles. Measurement of the production of ROS *in vitro* is an important step in evaluating the toxic potential of nanoparticles. ROS vary (e.g., peroxides, hydroxyl radicals, etc.), so there are several *in vitro* assays; only a few will be briefly described here.

Fluorescent dye production (e.g., 2, 7-dichlorofluorescein), through conversion of a non-fluorescent precursor by ROS, can be measured by fluorimetry (measurement of the light emitted after excitation by a specific-wavelength of light). The fluorescent emission is measured by several types of detectors, but phototubes (that convert light into an electrical current) are most commonly used. Because many nanoparticles alone are also fluorescent, a calibration using particles without dye must be performed in order to quantify the ROS production by air pollutants.

Electroparamagnetic resonance (EPR) is a technique that allows the measurement of specific ROS molecules (e.g., hydrogen peroxide vs. superoxide anion). EPR is a more specific method of analysis than the fluorescence technique. EPR has some pitfalls, including physical and chemical interference by particles. The technique must first be applied to cell-free systems to measure the ROS production by the particles alone.

Particle size and composition affect ROS production. For example, ultrafine (under 0.1 μm diameter), but not fine (2.5 μm to 0.1 μm in diameter) titanium dioxide (TiO_2), has been seen to stimulate ROS production by lung epithelial cells. Other cells found in the lung, such as macrophages and neutrophils, also produce measurable ROS in what is called the *phagocytic burst,* as when they are stimulated by engulfment of some particles. Although such *in vitro* tests are complicated, they are helpful in evaluating the factors that modify the toxicity of particles in the body.

Inflammation

Inflammation, or more specifically *inflammatory potential,* can be measured by detecting substances liberated by cells upon contact with nanoparticles and other air pollutants. Inflammation can be produced by elevated concentrations of particulate matter and gases, as well as small to modest exposures to *aeroallergens* (i.e., airborne substances, primarily of biological origin, that stimulate allergic responses). In most healthy individuals an inflammatory response is transient, but in sensitized subjects (e.g., asthmatics) the response may be overwhelming. The general sequence of air pollutant inflammation in the bronchial airways is shown in **Figure 9–8**. As shown in the figure, inflammation is a cascade-type process. The initial responses can be amplified by the recruitment of cells that respond to released mediators and produce additional inflammation. If the resulting cell and tissue damage is overwhelming, irreparable injury and/or disease production may result. Inflammation can be either an appropriate defensive response or a debilitating event.

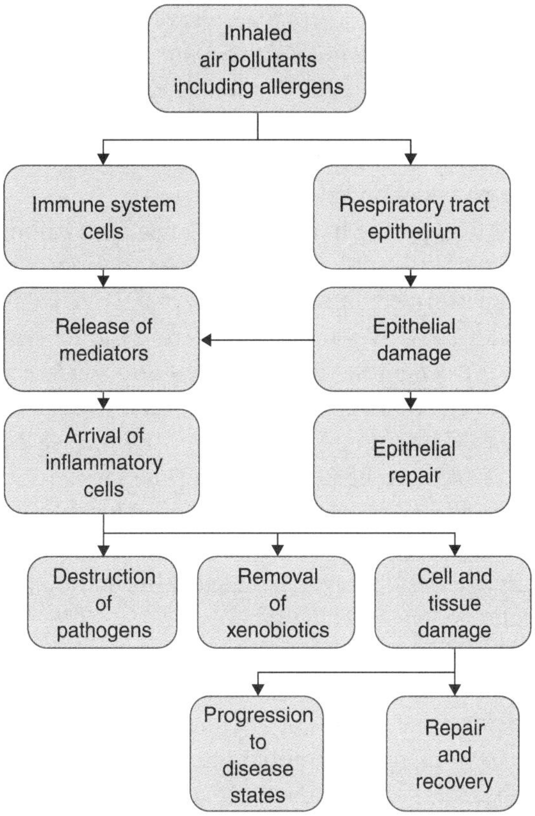

Figure 9–8 The sequence of air-pollutant induced inflammation in the respiratory tract. Note the potential for a cascade (arrival of inflammatory cells) that can amplify the tissue damage.
Source: The University of California Air Pollution Health Effects Laboratory, with kind permission.

The detection of inflammation associated with exposure to air pollutants in *in vitro* systems is based on measuring the mediators generated by inflamed (also called activated) cells. Such mediators include *cytokines* and *chemokines,* which are proteins made and secreted by a variety of cells. The cytokines and chemokines, and other mediators, serve to both promote and/or control inflammation. For *in vitro* detection of mediators, cultured cells are exposed to air pollutants (such as gases and/or particles) and then the culture is collected, centrifuged (to remove cells and particles), and analyzed for cytokine (or other mediator) proteins. The enzyme linked immunosorbent assay (ELISA) involves exposing the centrifuged serum to an antibody that attaches to the specific cytokine protein under consideration. The antibody-bound protein is then exposed to a second "indicator" antibody that attaches to the protein. The product is then measured by colorimetry (color detection), fluorescence emission, or detection of a radioactive label. A complicating factor is that some of the cytokine proteins may bind to the particles being tested, and thus escape detection. Although ELISA is widely used, care must be taken to account for the losses of antibodies to particles. The use of cultured cells (e.g., macrophages) that engulf particles and thus prevents the attachment of released mediators to particle surfaces is one of the ways to reduce mediator losses in the ELISA technique.

Additional Comments

In vitro toxicity testing of air pollutants is an important tool for estimating their potential health effects. In such tests it is vital to compare the results using tested particles (or gases) with *positive controls* (e.g., silica particles known to be highly toxic), and *negative controls* (non-toxic particles). In properly-conducted *in vitro* tests, various air pollutants can also be ranked according to their potential toxicity.

Prior to exposing humans (or human populations) to the tested pollutants, *in vivo* (live animal) tests must also be performed. *In vitro* tests are mainly valuable for eliminating highly-toxic particles from further consideration. *In vitro* tests do not currently provide a final safety evaluation; however, some scientists envision this as a viable possibility for the future (see NRC, 2007). Based on the complexity of the human system and the millions of interactions and metabolites present in the body at any given time, heavy reliance on *in vitro* testing will definitely be in the more distant future.

Nanoparticles

Much has been learned about the potential toxicity of nanoparticles as a result of *in vitro* testing. Particle size *per se* is an important modifier of toxicity. As previously described, the surface area of a given mass of particles is inversely proportional to their diameters. The total surface area of a given mass of 0.01 μm diameter particles is 10,000 times larger than 1.0 μm particles. Surface area, but not particle mass, is a key factor in inducing ROS and inflammatory processes. However, surface area alone is not the entire toxicity story. **Table 9–7** lists some of the known and/or suspected properties that influence the toxicity of inhaled nanoparticles; much of this knowledge has been obtained using *in vitro* methods.

Table 9–7 A list of some potentially important characteristics that influence the toxicity of nanoparticles.

Characteristic	*Comment*
Size	Influences dose and cell uptake
Shape	Influences deposition pattern and cell uptake
Surface area	Influences cell responses
Number	Influences toxic impact on cells
Porosity	Increases particles as carriers of other substances
Dissolution rate	Influences persistence and systemic distribution
Chemical reactivity	Determines cell responses
Chemical composition	Influences cell responses
Antigenicity	Ability to trigger an immune response
Surface coating	Influences cell responses and presentation of other substances
Presence of co-pollutants	Modifies cell and other responses

Data from Madl and Pinkerton (2009).

IV. ANIMAL STUDIES

Why Are Animal Studies Performed?

Animal studies form a bridge between *in vitro* studies and human studies. Animal studies provide information on the responses of complex intact mammals after exposure to xenobiotics (e.g., by air pollutant inhalation). Cell cultures are inadequate for understanding many types of effects including birth defects; fertility; the transition from DNA damage to cancer; multi-organ diseases (e.g., cardiovascular disease, hypertension, diabetes, and others); learning, sleep, and behavioral disorders; brain disease (e.g., Parkinson's and Alzheimer's disease); and many other diseases (such as kidney disease, osteoporosis, etc.). As an example, consider a chemotherapeutical company that can easily create 10,000 promising new chemical agents for the treatment of lung disease. Considerations based on the chemical structures and how they will most likely interact with normal cells and/or produce toxic metabolites can reduce the number of promising agents to 1,000. Initial *in vitro* studies on cell killing, ROS production, and DNA damage can reduce the list of agents to ten. The ten chemicals enter animal testing to look for direct toxic effects on the lung; ability to control bronchoconstriction; tolerance on repeated inhalation administration; production of birth defects; carcinogenicity; and a large number of other issues. Perhaps three of the promising drugs will be acceptable for initial human studies which will investigate metabolism of the agent, the acceptable safe dosage range, the methods of administering in an aerosol form, and the side-effects. Typically, one or two formulations may progress to being tested on sick humans who have had treatment failures with existing medications. After review by a government agency (such as the U.S. Federal Drug Administration, FDA), a formulation may be approved for treating the lung disease in adults. Further animal studies must be performed to look for effects on post-natal lung development and toxicity to young animals. If all goes well, the FDA may also approve the use of the new treatment in children. Without the controlled laboratory animal studies, it would be essentially impossible to introduce new medical treatments for humans, pets, livestock, or other animals. Similar considerations must be applied to the introduction of any new chemicals into the environment, including air pollutants from new fuels, novel pesticides, and engineered nanoparticles.

Rationale for Animal Studies

Why do animal studies actually work for protecting humans? Genetically, all mammals are strikingly similar, which produces basic similarities in metabolism, anatomy, disease processes, reproduction, behavior, and healing processes. There are some, although minor, important species differences in these characteristics, so animal testing must be performed on more than one species in order to assess the implications to humans. Rodents (especially mice and rats) are the backbone of animal research; but studies in higher mammals, such as dogs, sheep, and primates are essential for definitive information because they are more similar to humans. A major goal of animal studies is to eliminate the use of highly-toxic substances in order to prevent unacceptable harm in human populations. The use of animals in research has a long successful history of serving the interests of humans and other animal species. All laboratory animal studies must be approved by an ethics committee (see Chapter 12) in order to eliminate (or reduce) discomfort and distress. There must also be a significant benefit to humans and/or the ecosystem to justify animal studies.

Main Species Used in Inhalation Toxicology

Rodents

The *Rodenia,* especially rats and mice, are usually unwelcome, but familiar, companions of humans. They have shared our cities, homes, food, and many of our diseases for millennia. This familiarity and similarity, along with their availability, easy maintenance in the laboratory, and low cost, has made them important subjects for biomedical research. Some of the earliest research with rats was on nutrition and poisoning. Research with rats and mice, which began nearly a century ago, has established the minimum daily dietary requirements in humans for fats, carbohydrates, proteins, minerals, micronutrients, and vitamins. Humans owe most of what we know about our nutrition to laboratory rodents. In addition, research with rats and mice has identified significant poisons, mutagens, teratogens, and carcinogens. Less well known are the contributions of rodents to behavioral studies. Rats and mice can be taught to perform complex tasks, and deterioration in their learning and performance can identify substances that harm the brain, spinal nerves, and peripheral nervous system.

In recent decades, the development of *transgenic strains,* particularly in mice, has greatly amplified the utility of rodents in biomedical research. As genes and gene defects that predispose humans to specific diseases are discovered, those key genetic characteristics can often be produced in rodents. There are a large number of strains of transgenic mice and rats currently available to researchers. These strains include those predisposed to cancer, diabetes, bronchitis, asthma, atherosclerosis, hypertension, osteoporosis, glaucoma, cystic fibrosis, aggression, alcoholism, drug addiction, Alzheimer's disease, and obesity, to name only a few. Most of these rodents do not require special husbandry practices, but some are quite fragile and need special diets, housing, and other care.

Not all of the needed animal research can be performed using rodents. Their small size, short lifespans, and simple lung anatomies are a few of their limitations. Many techniques for studying lung function cannot be performed on tiny animals without sedation and surgical interventions. Laboratory mice and rats have a natural lifespan of about two to four years, depending on strain, diet, and quality of care. This span is inadequate for studying the long-term effects of air pollutants. Some diseases, such as cancer, fibrosis, and emphysema, in humans require 10 to 20 years to manifest. Another problem related to air pollution research with rodent lungs is the lack of a complex deep lung structure as seen in humans and other large mammals.

Other Species

The most important large animal species used in toxicological research are those which have previously established relationships with humans, so dogs, sheep, pigs, horses, rabbits, and cattle provide valuable information on the health effects of air pollutants.

Dogs may be the oldest of domesticated animals, as their bones have been found in excavations dating to the Stone Age. Their cooperative nature, trainability, and lung structure similar to that of humans has earned them an important place in air pollution studies. Today, purebred animals from *Class A suppliers* (suppliers that raise animals on their premises) have largely replaced strays and pound animals in the laboratory. The techniques for performing inhalation exposures using dogs have been developed over the past 50 years, as well as the procedures for maintaining their health in research settings. In a typical study, purebred Beagles are trained to wear custom-fitted latex masks instrumented with sensors (for oxygen, carbon dioxide, air flow rates, and air pollutants) and then exposed to air pollutants at rest or exercising on a treadmill. Most often, the exposures are to low levels of air pollutants. After the exposure, the dogs are monitored to follow their recovery before being returned to their kennel. Dogs are seldom used in studies that require sacrifice, as they are valuable, and after complete recovery they can be enrolled in additional studies. Public pressure against the use of dogs in research has led to the increased use of purpose-bred ferrets in air pollution studies. *Ferrets* are very cooperative and trainable, so they adapt well to laboratory procedures. The ferret's lungs are also anatomically similar to those of humans, and they seem to enjoy studies that require exercise during exposure to low levels of air pollutants.

Non-human primates have the greatest similarity to humans of all of the laboratory animals. Availability, cost, and difficulty in adapting to handling in the laboratory all limit their use in inhalation studies. Primates can also transmit several serious viral illnesses to humans, so they pose risks to their handlers. Primates are usually exposed to air pollutants while they are sedated and wearing masks or "helmets" that enclose their heads. Primates also require spacious housing with structures on which they climb, along with other "enrichments," such as toys for their entertainment. They are used in critical studies that cannot be accomplished using other species. Such studies include the testing of protective treatments (e.g., to test immunizations against aerosolized bioterrorism agents) that will be used on human populations, if needed. Whenever possible, primates are retired to live out their natural lives after they are used in laboratory investigations.

A variety of other large and medium-sized species, including horses, cattle, sheep, goats, pigs, and rabbits have been used in inhalation-toxicology research. These species are easy to work with, and they seldom require sedation in the laboratory. Each species has its own unique advantages for understanding the effects of inhaled air pollutants. *Horses,* like humans, spontaneously acquire a form of pulmonary emphysema. In addition, they can be quite valuable (e.g., as pets or as race horses), so the aim of many studies is to acquire information for their own benefit. Horses have some specific requirements, such as outdoor housing and lack of head-restraints (they require freedom of movement of their heads) during air pollutant exposures. Comfortable masks are typically attached to flexible tubes that supply them with airborne particles and gases. *Sheep* are very tolerant of procedures such as endotracheal intubation, which permit inhalation exposures that bypass the nasal region and allow the recovery of *lavage* fluids. Lavage is a procedure in which a small quality of physiological saline is placed in the lung and then recovered for biochemical and cellular studies. It is not tolerated by other mammals unless they are sedated or treated with local anesthetics (such as lidocaine). *Rabbits* are tolerant of serial blood-sampling procedures, and they efficiently manufacture new blood. Rabbits are thus of great value in immunological studies. Neonatal rabbits are a good model for premature human babies, so they are used to study inhalation treatments for human neonates.

One cannot overstress the ethical obligations researchers have to their animal subjects; the animals deserve comfortable and meaningful lives. Disrespect for laboratory animals is as offensive to most scientists as it is to most members of the general public. Failure to adequately protect animal subjects can lead to allegations of scientific misconduct and serious consequences.

V. HUMAN CLINICAL STUDIES

Human clinical studies bridge the gap between laboratory animal investigations and epidemiological studies. Clinical studies are controlled studies that involve relatively few human volunteers, in contrast to the larger numbers involved in epidemiological investigations. The focus of clinical toxicology studies is typically on either the development or the control (as in drug studies) of disease. The many diseases and conditions associated with the inhalation of air pollutants include: asthma; bronchitis; chronic obstructive pulmonary disease; cardiac dysfunction; increased susceptibility to infections; degraded athletic performance; learning disabilities; and nervous system disorders. The human subjects range from healthy to diseased and from children to elderly adults. Human studies can only be ethically conducted after laboratory animal studies have been performed that insure human safety (see Chapter 12 on ethics).

Pulmonary Function

Pulmonary function measurements are a mainstay in clinical studies of air pollutants. Such measurements can provide highly reliable information, and most can be performed safely on children and on exercising subjects. Another important consideration is that reversible pulmonary function changes can occur after exposure to inhaled air pollutants. Repeated pulmonary function measurements can be made pre-exposure and post-exposure to air pollutants in order to detect early and delayed responses and to follow a recovery period. Pulmonary function tests are varied and include simple measurements of lung volumes and air flow rates, as well as procedures that require blood sampling or radioisotope administration. A small sample of the available pulmonary function measurements includes:

- *ventilatory exchange*—breathing frequency (also called respiratory rate); tidal volume (the volume of air inspired or expired in a single breath); and minute volume (the total volume of air inspired or expired in one minute)
- *lung volumes*—vital capacity (the volume of air moved when a maximal inspiration is followed by a maximal expiration); total lung capacity (the vital capacity plus the remaining lung volume after a maximal expiration); functional residual capacity (the remaining lung volume after a maximal expiration); inspiratory capacity (the volume of air that can be inspired after a normal exhalation); and the inspiratory reserve volume (the volume of air that can be inspired after a normal inspiration)
- *respiratory mechanics*—forced expired volume over time (e.g., FEV_1 is the maximum volume of air that can be expired in 1 second); lung-flow resistance (which can be measured in various ways); lung and chest compliance (which measures the "stiffness" of the lungs and thorax); and flow-volume relationships (the flow rates at various lung volumes during normal, or maximum breathing efforts)
- *distribution of ventilation*—measurements of the evenness of filling of the lungs with inspired air (These measurements use an inhaled radioactive gas, such as ^{133}Xe, followed by scanning of the lungs with a radiation detector.)
- *diffusion capacity*—evaluates alveolar membrane thickness, pulmonary blood-capillary status, and evenness of distribution (e.g., as measured by the uptake of small concentrations of carbon monoxide in a single breath)
- *blood gases*—the efficiency with which gases, such as oxygen and carbon dioxide, are exchanged between the alveolar air and blood (These tests require blood sampling.)
- *bronchial provocation*—bronchial constriction assessments after inhaling an irritant substance

Many additional pulmonary function measurements are available, but these examples are quite sensitive when used for detecting the major potential effects produced by inhaled air pollutants. Some of these tests (e.g., those requiring radioisotopes) are not suitable for use in children unless they are also *required for diagnostic purposes.* Most pulmonary function measurements can be made in exercising subjects using a treadmill or a bicycle ergometer (a bicycle that has a variable resistance to pedaling). Note that most of these pulmonary function measurements require cooperation of the subjects, thus they are often difficult to perform in laboratory animals.

Cardiac Function

Several air pollutants in sufficient concentrations can alter cardiac function (see Chapters 1 and 10). In fact, cardiac patients are among the known susceptible sub-populations who have experienced morbidity and mortality during major air pollution episodes. Heart disease is a leading cause of death in the United States and other developed nations, which underscores the need for including cardiac function measurements in clinical investigations that assess the effects of air pollutants. Cardiac function measurements are usually based on monitoring electrical signals (produced by the beating heart) acquired by an array of electrodes placed on the chest, although, other techniques exist. Tracings of this electrical activity, called *electrocardiograms,* are analyzed in order to detect abnormalities. Trained cardiologists can detect skipped beats, premature beats, fibrillations (quivering and ineffective heart contractions), and ischemia (inadequate oxygenation of the heart muscle). Clinical studies of the effects of air pollutants on cardiac functions cannot always be safely conducted in people with severe cardiac disease. However, relatively-healthy subjects can be studied safely by cardiologists, provided resuscitation, defibrillation, and artificial ventilation techniques can be rapidly performed, if needed. Other techniques, including measurements of blood pressure, blood-clotting factors (e.g., platelet counts in the blood), and exercise performance are also used in clinical studies of inhaled air pollutants. As can be appreciated from the foregoing material, trained physicians are essential during the performance of most clinical studies.

Behavioral Studies

Some xenobiotic substances are capable of affecting behavior. Ozone exposure has been shown to decrease running performance in rats and humans. Lead (Pb), mercury, manganese, and solvents have well-known neurotoxicities. Some pesticides and some chemicals in consumer products are also implicated in central nervous system disorders. Odorous substances can alter behavior in both positive (e.g., perfumes) and negative ways. Drs. Bernard Weiss and Alice Rahill (1995) reviewed the application of behavioral measures in inhalation toxicology. For human volunteers and laboratory animals, a variety of behavioral studies can be conducted that measure intelligence, memory, vigilance, spatial discrimination, coordination, and mood. Such measures have not been extensively applied to air pollutants, but there is a need for progress in understanding the behavioral effects of air pollutant exposures on populations.

VI. EXPOSURE METHODS

Overview

Toxicology investigations using air pollutants require specialized equipment. The exposure system must deliver the desired aerosol particles and/or gases in a manner that can be related to natural exposures. Furthermore, unwanted pollutants must be suppressed, and the environmental temperature and humidity must be within specific ranges. Other potential stressors, such as lighting, vibration, noise and discomfort need to be considered. The subjects, be they cells, laboratory animals, or human volunteers, should be exposed in a manner that ensures their responses to the intended pollutant exposure can be accurately measured. Special considerations, such as biocontainment, apply when the air pollutants are infectious.

Inhalation System Requirements

An air pollutant exposure system for laboratory animals or humans consists of several modules (**Figure 9–9**) including: (1) an air cleaning and conditioning module that removes unwanted pollutants and adjusts the temperature and humidity of the air; (2) air pollutant generators; (3) an exposure-delivery module; (4) characterization equipment for the pollutants and environmental conditions; and (5) an exhaust module that removes the pollutants and cleans the air before it is exhausted into the environment. The design details will depend on many factors including the species and numbers of subjects; the duration of the exposure; and the toxicity of the studied air pollutants.

In Vitro *Exposures*

In vitro exposures to air pollutants are less complex than whole-animal exposures, but they require careful consideration of how the pollutants are presented and how the desired doses are delivered. The usual problem is how to maintain and support the viability of exposed cells (or sub-cellular entities) during the experiment. A primary consideration is to not overdose or stress the *in vitro* preparation to the extent that the results are unrealistic.

Laboratory Animal and Human Exposures

Inhalation Exposures

Exposures of laboratory animals or human volunteers to air pollutants face ethical, scientific, and practical issues that must be carefully considered. The subjects are dependent on the experimenters to provide breathable air and comfortable environmental conditions. The study protocols must be approved by an independent ethics committee, such as an Institutional Review Board (for human volunteers) or an Animal Care and use Committee (for live vertebrate animal subjects), as described in Chapter 12. The mode of air pol-

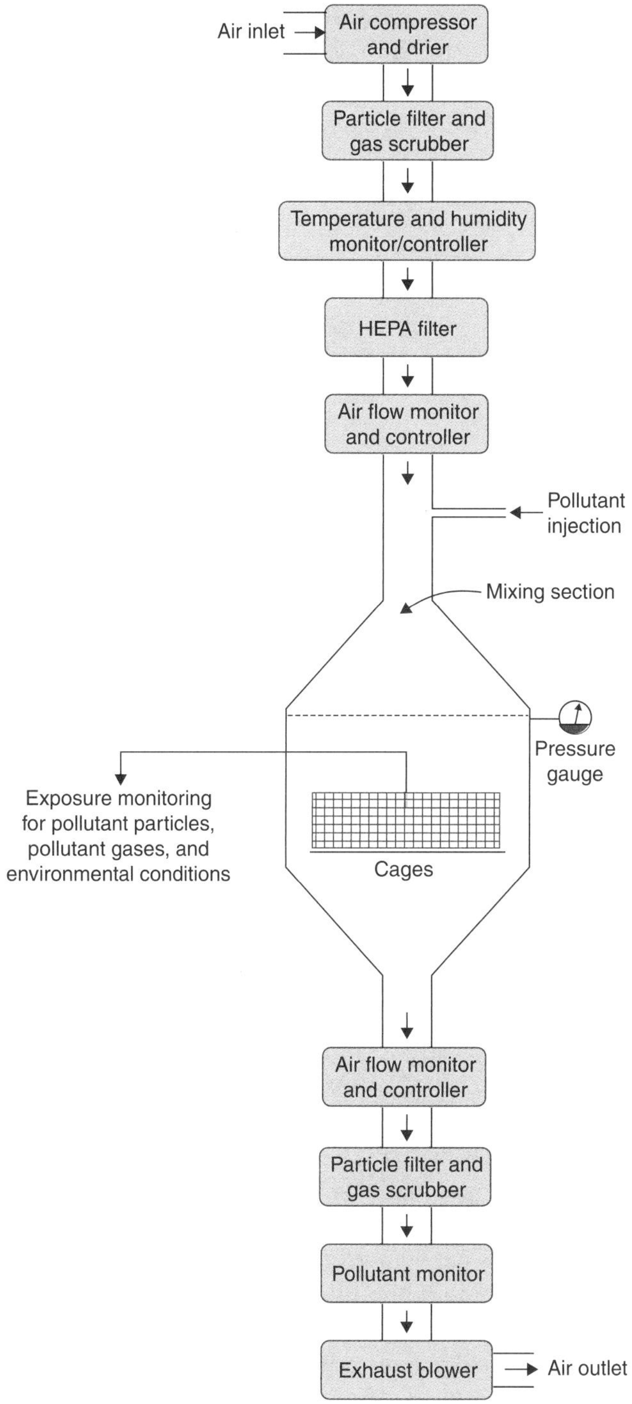

Figure 9–9 The many components of an inhalation chamber exposure system.
Source: The University of California Air Pollution Health Effects Laboratory, with kind permission.

lutant delivery selected can range from whole-body to partial-lung exposures (**Table 9–8**). Whole-body chambers are best for *chronic* exposures of animals (because the chambers also serve as living quarters) and for other than brief exposures of human volunteers. In some studies, the chambers are large enough to permit the investigators to enter in order to monitor responses of the subjects during the study. This is particularly important for human subjects who require medical surveillance and/or supervision during exercising (e.g., on treadmills or bicycle ergometers). However, chambers have disadvantages: they are expensive, may require large quantities of pollutants and throughput air, and expose the subjects' skin (or fur) and eyes to pollutants. Exposures via head-only enclosures, masks, nasal or oral tubes, or catheters (for lung-only or partial-lung exposures) are efficient (they use less exposure material), and they deliver more precise doses than chambers. The disadvantages include subject discomfort and sometimes the need for anesthesia.

Instillation Methods

An alternative to inhalation exposure is to instill a liquid (such as physiological saline) that contains the pollutants into the nose (where it is subsequently inhaled), or directly into the trachea (which requires heavy sedation or anesthesia). Although instillation is relatively inexpensive, it suffers from several artifacts. It is not as realistic as inhalation is with respect to the distribution of the pollutants throughout the respiratory tract, and the instilled fluid can cause local tissue damage that complicate interpretation of the results. Also, pollutant-fluid interactions can change the physical and chemical characteristics of the exposure material. Instillations are most valuable for screening studies (where a large number of pollutants must be compared with respect to toxicity), or mechanistic studies (where biochemical interactions, including metabolism and pathways of toxicity are studied in detail). Instillation exposures are not as useful as inhalation studies for simulating whole-animal effects of inhaled air pollutants.

VII. UNSOLVED PROBLEMS IN AIR POLLUTION TOXICOLOGY

Air pollution toxicology is still in an early stage of development. The basic tools used for *in vitro,* laboratory animal, and human clinical toxicology studies are in place, but there are gaps in knowledge when it comes

Table 9–8 Air pollutant exposure methods, their advantages and disadvantages.

Exposure Method	*Advantages*	*Disadvantages*
Whole-body (chambers)	Good for chronic exposures Subject comfort Control of temperature, humidity	Requires large quantities of air and pollutants Exposes entire body Non-uniformity of exposure
Head-only, Nose-only, and Mouth-only	Good for brief exposures Does not contaminate entire body Uses small amounts of air pollutants	Less comfortable than chambers Requires more manpower than chambers Equipment must often be custom-made
Lung-only and Partial Lung-only	Exposure is confined to lung as nose is bypassed Precise placement of dose Uses very little exposure material	Requires sedation or anesthesia Requires technical expertise Invasive and stressful
Instillation (of saline plus pollutants)	Precise dosing Can perform quickly Instillate can be recovered for study	Unrealistic route of exposure Can damage tissues Requires anesthesia Stressful to subjects

to interpreting and integrating the data. The following list highlights some of the unsolved problems, which can be also interpreted as opportunities for future researchers:

- *chronic vs. acute effects*—With few exceptions (e.g., lifetime studies on dogs) inhalation toxicology investigations are acute, or less than a few years in duration; laboratory animals do not live as long as humans, and extrapolating data to longer-term situations is not scientifically well-grounded. In addition, there is some evidence short-term adverse effects protect mammals from long-term adverse effects. This is true for ozone, where brief pre-exposures can be protective for even relatively high-dose subsequent exposures. Also, early exposure of children to infections and some air contaminants appears to help prevent them from developing asthma. What is known about the health effects of chronic exposures to air pollutants comes from epidemiological studies. Toxicology studies are needed to identify the roles of specific components and characteristics of air pollutants over the lifespan.
- *individual differences*—Most toxicology studies involve the use of small groups. Groups of 10, 20, or 30 subjects may be exposed to various levels (including a clean air group) of an air pollutant and statistical tests used to document group changes. This design is very powerful for eliminating spurious (i.e., chance) findings, but it obscures the individual responses. An air pollution episode in a city with 10 million people might be associated with an increase in the normal daily death rate of 200 people by only 0.2 percent. This death rate increase represents less than 1 person in a population of 10 million. It is not known who that person is or which air pollutant (if any) caused their death that particular day. It is known that individuals in certain sub-populations (the ill, elderly, etc.) are more likely to be adversely affected in an air pollution episode, but even within these groups great differences in susceptibility exist. If the most vulnerable individuals could be identified, then cost-effective individual actions could be taken to protect them.
- *extrapolation*—The main goal of extrapolating toxicology data is to protect humans and other valuable species. Data from cell cultures and laboratory animals has been, and will continue to be, essential for public health protection. However, there are difficulties and uncertainties involved with extrapolation of data from cell cultures and laboratory animal

species to human exposures, which are often orders of magnitude lower. There are even difficulties in extrapolating toxicology data from one individual to others. Advances in understanding air pollutant uptake, distribution, metabolism, storage and excretion are needed for the extrapolation of toxicology data from one species to another.

- *mixtures*—Most toxicology data is on the effects of exposure to substances one at a time. These types of studies have clearly identified safe and unsafe levels of thousands of substances. Under realistic conditions, humans are not exposed to air pollutants, or other toxicants, individually. Each breath of normal air can be expected to contain millions of different substances if one includes concentrations of a few molecules or more. More practically, one can expect to see several potentially important air pollutants elevated together during an air pollution episode. Interactions between a few combinations of particles and gases have been studied by toxicologists, but no clear picture has emerged from such studies. In some cases the combined pollutants have been less toxic (i.e., found to be protective; an *antagonistic effect*) than the individual pollutants alone. In other cases the combination has been more toxic than the sum of the individual components (i.e., the combinations had a *synergistic* effect). Additional studies on pollutant combinations including complex mixtures are needed. A neglected, but important type of toxicology study involves starting with a complex mixture and looking at the effects of *reduction* of individual components one-by-one. These types of studies would be useful for designing effective control strategies for individual air pollutants.
- *low doses*—Toxicologists prefer to study moderate to high doses of substances because such doses produce detectable effects in small groups of subjects. The lower the dose, the larger the group size needed to study a pollutant. Extrapolation of moderate or high doses to low doses is risky at best. In recent years, many instances of the protective effects of low doses of substances that are toxic at higher doses have been demonstrated (see Calabrese, 2008). It is quite possible some low doses of common air pollutants are required in order to maintain healthy defenses against potentially harmful exposures to high doses. The beneficial effects of low doses (*hormesis*) have been seen in a large number of toxicology studies. Additional research is needed to evaluate the effects, both *adverse* and *beneficial,* of inhaled air pollutants.

VIII. SUMMARY OF MAJOR POINTS

Toxicology studies are essential to the understanding of air pollutants and to the development of rational control strategies. Toxicology is still based on Paracelsius' proclamation made over 400 years ago that "All substances are poisons; there is none which is not a poison. The right dose differentiates a poison from a remedy."

For air pollutants the dose is not easy to define, as it involves the air concentration, breathing rate, airway anatomy, internal metabolism, distribution in the body, and other factors that differ from person-to-person, and even from time-to-time in an individual. Some groups, such as small children, can be expected to receive greater doses from air pollutants per kilogram of body mass; small mammals require more oxygen per unit of body mass in order to maintain their body temperature.

Inhalation is a route of exposure that exposes the entire body to pollutants; especially those that enter the blood. Therefore, the effects of inhaled air pollutants are not limited to the respiratory tract. *Toxicokinetic*s, and toxicokinetic modeling, focuses on the absorption, distribution, metabolism, storage, and excretion (ADMSE) of xenobiotic substances. The metabolism of potentially toxic air pollutants usually (but not always) serves to increase their elimination rates (e.g., in urine and bile) from the body, but some substances are stored in the body from months to years. Metabolism generally decreases the toxicity of inhaled air pollutants, but this is not true in all cases.

Air pollution toxicology is a specialized field that conducts *in vitro,* laboratory-animal, and human-volunteer studies. Each type of study adds to the knowledge base, but none can provide the necessary information alone. *In vitro* studies are essential for screening purposes and for discovering the biological mechanisms of injury and disease. Animal studies are critical for understanding whole-animal effects of pollutants, including cancer, heart disease, asthma, birth defects, and others. Human clinical studies are needed, especially for developing information to set safe levels of air pollutants, and to protect potentially sensitive sub-groups.

The methods for exposing subjects to air pollutants in the laboratory are quite varied and include chambers, masks, and indwelling tubes (e.g., lavage tube or catheter). Each method has its important advantages and disadvantages, so no one method can be said to be better than the others.

Much is known about the health effects of air pollutants, but several stubborn uncertainties remain. Little is known about the effects of chronic exposures and how acute exposures influence the susceptibility to subsequent exposures. Individuals differ substantially in their responses to air pollutants, but the reasons for such diversity are only beginning to be understood. The effects of low doses of air pollutants, which are what we experience today, are also poorly understood. Many substances exhibit *hormesis* (i.e., beneficial effects at low doses), but the implications to air pollution toxicology are not known. The effects of realistic mixtures of air pollutants are poorly understood in relation to the effects of individual substances. Clearly, there is much work to be done by air pollution toxicologists. Such work has importance beyond understanding air pollutants; it also provides new basic knowledge on complex living things and how they function in their day-to-day lives.

IX. QUIZ AND PROBLEMS

Quiz Questions

(select the best answer)

1. A basic principle in toxicology is:
   a. man-made chemicals are more toxic than natural substances.
   b. all substances are poisonous if the dose is large enough.
   c. some substances are harmless regardless of the dose.
   d. the pH makes the poison.
2. In Equation 9–1 what does "*E*" stand for?
   a. effectiveness
   b. effort
   c. effluent
   d. efficiency
3. Phase 1 and Phase 2 biotransformation reactions:
   a. protect the body by forming water-soluble compounds.
   b. are catalyzed by enzymes.
   c. can sometimes increase the toxicity of an absorbed pollutant.
   d. All of the above are true.
4. Which of the following elimination pathways for xenobiotic chemicals occur in the lungs?
   a. Exhalation and metabolic conversion
   b. Biliary excretion and metabolic conversion
   c. Storage in alveolar epithelial cells
   d. None, only the liver and kidneys eliminate xenobiotic chemicals
5. *In vitro* toxicology studies of air pollutants:
   a. are conducted in mice, rats, and other mammals.
   b. are conducted in medical clinics.
   c. are no longer being conducted.
   d. None of the above are true.
6. Laboratory animal studies of air pollutants:
   a. are conducted in rodents but not larger mammals.
   b. exclusively use non-human primates because of their cooperative nature.
   c. look for birth defects, cancer, and other effects that cannot be studied in cell cultures.
   d. must be conducted by physicians in clinical settings.
7. Human clinical studies of air pollutants:
   a. are illegal for ethical reasons.
   b. cannot be conducted on people with asthma or other diseases.
   c. do not require the subject's consent if the risks are small.
   d. must be conducted with the consent of the subjects or their legal guardians.
8. What type of toxicology study is preferred when one is performing *initial screening* tests on previously unstudied nanoparticles?
   a. *In vitro* testing
   b. Pulmonary function tests on dogs and/or cats
   c. Human clinical studies on asthmatics
   d. Human clinical studies on healthy adults
9. Among the unsolved problems in the study of air pollutants are:
   a. evaluating the lethality of particles to lung cells.
   b. evaluating the lethality of gases to lung cells.
   c. evaluating the toxicity of complex mixtures of air pollutants.
   d. evaluating the potential capacity of air pollutants to prevent aging.
10. For regulatory purposes (i.e., establishing ambient air quality standards) what types of tests are considered?
    a. *In vitro* tests
    b. Laboratory animal inhalation tests
    c. Human clinical tests
    d. All of the above are true.

Problems

1. Referring to Figure 9–1 estimate how much more air per gram of body mass is inhaled per minute by a mouse than by an adult human. Which of the following is closest to your answer?
   a. 10 times more
   b. 3 times more
   c. 2 times more
2. An adult and a 4-year-old child are running to get out of a dust cloud. The air concentration is 10 milligrams per cubic meter ($1m^3$ = 1,000 liters), they are in the cloud for 10 minutes, and the dust particles have a deposition efficiency of 0.9 in both the adult and the child. Using Equation 9–1 and Table 7–6 find the following:
   a. How many milligrams will deposit in each person?
   b. What is the deposition normalized to body mass for each person?
3. How important are individual differences in body size, state of health, and genetic characteristics in determining the absorption, distribution, metabolism, storage, and excretion (ADMSE) of xenobiotic substances? Give three to five examples. Limit your discussion to two pages or less (double-spaced).
4. Design and discuss a series of studies on a novel inhaled medicine that could be used to reverse severe bronchial constriction in newborn children.
5. If no clearance was occurring in Figure 9–7, what would the final lung burden by at the 30-day time point for each of the dosing regimens?
6. Read one of the listed references by Calabrese, and discuss how it might apply to the setting of air pollution standards.

X. DISCUSSION TOPICS

1. Small children seem to be more susceptible to adverse effects than adults for some types of outdoor air pollutants. What factors might explain this phenomenon?
2. Why are occupational air quality standards different from the U.S. EPA's National Ambient Air Quality Standards? Refer to Table 9–3 for examples.
3. How important are laboratory animals in the study of air pollutant toxicology? Can animal models be completely replaced? If so, when might this occur? If not, why not?
4. Review section VII on unsolved toxicology problems and discuss which problems are most likely to be solved in the next 10 to 20 years.

References and Recommended Reading

ACGIH®, *2010 TLVs® and BEIs®,* American Conference of Governmental Industrial Hygienists, Cincinnati, OH, 2010.

Ballantyne, B., Marrs, T. C., and Syversen, T., eds., *General and Applied Toxicology, 3rd Edition,* John Wiley & Sons, West Sussex, UK, 2009.

Calabrese, E. J., Hormesis: Why it is important to toxicology and toxicologists, *Environ. Toxicol. Chem.,* 27:1451–1474, 2008a.

Calabrese, E. J., Hormesis and medicine, *Brit. J. Clin. Pharmaco.,* 66:594–617, 2008b.

Caldwell, J., Conjugation reactions in foreign–compound metabolism: Definition, consequences, and species variations, *Drug Metab. Rev.,* 1982, 13:745–777.

Clark, R. P., Safety cabinets, fume cupboards and other containment systems, in *Bioaerosols Handbook,* Cox, C. S. and Wathes, C. M., eds., Lewis Publishers, Boca Raton, FL, 1995, pp. 473–504.

Eaton, D. L. and Klaassen, C. D., Principles of toxicology, in *Casarett and Doull's Essentials of Toxicology,* Klaassen, C. D. and Watkins III, J. B., eds., McGraw–Hill, New York, 2003, pp. 6–20.

Donaldson, K. and Borm, P., eds., *Particle Toxicology,* CRC Press (Taylor & Francis Group), Boca Raton, FL, 2007.

Gallo, M. A., History and scope of toxicology, in *Casarett and Doull's Essentials of Toxicology,* Klaassen, C. D. and Watkins III, J. B., eds., McGraw–Hill, New York, 2003, pp. 3–5.

Gardner, D. E., Crapo, J. D., and McClellan, R. O., *Toxicology of the Lung, 3rd Edition,* Taylor & Francis, Philadelphia, PA, 1999.

Gehr, R. and Heyder, J., eds., *Particle-Lung Interactions,* Marcel Dekker, Inc., New York, 2000.

Guyton, A. C., Analysis of respiratory patterns in laboratory animals, *Am. J. Physiol.,* 150:78–83, 1947.

Guyton, A. C., ed., *Textbook of Medical Physiology, 8th Edition,* W. B. Saunders, Philadelphia, PA, 1991, pp. 186.

Hodge, H. C. and Sterner, J. H., Tabulation of toxicity classes, *Am. Ind. Hyg. Assoc. Q.,* 10:93–96, 1949.

Holgate, S. T., Samet, J. M., Koren, H. S., and Maynard, R. L., eds., *Air Pollution and Health,* Academic Press, San Diego, CA, 1999.

ICRP (International Commission of Radiological Protection), *Report of the Task Group on Reference Man, No. 23,* Pergamon Press, New York, 1975.

Klaassen, C. D., ed., *Casarett and Doull's Toxicology: The Basic Science of Poisons, 7th Edition,* McGraw–Hill, New York, 2008.

Kodavanti, U. P., Costa, D. L., and Bromberg, P. A., Rodent models of cardiopulmonary disease: Their potential applicability in studies of air pollutant susceptibility, *Environ. Health Persp.,* Suppl. 1, 106: 111–130, 1998.

Kodavanti, U. P. and Costa, D. L., Rodent models of susceptibility: What is their place in inhalation toxicology?, *Respir. Physiol.,* 128:57–70, 2001.

Lambert, L. B., Singer, T. M., Boucher, S. E., and Douglas, G. R., Detailed review of transgenic rodent mutation assays, *Mut. Rsh. Rev. Mut. Rsh.,* 590:1–280, 2005.

Lenaz, G. and Strocchi, P., Reactive oxygen species in the induction of toxicity, in *General and Applied Toxicology, 3rd Edition, Volume 1,* Ballantyne, B., Marrs, T. C., and Syversen, T., eds., John Wiley & Sons, West Sussex, UK, 2009, pp. 367–410.

Lewis, R. J., *Sax's Dangerous Properties of Industrial Materials, 10th Edition, Volume. 1 & 2,* John Wiley & Sons, New York, 2000.

Lidstedt, S. L. and Schaeffer, P. J., Use of allometry in predicting anatomical and physiological parameters in mammals, *Lab Anim.,* 36:1–19, 2002.

Lipfert, F. W., *Air Pollution and Health: A Critical Review and Data Sourcebook,* Van Nostrand Reinhold, New York, 1994.

Lippmann, M., Cohen, B. S., and Schlesinger, R. B., *Environmental Health Science: Recognition, Evaluation and Control of Chemical and Physical Health Hazards,* Oxford University Press, New York, 2003.

Madl, A. K. and Pinkerton, K. E., Health effects of inhaled engineered and incidental nanoparticles, *Crit. Rev. Toxicol.,* 39:629–658, 2009.

McClellan, R. O. and Henderson, R. F., eds., *Concepts in Inhalation Toxicology, 2nd Edition,* Taylor & Francis, Washington, DC, 1995.

NCRP (National Council on Radiation Protection and Measurements), *Deposition Retention and Dosimetry of Inhaled Radioactive Substances,* NCRP SC 57–2 Report, National Council on Radiation Protection and Measurements, Bethesda, MD, 1997.

NRC (National Research Council), Committee on Toxicity Testing and Assessment of Environmental Agents, *Toxicity Testing in the 21st Century: A Vision and a Strategy.* Washington, D.C.: The National Academies Press, 2007.

Parkinson, A. and Ogilvie, B. W., Biotransformation of xenobiotics, in *Casarett and Doull's Toxicology: The Basic Science of Poisons, 7th Edition,* Klaassen, C. D., ed., McGraw–Hill, New York, 2008, pp. 161–304.

Phalen, R. F., *Inhalation Studies: Foundations and Techniques, 2nd Edition,* Informa Healthcare, New York, 2009.

Shen, D. D., Toxicokinetics, in *Casarett and Doull's Toxicology: The Basic Science of Poisons, 7th Edition,* Klaassen, C. D., ed., McGraw–Hill, New York, 2008, pp. 305–325.

Stone, V., Johnston, H., and Schins, R. P. F., Development of *in vitro* systems for nanotechnology: Methodological considerations, *Crit. Rev. Toxicol.,* 39:613–626, 2009.

Timbrell, J. A. and Marrs, T. C., Biotransformation of xenobiotics, in *General and Applied Toxicology, 3rd Edition, Volume 1,* Ballantyne, B., Marrs, T. C., Syversen, T., eds., John Wiley & Sons, West Sussex, UK, 2009, pp. 89–126.

U.S. EPA (U.S. Environmental Protection Agency), *National Ambient Air Quality Standards* (NAAQS), http://www.epa.gov/air/criteria.html (accessed October 3, 2010).

Weiss, B. and Rahill, A., Application of behavioral measures to inhalation toxicology, in *Concepts in Inhalation Toxicology, 2nd Edition,* McClellan, R. O. and Henderson, R. F., eds., Taylor & Francis, Washington, DC, 1995, pp. 505–532.

Chapter 10

Epidemiology and Air Pollution

LEARNING OBJECTIVES

By the end of this chapter the reader will be able to:

- define the term "epidemiology" and explain its role in understanding the effects of air pollutants
- discuss some of the epidemiologic associations between adverse health outcomes and air pollution episodes
- discuss the issues involved in establishing cause and effect relationships
- define "susceptible sub-populations" in relation to the effects of air pollutants

CHAPTER OUTLINE

I. INTRODUCTION: WHAT IS EPIDEMIOLOGY AND WHY IS IT IMPORTANT?

Definition and Scope

Epidemiologic studies are instrumental in establishing air quality standards because they describe the effects of actual levels and types of air pollutants on human populations under environmental exposure conditions. Epidemiology, a mainstay of public health (see **Exhibit 10–1**), has its origins in studies of widespread infectious diseases (i.e., epidemics). The field has evolved to include the study of essentially *all factors* that have a significant impact on human health including, nutritional, genetic, cultural, socioeconomic, and environmental aspects. Some epidemiologists also include non-human subjects in their studies, but the main focus of epidemiology is on *human populations.*

There are many definitions of epidemiology, but these two provided by Dr. Robert Friis (2008) succinctly capture the essence:

1. Epidemiology originates from the Greek words epi (upon) + demos (people) + logy (study of).
2. Epidemiology is concerned with the distribution and determinants of health and diseases, morbidity, injuries, disability, and mortality in populations. Epidemiologic studies are applied to the control of health problems in populations. The key aspects of this definition are determinants, distribution, population, and health phenomena (e.g., morbidity and mortality).

A difference between epidemiology and either clinical medicine or experimental biology is that the focus of epidemiology is on groups rather than on individual subjects. However, epidemiologists often acquire diverse data, such as clinical measurements on their subjects, exposure assessments, corroborative laboratory animal toxicology assessments, and other types of controlled laboratory data. Air pollution epidemiologists apply a variety of methods for identifying and understanding threats to human health for the ultimate purpose of reducing, or even eliminating, those threats.

The early practice of epidemiology included identifying causal factors for many infectious diseases. Dr. John Snow, considered the father of modern epidemiology, demonstrated the power of observational methods by identifying a contaminated water source in association with an 1854 outbreak of cholera in London. The actual cause of cholera (a rod-shaped bacterium, *Vibrio cholerae*) was not discovered until sometime later, but Snow's study was pivotal in the prevention of future disease outbreaks, as well as in advancing epidemiology itself. His work also demonstrated the power of a "natural experiment" in which a change in the cause alters the outcome.

Air Pollution Studies

Everyone is continuously exposed to air pollutants, but the use of effective control strategies has steadily improved air quality. Since the modern relatively low levels of air pollutants affect only a small fraction of the total exposed population, it is necessary to study large groups using sophisticated statistical methods in order to discover adverse health effects. Furthermore, the large majority of health effects produced by air pollutants are not unique, so there is a significant "background" level of the same effects that are produced by other risk factors (e.g., smoking, diet, and aging). In reality, air pollution is a contributor, along with other factors, to modern diseases (see **Exhibit 10–2**). Epidemiologists must, therefore, often detect the relatively small disease signals produced by modern levels of air pollutants. Such detection requires the use of biostatistical techniques and sophisticated study designs.

Epidemiology is one of the indispensible tools used in modern air pollution research. Because the scope of such epidemiologic research is both large and complex, this chapter will introduce only a few important concepts and then present samples of some major epidemiologic findings on the effects of air pollution on human health. The goal is to provide both an understanding of some basic epidemiological methods and issues and an appreciation for how they are applied to air pollutant exposures. Because the epidemiology of air pollution is rapidly evolving, air pollution scientists refer to the U.S. Environmental Protection Agency's periodic extensive reviews of the literature to stay up-to-date.

II. IMPORTANT CONCEPTS IN EPIDEMIOLOGY

Overview of Statistical Techniques and Concepts Used by Epidemiologists

Epidemiology is a mathematically-intensive field. There are two main classes of epidemiological studies:

- *descriptive*—involving the gathering, summarizing, and presenting of data taken from a sample

Exhibit 10–1 The Story of Public Health

"Epidemiology is a science; public health is a mission that is implemented through societal action." (Savitz, et al., 1999)

Public health is an important tool designed to protect and advance the health and well-being of society. It includes the *detection, analysis, and control* of diseases (both those caused by infections and those caused by other factors). The basic scientific method of discovery used in public health is epidemiology (e.g., see Friis, 2008). Once a problem has been detected by epidemiologists, analysis of causal factors and initiation of control measures can follow. The biological sciences, social sciences, engineering, law, and medicine are among the many tools used in public health practice for identifying and controlling a public health problem.

For much of history, the problem of combating infectious diseases was the major challenge faced by public health officials. Most notable have been the great plagues of the 6th and 14th centuries, which killed an estimated 175 million people worldwide. The causal organism, *Yersinia pestis,* is a bacterium that is capable of infecting both humans and rodents. It is transmitted by flea bites (causing bubonic plague) and through inhalation (causing pneumonic plague). Rodent and flea control (using rodenticides and insecticides), isolation of infected persons, and more recently the use of antibiotics are used to fight the plague diseases. The use of a plague vaccine is generally limited to laboratory and field workers who work directly with *Y. pestis,* or work in areas where infection risk is significant.

In the 16th, 17th, and 18th centuries, advances in methods of data collection and analysis laid the foundations for the scientific study of populations (i.e., demographics). Accurate data on diseases, occupations, educational attainment, and population sizes and locations became available to public health officials.

In the late 18th century, the English physician Edward Jenner (1749–1823) introduced vaccination as a method to control smallpox, which was produced by a virus and spread by airborne droplets and direct contact with infected items. Vaccination was responsible for the complete eradication of the disease, as announced by the World Health Organization in 1979. Today, public vaccination programs are a major public health tool. Concurrent with efforts to control infectious diseases, public health efforts in improving diet, food and water quality, and basic sanitation have contributed to steady increases in lifespan throughout the world.

In the late 18th and early 19th centuries, the discovery of microbes and advances in microbiology provided public health practitioners with powerful tools to fight infectious diseases. Thus, along with mass immunizations, laboratory testing of water, milk and food supplies are now routinely performed in many nations. The growth of public and voluntary health agencies, school health instruction, maternal health care, and public health nursing began to meet the needs of the growing populations in industrialized countries.

Worldwide cooperation in public health eventually led to the founding of the World Health Organization (WHO) within the United Nations in 1942. The WHO focuses on infectious diseases (such as tuberculosis, acquired immune deficiency syndrome, flu epidemics, and malaria), the availability of clean water, sanitation, and health education. Recently, the WHO has published world-wide air-quality standards.

In various nations, public health efforts range from being highly centralized to being administered locally. New challenges continue to confront public health agencies. Modern challenges include:

- the AIDS epidemic;
- emerging new (and old) infections;
- the threat of terrorism (including biological, chemical, nuclear, and psychological);
- starvation and malnutrition;
- overpopulation;
- environmental pollution; and
- civilian casualties of war.

The above, as well as unknown future challenges, must be faced by the next generation of public health officials.

---

population (e.g., daily asthma medication usage in a public school);

- *analytic*—largely based on probability theory, analytic techniques are used for the discovery of *causal relationships,* or reasons for the observed findings. A variety of methods are used, including quantifying risk levels, *hypothesis testing,* and the use of *natural experiments.*

Descriptive statistics can be used to express the nature and characteristics of distributions of population and exposure data by the use of measures of *central*

Exhibit 10–2 Multiple Causation of Diseases Associated with Air Pollutants

Essentially all diseases and conditions that are associated with modern environmental air pollutants have multiple causes. Because air pollution can be a contributing factor to common diseases, its control may significantly relieve any specific health problem.

Asthma is a good example of a multi-causal disease for which high levels of air pollutants can contribute to its periodic exacerbation. A key factor in the diagnosis of asthma is a hyperresponsiveness of airways characterized by excessive narrowing of the air passages. The reasons why particular individuals become asthmatic are still unclear. The *risk factors,* none of which are essential for developing the disease, include a genetic predisposition, early childhood events, obesity, and heavy exposures to allergens and irritants. Oddly, early exposure to some airborne allergens and viral and parasitic infections protect children from developing the asthma.

Once a person has developed the disease, episodes of wheezing and difficulty in breathing can be triggered by a large variety of factors. The known and/or strongly suspected risk factors that can trigger an episode include the following:

- food and airborne allergens (including pollens and spores);
- smoke from tobacco, wood, and other burning substances;
- infections;
- temperature extremes;
- medications, including aspirin, and penicillin;
- psychological stress;
- laughter;
- exercise;
- organic vapors, including paint thinners and gasoline; and
- sufficiently elevated levels of many types of airborne particles and gases.

An asthma attack often occurs when the *total* weight of one or more risk factors exceeds the level that can be tolerated. This is similar to the situation of heavy repetitive lifting with improper body mechanics, where the combined weight, repetitive trauma, and poor body mechanics all lead to the eventual harm. Current treatments for controlling existing asthma include avoiding precipitating factors and using prescribed medications. Prevention, focusing on the identification and control of risk factors, must be done on a case-by-case basis. However, effective methods for preventing the disease altogether are still elusive.

---

tendency (e.g., means, modes, and medians) and measures of *variability* (e.g., standard deviations, standard errors, and ranges, such as 95 percent confidence intervals). Descriptive statistics can also be used to demonstrate associations (i.e., correlations) between two and/or among several data sets. An example is the well-known associations between major air pollution episodes and increases in human mortality.

Statistical data (also called *exposure variables* and *outcomes*) can be either *dependent,* which is the outcome of interest (e.g., daily doses of asthma medications used by school children), or *independent,* which is an exposure variable that is believed to modify the outcome (e.g., concentration of fine particles in the air, pollen counts, or average daily air temperature). In air pollution epidemiology, statistical techniques are used to establish quantitative relationships between the exposure variables and the outcomes. For example, asthma medication usage in allergic Atlanta school children (outcome, dependent variable) may be observed to increase by a stated percent when the grass pollen count (exposure variable, in grains per m^3 of air) increases by a specified amount. Thus, in this example an *association* is sought between medication usage and airborne pollen counts. The association is an observed mathematical relationship between the independent and dependent variables. It is not necessarily a cause-and-effect relationship, as other exposure variables (e.g., mold spores, high temperature, or high levels of ozone) may be the actual cause or components of a causal combination that increases medication use. In this example, pollen count could have been a *surrogate* exposure variable that varies with the real causal agents. Surrogates will be described later in this chapter.

Associations can be confused with true causal relationships, especially because such terms as "link," "connection," "statistically significant relationship," or "significant relationship" are commonly used to describe associations. Nonetheless, identifying associations between disease and potential causal factors is an important step in the process of establishing true causality.

Various Meanings of "Significance"

When a claim is made that a result is "significant" it is important to understand the basis of the claim. In conversational usage "significant" means "important," "meaningful," or "large." In the biomedical realm there are three primary meanings:

- *statistically significant*—likely to be valid but with a stated probability of being invalid (e.g., the probability that the result is due to chance alone);
- *biologically significant*—having an effect on a biological system (e.g., a person, organ, cell, etc.) that modifies the form or function of that system in a non-trivial manner; and
- *clinically significant*—of medical importance, either with respect to deviation from the normal range, or affecting the health status of an individual or a population.

Figure 10–1 shows a hypothetical dose-response curve with some important doses and outcomes indicated. Such a curve applies to either an individual or a population.

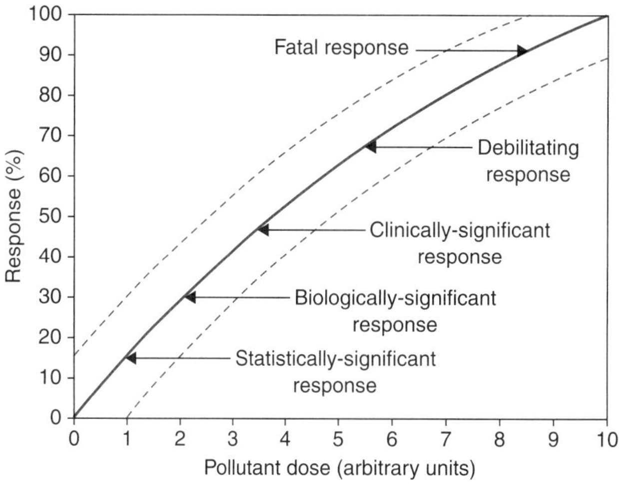

Figure 10–1 A hypothetical dose-response curve with a 95% confidence band. The statistically-significant response (at 1 dose unit) is where the lower 95% confidence band is just above the zero response. Biologically-significant, clinically-significant, debilitating, and fatal responses (and doses) are also depicted.
Source: The University of California Air Pollution Health Effects Laboratory, with kind permission.

Statistical significance is measured in terms of the probability that an observed value is, or is not, a member of a given (unexposed or control) population of values. Such significance is the basis of *hypothesis testing,* which is about ruling out chance as producing the outcomes (effects). Usually a value (e.g., average daily medication usage) of an exposed or treated group is compared with the normal or average values of an unexposed or untreated group, and the probability that the two group values do, or do not, differ is calculated. A permissible small error in claiming that the treated group's value is not normal (e.g., the medication use is increased) is selected prior to starting a study, and a claim is made as to whether or not the treated group differs from the normal group. Usually a permissible error of 5 percent or less must be observed before statistical significance is claimed, in which case it is stated that the group difference is significant with a significance level (p) of 0.05 or less. Thus, there is only a 5 percent chance that the claimed difference is due to chance, which implies that there is a 95 percent chance that the claimed difference is valid. Other than this brief description, hypothesis testing is beyond the scope of this chapter, but it is important to understand that statistical significance is based on mathematical probability calculations.

Statistical significance does not necessarily imply biological or clinical significance. An observed change in a population may be real but of little consequence. For example, public school children gain height at an average rate of about 5.6 cm (2.2 inches) per year, i.e., about 0.15 mm per day. If we measure the height of 10,000 nine-year-old children on a given day we might get an average of 120 cm with a standard deviation (SD) of 10 cm. If the distribution of heights is approximately normally distributed, the uncertainty, or standard error of our mean value of height (the standard deviation of the mean, SEM, is ± 1 mm) would be:

$$\text{SEM} = \text{SD}/(\text{number of subjects})^{1/2}. \qquad \text{(Eq. 10–1)}$$

If we re-measure the children two weeks later we may find that they have grown on the average 2.1 mm. The difference in height is highly statistically significant ($p < 0.05$) as it is greater than ± 2 standard errors of the mean. However, the difference in height is neither biologically nor clinically significant. *The point is that statistical significance alone does not necessarily have any practical importance with respect to fitness.* If we re-measured the children one year later, their increase in height would be statistically significant, clinically significant, and biologically significant.

Epidemiologic studies frequently use an alternate method to express the use of a significance level of 5 percent in making a claim that an effect has been detected. This is done by providing a 95 percent range (also called a *confidence interval*) around a measured mean value. For example, if an increased daily death rate of 0.5 percent (the relative risk, RR=1.005) is seen during an air pollution episode, the 95 percent confidence interval might be 1.001 to 1.009. This means that the investigators can claim that the true increase in death rate is between 0.1 and 0.9 percent, with a 5 percent probability of being wrong in making the claim. Both relative risk *(RR)* and confidence interval *(CI)* will be described in more detail later in this chapter.

Biological significance implies that a measured change is likely to alter the fitness of the subject. The biologically-significant change varies with the situation. For example, a 10 percent change in running speed can make the difference of escaping or not escaping a dangerous situation. By contrast, in athletic events a 0.01 percent change in running performance can separate first and second place competitors. For many biological parameters, a 10 percent change can make a real difference in fitness. Epidemiologists may design a study (e.g., of an air pollution episode) that will detect a ± 10 percent difference in blood pressure, with a 5 percent confidence level, in order to claim that the exposure had a biologically-significant effect on blood pressure. If the study has a positive finding, one might expect similar exposures of other populations to also produce a change in the blood pressure of 10 percent or more.

Clinical significance is used by physicians to make decisions on whether or not to recommend medical intervention, or to diagnose a medical condition. The significance of a given difference from the normal population or even a change from a baseline (i.e., previous) measurement in a given patient varies greatly for different medical conditions. Wide population variations in body height (e.g., ± 10 percent) are normal, but small percentage changes in an individual's body temperature (e.g., ± 2 percent) are significant. Clinical significance implies a statistically significant difference from some relevant normal value.

Summing up, statistical significance is a mathematical concept that depends on measurement accuracy and precision, number of subjects on which measurements are made, and population variability. It may or may not have any practical value. Biological and clinical significance, on the other hand, generally relate to fitness, health, and well-being. Strictly speaking, air pollution epidemiologists first should establish if an effect has statistical significance, and then they, or others, seek to determine whether or not it is also biologically or clinically significant. From a public health perspective, an effect need not reach clinical significance in order to be of concern.

Cause and Effect Relationships (Causality)

Although epidemiological data alone cannot usually *definitively prove* that a *cause-and-effect relationship* exists between independent and dependent variables, such data can strongly infer that a causal relationship exists. In some cases the epidemiological data alone are strong enough to permit a reasonable judgment that a causal relationship should be assumed. Thus, some interventions to protect public health, such as removing lead (Pb) from gasoline and controlling cigarette smoking, have been initiated primarily on evidence from epidemiologic studies.

Before proceeding to a discussion on how epidemiologic data can be used to support a cause-and-effect relationship, it is useful to examine some general scientific criteria for supporting the claim that an independent variable A (the putative cause) actually causes B (the observed effect). In a sense, the following criteria define "cause" and "effect" and how they are related to one another:

- A proceeds B (i.e., temporality).
- If A occurs, B occurs (i.e., repeatability).
- If A does not occur, B does not occur (the purpose of controls).
- The strength of A influences the magnitude of B (i.e., a dose-response relationship).
- A is not just a surrogate marker for A^*, the true cause (i.e., absence of confounding).
- The relationship between A and B are not due to chance alone (i.e., statistical significance).
- The cause and effect relationship makes sense (i.e., plausibility).

In truth, the general problem of establishing causality has not been solved, as the concept has different meanings in physics, philosophy, law, medicine, religion, and public health. One person may accept the evidence for causality in a given situation, while another may not. A determination of causality is in essence a decision, conclusion, or a verdict by an observer.

For the purpose of exploring causality as related to epidemiology and air pollution, it is useful to examine a seminal paper by Sir Austin Bradford Hill (1897–1991), who was a Professor Emeritus of Medical Statistics at the University of London at the time the paper was published. Professor Hill postulated that respiratory illness is associated with the inhalation of environmental dust and then asked: "In what circumstances can we pass from this observed *association* to a verdict of *causation?*"

To answer his own question, Hill suggested that the following nine items should be considered before making a decision on causality. Although these items are frequently referred to as "Hill's criteria," he cautioned that none of the items are absolutely necessary for a claim of causality, so we will refer to them only as "considerations."

- *Strength*—First on Hill's list is "the strength of the association." Hill noted that the *relative risk* (i.e., *RR,* which is a ratio of disease occurrence in exposed groups to groups with no exposure) of heavy smokers developing lung cancer is nearly 30 times that for non-smokers. A relative risk of greater than 5 (e.g., a *RR* of 200 was seen for scrotal cancer in English chimney sweeps) is strong evidence for causality. Some scientists set a lower limit of a relative risk of 2- or 3-fold for a convincing causal association, while others would accept a much lower fold increase, especially when it is seen in multiple well-conducted studies that are accompanied by corroborative evidence.
- *Consistency*—Is the association *repeatable* (e.g., by the original investigator) and *reproducible* (e.g., by other investigators in different locations, times, and circumstances)?
- *Specificity*—Does the specific event (e.g., dust exposure) lead to specific outcomes (e.g., lung cancers) that are not common in cases where the event is absent? Lung cancer, for example, is not common in people who do not have significant exposures to carcinogens or other factors that can produce or promote cancer development. Hill cautioned that many diseases have multiple causes (e.g., lung cancer can be caused by smoking and radiation exposure, as well as some inhaled dusts), so specificity must not be over-emphasized in making judgments about causality.
- *Temporality*—As previously stated, the cause must come before the effect. However, some diseases, such as lung cancer, may occur decades after the initial exposure (e.g., cigarette smoking or exposure to asbestos), so near-term temporality is not essential.
- *Biological gradient*—Is there a dose-response relationship? Evidence that a greater exposure leads to a greater response is an important finding. Yet, a dose-response relationship for humans is often difficult to obtain in many cases.
- *Plausibility*—Is there a reasonable biological explanation? The deposition of known carcinogens in the lung followed by lung cancer is biologically plausible. However, sometimes a state of biological ignorance (e.g., as was the case for infections before the discovery of bacteria) can prevent the claim of plausibility.
- *Coherence*—Does the proposed causal relationship conflict with what is known outside of the epidemiological association, or alternatively does other evidence (e.g., observing microscopic changes in dust-exposed lung cells that imply they are cancerous) support the claim of causality? Reproducing the event in exposed laboratory animals would also provide coherence, but in some cases the response in humans is a unique one.
- *Experiment*—For example, does preventing the exposure (e.g., using dust masks or other controls) prevent the effect (e.g., lung cancer)? If such experiments (called "natural experiments" by epidemiologists) have been done, the outcomes are very important to determinations of causality.
- *Analogy*—Have exposures to similar agents led to similar outcomes? For example, have exposures to other dusts also produced lung cancers?

As can be seen, Hill's considerations can be challenged individually in specific cases. It is the overall weight of the available evidence that is important. Also, if other types of confirmatory studies (e.g., controlled investigations in which suspected agents are tested one-by-one in realistic doses) are done, the case for a verdict of causality is strengthened. Ideally, confirmatory clinical studies using human volunteers would be performed. When death or serious illness are possible, such studies cannot be done ethically in people for obvious reasons (see Chapter 12 on ethics). Fortunately, good laboratory animal models do exist for most human diseases (see Chapter 9 on toxicology). Assigning causality can be considered as belonging to the branch of analytic epidemiology, as it is neither purely descriptive, nor purely inferential in a statistical sense.

The report of the Advisory Committee to the Surgeon General of the Public Health Service, *Smoking and Health* (U.S. DHHS, 1964) addressed the issue of causality under the rubric "Criteria for Judgments." The Committee stated: "when coupled with the other data, results from the epidemiologic studies can provide the basis upon which judgments of causality can be made" and "statistical methods cannot establish proof of a causal relationship in an association." Although the Committee did not use epidemiologic data alone to establish causality, it considered such data as important to making the judgment of causality. The conclusion of the committee's analysis was: "Cigarette smoking is a health hazard of sufficient importance in the United States to warrant appropriate remedial action."

In its 2009 evaluation of the health effects of particulate air pollution, the U.S. EPA established five levels of determination of causality. The levels were based in part on Hill's considerations. Here is an abbreviated summary of EPA's levels of causality:

- *"Causal relationship"*—Well-conducted studies using realistic exposures have been replicated, and chance, bias, and confounding can be "ruled out with reasonable confidence."
- *"Likely to be a causal relationship"*—Evidence is similar to that for a causal relationship, "but important uncertainties remain." Uncertainties could include possible confounding of the effects of the studied pollutant (e.g., air pollutant particles) by co-pollutant exposures (e.g., gaseous air pollutants); limited or inconsistent information on mode of action; and limited or no human data to support existing high-quality animal studies.
- *"Suggestive of a causal relationship"*—Evidence suggests a causal relationship at realistic exposure levels, but "chance, bias, and confounding cannot be ruled out."
- *"Inadequate to infer a causal relationship"*—The available studies lack the "quantity, quality, consistency, or statistical power" on which to base a decision.
- *"Not likely to be a causal relationship"*—The evidence from several studies suggests that a causal relationship is unlikely. The studies covered the range of likely environmental exposures and considered "susceptible or vulnerable subpopulations." The available studies were consistent in showing no adverse effects.

The U.S. EPA uses these levels of determination of causality in deciding whether, or not, to establish air quality criteria to protect public health. Interestingly, Hill's first consideration, the *strength* of the associations, is not included. That may relate to EPA's mandate to err on the side of caution.

Relative Risk

In the previous section, Hill's first consideration, the strength of an association, can be expressed mathematically as a *relative risk, RR. RR* is the ratio of the risk of an event (e.g., a disease) associated with a treatment (e.g., a specific air pollutant exposure) to the risk of no treatment (i.e., no exposure). *Incidence,* which is the rate of development of a disease per unit population per unit time, can be used to calculate *RR.* The time span can vary (e.g., an hour, days, years, or a lifetime). *Prevalence* refers to the number (not rate) of existing cases of a condition in a defined population at a specific time. If I_o is the incidence of the event in an unexposed population and I_e the incidence in an exposed population:

$$RR = \frac{I_e}{I_o}. \qquad \text{(Eq. 10–2)}$$

As an example, assume that the lifetime risk of getting lung cancer after 30 years of heavy cigarette smoking is 1 for every 20 smokers, and the risk in non-smokers is 1 in 600. The *RR* would be:

$$RR = \frac{1/20}{1/600} = \frac{600}{20} = 30. \qquad \text{(Eq. 10–3)}$$

So the long-term smokers are 30 times more likely to develop lung cancer than the non-smokers. Another way of stating this is that long-term heavy smoking increases the risk of getting lung cancer by about 3,000 percent. This is an example of an extremely strong association, which would certainly justify encouraging people to not smoke cigarettes. However, examples of such strong associations are uncommon in modern air pollution epidemiology.

The very strong relative risk example for smokers can be looked at two ways. On the positive side, 95 percent of the heavy smokers will not get lung cancer, as the lifetime risk is only 1 in 20. On the negative side, the risk of lung cancer in heavy smokers is enormous when compared to that in non-smokers. It should be noted that heavy cigarette smoking also carries other substantial risks (e.g., emphysema, heart disease, and other serious conditions) in addition to cancer.

How big must the *RR* be before it is considered to be important? This key question has no single answer, as it depends on the specific circumstances and who is asked. People are willing to voluntarily accept behaviors that have relative risks that are substantial, if the risk is familiar, under their control, and there are offsetting benefits. An example is driving a small automobile (instead of a large one) to and from work for several years, that had a risk of dying (as of 1979) in an accident of about 1 in 400. For a large number of people, the benefits (mainly economic) in this case, outweigh the excess risk associated with colliding with a larger vehicle. In contrast, regulatory agencies consider an excess cancer risk of only 1 in 100,000 (an *RR* of 1.00001) as significant. The logic is that if a chemical (say a pesticide) is associated with an increase in the cancer rate of 1 in 100,000, and that there are 300 million people at risk, then there may be 3,000 additional cancers. So, from the point of view of an individual, the risk associated with the pesticide is negligible but to the regulator it may be significant.

Similar considerations apply to air pollution standards. Regulatory decisions are influenced by relative risks of mortality on the order of 1.05 or less. When someone asks "Are current air pollution levels dangerous to me?," the answer may be "no," unless they are one of the unlucky (or health-compromised) few for which it is harmful. If one asks a regulator "Are current levels of air pollution dangerous?," the answer may be "yes," if the exposed population is large. It should be noted that breathing air pollution is involuntary and generally not associated with any direct benefits. Indirect benefits of air pollution may include the economic advantages associated with jobs, cost of goods, and services associated with the sources (e.g., electric power plants, transportation, farming, and manufacturing). In any case, knowing the *RR* places the risk into a perspective that can be objectively evaluated.

Odds Ratio (OR)

In cases when the incidence rates are not known or available (e.g., when cases are rare) to calculate a *RR*, the *odds ratio* (*OR*) is often used as an estimate of risk. When the health outcome (e.g., disease) is infrequent the *OR* is an effective estimate of the *RR*. Case-control studies compare the odds of an exposed group (cases) having the outcome (e.g., disease) with the odds of an unexposed (controls) group also having the outcome. The ratio of odds of the exposed group being diseased to the odds of the unexposed group having the disease is the *OR*. From the cigarette smoker's lung cancer example above the odds ratio would be

$$OR = \frac{1/19}{1/599} = \frac{599}{19} = 32. \qquad \text{(Eq. 10–4)}$$

Attributable Risk (AR)

The effect of a risk factor, such as exposure to elevated levels of air pollution, can also be expressed as an *attributable risk (AR)*. For example, if the yearly risk of death in people not exposed to urban air pollution per 100,000 people is R_o, and the yearly risk of death from the urban air pollution exposure per 100,000 people is R_{air}, then the attributable mortality risk, AR_{air} from air pollutants is

$$AR_{air} = R_{air} - R_o. \qquad \text{(Eq. 10–5)}$$

The attributable risk is useful in that it isolates the risk associated with a particular potential cause. If several risk factors are compared with respect to their *ARs*, decisions can be made on their relative importance. Decisions that are based on such ranking of various risks are useful for assigning priorities for spending the always limited funds available for protecting public health.

Confidence Interval

The confidence interval, CI, for an estimate of a value (e.g., a mean or an *RR*) is a range within which the true value is claimed to lie (**Figure 10–2**). By convention, the *CI* is usually such that there is a 95 percent probability that the true value lies within its range. In the case of a relative risk, if the 95 percent confidence interval does not include "1" then one concludes that there is a statistically-significant risk, with only a 5 percent chance that the conclusion is wrong. Confidence intervals are especially useful in epidemiology when several studies of a given health effect are being compared or combined (i.e., in a *meta-analysis*). The *CI* range depends on several factors, but the variability of the effect in the population and the number of subjects studied are primary determinants.

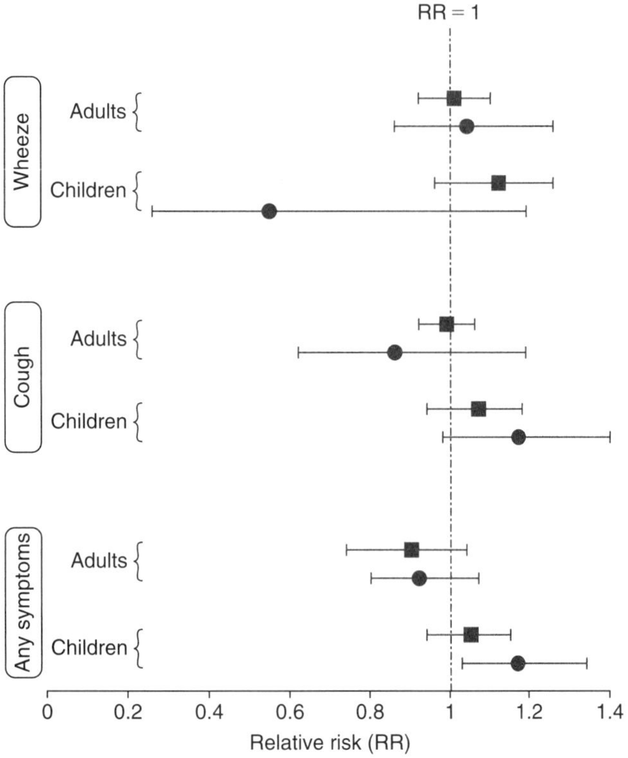

Figure 10–2 Relative risks of asthmatic adults and children along with 95% confidence intervals associated with an increase of 10 μg/m³ for $PM_{2.5}$ (fine particles, dark circles) $PM_{10\text{-}2.5}$ (coarse particles, dark squares). Plotted from data of Mar et al. (2004, their tables 6 and 7).
Source: The University of California Air Pollution Health Effects Laboratory, with kind permission.

Loss of Life Expectancy

Another way of looking at risks is to examine the extent to which they shorten lifespan. If a 90-year-old person dies prematurely, the loss of expected lifespan may be only a few days or weeks. If an infant dies prematurely, the loss of lifespan could be 85 years. The importance for air pollution epidemiology is that an observed excess mortality rate in the infirm and very old is not as tragic as it is in the young. Although one cannot confidently define how much loss of lifespan is significant, more than 30 days has been proposed by Dr. Bernard Cohen. Loss of life expectancy in years, *LLE,* is defined as

$$LLE = S - D, \qquad \text{(Eq. 10–6)}$$

where S is the expected lifespan and D is the age at which one dies. For example, if the expected lifespan S, of an individual is 75 years and they die at D, age 60, then the *LLE* is 15 years. A large number of risk factors, voluntary and involuntary, influence *LLE.* The largest factor is the country in which one lives. Poverty, lack of technology, low educational levels, and poor public health and medical services are associated with national life expectancies that can be less than 50 years. In technologically-advanced countries, life expectancies are typically 80 years or more. Even in the United States, poverty is associated with an *LLE* of about 10 years when compared to those above the poverty level. Air pollution levels in the United States were estimated by Dr. Bernard Cohen in 1991 to produce an average *LLE* of 0.21 years (77 days). Since that time, air quality has improved. Even so, some more recent epidemiological evidence indicates a larger *LLE.* Dr. C. Arden Pope and colleagues estimated that the *LLE* associated with an increase of 10 μg/m³ of fine particulate material ($PM_{2.5}$) was 0.61 years (223 days). **Table 10–1** shows some risk factors and their associated effects on *LLE* and lifespan. The table is useful for prioritizing research and/or efforts intended to improve public health.

Confounders and Surrogates

Epidemiological studies differ from controlled laboratory studies in several ways. Because most epidemiological studies are observational, the independent variables (e.g., exposure to specific air pollutants) are not manipulated. This stands in contrast to laboratory experiments in which the independent exposure variables can be varied, and even withheld, as the investigator decides. In the laboratory, each suspected chemical agent can be studied individually in its pure state as opposed to the inevitable mixtures found in the real world. Furthermore, in epidemiological studies, matched control groups that are unexposed may be difficult to find. In the laboratory, one may use control groups that do not differ in any significant way other than the absence of the studied intentional exposure. As a result, epidemiological studies are subject to more uncertainty, including *confounding,* than are controlled studies. According to Gary Friedman, Director of Kaiser Permanente's Division of Research, it is the strength of associations between potential causal factors and a disease, not the statistical significance of the associations, which should be the basis of judging whether a potential confounder exists. Confounders are easier to define than to eliminate.

Table 10–1 Loss of life expectancy (LLE) associated with various factors, and lifespans in selected nations. *A minus LLE is a gain in life expectancy.*

Occupation/Activity/Status	*LLE*
Alcoholic	11 years
Poverty	10 years
Smoker	6.6 years
Fire fighting	3.5 years
Coal mining	3.2 years
Police work	2.8 years
20% overweight	2.7 years
Construction	1.4 years
Unemployed	1.4 years
Oil & gas recovery	0.4 years
Parachuting	25 days
Professional boxing	8 days
Airline employee	–0.1 years
Agriculture	–0.7 years
Chemical manufacturing	–0.8 years
Wholesale trade	–1.3 years
Education services (e.g., teacher)	–3.1 years
Country of Residence	*Lifespan*
World average	67 years
Japan	82 years
European Union	79 years
United States	78 years
Russia	66 years
S. Africa	49 years
Swaziland	32 years (highest incidence of HIV/AIDS in the world, 26%)

Data from Cohen and Lee (1979); Cohen (1991); and CIA (2009).

A *confounder* is an unwanted independent variable that distorts the results of a study. Examples of confounders in epidemiologic air pollution studies include the following:

- unmeasured *population variables,* such as socioeconomic status (SES) that vary with exposure to air pollutants, and thus distort the association between adverse health effects and air pollutant levels. (Poor people suffer many health inequalities *and* often live in more-polluted neighborhoods.)
- unmeasured *behavioral factors,* such as people taking excess medications, becoming anxious, or spending more time indoors (indoor air quality is generally poor) on days that have "smog alerts."
- unmeasured *exposure parameters,* such as cigarette smoking, ambient temperature, or exposure to a specific substance, that distorts the association between adverse health effects and the measured air pollutant. The measured air pollutant could be $PM_{2.5}$ (the mass of particles less than 2.5 μm in diameter per cubic meter of air), but the health effects could be caused (or influenced) by mold spores, ultrafine particle counts, co-pollutant gases (e.g., NO_2, SO_2, or O_3), or metals (e.g., nickel or vanadium) on particle surfaces.

Epidemiologists are aware of confounders and use a variety of methods to control their effects, including using multivariate techniques in which suspected confounders are modeled in the study design. Correcting for the effects of cigarette smoking, education, obesity, temperature extremes, and socioeconomic factors is commonly done in modern air pollution epidemiologic studies. However, it may be difficult to evaluate potential confounders in a population, especially when using records and historical data. Care must always be exercised when there are suspected gaps in relevant knowledge on a population. For example, burning coal or wood indoors with inadequate ventilation can result in exposures 100 to 1000 times higher than those outdoors. If such indoor exposures are not included in a study of the effects of outdoor air pollution, erroneous conclusions may be reached.

Surrogate independent variables are those that are assumed to affect the dependent variable (e.g., mortality), but in reality have no direct effect on the study outcome. Surrogates only co-vary with a variable that does alter the outcome. That is, the surrogate variable tracks the variable that produces the outcome. The possibility that measured air pollutants are surrogates is a major concern in air pollution studies. Considering particles as an example, the measured variable, particulate mass, may be a surrogate for a specific chemical component found in the particles. For all air pollutants, the measured variable may be a surrogate for unmeasured co-pollutants or weather-related variables. Measurement of the surrogate variable in a study will often produce an association, but eliminating the surrogate exposure (as by a regulatory action) will not necessarily change the health-related outcome. In order to illustrate the concept, consider a hypothetical example. In a study of daily death rates in the elderly, an investigator has acquired data on daily air conditioner use during hot weather. Sure enough, there is a significant positive association between deaths and air conditioner use; when the number of hours of air conditioner use increases, so do death rates in the subjects. The wrong conclusion would be that the air conditioners were responsible for the excess deaths, and that it is therefore reasonable to deny air conditioners to the elderly. The proper conclusion would be that the heat waves were deadly, and that the elderly should have air conditioners. Air conditioner use was a surrogate for high temperatures. Although this example is obvious, it demonstrates why it is important to understand the influence that surrogate variables can have on an air pollution study. Finding a *low relative risk* in a study should alert one to the possibility of surrogate confounding. **Table 10–2** lists some exposure parameters that are measured in air pollution epidemiology studies, along with confounders for which the measured variable may be a surrogate.

Table 10–2 Exposure parameters measured in epidemiological studies and confounders for which the measured variable may be a surrogate.

Measured Parameter	*Confounder*
PM_{10} and $PM_{2.5}$ mass	co-pollutant gases acidity oxidative capacity allergenicity fine-fraction silica catalytic activity indoor pollutants socioeconomic status particle count or surface area
Ultrafine particle count	co-pollutant gases organics
Ozone, NO_x	carbonaceous particles carbon monoxide environmental temperature
SO_2	acidity sulfates
Any air pollutant	socioeconomic factors behavioral factors medication side effects

In some cases a surrogate is intentionally used as a general indicator of air quality. For example, carbon dioxide, a byproduct of human respiration, is often used to indicate inadequate ventilation indoors and the potential build-up of other indoor air pollutants. The surrogate, carbon dioxide may not be producing any significant effects but taking steps to control it will control other air pollutants. Visibility can also be a surrogate for outdoor air health risks. When visibility is impaired, pollutants that directly affect health may be present, and improving visibility may improve public health. One of the key issues in air pollution research is to ensure surrogates are identified and not confused with causal variables.

III. TYPES OF EPIDEMIOLOGY STUDIES

Common Study Designs

In epidemiology, the study design is critically important, as it will determine the ultimate statistical strength and the way confounding is dealt with. A complete listing of all of the study designs used by epidemiologists cannot be provided here, as there are many variations that continually evolve to meet new challenges. Some examples of general study designs that are used in air pollution research include:

- *ecologic*—Both descriptive and analytic studies in which the units of analysis are groups of people as opposed to individuals. Ecologic studies have a variety of shortcomings, as subgroups may respond very differently from one another, exposures of individuals may differ from average exposures, and events occurring in only a few individuals may be masked.
- *cross-sectional (prevalence)*—Descriptive studies are those in which a group, or groups, are studied *at a given time* to discover relationships between exposures and current health measures.
- *case-control*—Analytic studies in which groups with an existing outcome (e.g., they have a specific disease) are compared to control groups (specific disease free) to discover potential causes.
- *cohort (incidence)*—Descriptive studies in which a group (cohort) or groups defined by exposure levels are followed *over time* to document their development of new cases of outcomes (e.g., specific diseases).

- *time series*—An analysis technique in which a given population is studied for health outcomes at specific intervals (e.g., daily) which are correlated with exposures measured at the same or previous time points.
- *intervention*—Analytic studies in which the effect of an *intentional change in exposure* of a population is evaluated to discover any associated changes in health outcomes. For example, specific disease rates are compared *before* and *after* air quality has improved.

These types of study designs are not mutually exclusive. One can perform cross-sectional ecologic studies, time series intervention studies, etc. Also, an air pollution exposure can be based on one or more central monitoring stations (i.e., ecological), but the effects can be measured in individuals, which makes the study "semi-ecological" or "semi-individual." **Table 10–3** presents a validity-ranking of several experimental designs with respect to their ability to support cause-and-effect relationships. The table is conceptually instructive, but it cannot be used to rank individual studies, as many factors influence their validity. For example, time-series studies in which confounders are indentified and controlled for can be especially helpful in establishing causality. Most often the impact of the overall epidemiological evidence will be greatest when supported by multiple study designs. Each study design has distinct advantages and disadvantages to consider. Together they often complement each other.

Table 10–3 Ranking of various study designs with respect to establishing cause-and-effect relationships.

Validity Ranking	*Study Type or Design*
1. Highest	Controlled experimental studies, including clinical trials and community trials
2.	Prospective cohort studies
3.	Retrospective cohort studies
4.	Case-control studies
5.	Time-series studies
6.	Cross-sectional studies
7.	Ecologic studies
8.	Individual case studies
9. Lowest	Anecdotal descriptions

Data from Friis and Sellers (2004); and Künzli and Tager (1997).

Contemporary, Prospective, and Retrospective Data Collection

Epidemiologists gather data on individuals and populations in various ways. *Contemporary data* (also called cross-sectional data) describe the characteristics of a subject or population only at the given time of measurement. Such data are used in cross-sectional studies in which groups are compared at a specific time point. Contemporary data are also used for selecting groups for further study. For example, contemporary data could be used to identify current smokers, persons with a history of a disease of interest, or people who have a particular socioeconomic status, occupation, or residential location. These groups can then be entered into another study design in order to discover exposure factors (such as air pollutants) that may have created or might modify their health status. *Prospective* (looking forward) data are gathered on subjects after they have been identified and placed into a group or groups. For example, a group of children with a history of diagnosed asthma could be followed to obtain their future daily medication usage, periodic pulmonary function test results, and hospital admissions. Such health-related data could be correlated with environmental measurements, including meteorological and air quality data. *Retrospective* (looking into the past) data are based on historical records (e.g., medical records or job titles) or recall surveys (e.g., smoking histories, residence histories, or travel histories). *Recall surveys* are not always provided directly by the subjects (especially if they are deceased or missing); relatives, friends, coworkers, or other informants can supply primary or confirmatory information on the studied subjects. Recall survey questions must be carefully prepared in order to include information about any potential confounders. Note that the gathering of personal data, whether it be contemporary, prospective, or retrospective, involves an ethical obligation to protect the subject's confidentiality (see Chapter 12 on ethics).

Study designs based on the method of data collection include the following examples, some of which were previously indentified:

- *cross-sectional (prevalence) study*—Subjects are first selected, and then data on their health status and/or exposure status are compiled. A goal could be to relate their current health status to their air pollution exposure (either past or present). Early epidemilogists Lester Lave and Eugene Seskin (1970) used this study design to find associations between mortality, along with several morbidities, and several air pollution exposure parameters.

- *prospective cohort study*—The study begins by placing individuals into groups (cohorts) on the basis of their exposure-related status (e.g., the city in which they reside). The health status of the groups are followed for a period of time, and associations between disease states and exposure are calculated. The seminal "Harvard Six Cities Study" by Douglas Dockery and colleagues (1993) found associations between death rates and air pollution in six eastern U.S. cities with differing air pollution characteristics (e.g., types and levels of particles and gases).
- *retrospective cohort study*—In order to overcome the time and cost required to conduct a prospective cohort study, the historical exposures of specified (by health status) groups are reconstructed. The current health status can then be associated with historical exposure records.

These are only examples that demonstrate the types of studies that use contemporary, prospective, and retrospective data. A particular study may use several types of data to achieve its objectives.

Time-Series Studies

In recent decades, longitudinal time-series epidemiologic study designs have been widely used in air pollution investigations. George Thurston and Kazuhiko Ito of New York University discussed the strengths and weaknesses of such studies (Thurston and Ito, 1999). In a longitudinal time-series study a single population is studied over time (which varies from days to years). The population data collected during the study includes "outcomes" (such as deaths, hospital admissions, emergency room visits, school/work absences, etc.) and "exposures" (such as air temperature, relative humidity, and ambient air pollutant monitoring data from stationary monitors). Sophisticated statistical techniques are then used to ascertain the effects on health of each of the exposure variables. Both the outcome data and the exposure data are often highly variable, so the task is to isolate and quantify meaningful signals. For example, daily deaths vary with month, day of the week, ambient temperature, relative humidity, and other variables, such as levels of ozone, sulfur dioxide, particulate material, etc. Depending on the objectives of a particular study, any of these variables can be confounders. In addition, many of the exposure variables co-vary, in that they go up and down together (e.g., ozone and temperature, which are both driven by sunlight). In a good study, confounding and *bias* (e.g., a systematic error in measurement; poor selection of subjects; lying by subjects; or *publication bias* by reporting only positive findings) are modeled and corrected, leaving valid measures of the effects of air pollutants. The measure of the effect of a pollutant is sometimes called an *effect size,* such as the percent increase in deaths per unit increase in exposure (e.g., a 0.1 ppm increase in ozone concentration). Time-series studies, which can be very powerful, have the following advantages:

- Many serious individual-level confounders, such as smoking, occupation, residence location, socioeconomic status, and education, are automatically controlled for because they do not usually change from day to day.
- The population serves as its own control, because the health outcome *changes* are compared on days with different pollutant exposure levels.
- The statistical power can be excellent, as large numbers of people and several air pollution data points are studied at many time points.
- Since large populations are studied, they include sensitive, even critically-ill individuals that are otherwise not accessible.
- The exposures occur in their full complexity.
- The study conclusions can be independently verified, and even re-analyzed, if the original data sets are shared with other investigators. Data sharing can be a contentious activity, as investigators sometimes see their data as property that can be used to support future research proposals. Also, there may be concerns related to protecting the confidentiality of subjects.

Time-series studies also present some statistical challenges, including:

- They are usually "ecologic" in that individual outcomes and exposure data are not collected. This can be overcome (at additional expense) by the use of individual wearable portable air pollutant monitors, examination of individual health data (e.g., individual diaries and medical records), and inclu-

sion of individual time-activity patterns (e.g., time spent at work, home, commuting, etc.).
- Long-term confounding trends (e.g., seasonal effects on health) and correlations among the exposure variables (e.g., co-varying air pollutants) require sophisticated analyses and corrections in order to prevent misleading results.
- Random errors in the data can hide some valid associations, which usually leads to an underestimation of the effects of some specific air pollutants and of the effects on small sensitive sub-groups.

Nevertheless, such time-series study methods provide valuable data that is used by regulators in setting air pollution standards. The use of time-series studies on the effects of environmental ozone is presented later in this chapter.

IV. AIR POLLUTION EPIDEMIOLOGY

Overview

Epidemiologic investigations of the effects of air pollution on health have evolved from the early studies (when air pollution levels were high) to modern studies (in which air pollution levels are relatively low). Steady improvements in air quality, which led to improvements in public health, have challenged epidemiologists to develop more sophisticated study techniques that detect more subtle effects. Some of the new epidemiologic techniques are sufficiently sensitive to uncover statistically-significant associations for increases in relative risks well below 1 percent. Understandably, such small associations prompt discussions about the roles of confounders, and the costs of controlling the risk in relation to the need for funds to address other public health problems. Nevertheless, modern epidemiological studies have presented evidence that even low-levels of some air pollutants can produce adverse health effects that cannot be ignored by regulators.

Early Epidemiological Studies

As was described in Chapter 1, the great air pollution disasters in the Meuse River Valley, Belgium (December, 1930), Donora, PA (October 1948), and London (December, 1952) convincingly linked human mortality and morbidity to severe air pollution episodes. The excess deaths and attributable illnesses were measured using relatively-crude epidemiological techniques. In the early studies, excess deaths and other adverse effects were counted, categorized, and temporally related to the rise and fall of a few crudely measured air contaminants (Figure 1–5). Such quantification had a significant impact on the public and on government officials, as it provided both "body counts" and a culprit, specifically combustion-related air pollution. Equally important was the impact of these early studies on the scientific community, which often equates quantification with knowledge. This long-standing scientific view was captured by Sir William Thomson (Lord Kelvin, 1824–1907), English physicist and mathematician.

> ". . . when you cannot express it in numbers, your knowledge is of a meager, and unsatisfactory kind."

In an attempt to understand the air pollution disasters, epidemiologists, physicians, and chemists initiated a variety of new types of studies. A particularly good example of the interest in quantitation in the early decades following the 1952 London episode is found in a key paper by Lester Lave and Eugene Seskin (1970). The focus of their review of existing studies, and presentation of new epidemiologic data, was on sulfur dioxide (SO_2) and particulate matter. Among the quantitative epidemiologic study design models that were described was a linear model:

$$MR = a_0 + a_1 P + a_2 S + e, \qquad \text{(Eq. 10–7)}$$

where MR is the mortality rate (e.g., per 100,000 population), a_0, a_1, and a_2 are to-be-determined coefficients, P is a measure of a specific air pollutant, S is a measure of socioeconomic status (SES), and e is an error term. Such a model, although an elementary one, allows a determination of the background mortality rate (a_0), and the strengths of the effects of air pollution measures (a_1) and the effects of the SES (a_2). With such an equation, the strength of specific air pollutants as modifiers of mortality could be separated out from other factors using the gathered data. The quantitative analyses produced several results, including an analysis of the effects that *improvements* in air quality and SES would have on reducing mortality and morbidity rates. For example, a 10 percent decrease in particulate air pollution in polluted English Communities was projected to decrease the total death rate by 0.5 percent and the infant death rate by 0.7 percent. A 10 percent decrease in airborne sulfates (which is a component of combustion-derived particles) was projected to decrease the total death rate by 0.4 percent and the infant death rate by 0.3 percent. In comparison, a 10 percent decrease in the fraction of poor families was projected to decrease the total death rate by 0.2 percent and the fetal death rate by 2 percent. The analyses indicated that poor birth outcomes were more strongly associated with poverty than with air pollution. Clearly, the quantitative modeling approach was capable of separating the adverse effects of specific pollutants from socioeconomic effects. It also introduced the important concept that cleaner air and less poverty could produce predictable improvements in public health. After this paper, the quantitative methods used by epidemiologists rapidly advanced in sophistication. In order to demonstrate these advances, two selected air pollutants, ozone and particulate matter, will be examined in some detail.

Recent Epidemiological Studies of Ozone and Particulate Material

Ozone

Ozone (O_3) is a powerful oxidant gas that rapidly reacts on contact with many substances including those that are found in living tissue. At high concentrations it rapidly produces cell death, and at low concentrations it damages proteins, lipids (e.g., in cell membranes), and a variety of important biomolecules (e.g., those in mucus, pulmonary surfactant, and the interior of cells). As a result, at various concentrations ozone is capable of producing a range of potentially-adverse health effects. Ozone can also produce *adaptation* or *tolerance* to subsequent exposures to it and other oxidants. As for all substances, the effects of ozone depend on the dose delivered to tissues. The airborne concentration (C) of ozone is more important than exposure time (T) in producing adverse health effects (E). The relationship $E = C^2T$ has been proposed for ozone. As a secondary pollutant (formed in the air by the action of sunlight) it is difficult to control in the environment. Current 8-hour averages are below 0.075 ppm, the 2008 National Ambient Air Quality Standard, in 50 percent of the U.S. cities. The motivation to pursue even more expensive emissions controls comes from a variety of studies, including epidemiological investigations.

Epidemiological studies indicate that sufficiently elevated ozone levels are associated with a variety of short-term (that occur within days of initial exposure) adverse health effects. These associations include:

- decreased pulmonary function;
- aggravation of asthma and other pre-existing lung diseases;
- increased hospital admissions for respiratory problems; and
- increased death rates.

The long-term (i.e., months or longer) effects of ozone have not been well-studied in humans, but laboratory animal studies at levels of 0.1 ppm and greater have shown the following effects:

- thickening of deep-lung epithelial surfaces (alveolar walls), which can decrease the efficiency of oxygen uptake in the lung;
- changes in cell populations of the lung in which thin alveolar (gas-exchange) cells are replaced by cuboidal cells;
- loss of ciliated cells that are responsible for removing mucus and trapped pollutants from the tracheobronchial airways; and
- increased susceptibility to respiratory tract infections.

Potential confounders in epidemiological ozone studies include elevated temperature; automobile exhaust components (nitrogen oxides, carbon monoxide, ultrafine particles, organics); numerous reaction products of ozone with other air pollutants; possibly increased outdoor exposures and exercise on sunny days; and elevated general pollutants on days with temperature inversions.

George Thurston and Kazuhiko Ito (1999) reviewed the epidemiological data on environmental (as opposed to occupational) ozone exposures. Most of the data came from longitudinal time-series studies, which essentially eliminates confounding from factors such as socioeconomic status, cigarette smoking, place of residence, and occupation (assuming that such factors did not change significantly during the study). Thus, population measurements and concurrent (or proximate) air pollution measurements could be analyzed in order to quantify associations between health status and ozone exposure. In the most basic form of time-series study, the same-day health measures and concurrent air pollution measures are correlated. However, same-day correlations between outcomes and exposures did not tell the whole story. The health effects of ozone may occur sometime (e.g., 1 to 3 days) after the air pollution exposure. Also, a few days of cumulative exposure may be required in order to produce measurable health effects, especially when ozone levels are modest. Therefore, some of the time-series studies also included a *lag* (e.g., 1, 2, 3, etc., days) between exposure and effect, and/or a *running average* (e.g., of the previous 2 or 3 days) of the air pollution measures. In some cases, the air pollution measure was an *increment,* such as the current day ozone level minus the running average of the previous 1, 2, or more days. In this case, the increment was not an actual pollutant level, but a *change* in the level, which complicates the interpretation. If an adverse health effect was associated with an incremental exposure after 3 days of relatively clean air, it is possible that the multiday clean air period produced a down-regulation of natural defenses, which intensified the subsequent ozone effects. In this case, the *level* of ozone might not have been as toxic as was a *change in the level.* This phenomenon implies that biological systems can have short-term loss of adaptation to environmental conditions, including air pollution levels.

Thurston and Ito organized their ozone review into the following topics: premature mortality; hospital admissions (HA); emergency department (ED) visits; and individual-level studies, such as those in which pulmonary function measurements in school children were correlated with near-term environmental ozone exposures.

Ozone and Mortality

When ozone exposure is correlated with mortality (e.g., total daily deaths, or daily cardiovascular/pulmonary deaths) several of the previously mentioned potential confounders must be considered. Ozone levels tend to be higher on hot days when there is significant sunshine. Extreme temperatures, both high and low, produce increases in daily death rates as shown in **Figure 10–3**. Also, when ozone levels are high, there are changes in other air pollutants as a result of concurrent factors (confounders) such as temperature inversions and altered air chemistry (see Chapter 3). These and other confounders can only be taken into account by careful design of the time-series studies.

Considering only the studies that corrected the mortality data for season, temperature, and co-pollutant effects, elevated ozone exposures were still shown to have consistent associations with mortality. The associations could be expressed in terms of an *effect size,*

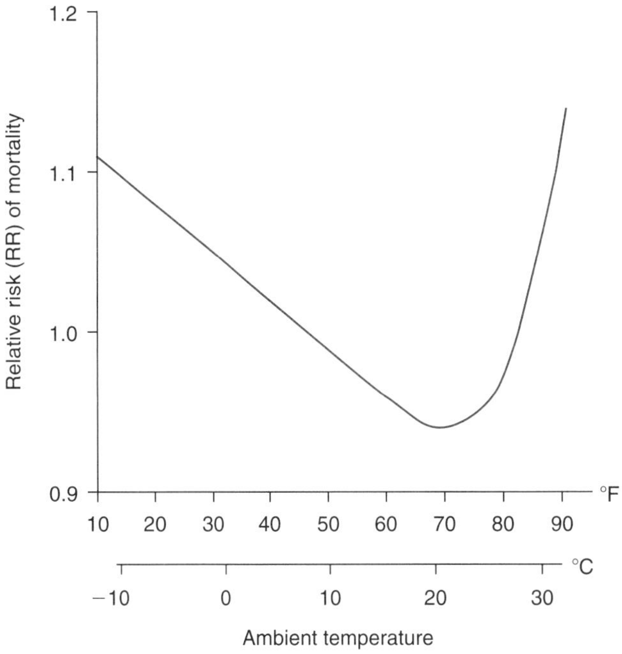

Figure 10–3 The effect of ambient temperature on death rate in six northern U.S. cities. Data from Curriero et al. (2002).
Source: The University of California Air Pollution Health Effects Laboratory, with kind permission.

which is the increased relative risk (*RR*) of death (or other outcomes) per unit increase of 0.1 ppm (parts per million) of ozone. When five well-designed studies were examined, the average effect size was an *RR* of 1.07 per 0.1 ppm of ozone (**Figure 10–4**). That is, on average, an increase of 0.1 ppm of ozone was associated with a 7 percent increase in the death rate. When the data were corrected for the concurrent particle concentration, the effect size for ozone decreased slightly to 1.06 (i.e., 6 percent), indicating the probability of an independent effect of ozone on mortality.

Ozone and Hospital Admissions

People are admitted to a hospital for a variety of reasons, which fall into two general categories: "unscheduled" (e.g., emergency admissions) and "scheduled" (i.e., by prior appointment). Furthermore, the disease diagnosis category is recorded upon admission, which permits it to be included in a study. Relevant diagnoses for ozone-related admissions include pneumonia, COPD (chronic obstructive pulmonary disease), and asthma,

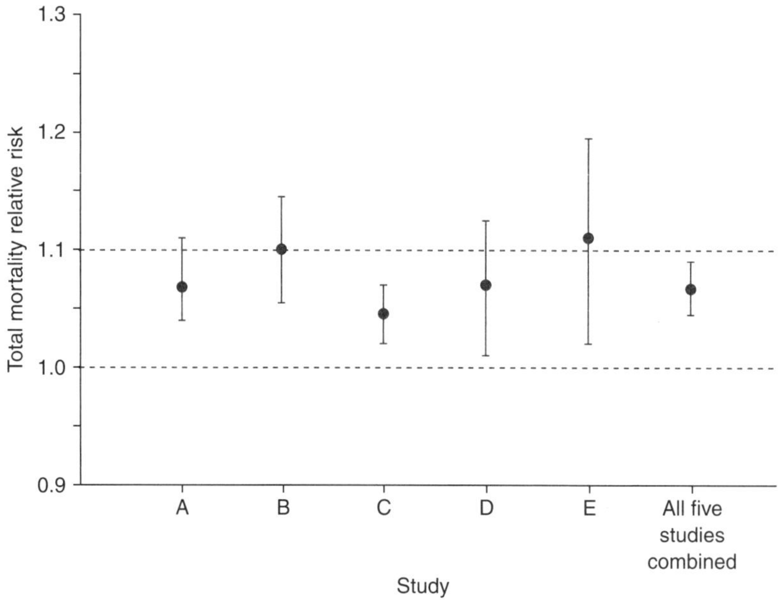

Figure 10–4 Relative risks of mortality and 95% confidence intervals associated with an increased ozone level of 0.1 ppm as seen in several studies. Data from Thurston and Ito (1999).
Source: The University of California Air Pollution Health Effects Laboratory, with kind permission.

but not accidents or drug overdoses. Most unscheduled hospital admissions (which is a category of interest in time-series ozone studies) are preceded by an emergency department (ED) visit. The majority of people that visit the ED are not admitted to the hospital, so the epidemiological-study statistics are stronger for ED visits than for hospital admissions (as there are more ED cases). On the other hand, data on hospital admissions are more likely than ED data to be available in computerized form, which facilitates their inclusion in epidemiologic studies and their analysis.

Several epidemiologic studies in the United States, Canada, and Europe correlated respiratory hospital admissions with atmospheric ozone data. Thurston and Ito reviewed 27 such studies, all of which showed a trend increase with increased environmental ozone exposure; all but 11 were statically significant at the 5 percent level. The 27 studies were analyzed to obtain a composite *RR* associated, on average, with a 0.1 ppm increase in the maximum 1-hour ozone concentration. The *RR* for major respiratory and asthma admissions combined was 1.18 (i.e., an 18 percent increase in hospital admissions). In this case the meta-analysis increased the confidence in the determination of a causal relationship between ozone exposure and hospital admissions.

Ozone and Emergency Department Visits

As a result of variations in the study designs, ED visit studies were not combined in a meta-analysis in the review. For the seven studies that were reviewed, the results were mixed. Six of the studies demonstrated a statistically-significant association between elevated ozone levels (usually above 0.06 ppm) for asthma-related ED visits for one or more age groups. In some studies, children were more affected, but in other studies, the elderly (65 plus years in age) were more affected. The conclusion was that elevated ozone exposures were associated with ED visits, but there was no obvious effect of subject age.

Individual-Level Studies and Ozone Exposure

Individual-level studies differ from group-level studies (i.e., ecologic studies) in important ways. Because such studies are labor-intensive, individual-level studies must involve smaller numbers of subjects (typically less than 100). The individuals, who are mainly studied in a laboratory, are not likely to be members of the most susceptible sub-populations. Those subjects who suffer serious effects, such as death or conditions requiring hospitalization, are seldom available for laboratory study. On the other hand, individual-level studies can investigate some potentially susceptible ambulatory groups, such as the elderly, children, and those with some pre-existing diseases. When small groups are studied, each individual can serve as their own control, as periodic examinations (e.g., pulmonary function measurements) can be conducted before, during, and after an exposure.

With respect to ozone, the most extensive individual-level epidemiologic studies were conducted on children aged 7 to 17 years enrolled in summer camps. Six such camp studies in the United States and Canada conducted serial pulmonary function measurements in over 600 youths. The youths were exposed to varying levels of environmental ozone (along with other pollutants and temperature variations) at a camp for children with medically-managed asthma. In these studies, the measures of effect included FEV_1 (forced expiratory volume in the first second of a maximum expiration), peak air flow rate (also during a maximum expiration maneuver), and the number of uses per day of a doctor-prescribed asthma medication. Hourly ambient ozone concentration measurements were made at the camps during the studies. When the FEV_1 data were combined from the six studies, a statistically significant association was seen between a decrease in FEV_1 (measured in the afternoon) and ozone exposure. A typical value for FEV_1 in children is about 1,300 mL (milliliter). On average, an increase of 0.1 ppm ozone was associated with a decrease in FEV_1 of 50 mL, with a significance (p-value) of 0.0001. For asthmatics, the medication usage doubled for an increase of 0.1 ppm ozone, and the peak flow rate decreased about 30 percent. Such changes are clinically-significant, although not life-threatening, and they clearly indicate that ozone and/or its associated co-pollutants could degrade pulmonary function in school-camp children, and worsen the symptoms in asthmatic children. Individual-level studies provide evidence for a causal link between elevated ozone levels and adverse effects on pulmonary function. A reasonable general recommendation would be to not require outdoor exercise in children during ozone episodes, as exercise produces increases in the inhaled doses of ozone and other air pollutants.

Particulate Material

Particulate air pollution comprises a great variety of materials, and a broad range of particle sizes (see Chapters 1, 2, and 3). As a result, the particle-related studies performed by epidemiologists have been both varied and numerous. A February 2010 scientific database search on "partic*" and "epidem*" returned about 80,000 citations. A discussion of this vast literature is prohibitive. Here, we will consider mainly the particulate ambient air pollution size categories as regulated (or reviewed) by the U.S. EPA. Even then, the health-related studies that are presented here will, out of necessity, include only a small number of illustrative samples of those that have been published. For a more complete analysis and review, those interested can consult U.S. EPA reports (e.g., U.S. EPA, 2004, 2009).

Because the chemical composition of PM varies substantially by season, location, and numerous other factors, the results of epidemiological studies on particulate air pollution are highly variable, and sometimes even contradictory. The U.S. EPA has thoroughly reviewed the epidemiological studies conducted between 2004 to 2009 in its *Integrated Science Assessment for Particulate Matter* (ISA) (U.S. EPA, 2009). What follows here is abstracted from that report. The ISA covered epidemiological research on 19 specific health-related outcome categories for PM including:

- heart rate and heart rate variability
- cardiac arrhythmia (a deviation from the normal heart rhythm)
- ischemia (a deficiency of blood supply)
- vasomotor function (variations in the caliber of blood vessels)
- blood pressure
- cardiac contractility (heart muscle contraction)
- systemic inflammation (inflammation in the body)
- hemostasis, thrombosis and coagulation factors (blood flow and blood clotting)
- oxidative stress (in the heart or other organs and tissues)
- respiratory-related hospital admissions, emergency department visits, and physician visits
- cardiovascular-related hospital admissions and emergency department visits
- respiratory symptoms and medication use
- pulmonary function (a variety of clinically-relevant lung functions)
- pulmonary inflammation
- other types of pulmonary injury
- allergic responses, including asthma attacks
- host defenses (primarily, infection-related)
- central nervous system effects (related to brain activity)
- respiratory-related mortality

For each of these categories, the U.S. EPA made determinations of causality related to PM exposure. Causality determinations were based on the combined evidence from epidemiologic, toxicologic, and mechanistic (chemical pathways) data. What follows are only illustrative examples of the assessments. The five causality categories (that were previously defined in this chapter) used were:

- "causal relationship"
- "likely to be a causal relationship"
- "suggestive of a causal relationship"
- "inadequate to infer a causal relationship"
- "not likely to be a causal relationship"

Short-term Cardiovascular Effects and $PM_{2.5}$ Exposure

Emergency department visits and hospital admissions for cardiovascular diseases were consistently observed in PM-related epidemiological studies. There was strong evidence of both regional and seasonal variations. The largest risk effect estimates (up to about 2 percent per 10 μg/m^3 increase in $PM_{2.5}$) were seen in the Northeastern U.S. in the winter months. Ischemic diseases and traffic-related pollutants appeared to be driving the emergency department visits and hospital admissions.

Cardiovascular-related mortality increases ranging from 0.5 to 0.9 percent were associated with 24-hour average $PM_{2.5}$ concentrations above 13 μg/m^3, which is consistent with the hospital admissions and emergency department visits noted above. With respect to the composition of $PM_{2.5}$, metals and particulate carbon were implicated as primarily important for producing the effects. Michelle Bell and colleagues found statistically-significant hospital admission associations in older adults only for the vanadium, nickel, and elemental carbon components of $PM_{2.5}$. Confounding by these few chemical components is a likely explanation for the large regional and seasonal variations in $PM_{2.5}$ effects when PM mass (but not PM composition) was measured.

The epidemiological studies on blood pressure, cardiac arrhythmias (irregular heart beats), and markers of blood coagulation have not been consistent in demonstrating associations with exposure to $PM_{2.5}$ mass. However, animal studies and some human clinical investigations support such health effects for selected components of $PM_{2.5}$, especially nickel.

Although there are numerous positive epidemiologic studies associating $PM_{2.5}$ exposure with adverse cardiovascular outcomes, there are also conflicting studies, particularly in the Western U.S. (see Enstrom, 2005). The controlled laboratory animal and human clinical studies have been generally supportive of the positive epidemiological studies that associate $PM_{2.5}$ exposure with adverse cardiovascular outcomes. The U.S. EPA concluded that: "Together, the collective evidence is sufficient to conclude that a causal relationship exists between short-term $PM_{2.5}$ exposures and cardiovascular effects." A more cautious conclusion might be that some components (some metals and elemental carbon) of $PM_{2.5}$ in some regions of the United States have produced adverse cardiovascular outcomes, including increased mortality. It must be noted that PM composition may not be considered, by law, in establishing the National Ambient Air Quality Standards for PM. Thus, the U.S. EPA must establish PM standards based on particulate mass in given particle size ranges. Also, the EPA standards, which are national, must be met in all cities, and in all seasons without regard to chemical composition. This constraint placed on the EPA has led to controversy and challenges of its mass-based PM standards.

Short-term Cardiovascular Effects of $PM_{10-2.5}$ and Ultrafine PM

The coarse mass fraction of particulate air pollution, $PM_{10-2.5}$, is generally dominated by natural sources, but resuspended road dust, construction, farming, and some other human-related sources and activities also contribute. Because coarse particles settle out of the air rapidly (e.g., at velocities above 50 cm per hour), population exposures tend to be localized near $PM_{10-2.5}$ sources. The substantial heterogeneity in population exposure degrades the sensitivity of epidemiological studies, which typically rely on only a few fixed air monitoring stations. In addition, if the $PM_{10-2.5}$ mass is determined by subtracting the measured $PM_{2.5}$ mass from the measured PM_{10} mass, there will be an increased error (because errors in the two particle-mass measurements combine). Also, natural biological allergens in the coarse particle mode can strongly affect sensitive population groups. All of these factors produce substantial uncertainty in the epidemiological findings for coarse particles that are based only on average levels of particle mass.

As expected, the epidemiological associations between coarse particles and cardiovascular outcomes are more variable than the associations for fine particles. Some studies have indicated positive associations between coarse particles and hospital admissions and emergency department visits, but other studies have produced non-significant and/or no associations. Most coarse-particle studies have not adequately controlled for confounders, especially for the effects of biological aerosols (that produce infections and allergic responses) and co-present fine particles. Corroborating clinical and laboratory animal studies have also produced variable results. Rodents (which are commonly used laboratory animals) have not been good models for coarse particle effects because of the high nasal deposition in rats and mice (and thus poor exposure of deeper respiratory tract regions) for particles larger than a few micrometers in diameter. Therefore, after reviewing the data, the EPA classified the cardiovascular effects of $PM_{10-2.5}$ as being only "suggestive of a causal relationship."

Ultrafine particles (UFP), which are particles that are smaller in diameter than 0.1 μm, have been present in the troposphere in substantial numbers long before humans roamed the Earth. Natural UFP are generated by fires, other high-temperature processes, radioactive decay (e.g., radon decay), and chemical reactions among gases that produce solid or liquid reaction products (e.g., oxides of sulfur and some organic gases and vapors). Human-generated sources of UFP include fuel combustion and other high-temperature processes. These sources of UFP also produce a variety of co-pollutant gases.

Interest in the potential human health effects of UFP as a separate class of PM is relatively recent. This interest was largely stimulated by laboratory animal studies demonstrating that UFP could move through membranes of the body and have biological effects when inhaled in small mass quantities (but large particle numbers). **Table 10–4** shows, for various individual particle sizes, the number, mass and surface area, and the same measures for one microgram (μg) of particles. The smaller particles individually have smaller masses and surface areas than the larger particles, but when one considers a total mass of a μg of particles, the situation is reversed; the smaller particles have much greater total counts and total surface areas. The current concern is that ultrafine particles may have an increased toxicity (over that for fine or coarse particles), even in small mass concentrations, as a result of their collective surface areas as well as their ability to translocate from the lungs to other organs and tissues.

The epidemiologic data on the effects of ultrafine particles are mixed. One time-series study in Atlanta, GA found no associations between 24-hour levels of UFP and any cardiovascular outcomes, including cardiovascular disease, arrhythmia, congestive heart failure, ischemic heart disease, peripheral vascular disease, and cerebrovascular disease. Other studies have shown some UFP-associated sub-clinical changes in heart function, but there is insufficient evidence for causality. The EPA concluded that "evidence is suggestive of a causal relationship between ultrafine PM exposure and cardiovascular effects." However, caution is advised in implicating only the UFPs, as associated gaseous co-pollutants are found in substantial concentrations. For example, diesel engine exhaust has copious UFPs, but also can have significant concentrations of carbon monoxide, nitrogen oxide gases, sulfur dioxide, and organic gases and vapors. Each of these co-pollutants can produce adverse health effects in sufficient concentrations.

Short-term Respiratory Effects and PM

Epidemiologic studies of $PM_{2.5}$ and respiratory effects have demonstrated several consistently statistically-significant associations. Respiratory-related hospitalizations and emergency department visits were seen to

Table 10–4 Particle properties vs. diameter (D_p) for spherical particles with a density of 2 g/cm^3.

D_p (μm)	Mass (gm)	Surface Area (cm^2)	Number/μg (count)	Surface/μg (cm^2)
10	1.05×10^{-9}	3.14×10^{-6}	0.95×10^{3}	2.98×10^{-3}
1	1.05×10^{-12}	3.14×10^{-8}	0.95×10^{6}	2.98×10^{-2}
0.1	1.05×10^{-15}	3.14×10^{-10}	0.95×10^{9}	2.98×10^{-1}
0.01	1.05×10^{-18}	3.14×10^{-12}	0.95×10^{12}	2.98
Change Going From 10 μm to 0.01 μm D_p	Decreases by 9 orders of magnitude	Decreases by 6 orders of magnitude	Increases by 9 orders of magnitude	Increases by 3 orders of magnitude

increase by an average of 1 to 4 percent per 10 μg/m^3 increase in $PM_{2.5}$ in several cities. Similarly, positive associations have been seen between $PM_{2.5}$ and respiratory related mortality, with an effect estimate (*RR*) range of 1.7 to 2.2 percent per 10 μg/m^3 increase in mass concentration. Results have been mixed regarding $PM_{2.5}$ associations and measures of asthma exacerbations. Pulmonary function measurements are also mixed with respect to the effects of $PM_{2.5}$, but the majority of epidemiologic studies show positive effects. For example, in asthmatic children, an increase of 10 μg/m^3 is associated with a 1 to 3.4 percent decrease in FEV_1. Many of the positive associations in children and in adults are not necessarily driven by $PM_{2.5}$ mass concentration alone. As previously noted, recent studies implicate specific components (including nickel, vanadium, organic carbon, and elemental carbon), or specific sources such as wood-burning and vehicular traffic in producing effects. It is clear that $PM_{2.5}$ mass alone is not a strong causal agent in producing respiratory effects, and the EPA concluded that "a causal relationship is likely to exist between short-term $PM_{2.5}$ exposures and respiratory effects." Again, note that the EPA is constrained to regulate PM mass, not components of PM.

Coarse particles ($PM_{10-2.5}$) have been associated with hospital admissions and emergency department visits for respiratory conditions, more strongly in asthmatic children than in asthmatic adults. However, the strong correlations between $PM_{2.5}$ and $PM_{10-2.5}$ levels and the spatial heterogeneity of coarse particle exposures diminish the significance of the results. The coarse particle effects are confounded by fine particles. The EPA concluded that evidence for coarse particles is only "suggestive of a causal relationship" for respiratory effects.

Ultrafine particles, as measured by number concentration (e.g., thousands of particles per cm^3 of air), have been associated with respiratory-related hospital admissions and emergency department visits in Copenhagen (Denmark) and Helsinki (Finland), but not in Atlanta, GA. Oddly, the UFP concentrations in the European cities were between 6,200 and 8,200 particles/cm^3, while those in Atlanta were 38,000 particles/cm^3. This raises the possibility that confounders, such as the aforementioned co-pollutant gases, and/or weather phenomena were important. The EPA evaluated the evidence for causality of UFP as only "suggestive of a casual relationship."

Comments on PM Effects

The health effects of particulate material in all size ranges is an issue that continues to concern regulators. It is also a complex problem, given the current crude state of measurement of composition, the heterogeneity of exposures, and the seasonal and geographical variations in composition.

Peter Valberg and Ann Watson (1998) presented an alternate hypothesis that challenged the assumption that PM inhalation actually causes the observed associations. Non-PM factors, such as changes in weather and behavior might be causing (or at least influencing) the health associations in the epidemiology studies. A new technique, Geographic Information Systems (GIS), will help resolve some of these issues (**Exhibit 10–3**).

Exhibit 10–3 Geographic Information Systems (GIS)

One persistent problem in epidemiological research is matching the *locations of people* exposed to air pollutants with the *pollutant concentrations* at those same locations. Lists of zip codes, street addresses, and census data have been used, along with published locations of air sampling stations, to make the needed matches. The relatively recent availability of satellite global positioning systems (GPS), inexpensive computational capacity, and new software for data linking has come to the rescue.

What exactly are the new Geographic Information Systems (GIS)? Such systems are a set of powerful techniques for the acquisition, storage, retrieval, analysis, and display of three-dimensional geographical data. Positional data can be obtained from a variety of external digital sources (e.g., digitized maps that are in the public domain). The geographic data can be edited as needed using software and correlated with similarly digitized health-related data sets. Retrieval and display are performed using available software. The user has the capability of tailoring the process as needed for their study.

Epidemiologists have adopted GIS technology as an essential tool for promoting public health, as by mapping the geographical spread of diseases during epidemics. From an air pollution perspective, automated matching of locations of exposed individuals (and populations) with interpolated air-quality data is an important task that can be efficiently performed using GIS technology. Many degree programs already offer training in Geographic Information Systems.

V. POTENTIALLY SUSCEPTIBLE SUB-POPULATIONS

Natural variability in the responses of individuals in a population to *any change* in the environment is expected. Such variability produces a broad composite population dose-response curve (**Figure 10–5**). Human populations are so diverse that they must often be broken into sub-populations, each with its own dose-response relationship to air pollutants. The potentially susceptible sub-populations (with respect to the effects of air pollution responses) are sometimes classified into two general groups:

- those who are likely to be *more sensitive* to a given air concentration—including the ill, elderly, very young, undernourished, etc.
- those who have *greater exposures* and/or fewer options for avoiding exposure, or for seeking medical treatment—including the poor, uneducated, exercising school children, exercising outdoor workers, and those living or working near substantial sources of air pollutants (e.g., near heavy traffic, incinerators, and significant industrial and agricultural operations).

Nationwide environmental air standards apply to all segments of the population, so ideally, they would also protect all of the major susceptible sub-populations. Such sub-populations are most-reliably identified by epidemiological methods, but clinical, laboratory, environmental, and even mathematical modeling studies can also be used to identify, confirm, or suggest susceptibility. Data collected from hospital admissions, emergency department visits, death certificates, and questionnaires often contain data that help to identify those sub-populations that are the most strongly affected by air pollutants. Some clinical studies focus on subgroups, such as asthmatics, diabetics, cardiac patients, cigarette smokers, the elderly, and exercising subjects in order to identify susceptibility factors. Laboratory animal studies that use genetically-modified and/or disease models (see Chapter 9) can also point to possible susceptible human populations. Environmental air concentration measurements are used to identify locations (e.g., those near heavy sources of air pollutants) where the exposures of some people are elevated in relation to average population exposures. Mathematical modeling also contributes to identifying susceptible groups by simulating how their particular anatomical and physiological characteristics might influence their doses.

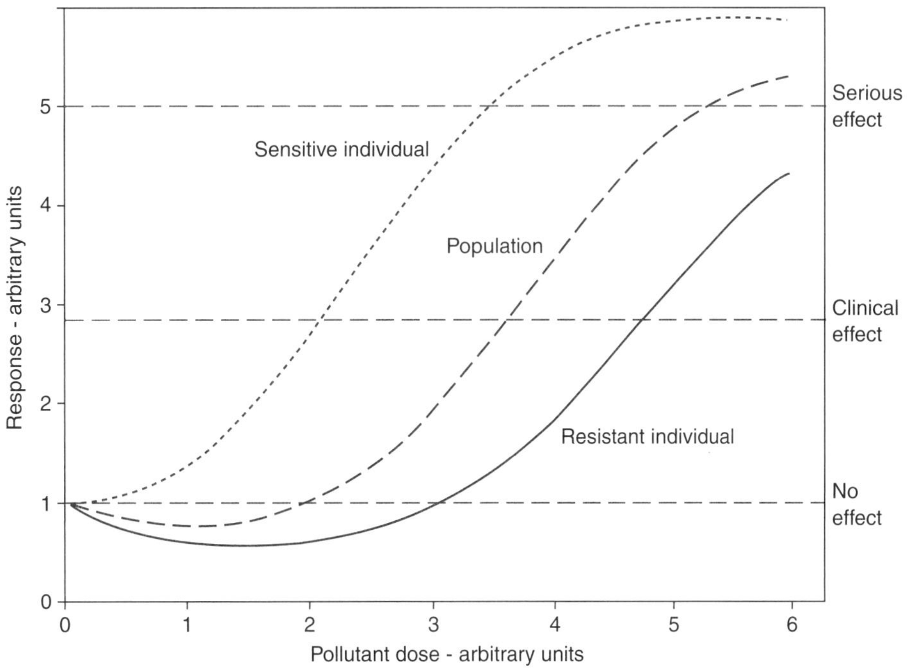

Figure 10–5 Dose-response curves showing those for a sensitive individual, a resistant individual, and the entire population. At low doses, a "hormetic" (i.e. beneficial) effect is shown where the curves fall below a "no effect" level. *Source:* The University of California Air Pollution Health Effects Laboratory, with kind permission.

Thus, data from several disciplines contribute to identifying potentially-susceptible sub-populations.

The following groups have been identified as more susceptible, or suspected of being more susceptible, to air pollutants than the general population:

- the elderly, i.e., age 65 years and older;
- young children from birth to 3 years of age;
- persons with significant respiratory and/or cardiovascular diseases;
- persons with transient respiratory-tract infections;
- diabetics and obese individuals;
- school children and workers who routinely exercise outdoors during air pollution episodes;
- persons exposed in locations near significant air pollution sources, such as heavy-traffic or poorly-controlled industrial and agricultural operations;
- the disadvantaged (e.g., by poverty, low educational levels, and other socioeconomic status factors);
- persons genetically predisposed to illness; and
- possibly those who are exposed *in utero.*

The list of potentially susceptible sub-populations is still evolving as new research results are published. New air pollution standards are established after consideration of the epidemiologic and other literature, which may not adequately represent all of the potentially susceptible sub-populations.

It should be emphasized that nation-wide or even city-wide air standards cannot protect absolutely everyone from harm. Even background levels (e.g., those that would be present if anthropogenic contributions were zero) of natural sources such as microbes, pollen-producing plants, and windblown dust will adversely affect select individuals in a population. Therefore, some very sensitive people may need to resort to avoiding outdoor exposures during air pollution episodes and/or to using doctor-prescribed medications for their protection. Deciding on how to adequately protect sensitive sub-populations and individuals presents a formidable challenge to regulators and other health professionals.

VI. SUMMARY OF MAJOR POINTS

Epidemiology, the study of the health of human populations, is critically-important for identifying the levels of air pollutants that significantly contribute to disease and death in people. *Biostatistical techniques* are a powerful and evolving tool used in epidemiological studies to establish statically-significant associations between measures of air pollution and adverse health outcomes. Epidemiology can contribute to a judgment that a *cause-and-effect* relationship exists between an environmental exposure and an adverse health outcome. Although confirmatory data from controlled (e.g., laboratory) studies are important for causal determinations, strong epidemiological evidence (as exists for lung cancer and cigarette smoking) alone can trigger regulatory actions. Epidemiology can identify the *relative risk* (*RR*) associated with an air pollutant exposure or other environmental factors. A value for *RR* of 1 implies no additional risk, a value of 2 indicates a doubling of risk, and a value of *X* indicates an *X*-fold increase in risk. Cigarette smokers have a *RR* of 30 for developing lung cancer. However, for the effects of air pollution, a *RR* of 1.01 (a 1 percent increase in risk) or sometimes less for a serious adverse effect can stimulate regulatory action. Epidemiological studies are affected by *confounders,* which are uncontrolled variables that influence the outcome being studied, and by *surrogates,* which co-vary with the causal variables. For environmental air pollution studies, confounders include socioeconomic status (SES), indoor air pollutants, unmeasured outdoor air pollutants, temperature extremes, and other factors. Surrogates for air pollutants may also include particle mass, particle count, and visibility.

A large variety of study designs are used by epidemiologists to investigate the effects of air pollutants on health. Such designs include *ecologic, cross-sectional, case-control, cohort, time-series,* and *intervention.* Some epidemiologic studies combine these and other study designs in order to identify the effects of air pollutants. The data collected in an epidemiology study may be *retrospective, contemporary,* and/or *prospective. Time-series* methods are commonly used in air pollution studies because they control for many important population-related confounders such as SES, occupation, and residential location.

Air pollution studies have evolved in their designs since the time of the great air pollution disasters of the 20th century. Modern studies are usually very sophisticated with respect to identifying, and controlling, known or suspected confounders. They also contribute to identifying *sensitive sub-populations,* and they can establish quantitative relationships between changes in air pollutant levels and changes (usually increases) in adverse health effects. Such quantitative relationships inform regulators on the benefits of tightening air standards (e.g., the U.S. EPA National Ambient Air Quality Standards, NAAQS).

Ozone and particulate matter (PM) serve as examples of modern problematic air pollutants that have been extensively studied by epidemiologists. Ozone is a specific chemical (an oxidizing gas), but PM is not chemically-specific, as its composition is variable in both time and location. Accordingly, PM is studied in particle size ranges: $PM_{2.5}$ (fine particles); $PM_{10-2.5}$ (coarse particles); PM_{10}; and UFP (ultrafine particles). Currently, UFP is not regulated in the United States. Evidence is strong and growing that indicates PM composition, as opposed to just particle size and mass, is important for producing adverse health effects. As the composition of PM varies substantially from city to city, mandated national standards based on particle size and mass have generated challenge and controversy. Similarly, PM varies in composition from nation-to-nation, currently making international standards problematic.

The identification of potentially-sensitive sub-populations with respect to the adverse effects of air pollutants is continually evolving. To date, several sensitive sub-populations have been identified including the elderly, the ill, children, and those with low SES.

The understanding of the adverse effects of air pollution exposure has been significantly advanced by epidemiologists. Yet, many uncertainties remain, so the opportunities for those who pursue careers in epidemiology should be strong for the foreseeable future.

VII. QUIZ AND PROBLEMS

Quiz Questions

(select the best answer)

1. Epidemiology is important for establishing air pollution standards because:
   a. it provides information on actual air pollutant exposures of human populations.
   b. it is the best way to identify the cellular mechanisms of toxic actions of specific chemical constitutes of air pollutant mixtures.
   c. any epidemiologic association is proof of a cause-and-effect relationship.
   d. All of the above are true.
2. Which consideration does *not* help in scientifically establishing a cause-and-effect relationship?
   a. The cause precedes the effect.
   b. The strength of the cause affects the strength of the effect.
   c. The effect must immediately follow the cause.
   d. If the cause is removed the effect does not occur.
3. Hill's considerations for establishing a cause-and-effect relationship:
   a. must all be fulfilled.
   b. must be confirmed by animal studies.
   c. need not all be fulfilled.
   d. are not relevant.
4. Relative risk (*RR*) is:
   a. the same as the confidence interval (*CI*).
   b. a ratio of two risks.
   c. a difference between two risks.
   d. a measure of loss of life expectancy (*LLE*).
5. With respect to sensitive sub-populations for the adverse effects of air pollutants, which set of groups are included?
   a. the very young, the very old, and those suffering from cardiovascular diseases
   b. the very young, the very old, and people who have high socioeconomic status
   c. the very old, the obese, and children who are exercising indoors
   d. cardiovascular patients, small adults, and people who have below normal lung capacity
6. Particulate matter (PM) is:
   a. regulated by the U.S. EPA in particle size ranges.
   b. variable in composition from city to city but *not* from season to season.
   c. is not harmful to public health.
   d. Only a. and b. are true.
7. Ozone is an air pollutant that:
   a. is mainly a "primary" pollutant.
   b. is mainly a "secondary" air pollutant.
   c. is neither a "primary" nor a "secondary" pollutant.
   d. is associated with human mortality, but not human morbidity.
8. Time-series epidemiological studies:
   a. control for confounders such as socioeconomic status and occupation.
   b. involve small numbers of subjects in relation to other study designs.
   c. are not useful for studying the effects of gaseous pollutants.
   d. All of the above are true.

9. Retrospective epidemiological data:
   a. are a sub-category of prospective data.
   b. can be obtained using medical records but *not* questionnaires.
   c. cannot be used for deceased subjects.
   d. may be published if confidentiality is protected with respect to identifying the subjects.
10. Which of the following are confounders for ozone exposure?
    a. Ambient temperature
    b. Co-pollutant particles and gases
    c. Both a. and b. are true.
    d. None of the above are true.
11. What type of epidemiologic study design would be preferred for studying asthmatic children at summer camps?
    a. Retrospective cohort studies
    b. Time-series individual studies
    c. Ecologic studies
    d. Mathematical modeling studies

Problems

1. If the relative risk of an adverse effect associated with an air pollution episode is 1.01, and the exposed population is 1 million people, how many people are likely to experience the adverse effect?
2. If the PM_{10} level is 100 μg/m^3 of air, and the $PM_{2.5}$ level is 25 μg/m^3, what are the concentrations of a) fine particles and b) coarse particles?
3. If the average daily mortality in a city of 10 million people is 200 on days without air pollution episodes, and 202 on days with increases of 10 μg/m^3 of $PM_{2.5}$, what is the relative risk associated with the increased $PM_{2.5}$?
4. Given that a group of asthmatic children use an inhaled medication once a day ($RR = 1$) when air quality is good, and twice a day ($RR = 2$) when the ozone level is above 0.1 ppm, what 95 percent confidence interval on the relative risk is required for the finding to be statistically significant at a 5 percent significance level?
5. Using Figure 10–3 how many more deaths would you expect in a 90° heat wave, than at 70° when the death rate at 70° is 100 per day?
6. Review the 1998 reference by Valberg and Watson. Briefly discuss their suggestion that human *behavior* during PM episodes might produce health effects, as opposed to the direct effects of inhaling PM.

VIII. DISCUSSION TOPICS

1. Should air pollution standards in Mexico be as stringent as those in the United States or in England? If so, why? If not why not?
2. Are current air pollution standards in your city currently adequate? Discuss the pros and cons of tightening the current air standards.
3. Discuss the considerations involved in designing, conducting, and publishing an epidemiologic study on the effects of ozone in AIDS patients.
4. Should air quality standards be based only on adverse health effects that reach clinical significance? Alternatively, is *biological significance,* or just *statistical significance,* strong enough to support a proposed air quality standard?

References and Recommended Reading

Baker, D. and Nieuwenhuijsen, M. J., eds., *Environmental Epidemiology: Study Methods and Application,* Oxford University Press, New York, 2008.

Basu, R. and Ostro, B. D., A multicounty analysis identifying populations vulnerable to mortality associated with high ambient temperature in California, *Am. J. Epidemiol.,* 168:632–637, 2008.

Bell, M. L., Ebisu, K., Peng, R. D., Samet, J. M., and Dominici, F., Hospital admissions and chemical composition of fine particle air pollution, *Am. J. Resp. Crit. Care Med.,* 179:1115–1120, 2009.

CIA (Central Intelligence Agency), https://www.cia.gov/library/publications/the-world-factbook/ (accessed November 16, 2009).

Clancy L., Goodman, P., Sinclair, H., and Dockery, D. W., Effect of air pollution control on death rates in Dublin, Ireland: An intervention study, *Lancet,* 360 (9341):1210–1214, 2002.

Cohen, B. L., Catalog of risks extended and updated, *Health Phys.,* 61:317–335, 1991.

Cohen, B. L. and Lee, I. S., A catalog of risks, *Health Phys.,* 36:707–722, 1979.

Costa, D. L., Air pollution, in *Casarett and Doull's Essentials of Toxicology,* Klaassen, C. D. and Watkins, J. B. III, eds., McGraw–Hill, New York, 2003, pp. 407–418.

Curriero, F. C., Heiner, K. S., Samet, J. M., Zeger, S. L., Strug, L., and Patz, J. A., Temperature and mortality in 11 cities of the eastern United States, *Am. J. Epidemiol.* 155:80–87, 2002.

DHHS (U.S. Department of Health and Human Services), *Smoking and Health: Report of the Advisory Committee to the Surgeon General of the Public Health Service,* U.S. Public Health Service, Atlanta, GA, 1964.

DHHS (U.S. Department of Health and Human Services), *The Health Consequences of Smoking: A Report of the Surgeon General,* U.S. Department of Health and Human Services, Atlanta, GA, 1973.

DHHS (U.S. Department of Health and Human Services), *The Health Consequences of Smoking: Cancer; A Report of the Surgeon General,* U.S. Department of Health and Human Services, Centers for Disease Control and Prevention, Atlanta, GA, 1982.

Dockery, D. W., Pope, C. A., Xu, X. P., Spengler, J. D., Ware, J. H., Fay, M. E., Ferris, B. G., and Speizer, F. E., An association between air pollution and mortality in six United States cities, *N. Engl. J. Med.* 329:1753–1759, 1993.

Enstrom, J. E., Fine particulate air pollution and total mortality among elderly Californians, 1973–2002., *Inhal. Toxicol.,* 17:803–816, 2005.

EPA (U.S. Environmental Protection Agency). *Air Quality Criteria for Particulate Matter, Vol. I and II, EPA/600/P–99/002 aF–bF,* U.S. Environmental Protection Agency, Research Triangle Park, NC, 2004.

EPA (U.S. Environmental Protection Agency), *Integrated Science Assessment for Particulate Matter: Second External Review Draft,* EPA 600/R–08/139B, U.S. Environmental Protection Agency, Research Triangle Park, NC, 2009.

Friedman, G. D., *Primer of Epidemiology, 4th Edition,* McGraw–Hill, New York, 1994, pp. 210, 194–224.

Friis, R. H. and Sellers, T. A., *Epidemiology for Public Health Practice, 4th Edition,* Jones and Bartlett, Sudbury, MA, 2008.

Gordis, L., *Epidemiology, 4th Edition,* Saunders (Elsevier, Inc.), Philadelphia, PA., 2009.

Hill, A. B., The environment and disease: Association or causation?, *Proc. Roy. Soc. Med., Occup. Med.,* 58:295–300, 1965.

Holgate, S. T., Samet, J. M., Koen, H. S., and Maynard, R. L., eds., *Air Pollution and Health,* Academic Press, San Diego, CA, 1999.

Künzli, N. and Tager, I. B., The semi-individual study in air pollution epidemiology: A valid design as compared to ecologic studies, *Environ. Health Perspect.,* 105:1078–1083, 1997.

Lave, L. B. and Seskin, E. P., Air pollution and human health, *Science,* 169:723–733, 1970.

Lipfert, F. W., *Air Pollution and Community Health: A Critical Review and Data Sourcebook.* Van Nostrand Reinhold, New York, 1994.

Mar, T. F., Larson, T. V., Stier, R. A., Clairborn, C., and Koenig, J. Q., An analysis of the association between respiratory symptoms in subjects with asthma and daily air pollution in Spokane, Washington, *Inhal. Toxicol.,* 16:809–815, 2004.

Metzger, K. B., Tolbert, P. E., Klein, M., Peel, J. L., Flanders, W. D., Todd, K., Mulholland, J. A., Ryan, P. B., and Frumkin, H., Ambient air pollution and cardiovascular emergency department visits, *Epidemiol.,* 15:46–56, 2004.

Norman, G. R. and Streiner, D. L., *Biostatistics: The Bare Essentials, 3rd Edition,* B. C. Becker, Inc., Hamilton, Ontario, 2008.

Phalen, R. F., *The Particulate Air Pollution Controversy: A Case Study and Lessons Learned,* Kluwer Academic Publishers, Boston, MA, 2002.

Pope, C. A. III, Review: Epidemiological basis for particulate air pollution health standards, *Aerosol Sci. and Technol.,* 32:4–14, 2000.

Pope, C. A. III, Ezzati, M., and Dockery, D. W., Fine-particulate air pollution and life expectancy in the United States, *N. Engl. J. Med,* 360:376–386, 2009.

Savitz, D. A., Poole, C., and Miller, W. C., Reassessing the role of epidemiology in public health, *Am. J. Public Health,* 89:1158–1161, 1999.

Taubes, G., Epidemiology faces its limits, *Science,* 269:164–169, 1995.

Thomson, W. (Lord Kelvin), *Popular Lectures and Addresses, Vol. I, Constitution of Matter,* Macmillan, London, 1889.

Thurston, G. D. and Ito, K., Epidemiological studies of ozone exposure effects, in *Air Pollution and Health,* Holgate, S. T., Samet, J. M., Koren, H. S., and Maynard, R. L., eds., Academic Press, San Diego, CA, 1999, pp. 485–510.

Thurston, G. D. and Kinney, P. L., Air pollution epidemiology: Considerations in time-series modeling, *Inhal. Toxicol.,* 7:71–83, 1995.

Valberg, P. A. and Watson, A. Y., Alternative hypotheses linking outdoor particulate matter with daily morbidity and mortality, *Inhal. Toxicol.,* 10:641–662, 1998.

Wynder, E. L., Invited Commentary: Response to *Science* article: "Epidemiology faces its limits," *Am. J. Epidemiol.,* 143:747–749, 1996.

Chapter 11

Risk Assessment

LEARNING OBJECTIVES

By the end of this chapter the reader will be able to:

- define and distinguish "risk," "risk assessment" and "risk communication"
- discuss the four main stages of the risk assessment process and how uncertainties are addressed at each stage
- identify the major differences between how carcinogens and non-carcinogens are evaluated in the risk assessment process
- interpret and discuss U.S. EPA risk assessment data in the Integrated Risk Information System (IRIS) at http://www.epa.gov/IRIS/
- discuss the barriers to effective risk communication and the methods for overcoming these barriers

CHAPTER OUTLINE

I. INTRODUCTION

Risk assessments provide the major rationale for setting air quality standards for protecting the general public and workers. In order to understand how risk assessments are performed it is helpful to understand risk itself.

Two of the many special characteristics that set humans apart from other species are our propensity to take risks and our ability to look to the future. We have taken sail to the uncharted perils of the ocean in search of unknown riches. We continue to explore the inhospitable vast expanse of outer space for the promise of adventure and knowledge. We invest our money and time on education, careers, and even the work of others (e.g., stock market) for future prosperity. In addition, we routinely propel ourselves at dangerously high speeds, well beyond our natural capacity, for both excitement and perhaps an attempt to slow the natural progression of time. Risk taking is in our nature.

In general, if the potential benefit to us and others is great enough, then it may be worth taking even a substantial risk. With assisted travel via automobiles, we assume the benefit of getting from point A to point B with alacrity far outweighs the risks of our own bodily injury or even the possibility of community illnesses from the air pollution it produces. However, to assess these sometimes forbidding risks we must be able to objectively measure them and weigh them against the associated benefits. This is a central objective of the field of risk assessment.

What is Risk?

Definitions and Examples

Risk can be defined as the probability and magnitude of an adverse event within a specified population and within a given time.

For example, the risk of a college student dying in an automobile accident may be one in 40,000 per year. In the liability insurance community the adverse event is often translated into a monetary loss. With respect to business investments risk can also be translated into monetary gains or losses. Assigning a measurable value (e.g., amount of money) simplifies the process of *risk assessment,* which can be defined as the systematic process of determining the probability and magnitude (e.g., a dollar amount) of an undesired event. Overall, risk assessment is aimed at predicting future outcomes, which we should all know is not an exact science. If it were, then there would not be much need for insurance policies, competitive games, or gambling.

In contrast to the insurance or business models, it is more difficult to assign an objective scale of loss regarding the severity of health-related outcomes. For example, what is the value of a shortened lifespan, an asthma attack, or reduced quality of life associated with a chronic respiratory-tract illness? These are all important factors people value and expect professionals to evaluate and control. On the other hand, employment and economic stability are equally important factors that directly affect the quality of our lives. In the United States, a portion of the money, goods, and services that flow through a community are used to help educate children, improve public health, and increase personal safety. The wealth of a community need not be measured by a median income; instead, we can just as easily measure the wealth by the quality of the schools, law enforcement, health care, public services (e.g., parks), and access to public health services within the community. Thus, gainful employment is a fundamental building block, which should not be overlooked, for a healthy and safe community.

Balancing Risks

The following is a brief list of consequences and benefits that should be considered in the risk assessment process:

- death rates (mortality) and lifespan
- disease (morbidity)
- permanent disability
- impairment of the senses and discomfort (e.g., eye irritation)
- psychological health (e.g., freedom from excessive worry and noxious odors)
- suppression of the immune system
- reproductive health and fertility
- birth defects
- economic benefits
  - employment
  - affordable energy (e.g., automobile fuel and electricity)
  - access to quality goods, services, and housing
- public health benefits
  - sanitation
  - affordable foods and nutrition
  - disease prevention

It must be noted that any major decision, such as a regulation regarding the control or removal of an air pollutant, may also involve some immediate loss of jobs and other negative economic impacts. To best serve society, both the benefits and consequences of such decisions must be considered. In addition, feasible alternatives should be available for comparison and as part of the decision process. For example, there should not be a decision to ban an industrial chemical or process unless viable alternatives have also been evaluated in an equivalent manner on the basis of risks, costs, and benefits to society. It is much easier to make a decision when no alternatives have been evaluated, especially because it may appear to err on the side of caution. However, scientists and regulators have an ethical responsibility to ensure the best possible choice (at the time) is made for the overall betterment of society. The environmental risk assessment process has its challenges and controversies, which are discussed in this chapter. The underlying goal here is to introduce the basic risk assessment process used in air pollution science and to describe some of its pitfalls and uncertainties.

Quantifying Risks

We typically present risks as proportions, such as the percentage (or probability) of cancers associated with certain personal behaviors or environmental conditions. As can be seen in **Table 11–1** the combined cancer risks associated with smoking and poor diet far outweigh those associated with environmental pollution. **Figure 11–1** provides a graphical representation of the relative percentages of cancer deaths attributed to various environmental and behavioral factors. Personal behaviors such as poor diet, tobacco smoking, alcohol consumption, and unprotected sexual activity are strongly associated with cancer, much more than environmental pollution. However, even modern levels of air pollution clearly have a negative impact on the health and longevity of susceptible groups of individuals (e.g., young children, the elderly, and those with cardiovascular or lung diseases). Risk can also be expressed as the estimated number of deaths or disease cases per million person lifetimes, as illustrated in **Table 11–2**.

Communicating Risks

One effective way to communicate risk to the general population is the use of lost days of life associated with an activity, occupation, or exposure (**Table 11–3**). One problem with this, and most other representations of risk, is that they do not accurately predict the outcome for an individual. For example, smoking a pack of cigarettes per day may not result in lung cancer nor reduce the lifespan for some, but may greatly reduce the lifespan of a susceptible or unfortunate person. Still, it is known that lung cancer within a population can be greatly reduced if substantially fewer people smoke.

If effectively communicated, clear representations of risk can be used to modify high risk behaviors, and benefit society as a whole. *Risk communication,* which is the exchange of information about risks between those responsible for assessing, minimizing, and regulating risks, and those who may be affected by the outcomes of those risks, is an essential step in the risk assessment process. How risks are presented to the public will influence personal perceptions and judgments regarding those risks. Without effective risk communication, the risk assessment process can be an expensive exercise in futility.

Table 11–1 Percentage of cancer deaths attributed to various environmental and behavioral factors.

Attributable Factor	*Percent of Cancer Deaths (United States, 1981)*	*Percent of Cancer Deaths (United Kingdom, 1998)*
Tobacco	25–40	29–31
Diet	10–70	20–50
Infections	About 10	10–20
Ionizing and UV radiation	2–4	5–7
Occupational exposures	2–8	2–4
Environmental pollution	1–5	1–5

Data from Doll and Peto (1981); Doll (1998, 1999).

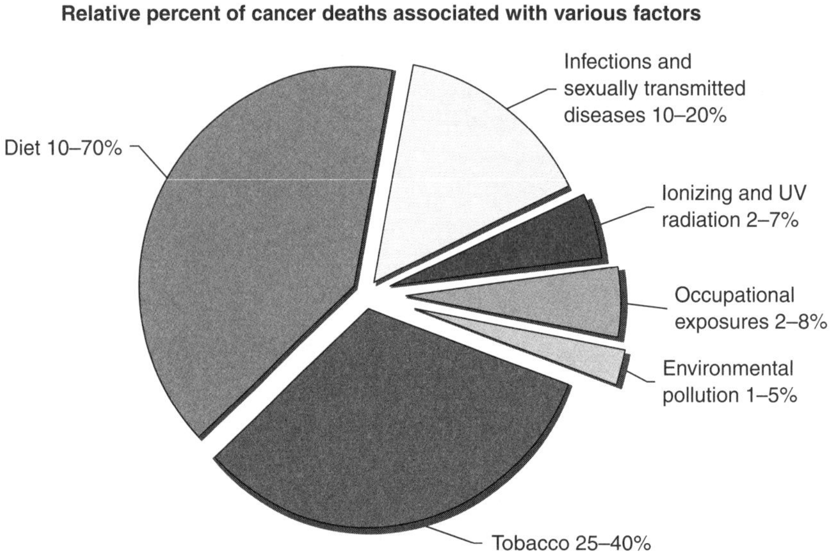

Figure 11–1 Graphical representation of the relative percentages of cancer deaths attributed to various environmental and behavioral factors, information from Doll and Peto (1981); Doll (1998,1999). Average percentages are approximated; however, the percentages do not add up to 100% because overlap exists among the various factors. *Source:* The University of California Air Pollution Health Effects Laboratory, with kind permission.

Table 11–2 Activities associated with a one-in-a-million lifetime risk of death.

Activity	*Cause of Death*
Smoking 1.4 cigarettes	Cancer, heart disease
Living 2 months with a cigarette smoker	Cancer, heart disease
One chest x-ray	Cancer (radiation)
Eating 100 charcoal-broiled steaks	Cancer (benzopyrene)
Eating 40 tablespoons of peanut butter	Liver cancer (aflatoxin B_1)
Drinking chlorinated water (Miami, Florida)	Cancer (chloroform)
Living 2 months in Denver, Colorado	Cancer (cosmic radiation)
Living 2 days in New York, New York	Air pollution
Living 5 years adjacent to a nuclear plant	Cancer (radiation)
Traveling 10 miles by bicycle	Accident
Traveling 300 miles by car	Accident
Traveling 1000 miles by jet airplane	Accident
Traveling 6000 miles by jet plane	Cancer (cosmic radiation)

Data from Wilson (1979).

Table 11–3 Loss or gain in life expectancy.

Factor	*Average Loss or Gain in Life Expectancy*
Being an alcoholic	Loss of 11 years
Living in poverty	Loss of 10 years
Being 45% overweight	Loss of 9 years
Having a risky occupation (Fireman)	Loss of 3.5 years
Having less than 8 years education	Loss of 2.1 years
Being unemployed	Loss of 1.4 years
Driving a small/compact car	Loss of 2 months
Eating 1 tbsp peanut butter per day	Loss of 1 day
Being a college graduate	Gain of 2.4 years
Being an educator	Gain of 3.1 years
Being a corporate executive	Gain of 4.7 years

Data from Cohen and Lee (1979); and Cohen (1991).

Early Beginnings of Formal Risk Assessment

The earliest beginnings of the risk assessment process in the United States were in the 1930s within the occupational medicine and industrial hygiene professions, with the development of the dose-response relationship and the American Conference of Governmental Industrial Hygienists (ACGIH®) threshold limit values (TLVs®). The TLVs® refer to airborne chemical concentrations for which it is believed that nearly all workers may be repeatedly exposed on routine or normal workdays, over a working lifetime, without adverse health effects. Note that the ACGIH® does not claim that *all* workers can be protected in the workplace. Additional precautionary measures, at the individual level, may be necessary to protect sensitive or susceptible persons.

The general risk assessment approach is straightforward, still used today, and can be simplified into the following steps:

1. Determination of an acceptable exposure threshold (TLV®) using clinical, toxicological (dose-response), and epidemiological data
2. Determination of human exposures
3. Comparison of human exposures to the TLV®

In the public arena, risk assessment is used by government agencies in the establishment of regulatory standards for safe drinking water, air, food, medicines, and consumer products. The 1958 *Delaney Amendment* to the Food, Drug, and Cosmetic Act was one of the first regulatory applications of risk assessment in an attempt to reduce the cancer-producing potential of food additives. Initially, the "Delaney Clause" was aimed at reducing the cancer risk to less than 1 in 100 million person lifetimes, represented as 1.0×10^{-8}. This was later changed to a risk criterion of 1 in a million or 1.0×10^{-6}, a level that was also adopted by the U.S. EPA for many carcinogens. By comparison, about 1 in 5 people will develop some form of cancer at some time in their life. Environmental and occupational health professionals also use risk assessment to minimize the risks of chemical, physical (e.g., noise, vibration, or radiation), or biological (e.g., viruses and bacteria) exposures for workers and the general public.

The Hazard Quotient

Evaluating a potential occupational hazard at the final step (i.e., comparison of human exposures to the TLV®) is often accomplished by simply dividing the exposure by the TLV®. When the units of measure are the same (e.g., mg/m^3 of air), then the resulting numerical value will be a unitless proportion centered on the value 1.0. In the environmental arena, this measure, sometimes termed the *hazard quotient* (*HQ*), is used to characterize the risk associated with an exposure:

$$HQ = \text{exposure concentration/TLV} \qquad \text{(Eq. 11–1)}$$

Because exposure concentrations vary with time, location, and environmental conditions, a number of measurements, determined statistically, are required to adequately represent the actual exposures. Thus, the exposure concentration is never a single value and is commonly reported as being within a 95th percentile *confidence interval* (CI). If the data are normally distributed, the CI is the sample mean ($\bar{x}$) plus or minus about two standard errors for large sample sizes (n). The CI is an estimate of the total population mean exposure (μ) and can be expressed as

$$\bar{x} - t_{(0.05,df)}SE \le \mu \le \bar{x} + t_{(0.05,df)}SE, \qquad \text{(Eq. 11–2)}$$

where $t_{(0.05,df)}$ is the t distribution value for n–1 degrees of freedom at the 95 percent confidence level. For large sample sizes greater than about 120 the $t_{(0.05,\infty)}$ is equal to about 1.96. For smaller sample sizes near 10 the $t_{(0.05,9)}$ is about 2.3. A *Student's t distribution* table is used for varying sample sizes (degrees of freedom) and critical probability values other than the 95th percentile.

The standard error (*SE*) for a sample size (n) is defined as

$$SE = \sigma_{n-1}/\sqrt{n}, \qquad \text{(Eq. 11–3)}$$

where σ_{n-1} is the standard deviation for n–1 degrees of freedom. The standard deviation is defined in Chapter 4 on Sampling and Analysis. The resulting confidence interval with lower and upper limits provides an appropriate measure of the population exposure. In contrast, the sole use of a mean or median measure of central tendency would be protective of only the lower half of the exposed population, which is not suitable for practical risk determinations. The goal is to protect a large majority of the population, with the understanding that the uncertainties and statistical probabilities will not effectively protect all people in the population. There is always a random chance that an untoward event, such as a disease, will occur in some fraction of the exposed population. Thus, we cannot protect everyone, even if the risk assessment is expertly performed. Just as with the occupational health setting, sensitive or susceptible persons may have to take additional precautionary measures, on an individual level, to protect their health and well-being.

The confidence interval for the exposure concentration can be used in the hazard quotient calculation, which will result in a *HQ* confidence interval. Interpretation of the *HQ* is based on whether the lower or upper confidence limits are below, above, or overlap the value 1.0. The following guidelines are useful in the interpretation of the hazard quotient:

- An upper limit of less than 1.0 indicates that exposures are below the acceptable threshold and of less concern.
- A lower limit greater than 1.0 indicates exposures exceed the identified acceptable level, which is probable cause for concern and remedial action, if the risk trade-offs are reasonable.
- When the lower and upper limits encompass 1.0 this indicates that a portion of population exceeds the identified acceptable level, which may be a cause for concern, especially for vulnerable or hypersensitive individuals within the population.

Lastly, when the *HQ* confidence interval overlaps or slightly exceeds 1.0 this does not imply that the population is overexposed or that there will be adverse health effects in exposed individuals. Margins of safety are built into the acceptable threshold values to account for hypersensitive individuals. Thus, the *HQ* is a measure of desired protection and not of a guaranteed outcome.

Air Pollution Risk Assessment

For chemical air pollutants, risk can be expressed as a function of exposure and toxicity. While toxicity is easily translated as the magnitude of an adverse event at a given dose, exposure can only be loosely associated with an increase or decrease in the probability of an outcome. Much of this lack of certainty is due to the lack of reliable data on health effects at low exposure concentrations (as occur in most modern daily air pollution exposures). The use of exposure has both advantages and disadvantages, which will be discussed in more detail.

Both a measurable exposure and a measurable toxicity are required to assess the risks associated with an air pollutant. Without one or the other being significant there is little risk involved. This relationship can best be conceptualized by the formula

$$Risk = f(exposure \times toxicity). \qquad \text{(Eq. 11–4)}$$

As an example, a small exposure to a highly toxic contaminant may have the same risk associated with it as a large exposure to a slightly toxic material. Zero exposure to even a highly toxic substance possesses no risk. Conversely, moderate exposure to a relatively nontoxic substance may also pose no significant risk. This simplified model provides a convenient means of assessing risk for known exposures and toxicities. One major advantage of this model is that exposure is both concrete and measurable in principle. However, a major drawback stems from the inherent gaps in the data for how toxicity influences risk. Most of our toxicity data are for large exposures in laboratory animals or from major releases in the environment, whereas our everyday environmental exposures are often orders of magnitude smaller. The model must make the assumption that toxicity can be extrapolated from high to low exposures. Unfortunately, with extrapolation comes uncertainty, which is a continual area of concern, debate, and controversy in the field of air pollution risk assessment. Risk assessment is a continually evolving process. More information on the future directions and goals of risk assessment can be obtained in the National Research Council's (NRC) *Science and Decisions: Advancing Risk Assessment* (2009), which is discussed in **Exhibit 11–1**.

Today, the above mentioned formula has little use in risk assessment, because many other factors must be considered in the risk assessment process. These factors include the quality and quantity of the available data, strength of evidence in multiple animal species (i.e., concordance), extrapolation of results from animals to humans, the variable sensitivity of individuals, the effects of co-stressors, and metabolic differences between age groups, genders, and species. Most air pollution risk assessment follows the model developed by the NRC in 1983 with its publication of the "Red Book," *Risk Assessment in the Federal Government: Managing the Process.* According to this guideline report, *risk assessment* is the evaluation of information on the hazardous properties of substances, the extent of human exposure to them, and on the characterization of the resulting risk. The general

Exhibit 11–1 Does the red book risk assessment paradigm need to be revised?

Risk assessments have been valuable for evaluating a large variety of substances for their potential toxic effects. Yet there are many reasons to take a critical look at this process. For example, the current practice of performing assessments on one chemical at a time, as outlined in the venerable *Red Book* (NRC, 1983), was recently evaluated by the National Academies of Science. The resulting report *Science and Decisions: Advancing Risk Assessment,* sponsored by the U.S. EPA, takes a critical look at the past, present, and future of risk assessment (NRC, 2009). One reviewer of the report commented that it "will be a standard reference and textbook for many years to come" (Brewer, 2009).

The report made recommendations on two broad elements involved in performing risk *analyses* (the term used by the committee). First, the supporting *technical* analysis itself should be changed to better address the uncertainties and the substantial variability associated with dose-response curves. As is inherent or unavoidable in the current assessment process, greater uncertainty and variability do not necessary equate to greater risk. Second, the usefulness of the risk analysis process should be improved in order to meet the needs of decision-makers responsible for protecting public health. The assessment should be based on a *decision* and *viable options,* instead of exclusively on the risk associated with chemicals one at a time.

The new NRC report and its recommendations were necessary to address several problems. One, the process is so bogged down as to be dysfunctional in some instances. Risk assessments often take more than a decade to complete, largely due to the collection and analysis of vast amounts of non-critical or poorly relevant data. In addition, the current process does not adequately account for biases, misunderstandings, and practical realities among the various stakeholders. These include unrealistic public fears, political pressures, government agency responsibilities, industry capabilities, and economic realities. As a result, the process is flawed with respect to its ability to protect public health. The information provided for a specific chemical is often focused on potential harm, while neglecting the benefits associated with its use, or whether or not there are reasonable substitutes.

In summary, risk assessments should be based on the specific needs of decision-makers, taking into account the relative merits of decision options for managing risk. For example, a fuel additive that will improve automobile efficiency and decrease air pollution emissions requires a more careful analysis than simply an evaluation of its ability to produce harm at improbable doses. Again, as Paracelsus wisely pronounced, the dose makes the poison. Are the current risk assessments so one sided that the assessors and decision-makers see only the poison but not the bigger picture as it relates to public health?

process is a systematic approach divided into four stages (**Figure 11–2**):

1. Hazard identification
2. Hazard assessment or dose-response assessment
3. Exposure assessment
4. Risk characterization

Donald Barnes (1994) effectively interpreted these stages into the following rationale:

1. Is this stuff toxic?
2. How toxic is it?
3. Who is exposed to this stuff? How long? How often?
4. So what?

The process is fairly straightforward but filled with numerous unknowns including gaps in available information and other uncertainties in every stage. Ultimately, there would be no need to weigh and debate the risks or the probabilities of unfavorable consequences if we were certain of the outcomes.

Who Uses Risk Assessment?

Risk assessment is used by regulatory agencies and industry for a number of purposes. The U.S. Environmental Protection Agency (U.S. EPA) uses risk assessment to establish community air pollution exposure limits, emission standards, and to establish priorities for research and regulatory activities. Industries use risk assessment to protect its workers and to establish acceptable thresholds for air emissions of pollutants not regulated by the government. The overwhelming reality is that the U.S. government regulates hundreds of common air pollutants, while industry must deal with tens-of-thousands of chemicals on a routine basis. Thus, risk assessment is an important tool for companies wishing to minimize environmental and occupational liabilities (e.g., catastrophic releases and community exposures).

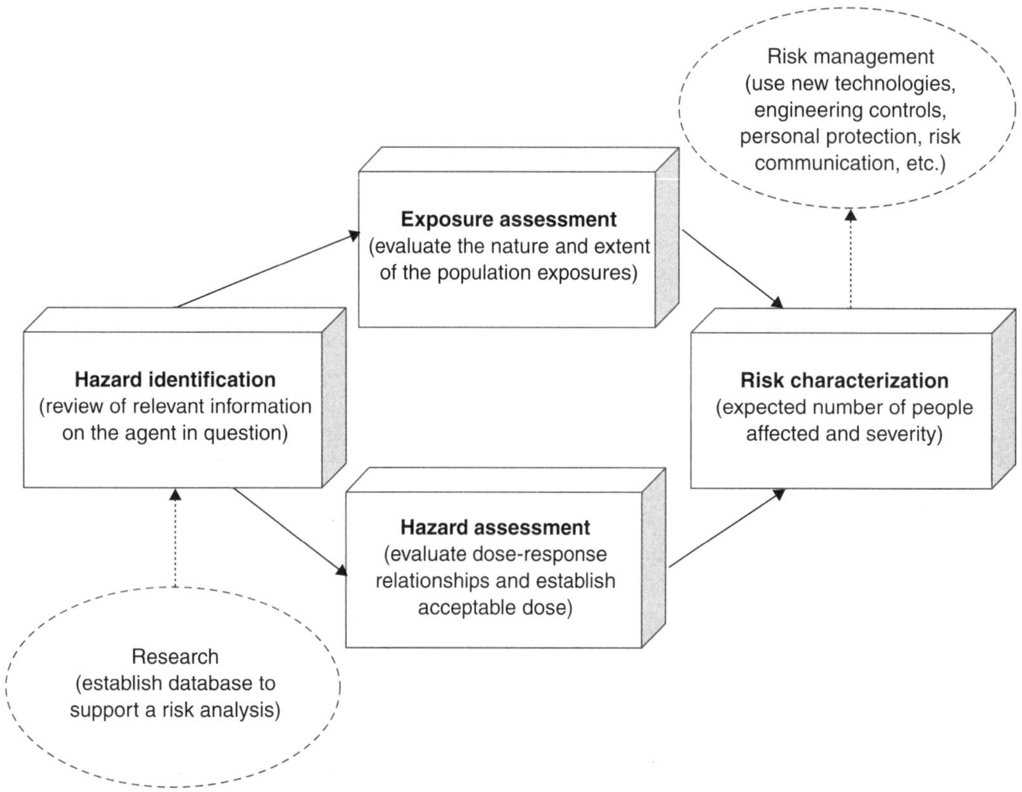

Figure 11–2 The risk assessment process.
Source: The University of California Air Pollution Health Effects Laboratory, with kind permission.

Decreasing liability is an important business strategy for minimizing losses, improving an organization's public image (i.e., "brand" identification), and increasing shareholder value for its investors. Managing risks and community exposures to air pollutants makes "good business sense."

Importantly, both the U.S. EPA and Occupational Safety and Health Administration (OSHA) provide incentives and reduced regulatory involvement for organizations that are effectively managing risks. The U.S. EPA provides compliance incentives, in the form of reduced penalties and reduced inspections, for organizations that develop *environmental management systems* aimed at identifying, disclosing, and correcting environmental hazards. In a similar manner, OSHA has the *Voluntary Protection Program* (VPP Star) for organizations that develop safety and health management systems that include risk assessments. The VPP Star provides organizations with special recognition and permits some level of self-evaluation and autonomy regarding compliance issues. Keeping the regulators happy also makes "good business sense."

Additional Considerations for Carcinogens

Carcinogens (i.e., substances that cause or promote cancer outcomes) pose a unique problem in risk assessment and are an area of continual debate in the scientific community. Due to the long latency periods (e.g., 20 to 30 or more years for cigarette smoke and asbestos) and unpredictable outcomes associated with exposure to potential carcinogens, acceptable thresholds cannot be easily established. Thus, there is great uncertainty at low doses of carcinogens. Note that asbestos-related cancers are quite rare, and that only about 1 in 20 smokers (~5 percent) develop lung cancer. Instead of using a hazard quotient that is dependent on an acceptable threshold dose, cancer risk is often expressed as an incremental probability of an exposed person developing cancer over a lifetime, known as the *excess cancer risk* (ECR). Application of the ERC is detailed later in this chapter.

II. HAZARD IDENTIFICATION

Hazard identification is a critical starting point in the risk assessment process that is aimed at identifying the potential adverse health effects (e.g., acute or chronic diseases) in an exposed population. It is a qualitative evaluation of the available toxicological, epidemiological, biological, and chemical data for the purpose of hazardous classification. It addresses the basic question of "Is this stuff toxic?" and substantiates the need for further risk assessment.

Many scientific organizations and government agencies are involved in the identification and classification of hazards. Detailed reports and databases are provided by the following (small sample of numerous) organizations:

- The Agency for Toxic Substances and Disease Registry (ATSDR) under the Centers for Disease Control and Prevention (CDC) produces detailed toxicological profiles on hazardous wastes (http://www.atsdr.cdc.gov).
- The ACGIH® publishes the *Documentation of the Threshold Limit Values and Biological Exposure Indices* (http://www.acgih.org).
- The International Agency for Research on Cancer (IARC) under the World Health Organization (WHO) publishes the *IARC Monographs on the Evaluation of Carcinogenic Risks to Humans* (http://www.iarc.fr).
- The International Programme on Chemical Safety (IPCS) publishes the *Concise International Chemical Assessment Documents* and *Environmental Health Criteria* on selected hazards (http://www.who.int/ipcs/en).
- The National Institute for Occupational Safety and Health (NIOSH) publishes criteria documents on occupational hazards (http://www.cdc.gov/niosh).
- The U.S. National Library of Medicine under the National Institutes of Health (NIH) provides the Toxicology Data Network (TOXNET) with integrated databases on toxicology, hazardous chemicals, environmental health, and toxic releases (http://toxnet.nlm.nih.gov).
- The U.S. EPA publishes the *Integrated Science Assessments* for air pollutants, the *Integrated Risk Information System* (IRIS), and *Toxicological Reviews* on environmental pollutants (http://www.epa.gov/iris).

The primary goal of hazard identification is to determine the relative toxicity of the substance of interest, and the toxicological consequences associated with human exposure. Toxicity is often categorized using a qualitative classification scale based on the dose required to produce an adverse health effect. The earliest example of this, which is still used today, was introduced by Harold Hodge and James Sterner (1949) and is illustrated in **Table 11–4**.

Another example of toxicity classification comes from the U.S. Code of Federal Regulations (40 CFR §156.62) where the U.S. EPA established an acute toxicity category for pesticides based on various routes of exposure (oral, inhalation, dermal, and ocular). The four categories range from highly toxic (I) to practically non-toxic (IV), which constitutes a qualitative scale. The categories are used in the labeling of pesticides, but can also be used to estimate the toxic potential of air

Table 11–4 Toxicity rating.

Toxicity Rating	*Toxicity Category*	*Lethal Single Oral Dose*	*Probable Lethal Human Dose*
1	Extremely toxic	≤1mg/kg	a taste
2	Highly toxic	1–50 mg/kg	1 grain
3	Moderately toxic	50–500 mg/kg	1 tsp.
4	Slightly toxic	0.5–5 g/kg	1 oz.
5	Practically non-toxic	5–15 g/kg	1 pint
6	Relatively harmless	≥15g/kg	1 quart

Data from Hodge and Sterner (1949).

Table 11–5 U.S. EPA acute toxicity categories for pesticide products.

Hazard Indicators	*I Highly Toxic*	*II*	*III*	*IV Practically Non-toxic*
Oral LD_{50} (mg/kg)^A	≤50	>50 to 500	>500 to 5,000	>5,000
Dermal LD_{50} (mg/kg)^A	≤200	>200 to 2000	>2000 to 20,000	>20,000
Inhalation LC_{50} (mg/liter)^B	≤0.2	>0.2 to 2	>2 to 20	>20
Eye irritation	Corrosive; corneal opacity not reversible within 7 days	Corneal opacity reversible within 7 days; irritation persisting for 7 days	No corneal opacity; irritation reversible within 7 days	No irritation
Skin irritation	Corrosive	Severe irritation at 72 hours	Moderate irritation at 72 hours	Mild or slight irritation at 72 hours

$^A LD_{50}$ is the lethal dose (mg/kg body weight) that is fatal to 50% of the test animals.
$^B LC_{50}$ is the lethal concentration (mg/liter air) that is fatal to 50% of the test animals in a given time (usually four hours).
Data from Code of Federal Regulations 40 CFR § 156.62.

pollutants other than pesticides. The U.S. EPA Acute Toxicity Categories for Pesticide Products are provided in **Table 11–5**.

The primary sources of toxicity information useful in hazard identification include:

- epidemiological studies
- *in vivo* bioassays
- *in vitro* studies
- evidence of biological activity
- chemical structure and reactivity information

Epidemiological Studies

Epidemiological studies define the relationship between exposure and disease in a given human population. The primary advantages are (1) that large populations can be evaluated, which may increase the statistical power for the detection of health effects at low doses and (2) the results apply to diverse populations under actual exposures. The primary limitation is that it is difficult to establish a "cause-and-effect" relationship within a complex and uncontrolled environment. Thus, it is important to understand that epidemiology studies alone are not sufficient to firmly establish a causal effect. Evidence of biological activity and toxicity are also needed to make a proper determination. On the other hand, strong epidemiologic data alone can trigger new regulations. The application of epidemiology in air pollution science is discussed in Chapter 10.

In Vivo Bioassays

In vivo bioassays in toxicology (see Chapter 9) are studies conducted on live animals to assess the effects of *exposure.* In order to see an effect, the doses are typically, but not always, much higher than would be seen in the normal environment. However, they provide useful information on the potential harm and lethality associated with exposure to a suspected hazard. Chapter 12 on Ethics also covers the need for laboratory animal studies to precede human clinical studies, which are more definitive.

In Vitro Methods

In vitro methods in toxicology (see Chapter 9) are often conducted in cell culture dishes or test tubes. These tests are performed on cells, tissues, biochemicals, or unicellular organisms to assess changes in cellular structure, function, survival, or growth. A major drawback of *in vitro* toxicity tests is that they cannot adequately mimic the complex metabolism and biotransformation that occurs within the human body. Nonetheless, the combined available *in vitro, in vivo,* and epidemiologic evidence is effectively used to assess and identify potential hazards to human health.

Evidence of Biological Activity

There exists a wealth of information on the acute toxicity of pollutants in reports associated with accidental spills and releases into the environment, as well as accidental or deliberate poisonings in humans and animals. Medical journals often report the signs and symptoms, and health effects due to chemical exposures and poisonings seen in emergency room visits. The ecological effects of chemical spills and releases are available in U.S. EPA investigation reports and the scientific literature. These types of information are a critical component of the hazard identification process, as they are indicators of potential harm to humans, animals, and ecosystems. However, caution and common sense must be used when relying on such observational data. Every toxicologist and environmental professional should know that "the dose makes the poison"; even excessive consumption of water can be fatal. Consequently, a primary goal of hazard identification is to identify a *critical dose* and its potential impact on human health and the environment.

Chemical Structure and Reactivity Information

Environmental chemists and toxicologists often look at chemical structure and the presence of reactive sites (e.g., double or triple bonds) and functional groups to help determine the potential toxicity of a chemical pollutant. Strong acids (H^+ proton donors) and bases (H^+ proton acceptors) have corrosive properties that can damage tissues. Carboxylic acids (–COOH) are among the weaker acids, but they may cause harm if they accumulate in the body. Aldehydes (–HC=O) and ketones (–C=OC–) are reactive and easily converted to carboxylic acids. Additional reactive functional groups include amines ($-NH_2$ or –NH), some halogenated compounds (e.g., Cl or Br), and highly reactive acid chlorides (–C=OCl). Additional properties of interest to toxicologists when assessing potential hazard include:

- molecular weight (e.g., determines diffusion across membranes);
- water solubility (e.g., determines pollutant deposition, distribution, and elimination);
- lipid solubility (relates to ability to cross cell membranes and to accumulate in the brain and fat tissue);
- ionic or dipole charge (e.g., influences deposition upon inhalation and chemical reactions);
- volatility (e.g., determines deposition upon inhalation and route of elimination); and
- stability (e.g., with respect to temperature, pH, metabolic action, and oxygen; determines fate in the body).

Chemical structure is needed to predict the absorption, distribution, metabolism, storage, and elimination of air pollutants in the human body, also known as *toxicokinetics* (see Chapter 9).

Hazard Identification of Carcinogens

Different rating systems exist for the identification and classification of carcinogens. Several organizations and government agencies have developed their own simplified rating systems. The most prominent include the World Health Organization's International Agency for Research on Cancer (IARC), the U.S. EPA (EPA), and the American Conference of Governmental Industrial Hygienists (ACGIH®). Each uses a similar approach and general classification system based on the strength of evidence in laboratory animals and humans. For the purposes of considering air pollution, we will focus on EPA's classification system.

EPA classifies carcinogens into seven categories, A through E, based on the adequacy of evidence in humans and animals. The data come primarily from epidemiological studies in humans and *in vivo* bioassays in laboratory animals. The general classification system is outlined as follows and the decision matrix is illustrated in **Table 11–6**:

- A: human carcinogen
- B: probable human carcinogen
  - B1 indicates limited human evidence
  - B2 indicates sufficient evidence in animals and inadequate or no evidence in humans
- C: possible human carcinogen
- D: not classifiable as to human carcinogenicity
- E: evidence of non-carcinogenicity in humans

A large amount of scientific research and evaluation goes into assigning the chemical classification of carcinogenicity for each chemical that is considered. The end

Table 11–6 Generalized decision logic for U.S. EPA classification of human carcinogens.

	Animal Evidence				
Human Evidence	*Sufficient*	*Limited*	*Inadequate*	*No Data*	*No Evidence*
Sufficient	A	A	A	A	A
Limited	B1	B1	B1	B1	B1
Inadequate	B2	C	D	D	D
No Data	B2	C	D	D	D
No Evidence	B2	C	D	D	E

See text for abbreviations.

product allows air pollution professionals a means of assessing carcinogenic potential. Reliable classification becomes valuable when comparing available chemical alternatives for a process, and when designing control strategies in the *risk management* process (that follows risk assessment). A major drawback to these generalized classification systems is that they do not take dose into account. Thus, a *full* risk assessment is necessary when regulatory or risk management decisions need to be made regarding the feasibility of existing control technologies, or when evaluating available alternatives.

III. HAZARD ASSESSMENT

Introduction

When a hazard has been identified and determined to be a significant concern, one proceeds to *hazard assessment,* the next stage in the risk assessment process. Hazard assessment involves two primary tasks: (1) to establish a dose-response relationship and (2) to define an *acceptable* exposure threshold or dose.

The first task, establishing a dose-response relationship, is critical for future extrapolations of animal and *in vitro* data to humans as well as for drawing conclusions regarding the associations among exposure and harm. It should be established that increases in the exposure dose (e.g., the concentration in the air) will result in increases in the harmful response (e.g., lung damage). In actuality, what really needs to be known is whether decreasing the dose will in fact reduce potential harm. In order to establish this relationship a positive correlation (e.g., linear, sigmoidal, or exponential) of the response with increasing dose is needed. An example of an ideal sigmoidal, dose-response curve is illustrated in **Figure 11–3**.

Reliable extrapolation from the often high doses used in animal or cellular toxicity tests downward to typical air pollution exposures in humans is not possible if low dose-response information is absent. In addition, any available occupational dose-response data must be extrapolated downward (except in unusual cases) before

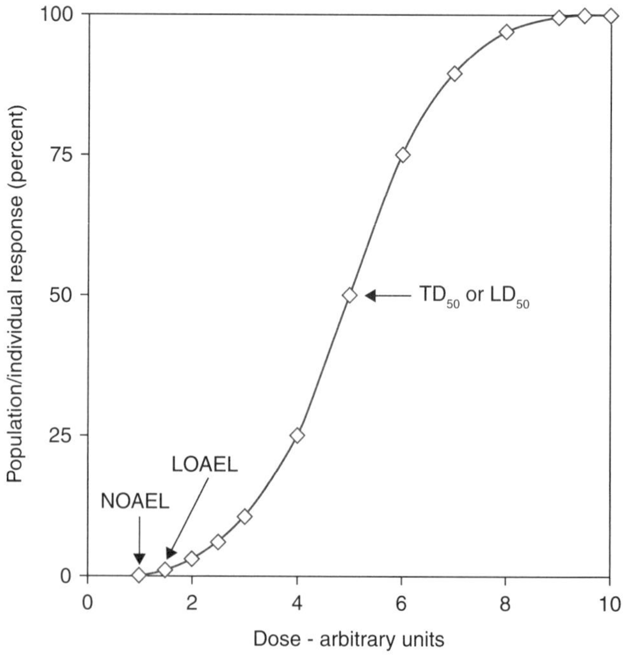

Figure 11–3 An example of an ideal dose-response curve. *Source:* The University of California Air Pollution Health Effects Laboratory, with kind permission.

it is applied to community levels to account for the large differences in daily exposure times (24 versus 8 hours), greater complexity of outdoor pollution, and the presence of vulnerable individuals. A healthy, adult worker with an 8-hour daily exposure is assumed to have at least 16 hours of no exposure between each workshift. In addition, many workers are exposed only 5 days of the week, which adds an additional 48 hours of no exposure time to a workweek. These breaks in exposure can effectively reduce the cumulative effect on worker health and allow for higher occupational exposures to pollutants. In contrast, community exposures are assumed to have no relief and also include a non-working population of children, elderly, and the infirm. Without well-established dose-response relationships it would be difficult to determine how to best protect these vulnerable segments of the population using the available toxicological data.

The second task in this process is to use the dose-response data and identified uncertainties to establish an *acceptable exposure level.* The primary steps are:

1. Interpret the dose-response data and define an exposure level (i.e., threshold dose) with little to no expected adverse effects;
2. Develop *safety factors* to account for unknowns, uncertainties, and susceptible individuals; and
3. Correct the threshold dose using these safety factors.

The acceptable exposure level is the threshold dose divided by the safety factors. This process is used most often with non-cancer hazards. There are some similarities between non-cancer and cancer assessments, but in general the approaches are quite different between the two. Both methods are discussed in more detail.

Non-Cancer Hazards

For non-cancer hazards the three steps introduced above can be applied, because an acceptable threshold can be established in most cases.

Step One: Dose-Response, NOAEL and LOAEL

The most common measures of no adverse effect are (1) the no observed adverse effect level (*NOAEL*) and (2) the no observed effect level (NOEL). The *NOAEL* is an exposure level at which there is no biologically- or statistically-significant increase in the severity of an adverse health effect in humans or animals.

On the dose-response curve (Figure 11–3) are doses that result in either no observable adverse response, or no observable response (of any type) in the test or subjects. In some cases there may be some measurable response (e.g., mild or transient irritation) at the administered doses in the study(s). Determination of a *NOAEL* or NOEL may not be possible, and instead the lowest observed adverse effect level (*LOAEL*) or lowest observed effect level (LOEL) can be used. The *LOAEL* is the lowest exposure level which causes a biologically- and statistically-significant adverse health effect outcome (e.g., persistent irritation or tumor) in humans or laboratory animals.

Because the *NOAEL* often ignores the dose-response curve and data, some scientists prefer use of a *benchmark dose* (BMD). The BMD is derived from a modeled dose-response curve and represents an upper confidence limit for a specified response level (typically, 5 percent or BMD_{05}). For instructional purposes, the discussion will refer to the *NOAEL* or *LOAEL.* Within the risk assessment process the NOEL, LOEL, and BMD can be substituted in place of these values.

Step Two: Safety Factors (UF and MF)

In step two, safety factors are formulated based on the quality of the data, missing data, uncertainties, and extrapolations. The U.S. EPA uses both an *uncertainty factor* (*UF*) and a *modifying factor* (*MF*). Each factor is unitless and used to adjust the *NOAEL* or *LOAEL.*

The uncertainty factors typically range from 10 to 1,000 and involve the product of several orders-of-magnitude (10-fold) adjustments. Each of the following factors is often associated with a 10-fold increase in the *UF*:

- inadequacies, contradictions, and gaps in the toxicity database;
- use of a *LOAEL* in the absence of a *NOAEL;*
- lack of chronic (lifetime) toxicity data (e.g., only acute or sub-chronic data are available);
- interspecies extrapolation from an animal model to humans (e.g., no human data); and
- unknown intraspecies variation (e.g., human variation and hypersensitive individuals).

A 3-fold uncertainty factor (i.e., UF = 3) is sometimes applied for lack of supporting data in additional animal species or lack of developmental and/or reproductive toxicity studies. Ultimately, uncertainty factors as high as 10,000 have been used in the risk assessment process, but most will fall in the range of 100 to 1,000.

The modifying factors typically range from one (no modification) to ten, mostly to account for concerns with study designs and exposure routes. For example, a toxicity study using rats may have administered the chemical in the food or through injection into the abdominal cavity. However, absorption via inhalation may be more rapid and complete, resulting in a need for a modifying correction to account for the expected body burdens in humans. It must also be noted that the *toxicokinetics* for most chemicals are strikingly different for inhalation versus ingestion. The blood supply from the gastrointestinal tract goes first to the liver, a major metabolic organ of the body, via the portal vein. Significant metabolism may occur before the blood reaches the rest of the body. With inhalation, chemicals may be directly transported, via the blood circulation (Figure 9–6), throughout the body, with or without significant metabolism. Inhalation can mimic an intravenous injection in some cases. Thus, a modifying factor may be used to account for inhalation in humans, when only oral dose data in animals are provided.

For pollutants evaluated by the U.S. EPA, the *UF* and *MF* values, and the rationale for their adoption, are available on the U.S. EPA's Integrated Risk Information System (IRIS) at http://www.epa.gov/iris/. Risk assessments listed in this database are primarily conducted on U.S. EPA listed *hazardous air pollutants (HAPs)*. As an example, a summary of the general information on ammonia (accessed September 15, 2010) is provided as follows:

- *Critical effect*: Increased severity of rhinitis (inflammation of mucous membranes of the nasal passages) and pneumonia with respiratory lesions
- *NOAEL*: 2.3 mg/m^3 (human equivalent adjusted concentration)
- *UF*: 30
  - A factor of 10 was used to account for sensitive individuals
  - A factor of 3 was used to account for the lack of chronic, developmental, and reproductive toxicity data, as well as the closeness of the *LOAEL* to the *NOAEL*
- *MF*: 1 (none)
- *RfC*: 0.1 mg/m^3 (which is discussed next in this section)

Step Three: Reference Concentration (RfC)

In step three, the U.S. EPA commonly uses a reference air concentration (RfC) in μg/m^3 to correct the threshold dose for inhalation hazards. In some cases a reference dose (RfD), normalized to body mass in units of mg/kg/day is used, but this is designed mostly for oral exposure hazards (e.g., drinking water). The RfC is defined as the *NOAEL* or *LOAEL* divided by the associated uncertainty and modifying factors:

$$RfC = NOAEL/(UF \times MF) \qquad \text{(Eq. 11–5)}$$

or

$$RfC = LOAEL/(UF \times MF). \qquad \text{(Eq. 11–6)}$$

The resulting *RfC* provides a means of evaluating the dose-response data and related uncertainties and corrections. The *RfC* is the adjusted acceptable human exposure threshold, and it is used in the final risk characterization, which aims to determine if the measured or predicted community exposures exceed the threshold level. The *RfC* can also be used to compare the toxicological risks between two or more pollutants. Risk management decisions, such as making a choice among various fuels, are easier to make if there are several alternatives, each with a *RfC* and exposure data, available for comparison. Keep in mind that decision-making is the usual motivation for initiating a formal risk assessment.

Table 11–7 provides examples of IRIS database risk assessment information for select hazardous air pollutants. Notice the effect of higher uncertainty factors on the *RfC,* as compared with the *NOAEL* or *LOAEL,* as well as the differences in critical health effects between the various pollutants.

A Common Mistake and Erroneous Assumption in Risk Assessment

It must be noted that dividing a *LOAEL* by a *UF* of 1,000 should not imply that the agent is 1,000 times more toxic in humans than in animals or isolated cells. There may in fact be no additional toxicity in humans. High *UF*s are more an indication of how uncertain scientists and regulators are on the potential health outcomes in humans, especially susceptible individuals. Thus, it would be poor judgment to conclude that a chemical with a low *RfC* and high *UF* must be highly toxic. In cases where it is necessary to evaluate and compare chemical alternatives it makes sense to evaluate the *NOAEL*s, *LOAEL*s, and *RfC*s equally. Failure to do so may result in the acceptance of a more toxic substance over one less studied.

Table 11–7 Examples of IRIS database risk assessment information for selected non-carcinogenic hazardous air pollutants (HAPs).

Chemical	*Critical Effect*	*Adjusted NOAEL/LOAEL*	*UF*	*MF*	*RfC*
Acetaldehyde	Degeneration of olfactory epithelium	NOAEL: 8.7 mg/m^3	1000	1	0.009 mg/m^3
Hydrogen chloride	Hyperplasia of nasal mucosa larynx and trachea	LOAEL: 6.1 mg/m^3	300	1	0.02 mg/m^3
Styrene	Central nervous system effects	NOAEL: 34 mg/m^3	30	1	1 mg/m^3
Toluene	Neurological effects in occupationally exposed workers	NOAEL: 46 mg/m^3	10	1	5 mg/m^3
Vinyl chloride	Liver cell polymorphism	NOAEL: 2.5 mg/m^3	30	1	0.1 mg/m^3
Xylenes (isomers)	Impaired motor coordination	NOAEL: 39 mg/m^3	300	1	0.1 mg/m^3

See text for abbreviations.
Database accessed September 15, 2010 from http://www.epa.gov/iris/.

Conversion of the RfC to RfD

In some cases it is necessary to convert a *RfC* in units of $\mu g/m^3$ into a *RfD* in units of mg/kg/day, which allows for corrections in body mass, ventilation rate, and exposure duration, discussed later in this chapter. The *RfD* can be calculated as

$$RfD = \frac{RfC \times V \times K}{W}, \quad \text{(Eq. 11–7)}$$

where

RfC = reference concentration of agent in air ($\mu g/m^3$)
V = ventilation rate (m^3/day; average sedentary male about 10 m^3/day)
K = unit conversion factor of 1.0×10^{-3} mg per μg
W = body mass or weight (kg; average male about 70 kg)

or

$$RfD = \frac{RfC \times V \times K \times F}{W}, \quad \text{(Eq. 11–8)}$$

where

F = exposure frequency (unitless) = number of days/year ÷ 365 days/year.

Equation 11–8 can be used to determine exposure dose when the exposure frequency is not continuous.

Cancer Hazards

For non-carcinogens, it is assumed that there is a threshold level below which adverse health effects are not likely to occur. In contrast, for carcinogens it is difficult to make this assumption. For many cancers, the disease appears to be due to an immune-system failure, instead of over-exposure to a carcinogen. Although humans are exposed to countless carcinogens on a daily basis, and not everyone develops cancer, but it is theoretically possible that one carcinogenic molecule can lead to a cancerous growth. This single-molecule theory does not describe how cancer normally progresses. The process is far more complex and there are numerous redundancies built into the human "self-repair" and immune systems to combat neoplastic changes (e.g., tumors). However, due to the severity of the associated diseases most regulators and health professionals work to significantly minimize the risks associated with cancer. The common and conservative assumption for carcinogens is that there is no "safe" or acceptable level of exposure or threshold. The goal established by the U.S. EPA is to reduce the cancer risk to less than one in a million per year or 1.0×10^{-6}. Many regulatory actions are based on an acceptable level of risk ranging from 1.0×10^{-4} (one in ten thousand) to 1.0×10^{-6}.

The assessment of dose-response data for carcinogens is complex and well beyond the scope of this chapter. The primary reasons for the complexity are:

- overall lack of cancer data at low exposure doses;
- long latency periods for the development of cancer; plus
- small acceptable risks assigned to carcinogens (e.g., less than one in a million excess cancers).

The U.S. EPA uses a number of linear and non-linear extrapolations, as well as benchmark models, to evaluate cancer risk. Due to the supposition that there is no acceptable threshold for carcinogens, many of the extrapolations are assumed to be linear with respect to risk down to a dose of zero. By default, and as a conservative measure, a linear model is commonly applied to carcinogens. In contrast, the dose-response model for non-carcinogens is assumed to be sigmoidal in shape (Figure 11–3), which represents the presence of hyper- and hypo-susceptible individuals, as well as a threshold *NOAEL* or *LOAEL*.

Because of the assumption that there is no exposure without an assumed risk, cancer assessments must establish an *acceptable* level of risk. In the occupational setting, this may be 1.0×10^{-4} or 1 in 10,000 workers over a lifetime, due to the intrinsic and accepted risks of the profession, and the fact that workers are paid to compensate for the additional risks. For example, firefighters have a dangerous and life-threatening job, which also exposes them to hazardous concentrations of air pollutants. In some cases there is also a formal increase in the base salary or (i.e., "hazard pay") to account for the inherent risks of a profession. In a community setting, the acceptable risk is often much lower and on the order of 1.0×10^{-6} or 1 in a million excess cancers. In both cases, a slope factor can be used to determine the level of acceptable risk.

The *slope factor* (SF) is commonly used by the U.S. EPA to assess cancer risk for both oral and inhalation hazards. The slope factor is related to the slope of a dose-response curve. Often, the linear portion of the low-dose section is interpolated using an upper 95th percentile confidence limit of the dose-response curve, which is designed to protect the most sensitive 5 percent of the population. The response in this case is the lifetime probability of excess cancers (e.g., 1×10^{-6}) and the dose is the exposure concentration (usually in μg/m^3 for inhalation). The slope, or rise-over-run, becomes the lifetime probability of excess cancer per unit exposure to the pollutant. For inhalation hazards the slope factor is often expressed as a *unit risk,* expressed as a lifetime probability of excess cancers per μg/m^3. The U.S. EPA commonly uses the terminology *inhalation unit risk* (IUR) and *air unit risk* (AUR) as a measure of the unit risk.

Because the measure of probability is unitless the slope factor units are in (μg/m^3)$^{-1}$, which are the reciprocal of those used for the *NOAEL* or *LOAEL* (e.g., μg/m^3). Thus, the risk assessment and characterization process differs between carcinogens and non-carcinogens.

The slope factor can be used to estimate the exposure concentration for different levels of cancer risk, by dividing the cancer risk by the slope factor. For example, if the inhalation unit risk for chloroform is 2.3×10^{-5} per μg/m^3, then the lifetime excess cancer risk of 1×10^{-6} would be about 0.04 μg/m^3. This is useful in establishing regulatory standards for the control of air pollutant emissions. **Table 11–8** provides examples of IRIS database risk assessment information on carcinogens for select hazardous air pollutants.

The slope factor can also be used to estimate the excess cancer risk for known lifetime exposures. For example, if a community's lifetime inhalation exposure concentration to chloroform was 0.1 μg/m^3, then we could predict, by multiplication of the slope factor and exposure concentration, a lifetime excess of 2.5 cancers per million persons. This is an example of a risk characterization, which compares exposures with acceptable risk. The next section discusses exposure assessment, which is then followed by risk characterization.

IV. EXPOSURE ASSESSMENT

Once a chemical contaminant has been identified as potentially hazardous (via hazard identification) and an acceptable threshold has been established (via hazard assessment) the next stage in the risk assessment process is to measure or estimate the population or ecosystem exposure. *Exposure assessments* can be performed for humans, animals, plants, and ecological systems. The primary aim of exposure assessment is to provide a reasonable estimate of community or population exposure for later comparison to the acceptable threshold of exposure, which completes the risk assessment process. Thus, it is important that the exposure

Table 11–8 Examples of IRIS database risk assessment information for selected carcinogenic hazardous air pollutants (HAPs).

Chemical Name (Disease)	*AUR or IUR per µg/m³ in air*	*Air Concentration at Three Specified Risk Levels*		
		1 in 10,000	*1 in 100,000*	*1 in 1,000,000*
Benzene (Leukemia)	2.2 to 7.8 × 10^{-6}	13 to 45 µg/m³	1.3 to 4.5 µg/m³	0.13 to 0.45 µg/m³
Chloroform (Liver cancer)	2.3 × 10^{-6}	4.0 µg/m³	0.4 µg/m³	0.04 µg/m³
Formaldehyde (Squamous cell carcinoma)	1.3 × 10^{-5}	8.0 µg/m³	0.8 µg/m³	0.08 µg/m³
Vinyl chloride (Liver cancer)	4.4 × 10^{-6} (lifetime–adult)	23 µg/m³	2.3 µg/m³	0.23 µg/m³

AUR = Air Unit Risk
IUR = Inhalation Unit Risk
Database accessed September 15, 2010 from http://www.epa.gov/iris/.

assessment units of measure are equivalent to those provided in the hazard assessment. As discussed earlier, the units for inhalation exposures are often µg/m³. Oral exposures are typically measured in mg/kg/day. For carcinogens, the slope factors would be in reciprocal units of $(\mu g/m^3)^{-1}$ and $(mg/kg/day)^{-1}$, respectively. Thus, units of µg/m³ or mg/kg/day suffice for both carcinogens and non-carcinogens.

Pollutant Sources (With an Emphasis on Air Pollutants)

Note that it is initially critical to identify non-background pollutant *sources* and expected *background concentrations* in the environment. The sources may contaminate air, water, soil, and food. In addition, the sources may be local, regional, or coming from long distances and across state or national borders. For air pollutant sources, they may be:

- *stationary,* e.g., an emission or smoke stack from an industrial process;
- *mobile,* e.g., an automobile or freight system;
- *point sources,* e.g., emission or exhaust stacks;
- *nonpoint sources,* e.g., fugitive emissions from agricultural areas or highways;
- *natural,* e.g., wildfires or windblown dust; or
- *anthropogenic,* e.g., coal-fired power plants.

Proper identification of the contributing pollution sources becomes important in the ensuing regulation and control of community exposures, especially if one source accounts for a majority of the air pollution concentrations. Identification of primary sources also aids in the recognition of those groups in the population who are most exposed and most at risk.

Routes of Exposure

Second, it is important to determine the key avenues or *routes of exposure* that need to be evaluated. The various routes of exposure include *inhalation, ingestion, dermal absorption,* and non-conventional routes such as *nursing, injection,* or *absorption through the eyes.* In air pollution, the primary route of exposure is inhalation. However, air pollutants can contaminate water sources, which could lead to the ingestion of contaminants in drinking water and food. Air pollutants may also contaminate a variety of surfaces and expose humans through dermal absorption following skin contact. In addition, air pollutants may contaminate the soil and expose humans via dermal absorption or through ingestion of agricultural products (plants and animals). Children may also eat dirt and non-food materials (this is called "pica" and often due to a nutritional-deficiency disorder), producing an exposure. All of these potential routes of exposure for a given pollutant will need to be evaluated. In most cases, the inhalation route is of primary concern in air pollution evaluations. To best illustrate this concept, the average person inhales about 10,000 liters of air per day versus about 3 liters of food and water combined per day (Figure 7–1). Dermal absorption typically accounts for only a small fraction of the total exposure, and is mostly a concern in occupational settings or with catastrophic releases of chemical contaminants into the environment.

Measurement of Exposure

The measures of exposure can be obtained using different methods. Exposure assessment can use data and estimates from:

- environmental monitoring;
- personal monitoring;
- historical exposure records; or
- mathematical models and projections.

Direct environmental monitoring of air concentrations is a common and preferred method for evaluating local or regional exposures. Data can be collected using direct-reading instruments or sampling methods that require subsequent laboratory analysis (see Chapter 4 on Sampling and Analysis). In order to obtain a *representative sample* of the population exposure, it is important to collect a statistically significant number of samples from different locations and within different time (e.g., day vs. night or seasonal) periods. Environmental monitoring data are often collected on a continual 24-hour basis and reported as annual averages. This provides an estimate of the average outdoor exposures but not the total environmental exposures. Most notably, the outdoor concentrations can vary widely by location, and most people spend a majority (about 90 percent) of their time indoors.

Reliable exposure estimates can be obtained from *personal exposure monitoring* methods, but they are time-consuming and costly. The air monitoring devices can be placed directly on individuals to monitor and/or sample their *breathing-zone.* There are distinct advantages to this method and most workplace exposure monitoring uses personal monitoring. However, community populations are much larger and more variable in nature (e.g., personal habits, cultural activities, and hobbies could produce additional exposures). The presence of multiple *indoor sources* of air pollutants must also be considered. Ultimately, it is difficult to obtain a representative sample of the general population using personal sampling. Thus, personal sampling is better suited for the evaluation of small subsets of the population (e.g., susceptible individuals such as asthmatics), or within the workplace.

Historical exposure data can be useful in estimating current exposures. Many government, university, and private organizations provide access to historical air pollution data from air monitoring stations. Caution must always be used when selecting past exposure data with high annual variability and trends. The reliability of the data collection methods must also be evaluated to ensure that proper *quality assurance and quality control (QA/QC)* methods were used to provide accurate and precise data. In addition, significant changes in pollution sources (e.g., in industrial processes, or vehicular traffic) may restrict the application of historical data.

Mathematical models and projections can also be used to estimate exposures. These types of methods use a combination of emission rates from identified pollution sources (e.g., vehicle exhaust emissions), physical and chemical transformations, transport mechanisms, historical data, meteorological patterns, and human activity pattern data to estimate expected exposure concentrations. High-speed computing has made exposure modeling an attractive and cost-efficient solution in the evaluation and control of air pollution. Some of the major problems are listed as follows:

- *propagation of error*—the error inherent in each data source used in the model add to produce a larger combined error, which reduces the predictive capability
- *complex environmental interactions*—little is understood about pollutant losses and the complex atmospheric interactions that may occur following emission and/or creation of many pollutants, which adds uncertainty to predictions
- *incorrect assumptions or estimates*—poorly-defined assumptions (e.g., minimal degradation in the environment) or estimated variables (e.g., average wind velocity) can have a significant outcome on the predictive capability of a model
- *lack of validation*—many models are not tested or validated to ensure that the predictions provide reliable estimates of the population exposures

As one can see, there is no single or simple solution to obtaining representative estimates of exposure for human populations. In most cases there is already a great deal of uncertainty at each stage in the risk assessment process, but an evaluation must still be made to support decisions. The ultimate goal is to make a scientific judgment or "educated guess" on whether or not the public is adequately protected against potential harm.

It is common practice to evaluate all available exposure data and to formulate a *reasonable maximal exposure (RME)* estimate, which is an estimate of the highest exposure. Assuming a worst-case scenario provides an additional margin of safety in the risk assessment process. As one can see, the exposure assessment process is not without uncertainties and unknowns.

Chronic Daily Intake (CDI)

Exposure concentrations in units of mass per volume of air (e.g., $\mu g/m^3$) are generally sufficient for air pollution risk assessments. In some cases it may be necessary to determine a measure of exposure dose in units of mg/kg/day, which allows for differences in body mass and exposure duration. The chronic daily intake (CDI) is a common measure of daily exposure dose that can be calculated from the airborne concentration using the following two equations:

$$CDI = \frac{C \times V}{W}, \qquad \text{(Eq. 11–9)}$$

where

C = concentration of agent in air (mg/m^3)
V = ventilation rate (m^3/day; average male about 10 m^3/day)
W = body mass or weight (kg; the average adult male is about 70 kg)

or

$$CDI = \frac{C \times V \times ET \times EF \times ED}{W \times AT}, \qquad \text{(Eq. 11–10)}$$

where

ET = exposure time = number of hours/24-hour period
EF = exposure frequency = number of days/year
ED = exposure duration = number of years
AT = averaging time = 365 days/year × lifetime years (e.g., 70 years).

If the average lifetime for a population was 70 years, then the AT would be equal to 365 days/year × 70 years or 25,550 days.

Equation 11–10 can be used to determine exposure dose when the exposure frequency is not continual. For example, the population exposure to an industrially produced contaminant may only occur 16 hours per day, or the industrial source may only operate five days out of each week.

There may also be a need to correct the exposure duration to account for *acute, sub-chronic,* and *chronic* exposure periods. Acute exposures are short-term exposures (e.g., hours to days), usually to higher levels of a contaminant. They are expected to have more immediate health effects such as irritation, sensitization, and neurological toxicity (e.g., central nervous system depression). The highly irritating and incapacitating effects of concentrated ammonia gas make it a good example of a chemical that may be evaluated and regulated based on its acute health effects. Sub-chronic exposures typically occur over periods of months to years and are likely to produce reproductive, teratogenic (e.g., birth defects), and even irreversible organ system toxicity. Lead is an environmental contaminant that can cause birth defects, infertility, learning deficits, and irreversible neurological damage following sub-chronic exposures over as little as a few months. Chronic exposures occur over years or a lifetime and are mostly associated with the development of cancer and irreversible organ system damage (e.g., "black lung" disease in coal miners). Highly organized organ systems, with a large functional reserve capacity, such as the liver, kidneys, and lungs are resilient to chemical damage, because their overall function is affected less by localized tissue damage. However, continual exposure can accumulate and cause catastrophic failure or an inability of the organ to function properly. Chronic chemical or physical insult may also lead to metaplastic changes (changes in cellular structure and function) that may eventually lead to neoplastic changes (significant changes in cellular growth) as part of the natural progression of many cancers. In most cases the exposure duration is accounted for in the hazard assessment process with the acceptable threshold concentration, reference concentration, or reference dose.

V. RISK CHARACTERIZATION

The final stage of the risk assessment process, *risk characterization,* involves the comparison of exposure assessment results with hazard assessment estimates to determine if the levels of exposure pose a risk to populations or ecosystems. When the units of measure are the same (e.g., $\mu g/m^3$) the simplest way to accomplish this task is to divide the exposure concentration by the acceptable threshold (e.g., reference dose). This provides a normalized and unitless comparison centered about the number one. If the exposure exceeds the acceptable threshold, then the ratio will be greater than one. If the exposure is below the acceptable threshold, then the ratio will be less than one. This method is used for non-carcinogens and the resulting ratio is often termed the *hazard quotient (HQ),* which was covered in some detail earlier in this chapter. In general, the following guidelines are useful in the interpretation of the hazard quotient:

- Values less than one indicate exposures are below the acceptable threshold and not an immediate concern.
- Values equal to one indicate a portion of the population may exceed the identified acceptable level, which may be a cause for concern, especially for vulnerable or hypersensitive individuals within the population.
- Values greater than one indicate exposures exceed the identified acceptable level, which is cause for concern and possible remedial action.

Non-Carcinogens

The U.S. EPA commonly uses the hazard quotient with non-carcinogen hazards. For inhalation exposures, the *HQ* can be defined as

$$HQ = C / RfC, \qquad \text{(Eq. 11–11)}$$

where

C = airborne exposure concentration (μg/m^3)
RfC = reference concentration (μg/m^3)

or

$$HQ = CDI / RfD, \qquad \text{(Eq. 11–12)}$$

where

CDI = chronic daily intake (mg/kg/day)
RfD = reference dose (mg/kg/day).

The measures of exposure concentration (C) or dose (CDI) are most often a projected *reasonable maximum exposure* (RME) or a 95th percentile upper confidence interval value, as discussed earlier in this chapter. Overall, hazard quotient values greater than or equal to one represent a risk of adverse effect within the exposed population or environment. Most regulatory objectives are aimed at ensuring community and environmental exposures are well below a *HQ* value of one.

Carcinogens

For carcinogens, the more common risk characterization measure is the *excess cancer risk (ECR),* which is the probability of an individual, within the greater population, of developing cancer over a lifetime. The *ECR,* sometimes termed the *excess lifetime cancer risk,* is determined from the product of the *slope factor* (*SF*) and exposure concentration (C). The slope factor (also termed inhalation unit risk or air unit risk) is a measure of the probability of an individual, within the population, developing cancer per unit intake (e.g., per μg/m^3) of a pollutant over a lifetime. For inhalation hazards, the *ECR* at a given environmental concentration is a product of the exposure concentration in units of μg/m^3 and slope factor or unit risk in units of (μg/m^3)$^{-1}$:

$$\textit{Excess Cancer Risk (ECR)} = C \times SF. \qquad \text{(Eq. 11–13)}$$

As discussed earlier, for environmental pollutants most regulatory actions are based on an acceptable *ECR* of one in one million exposed people or 1.0×10^{-6}. The intention is to ensure the general population is not exposed to environmental carcinogens at concentrations that would elevate the risk above 1.0×10^{-6}. In the United States, the predicted number of excess cancers for this criterion can be estimated as

$$\begin{aligned}\text{Predicted excess cancers} \\ &= ECR \times \text{total population} \\ &= (1.0 \times 10^{-6}) \times (300 \times 10^{6}) \\ &= 300 \text{ excess cancers.} \qquad \text{(Eq. 11–14)}\end{aligned}$$

Even as a conservative measure of risk this may pale in comparison to the half-a-million cancer deaths annually in the United States, but currently it is a common measure of cancer risk for air pollutants. If an air pollution exposure was at an *ECR* of 1.0×10^{-4} (1 in 10,000), then the predicted excess cancers would be about 30,000. Current U.S. EPA risk assessments address carcinogens within this range from 1.0×10^{-4} to 1.0×10^{-6}. The greatest potential problem is when these conservative estimates of risk are treated as actual losses. A value of these estimates of risk is that they can be used within the decision-making process when comparing alternative decisions, with a central goal of making the best decision for the general public. Whether or not these numbers have true meaning in relation to public health is debatable because all risks must be evaluated in a holistic manner. For example, decreasing the risk of a pesticide used to protect fruits and vegetables may increase the larger risks associated with a poor diet.

Lastly, the excess cancer risk does not take into account the severity or how treatable the cancer may be, which should be taken into consideration. However, at this point in time, regulators are less willing to allow the risks associated with carcinogens in contrast with other disease causing agents. Effective holistic and ethical risk communication (discussed later) become critical issues with carcinogens.

Cumulative Risk and Multiple Chemical Exposures

In some instances, the target organ and critical health effect may be the same for two or more pollutants. When the exposure is to a mixture of chemicals, additional analysis is needed. If an *additive effect* is anticipated, then it may be necessary to correct the *HQ* or *ECR* estimates for multiple chemical exposures or multiple exposure routes (e.g., inhalation, ingestion, and dermal). Caution must always be applied to ensure the cumulative effect is in fact an additive effect and not *synergistic* (i.e., greater than additive), such as with the combined effect of cigarette smoking and asbestos exposure where the risk of lung cancer is about 4 to 10 times what would be expected from a simple additive model.

If the critical or end-point health effect is expected to be additive, then a target organ-specific *hazard index (HI)* is often used to evaluate cumulative effect. For non-carcinogens, the *HI* is a simple sum of the target organ-specific hazard quotients:

$$HI = HQ_1 + HQ_2 + \ldots HQ_n. \qquad \text{(Eq. 11–15)}$$

The hazard index may also combine hazard quotients from multiple exposure routes to the same chemical contaminant. For example, for xylene (an aromatic hydrocarbon) the HQ_1 could represent oral exposure from food or water contamination, HQ_2 could represent dermal exposure from contaminated soils, and HQ_3 could represent inhalation exposure to xylene vapors. A new HQ_4 could also be added to the hazard index for exposure to ethyl benzene. Both xylene and ethyl benzene are aromatic hydrocarbons that can cause upper respiratory tract irritation and central nervous system depression. If the target organ-specific health effects are the same, and the effects are shown to be additive, then a hazard index can be used to account for multiple chemical exposures or exposure routes.

For carcinogens, the *cumulative excess cancer risk* can be evaluated in a similar manner by summing the target organ and cancer-specific excess lifetime cancer risks for equivalent pollutants and/or different exposure routes:

$$\text{Cumulative Excess Cancer Risk} = ECR_1 + ECR_2 + \ldots ECR_n. \qquad \text{(Eq. 11–16)}$$

For both non-carcinogens and carcinogens, cumulative risk is evaluated according to the target organ and when the critical health effect is the same and additive. Risks are evaluated separately or on an individual basis when these assumptions cannot be met. As an aside, mixtures can also be *antagonistic* (i.e., protective or less than additive) in their risks. In these cases the summation of risks will overestimate the actual risks.

VI. RISK COMMUNICATION

What is Risk Communication?

Change and improvement within a society, and on an individual level, requires two important events: (1) a decision to change and (2) action towards change. Both of these are part of the *risk management* process which aims to establish policies and procedures for minimizing unacceptable risks. *Risk communication* is the crucial intermediate step between risk assessment (the science) and risk management (policy). Risk communication is defined as the exchange of information about risks between those responsible for assessing, minimizing and regulating risks, and those who may be affected by the outcomes of those risks.

Why is Risk Communication Needed?

Unfortunately, there are numerous barriers between the objective estimates of risk projected by the experts and the perceptions of risk by the public, the media, special interest groups (e.g., non-profit organizations and activists), political factions, and policymakers. In many cases it is the failure of experts to disclose uncertainties and unknowns that have the greatest effect on public perceptions of risk. All too often the primary focus becomes "How do we convince the public that there is, or is not, a risk?" without attention to how the public will receive and process the provided information. Several factors can cause people to perceive risks as greater than judged by the experts, including:

- *lack of personal control*—for example, most people have a greater fear of flying over driving a car even though the risk of dying in a car is at least 200 times greater
- the *catastrophic nature* of the outcome
- *unfamiliarity* with the technology or "fear of the unknown"
- *unobservable or delayed effects*—for example, the possibility of delayed brain cancer from exposures to electromagnetic fields from cell phones and power lines has continued to be an area of public concern
- feelings of an *unfair distribution of risks,* especially for those living nearest a facility producing air pollution

Nuclear power is an excellent example of a technology perceived as having a much higher risk by the public than the experts estimate. For example, there is a "fear of the unknown" because most people are unfamiliar with nuclear radiation and do not know about the substantial radiation exposures we all receive from both natural sources and medical procedures (e.g., cosmic rays, coal-burning emissions, and medical x-rays). Such exposures do little harm and can produce substantial benefits. The potential catastrophe of a nuclear reactor accident stimulates exaggerated fears of pending doom. In addition, those living nearest a nuclear power plant are likely to feel a greater risk compared to those living farther away, even when this is often not the case. All of these factors can paint a dark picture for nuclear power, even in light of the economic and air pollution benefits of using clean nuclear energy sources. Lastly, the fears and distrust by the public may still remain even after it has been communicated by experts that the risks are insignificant in comparison to many other environmental hazards.

Table 11–9 provides an instructive example of how the perception of risk by U.S. college students can differ from the actual estimates of risk by experts. According to the average college student, nuclear power was ranked as the number one risk to the public. However, environmental experts ranked 19 other activities and technologies ahead of nuclear power. As another example, the college students greatly underestimated the risks associated with motor vehicles (students ranked it number 5; experts ranked it number 1) and swimming (students 30; experts 10). These examples clearly demonstrate that there has been a failure of risk communication.

Table 11–9 Perception of risk versus actual risks as determined by U.S. college students and experts. Ranking: 1 = most risk; 30 = least risk.

Technology or Activity	*College Students*	*Experts*
Nuclear power	**1 (highest)**	20
Motor vehicles (all)	5	**1 (highest)**
Motorcycles	6	6
Police work	8	17
Hunting	18	23
Food coloring	20	21
Skiing	25	**30 (lowest)**
Swimming	**30 (lowest)**	10

Data from Slovic (1987).

Humans are strongly influenced by emotions, which makes objective determinations difficult and is one reason why they disagree on both technical and non-technical issues. A real problem arises when we knowingly or unknowingly make sensational appeals to human emotions to influence decisions regarding risks. Even public health advocates have resorted to distortions of the truth in order to "scare" the public into making what they believe are better choices in life. In the 1980s a number of public health advertisements equated the effects of illicit drugs on the brain to an egg frying in a skillet. A more recent public announcement, this time regarding cigarette smoking, has gone as far as showing a mound of thousands of body bags piled high to tell the dangers of smoking. Both examples paint a strong picture of imminent death. With smoking, although it is a leading cause of illness and death, the "truth" is that only a small fraction (about 5 percent) of smokers die of lung cancer. This level of cancer risk is well hidden from the public. Such exaggeration of risk will backfire if the public loses confidence in health professionals. In contrast to the often temporary success of emotional "scare tactics," the advancement of air pollution science and public health should be dependent on objective holistic assessments, communications, and management of risks. Honesty and clarity work best.

Elements of Effective Risk Communication

The question still remains on how to communicate risks effectively and overcome the barriers involved with the public's accurate perception of those risks. Those responsible for assessing, minimizing, and regulating risks must present risks to the public in a manner that can be readily understood (i.e., clear, objective, and simplified), with respect to more familiar risks, and that does not mislead or influence decisions by appealing to emotion. Lastly, openness and trustworthiness are equally important in the risk communication process. The following are important guidelines for effective risk communication:

- *Avoid appeals to emotion*—Risk estimates that use total excess deaths, illnesses, and cancers can influence personal perceptions of risks, even when all valid alternatives to that risk have greater risks associated with them. The reality is that everyone eventually dies; hence the quality and length of life are topics that should be emphasized.
- *Frame information for objective decision making*—Provide information in a simplified manner that is relevant to the larger picture, such as the percentage of excess disease cases, potential loss of life span, or percent improvement/reduction in the living standard (e.g., gross national product).
- *Avoid using probabilities*—Probabilities are often difficult for people to accept, even if an extremely low risk probability is provided. Even one in a billion cancer risk can be considered unacceptable to some, especially if they have recently experienced a cancer-related death in the family. Cancer is a factor in about one in every four deaths in the United States or over half a million deaths annually, which is a probability of 0.25. However, the public may not perceive "0.25" much differently than a "1.0×10^{-6}" or "a one in a million risk of a cancer deaths" (which are equivalent to a probability of 0.000001). Probability, especially when exponents are involved, may have absolutely no meaning to the public.

Now assume one never picked up this book or studied risk assessment. Which risk value below is most likely to influence one's decision to consider a change in his/her behavior?

1. 0.00000025
2. 0.25
3. 2.5×10^{-6}
4. 2.5 in a million

- *Make risk comparisons*—Compare risk estimates with familiar risks or activities such as smoking, driving, riding a bicycle, eating a rich dessert, flying, medical x-ray, alcohol abuse, and unprotected sex. This also helps educate the public on the concept of risk and that it is a part of our daily lives.
- *Evaluate public health consequences and benefits alongside risks*—Explain the value of the pollutant in disease prevention (e.g., pesticides) or enhancing the availability of fresh produce and nutritional foods.
- *Evaluate economic costs and benefits alongside risks*—How many jobs will be lost or created; how many businesses will be forced to relocate operations overseas; will a decision increase energy costs, and how will our standard of living (e.g., gross national product) be affected in the process of implementing the environmental regulation?
- *Compare alternative technologies*—Make side-by-side comparisons on the basis of the risk estimates, public health benefits, and economic costs and benefits.
- *Safeguard openness*—Two-way communication with stakeholders and the community is important for ensuring trust, but non-expert, or biased, influence must not be allowed to disrupt the objective and scientific risk assessment process.

Suggestions for Meeting with the Public

The best ways to accomplish the objectives of public trust, openness, and scientific integrity include the following:

1. Establish a clear and realistic set of goals for the risk assessment. A well-designed agenda will compare risks and evaluate available alternatives for the improvement of the public health and well-being (economics and the environment included).
2. Provide early and sustained interaction and communication with the public and *stakeholders* (e.g., interest groups, industry, and affected communities).
3. Allow the public and stakeholders to voice concerns and make attempts to address each concern objectively and with sincerity.

4. Clearly identify those with special interests (e.g., political or personal agendas) and conflicts of interest (e.g., those who may profit or lose from any decision) involved in the risk assessment, communication, and management process. Limit or exclude participation by unqualified, non-experts with significant conflicts of interest.
5. Clearly identify the expertise and qualifications of all those involved in the risk assessment, communication, and management process.
6. Prepare a written technical report and include an assessment of potential bias, uncertainties, and assumptions used in the risk assessment process.
7. Establish a focus group with diverse interests (e.g., interest groups, community leaders, and affected industries) to review a draft of the written report, prior to its eventual release. Obtain feedback and recommendations from stakeholders.
8. Include an independent scientific review to help establish credibility.
9. Prepare a summary report or "white paper" for general release to the public.

Comments on Risk Assessments

Risk assessment is a systematic and objective methodology for estimating the health and environmental risks of exposure to pollutants. Scientists and regulators tend to take a cautious approach when assessing risk and make numerous assumptions or speculations in the process. There will always be a great deal of uncertainty associated with this or most other attempts to predict the future. The end result of a risk assessment is far from perfect, but it is an important step in our attempt to reduce risks and thereby improve the quality of human life.

VII. SUMMARY OF MAJOR POINTS

Risk assessment is the systematic process of determining the probability and magnitude of an undesired event. In air pollution science, it is aimed at predicting health outcomes related to air quality in the presence of substantial uncertainties and potential biases. The primary goal is to establish an objective and educated estimate of health risk, which can be used for comparison with familiar daily risks, and the risks associated with available alternative technologies. The risk assessment process can be broken down into four stages: (1) *hazard identification;* (2) *hazard assessment;* (3) *exposure assessment;* and (4) *risk characterization.* The first two stages are designed to identify whether or not an air pollutant poses a significant health or environmental hazard and to establish an acceptable threshold for exposure. For non-carcinogens, a non-linear dose-response relationship is assumed along with the notion that a no observed adverse effect level (*NOAEL*) or threshold exists. For many carcinogens, a linear dose-response relationship is assumed where there is no acknowledged "safe" dose. Nevertheless, an acceptable *threshold* (i.e., acceptable non-zero dose) is generally established in both cases. The third stage, exposure assessment, evaluates and estimates current or anticipated environmental exposures. In the last stage, risk characterization, the exposures are compared to the acceptable threshold and often evaluated as a *hazard quotient,* or *hazard index* for multiple exposures. Upper 95th percentile values above 1.0 indicate population exposures exceed the acceptable threshold.

Risk communication, the exchange of information about risks between those responsible for assessing risks and those affected by the outcomes of those risks, is the crucial intermediate step between *risk assessment* (the science) and *risk management* (policy). There are a number of identified barriers to effective risk communication and human perceptions of risk. Lastly, in order to effectively assess risk one must also holistically consider the public well-being, including the economic costs and benefits of potential regulatory actions. From the very beginning, risk assessors should establish a well-designed agenda that will compare risks and evaluate available alternatives for the improvement of public health and well-being (which includes both the environment and economic security).

VIII. QUIZ AND PROBLEMS

Quiz Questions

(select the best answer)

1. Risk assessment is used by government agencies in the establishment of:
   a. solid evidence of the health effects of a hazard at low doses.

b. regulatory standards for safe or acceptable water and air.
c. priorities for research and regulatory activities.
d. Both b. and c. are true.

2. The general stages of the risk assessment process include:
a. hazard identification, hazard assessment, dose assessment, and risk characterization.
b. hazard identification, dose-response assessment, exposure assessment, and risk characterization.
c. risk identification, hazard assessment, dose assessment. and risk characterization.
d. Both a. and b. are true.

3. Which of the following is designed to determine the relative toxicity and toxicological endpoints associated with potential exposure to an environmental contaminant?
a. Dose-response assessment
b. Hazard identification
c. Lethal dose assessment
d. Exposure assessment

4. Which of the following is a representative of a dose least likely to cause harm?
a. LD_{50}
b. TD_{LO}
c. LOAEL
d. NOAEL

5. What is a major drawback of the generalized carcinogen classification systems used by environmental agencies and organizations?
a. They do not take dose into account.
b. They do not evaluate the strength of evidence in humans versus animals.
c. They do not provide clear distinctions between human and animal carcinogens.
d. There are no major drawbacks.

6. In risk assessment, the primary goal of the dose-response or hazard assessment process is to:
a. determine the LOAEL.
b. determine the NOAEL.
c. establish an acceptable exposure level for the population or environment.
d. establish uncertainty and modifying factors.

7. Which of the following scenarios is likely to result in a 10-fold increase in the uncertainty factor (UF)?
a. Use of a LOAEL in the absence of an NOAEL
b. Availability of only acute toxicity data
c. Lack of human toxicity data
d. All of the above are true.

8. For carcinogens, the common and conservative assumption made by regulators and health professionals is:
a. there is a safe threshold.
b. there is no "safe" or acceptable level of exposure.
c. the immune system must be overwhelmed before there is a concern.
d. a risk of 1×10^{-4} is more acceptable.

9. The lifetime probability of excess cancers per $\mu g/m^3$ of air is an example of:
a. a unit risk.
b. an inhalation unit risk.
c. a slope factor.
d. All of the above are true.

10. When exposures to a non-carcinogen hazard may include multiple routes of exposure or the additive effects of similar hazards that affect the same target organ, then a ______________ is commonly used to evaluate the cumulative effect.
a. hazard quotient (HQ)
b. hazard index (HI)
c. cumulative excess cancer risk
d. synergistic effect

11. Which of the following is an example of a factor that can cause people to perceive risks as greater than how those same risks would be judged by an expert in the field?
a. Catastrophic nature of the outcome
b. Lack of personal control
c. Unfamiliarity with the technology
d. All of the above are true.

12. Which of the following is NOT an example of effective risk communication concerning a carcinogenic air pollutant?
a. Reporting the expected excess number of cancer deaths
b. Reporting the expected percentage of increase in cancer cases within the population
c. Reporting and comparing the risk estimates of alternative technologies
d. Reporting the economic costs of stricter regulation

Problems

1. The reasonable maximal exposure (RME) air concentration within an air basin, which represents a population exposure, for acetonitrile is 3.5 μg/m³.
   a. Use the IRIS database to determine the hazard quotient for the population inhalation exposures to acetonitrile.
   b. Is the population potentially overexposed to acetonitrile?
2. The 95th percentile confidence interval for a hazard quotient is 0.9 to 1.15. How would you interpret the hazard quotient and risk to the community?
3. The *LOAEL* for an air pollutant under study was determined to be 2.5 μg/m³, based on upper respiratory tract irritation. However, there was a general lack of available toxicity data and only acute toxicity testing was conducted in rats. No human data were available.
   a. Formulate an appropriate uncertainty factor (*UF*) based on the provided information. Justify your reasoning for each factor used to produce the *UF*.
   b. Why is a modifying factor (*MF*) of 3 appropriate in this case?
   c. Calculate the appropriate *RfC* for this pollutant; show your work.
4. Convert a *RfC* of 0.35 μg/m³ into an equivalent *RfD* in units of mg/kg/day, for an average male (70 kg) with an average ventilation rate of 10 m³/day.
5. Calculate the chronic daily intake for an average adult male exposed to a 0.12 mg/m³ air concentration of acrylonitrile.
6. The population surrounding an industrial plant is exposed to 0.25 μg/m³ (as a reasonable maximal exposure) of a pollutant for five days out of every week (about 261 days/year).
   a. What is the estimated chronic daily intake for adult males in the population?
   b. If the corresponding *RfC* is 0.2 μg/m³, is the population overexposed to the pollutant? Show your work.
7. The inhalation unit risk for chloroform is 2.3×10^{-5} $(\mu g/m^3)^{-1}$. Calculate the excess cancer risk for a population exposed to RME air concentrations of 4.88 μg/m³ (1 ppb) chloroform.
8. The inhalation unit risk for 1,1,1,2-tetrachloroethane is 7.4×10^{-6} $(\mu g/m^3)^{-1}$. Calculate the excess cancer risk for a population exposed to RME air concentrations of 6.87 μg/m³ (1 ppb) 1,1,1,2-tetrachloroethane.
9. From dose-response data, the inhalation unit risks for both chloroform (question 7) and 1,1,1,2-tetrachloroethane (question 8) are based on the development of liver-cell tumors in mice. Based on the results from questions 7 and 8, what is the estimated cumulative excess cancer risk for a combined exposure of 1 ppb of each pollutant?
10. Is it possible that setting $PM_{2.5}$ air pollution standards that protect the most sensitive individuals may in fact be harmful to the general population? Explain your answer.

IX. DISCUSSION TOPICS

1. From the information provided in Table 11–9, list and discuss the possible reasons why the U.S. college students perceived nuclear power as the number one ranked risk, when the experts ranked it 20th on the list.
2. Is it ethical to use powers of persuasion and appeals to human emotions and insecurities to bring about change? List and discuss the benefits and consequences of adopting a Machiavellian philosophy such as "the ends justify the means."
3. Discuss the value of scientific credibility as it relates to the evaluation and reporting of risks in our society? How does society perceive the value of science?
4. Is it possible that setting $PM_{2.5}$ air pollution standards that protect the most sensitive individuals may in fact be harmful to the general population? Does this issue apply to other air pollutants as well?
5. Go on the IRIS website at http://www.epa.gov/IRIS/ and evaluate the risk assessments for these industrial solvents: (a) styrene and xylene or (b) toluene and vinyl chloride. Make a table listing the critical effect(s): *NOAEL* or *LOAEL, UF, MF,* and *RfC*. Use the information and provide reasoning to determine why the final *RfC* values are similar or different from each other.
6. Find an aerosol cleaning product or chemical product in your home that provides the name of a possibly hazardous chemical ingredient. Use the resources provided at the beginning of Section II (Hazard Identification) to determine the relative toxicity (Table 11–4 or 11–5) and toxicological

endpoints (e.g., adverse health effects) associated with exposure to the chemical. Does the benefit of using the product outweigh the potential health consequences?

References and Recommended Reading

Barnes, D. G., Times are tough: Brother, can you paradigm?, *Risk Analysis,* 14:219–223, 1994.

Cohen, B. L., Catalog of risks extended and updated, *Health Physics,* 61:317–335, 1991.

Cohen, B. L. and Lee, I. S., A catalog of risks, *Health Physics,* 36:707–722, 1979.

Costa, D. L., Air pollution, in *Casarett and Doull's Essentials of Toxicology,* Klaassen, C. D., ed., McGraw–Hill, New York, 2008, pp. 1119–1156.

Doll, R., Epidemiological evidence of the effects of behavior and the environment on the risk of human cancer, *Recent Results Cancer Res.,* 154:3–21, 1998.

Doll, R., The Pierre Denis memorial lecture: Nature and nurture in the control of cancer, *European J. Cancer,* 35:16–23, 1999.

Doll, R. and Peto, R., The causes of cancer: Quantitative estimates of avoidable risks of cancer in the United States today, *J. Natl. Cancer Inst.,* 66:1191–1308, 1981.

Faustman, E. M. and Omenn, G. S., Risk Assessment, in *Casarett and Doull's Essentials of Toxicology,* Klaassen, C. D., ed., McGraw–Hill, New York, 2008, pp. 107–128.

Hodge, H. C. and Sterner, J. H., Tabulation of toxicity classes, *American Industrial Hygiene Association Quarterly,* 10:93–96, 1949.

Jayjock, M. A., Lynch, J. R., and Nelson, D. I., *Risk Assessment Principles for the Industrial Hygienist,* American Industrial Hygiene Association Press, Fairfax, VA, 2000.

Lippmann, M., Cohen, B. S., and Schlesinger, R. B., Risk Assessment, in *Environmental Health Science: Recognition, Evaluation, and Control of Chemical and Physical Health Hazards,* Oxford University Press, New York, 2003, pp. 396–420.

Nelson, D. I., Mirer, F., Bratt, G., and Anderson, D. O., Risk Assessment in the workplace, in *The Occupational Environment: Its Evaluation, Control, and Management, 2nd Edition,* Dinardi, S. R., ed., American Industrial Hygiene Association Press, Fairfax, VA, 2003, pp. 143–171.

NRC (National Research Council), *Improving Risk Communication,* National Academy Press, Washington, DC, 1989.

NRC (National Research Council), *Science and Judgment in Risk Assessment,* National Academy Press, Washington, DC, 1994.

NRC (National Research Council), *Science and Decisions: Advancing Risk Assessment, The National Academy Press, Washington, DC,* 2009.

Robson, M. and Toscano, W., eds., *Risk Assessment for Environmental Health,* John Wiley & Sons, San Francisco, CA, 2007.

Schmähl, D., Preussmann, R., and Berger, M. R., Causes of cancer—an alternative view to Doll and Peto (1981), *Klin Wochenschr,* 67:1169–1173, 1989.

Slovic, P. Perception of risk, *Science,* 236 (April 17, 1987):280–285.

Upton, A. C., Science and judgment in risk assessment: Needs and opportunities, *Env. Health Perspectives,* 102:908–909, 1994.

Wilson, R., Analyzing the daily risks of life, *Technology Review,* 81:40–46, 1979.

Chapter 12

Ethical Considerations: How They Apply to Air Pollution

LEARNING OBJECTIVES

By the end of this chapter the reader will be able to:

- define "ethics" and differentiate normative ethics from practical ethics
- describe the Tuskegee Syphilis Study and its impact on regulations covering research on human subjects
- describe the similarities and differences between regulations for animal research and for human research
- create a "Code of Ethics" for public health students

CHAPTER OUTLINE

I. INTRODUCTION

Why Bother?

If everyone just got along, respected one another, did not pollute the environment, and lent a helping hand to the needy, this chapter would not be necessary. Yet, human history as well as current events have shown that some regulation of human behavior is necessary for the safety, health, and well-being of the public. In addition, many people believe that nonhuman animals, ecosystems, and the environment are not adequately protected.

Ethical standards are established in a variety of ways. Public laws establish the rules for accepted general behavior. Governmental agencies set the rules for special problems such as construction standards, air and water quality, use of human and animal research subjects, and contaminant emissions from cars, power plants and factories. Professional societies and trade groups establish codes of conduct, including ethics requirements for their members. Local communities, clubs, families, religions, and even individuals also establish written and unwritten rules for behavior.

Thus, the ethical standards related to air pollution derive from many sources. Today, every person, professional group, social group, and business entity has several types of ethical constraints that bear on their activities that influence air quality. In the final analysis, it is the behavior of individuals who decide how to perform their daily activities that really matters.

Environmental laws are based on the public's expectation that the government will provide for safe, and even esthetically pleasing, environments, both indoors and outdoors.

As shown in **Exhibit 12–1** environmental ethics has a long history. Air pollution regulations have the force of law, and they are periodically revised. Such standards impact public health and welfare, both directly and indirectly through their economic effects. Because the direct (intended) effects are accompanied by their indirect (unintended) effects, new, or proposed, air quality regulations are usually controversial.

What Does "Ethics" Encompass?

Ethics has both formal and practical aspects. *Formal ethics* is a branch of philosophy that both studies the fundamental principles of moral behavior and describes the rules for acceptable behaviors. Philosophers who pursue ethics examine the history, variety, and basis for normative (i.e., proper) behaviors. They also take a lead in advancing their branch of philosophy, including holding conferences, publishing journals, and proposing changes in prevailing ethical standards. Specialists in ethics, i.e., *ethicists,* can be totally committed to the topic. The activities of ethicists can include analyses and critiques of timely ethical issues, sometimes with the intent of modifying the behavior of others. Ethics also interfaces with the law, and laws defining ethical behavior are continuously evolving. Such laws are examples of *practical ethics,* in that they define and regulate behavior.

One might assume that ethical standards have some firm basis, but it is difficult to identify universal ethical principles or standards. The simplistic concept that something is ethical only if it does no harm is impractical. Both time and place determine what behavior is ethical and what is unethical. Wartime battlefield ethical standards, for example, have little in common with the standards that underlie modern medical ethics: Secrecy, respect for persons, and considerations regarding acceptable risk of harm are worlds apart in these contrasting arenas. It is therefore difficult to identify any universally-accepted basis for defining what is proper and what is not. Even the best decisions usually have some undesirable consequences.

Rewards and punishments are used to encourage conformity with ethical standards. Some of these positive and negative incentives are externally imposed, but there are also internal guidelines. Individuals and organizations can take comfort when they know that they are conforming to high ethical standards and thus have nothing to hide. Success and status are linked to external perceptions of ethical behavior. In short, ethical standards do place constraints on behavior, but conforming to high standards reduces the risk of censure, or worse consequences.

Ethical standards change over time as society evolves, and as technology advances to present new challenges. Practical issues and contemporary concerns influence what is accepted behavior at any given time. In the past century, new areas of ethical concern have emerged including (1) those involving the use of human and laboratory animal research subjects and (2) protection of the environment. Humans and animals are widely used in air pollution studies to gather data for protecting people, domestic animals, wildlife, and ecosystems. Thus, ethical standards for protecting research subjects are relevant to this chapter, as are standards for protecting the general environment.

Exhibit 12–1 A brief history of environmental ethics.

The foundations of modern thinking about the relationship between industrial civilizations and the natural environment can be traced back over 200 years ago. A British scholar, T. Robert Malthus (1766–1834), argued in an essay that uncontrolled population growth was unsustainable without changes in the growth of food production (Malthus, 1798). According to Malthus:

> "Population when unchecked increases in a geometrical ratio. Subsistence increases only in an arithmetical ratio. A slight acquaintance with numbers will shew the immensity of the first power in comparison of the second."

Malthus' simple proposition stressed both the importance of (1) modeling future trends and (2) thinking about the impact of human population growth on human welfare. He was not directly concerned with the impact of population on the environment, but he helped to prepare the foundations for considering such problems.

More recently, thinkers and writers have developed the basic concept of Malthus into an environmental theme that includes the impact that humans have on nature. *Silent Spring,* a widely-read book by Rachel Carson (1963), drew attention to the ability of some useful pesticides (DDT, aldrin, and dieldrin) to concentrate in the environment. Such concentration, Carson argued, could have negative impacts on the environment (including bird populations) and on public health. Paul Ehrlich, a Stanford University ecologist, warned in his popular book *The Population Bomb* (1968) that human population growth has threatened the Earth's life-support systems.

NASA's Christmas 1968 photographic image of a fragile, beautiful Earth isolated in dark empty space made it evident that our planet has limited capacities to sustain life and to deal with environmental pollutants. Shortly after the image was published, the first Earth Day in 1970 and the creation of the U.S. Environmental Protection Agency (1970) firmly established a public commitment to the environment, and thus environmental ethics. The first administrator of the U.S. EPA, William D. Ruckelshaus, wrote in a press release two weeks after the formation of the agency:

> "So we shall be an advocate for the environment with individuals, with industry, and within government."

Thus, "the environment" achieved *legal status* and the protection of the U.S. government. Air quality is currently a major aspect of the legal commitment to protect the environment.

Topics related to environmental ethics were presented and debated in books, magazines, and the journal, *Inquiry,* throughout the 1970s. Philosophical issues emerged including whether or not nonhuman life had *intrinsic value* (i.e., was valuable for its own sake), rather than only *instrumental value* (i.e., value only in relation to its benefit to humans), and who might represent nonhuman living things in legal actions.

Publication opportunities were broadened by the founding of the journal, *Environmental Ethics,* by Eugene C. Hargrove in 1979; the establishment of the *International Society for Environmental Ethics* in 1990; the founding of the journals *Environmental Values* in England in 1992 and *Ethics and the Environment* in 1996; and the establishment of a second international association, the *International Association for Environmental Philosophy* in 1997.

For further information, in addition to the above sources, one can visit the *Internet Encyclopedia of Philosophy* (http://www.iep.utm.edu) and *The Stanford Encyclopedia of Philosophy* (http://plato.stanford.edu).

II. ETHICS AS A BRANCH OF PHILOSOPHY

Ethics, also called moral philosophy, is concerned with distinguishing right from wrong and good from bad. *Metaethics* (meaning "about ethics") addresses some of the larger questions such as "Is there any difference between right and wrong?" and "Are all moral standards just arbitrary?" Approaches to metaethics place philosophers into several camps, including the following:

- *Naturalists* maintain that moral concepts are matters of fact and that moral decisions can be based on scientific or factual investigations.
- *Cognitivists* maintain that moral judgments can be either true or false and can, in principle, be subjects of knowledge or cognition as opposed to emotion or volition.
- *Intuitionists* maintain that knowledge of right and wrong is self-evident, not requiring analysis.
- *Subjectivists* maintain that moral judgments are only about approval or disapproval by individuals or societies.

There are other ways to approach metaethics, but when it comes to deciding what should actually be done in a given situation, another branch of philosophy, *normative ethics,* applies. There are several normative theories, but they generally fall into the following two categories:

- *Consequentialist* (also teological) theory maintains that the morality (rightness or wrongness) of an action depends only on its consequences.
- *Nonconsequentialist* (also deontological) theory maintains that it is the motive, or conformity to an ethical rule or principle, that primarily determines the morality of an action.

There are several other normative theories that do not fall easily into these two categories, but nonetheless are used to distinguish moral from immoral actions:

- *Theological* theory maintains that it is the will of God, or divine command, that defines morality.
- *Utilitarian* theory maintains that affording the greatest happiness to the greatest number is the measure of morality of an action. There are debates on inclusion. Should only the happiness of humans be considered, or should other living things, and which ones, be included?
- *Situational* theory maintains that each situation may require a unique analysis of what is, or is not, moral.
- *Egoist* theory maintains that actions are right only if they are in the interest of the person performing the action (self-interest).

As one might expect, arguments about the morality of a given decision leads to invoking one or more of the above theories. There is no agreement on which theory is right or wrong as a measure of whether a given action (or decision) is itself right or wrong. Judgments must nevertheless be made, and there is seldom universal agreement on their correctness. In difficult decisions, precedent and personal and cultural norms are often the basis of judgments.

At times, the interests of one group (e.g., people with cancer) conflict with that of another group (e.g., laboratory animals in cancer therapy studies). This circumstance has been dealt with by a formal ethics review committee using the principle of maximizing the good (for cancer victims) and minimizing the harm (to research animals). In practice, an ethics committee will require adherence to good experimental design, use of the minimum number of laboratory animals to answer the scientific questions, and use of the best techniques for eliminating any pain and distress that might be experienced by the animals. If an independent ethics committee determines that the "good" significantly outweighs the "harm," the study may be approved, with monitoring and periodic review as the study progresses. For air pollution standards, the interests of susceptible individuals or groups (e.g., people with heart and lung disease) are counterbalanced by the interests of individuals or groups that must sacrifice economically in order to meet the standards.

III. HUMAN AND ANIMAL SUBJECTS RESEARCH ETHICS

From the perspective of this book it is useful to examine the ethical aspects of research with human and laboratory animal subjects. Such subjects are extensively used in toxicology and epidemiology research on the health effects of air pollutants. Air quality standards are dependent on the results of such research. Also, because the ethical standards in such research are highly developed and effectively monitored, they can provide insight on the potential future of the evolving topic of *environmental ethics.*

Historical Background

The late 1800s to the early 1900s was a remarkable period: radioactivity and x-rays were discovered; the practical automobile and airplane were developed; the industrial revolution began; medicine and public health rapidly advanced; the telephone revolutionized communication; and widely-available domestic and commercial electric power emerged. As a result, human life was radically and rapidly transformed, probably like never before. The industrial smokestack became a symbol of progress. The awesome power of science and technology was widely perceived to be in control of the future of humankind. This rapid change in lifestyle also generated a fear that poorly understood, yet powerful new forces were replacing the familiar traditional agents of change. There was concern over the possibility that the negative aspects of the scientific and industrial revolutions might outweigh their promised benefits. Some notable events illustrate this concern.

In 1818 Mary Shelley (1797–1851), young lover and future wife of the renowned poet Percival Shelley (1792–1822), wrote the novel *Frankenstein* while at a retreat with Percival and friends. In her horror story the protagonist, a young Dr. Frankenstein, who was obsessed with creating life, animated a creature (the "monster") made in the image of a large man. The doctor was shocked by and even hated his ugly creation, which led to a series of tragic events. The understandably traumatized monster turned murderous and eventually killed his creator's new wife. The monster fled to escape the nearly-insane Dr. Frankenstein's wrath. On its surface, the tale is one of science and ambition gone tragically wrong. The tale can be interpreted in at least two ways: (1) the horror associated with a parent (or creator) hating their offspring (or creation) and (2) the tragic consequences of the pursuit of science without due restraint and caution. It was the second interpretation that caught the public's imagination. Even the inventor Thomas Edison (1847–1931), who obviously embraced new technologies, produced a film, *Frankenstein,* in 1911. Later the English actor Boris Karloff (1887–1969, who was born as William H. Pratt), epitomized Dr. Frankenstein's monster in popular cinema (**Figure 12–1**). That, and other films, supported the public's fear of science, progress, and technology.

Fortunately, in the nearly two centuries since Mary Shelley wrote her famous horror novel, science and technology has not led to humankind's downfall. Human lifespans have doubled, infant mortality has declined, and access to mobility has increased. This fortuitous outcome is, in part, due to the basic ethical principles followed by most scientists. These principles include open communication of results, honesty in reporting, and an interest in improving the quality life and the general level of knowledge. There have been exceptions in adhering to these idealistic principles; for example, in cases of scientific fraud, in some corporate behavior, and in the interest of winning wars. Yet, the public's trust in science was still relatively high in the mid-1900s.

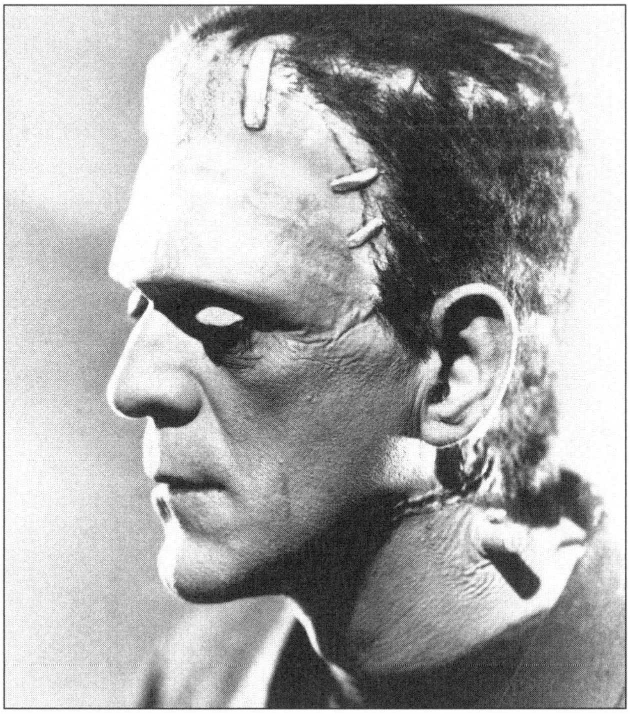

Figure 12–1 Boris Karloff as Dr. Frankenstein's monster from a 1931 film, *Frankenstein.*

Human Research Ethics

Although the public was somewhat sensitized to the potential abuses by scientists by sensational books and films, it was the shock of Nazi experiments during World War Two that proved that such abuses could be a rude reality. So-called "medical experiments" were performed on Jews and other prisoners. Many of the experiments were later determined to be "crimes against humanity." The human subjects were involuntarily subjected to extreme pain and even death. The studies included the effects of cold water immersion, oxygen deprivation, and new vaccines. The scientists (largely physicians) claimed that the studies were justified in that they sacrificed a few to benefit a much greater number. The trials of Nazi scientists at Nuremberg applied a set of principles (**Table 12–1**) in order to determine the guilt of the experimenters. Several of the investigators, were convicted and sentenced to death by hanging or to imprisonment. Only two of about 30 that were tried were acquitted. Even today, the *Nuremberg Principles* are central for defining "crimes against humanity" and the conditions for conducting ethical research with human and animal subjects. After the trials, the Nazi experiments were seen by many at the time as a bizarre Nazi anomaly that had been properly dealt with, perhaps once and for all time. For an analysis of the ethical aspects of the Nazi experiments, see the article "Too hard to face" by a Professor of Bioethics, Arthur Caplan (2005).

The view that civilized people could not engage in such studies was shattered when the U.S. press in 1973 exposed the infamous *Tuskegee Syphilis Study* see (Cobb, 1973). In October 1932, when the U.S. Public Health Service (PHS) began the study on about 400 Alabama black men, it was not known whether the accepted treatments for syphilis, including the administration of arsenic or mercury, were worse than the untreated disease. There was a perceived need to know the effects of "no treatment" on the course of syphilis. As a result, the subjects were not treated, and instead the

Table 12–1 The Nuremberg Code.

1. The voluntary consent of the human subject is absolutely essential. The duty and responsibility for ascertaining the quality of the consent rests upon each individual who initiates, directs, or engages in the experiment. It is a personal duty and responsibility which may not be delegated to another with impunity.[A]
2. The experiment should be such as to yield fruitful results for the good of society, unprocurable by other methods or means of study, and not random and unnecessary in nature.
3. The experiment should be so designed and based on the results of animal experimentation and a knowledge of the natural history of the disease or other problem under study [so] that the anticipated results will justify the performance of the experiment.
4. The experiment should be so conducted as to avoid all unnecessary physical and mental suffering and injury.
5. No experiment should be conducted where there is an *a priori* reason to believe that death or disabling injury will occur; except, perhaps, in those experiments where the experimental physicians also serve as subjects.
6. The degree of risk to be taken should never exceed that determined by the humanitarian importance of the problem to be solved by the experiment.
7. Proper preparations should be made and adequate facilities provided to protect the experimental subject against even remote possibilities of injury, disability, or death.
8. The experiment should be conducted only by scientifically qualified persons. The highest degree of skill and care should be required through all stages of the experiment of those who conduct or engage in the experiment.
9. During the course of the experiment the human subject should be at liberty to bring the experiment to an end if he has reached the physical or mental state where continuation of the experiment seems to him to be impossible.
10. During the course of the experiment the scientist in charge must be prepared to terminate the experiment at any state, if he has probable cause to believe, in the exercise of the good faith, superior skill, and careful judgment required of him that a continuation of the experiment is likely to result in injury, disability, or death in the experimental subject.

[A]Projects involving human subjects who are unable to give consent will require legally effective informed consent from guardians/conservators.

Reprinted from *Trials of War Criminals before the Nuremberg Military Tribunals under Control Council Law No. 10, Vol. 2, pp. 181–182.* Washington, D.C.: U.S. Government Printing Office, 1949.

natural course of the disease was followed and documented by the PHS physicians. Many believe that the original study was justified with respect to its intent, which was to determine the best treatment, if any, for the disease. The study continued for decades. What the press revealed almost 40 years into the study was shocking. In 1941, penicillin was discovered to be a cure for the disease, but the PHS physicians continued to withhold treatment from most of their research subjects. The decision to continue was based, in part, by concerns over the unknown long-term side effects of the new drug. This decision was widely perceived to be in violation of the Nuremberg Principles. Furthermore, the subjects were not informed that they were in a study without treatment for their disease. For a more thorough description of the Tuskegee Syphilis Study, see White (2000).

The U.S. Congress acted quickly and decisively. In 1974, regulations were instituted that required all government-funded human research to be performed only after approval by an ethics board, the *Institutional Review Board* (IRB). An IRB has specific requirements and responsibilities including:

- It must have at least five members with sufficient expertise and diversity to review research protocols, include more than one profession, include a non-scientist, and have at least one member who is not affiliated with the institution that it serves.
- It must review, require modifications as needed, and approve or disapprove the submitted protocols.
- It must conduct continuing review and monitoring and if necessary terminate or suspend the research.

- It must report to federal and institutional officials any serious non-compliance by the investigators or unexpected serious harm to the research subjects.

During IRB review of research protocols, all of the following criteria must be satisfied prior to approval:

- equitable selection of subjects (to protect disadvantaged populations such as minorities and prisoners)
- minimization of risks
- risks must be reasonable in relation to the importance of the project
- informed consent in the language of the subject is sought and documented
- monitoring, as needed to insure safety of the subjects
- protection of privacy and confidentiality of the subjects (However, a court order can force the release of information.)
- additional protection for fetuses, children, pregnant women, prisoners, mentally disabled, and other vulnerable individuals (Legal guardians or conservators must approve entry of some subjects into a study.)
- furthermore, the *consent form* in the language of the subject must have information on all foreseeable risks; how any injuries will be dealt with; and who to contact in the event of questions, problems, or injury
- a description of the research purposes, duration, procedures, and any benefits to the subject, others, or science
- a statement that participation is voluntary and that refusal to participate (or to withdraw at any time) will not adversely affect any benefits to which the subject is entitled
- a statement, if applicable, as to alternate treatments available to the subject if the study involves an experimental treatment for a disease or medical condition

This very brief description of some of the IRB's basic responsibilities illustrates the ethics review and monitoring considerations for human research. The review requirements have several ethical implications. First, ethical abuses (as occurred in the World War Two Nazi prison camps and in the Tuskegee study) will not be tolerated. Second, human subject rights include protection from unreasonable or unknown harms, access to information about potential risks, and power to refuse to participate in situations where known or suspected risks are involved. And third, the government is willing to establish formal requirements, with monitoring and sanctions to protect research subjects. These principles are more highly developed than those that govern air pollution exposures. People exposed to air pollutants seldom have the option of refusing to be exposed.

Animal Research Ethics

The long-standing relationships between humans and other animals include companionship, servitude (as for working animals), entertainment, herding (for food and other products), and use in science, product safety testing, and medicine. As a result, humans have developed respect and even affection for many species. The protections given to non-human research subjects are very similar to those provided for human research subjects. In several aspects the protections are even greater for laboratory animals.

The legal protection for animals in research predates that for human subjects. In England, antivivisectionists (who favored abolishing all live animal research) were involved in enacting the *Cruelty to Animals Act* of 1896. The act required investigators, who used vertebrate animals who might experience pain, to be licensed. Local laws against cruelty to animals, whether or not they are involved in research, also apply to laboratory animal studies. Animal protection in research, teaching, and testing has evolved significantly in the last 25 years in response to pressure from concerned groups and individuals. In the United States, federally-funded and other research is required to have approval from an *Institutional Animal Care and Use Committee* (IACUC). The IACUC is similar to the IRB (for human studies) with respect to composition and duties. The use of IACUCs to review animal research protocols is supplemented by a voluntary stringent accreditation system for research institutions. *The Association for Assessment and Accreditation of Laboratory Animal Care International* (AAALAC) (http://www.aaalac.org) publishes information including international regulations and resources that aid in promoting humane treatment of animals used in research, teaching, testing and training. Among the many criteria for accreditation by the AAALAC are stringent requirements related to housing, food, environmental quality, transportation, medical treatment, training of care personnel, and for some species, environmental enrichment.

Table 12–2 Abbreviated principles for the care and use of vertebrate animals in testing, research, and training.

1. The transportation, care, and use of animals should be consistent with the Animal Welfare Act and other applicable Federal laws, guidelines, and policies.[A]
2. Procedures involving animals should be designed and performed with due consideration of their relevance to human or animal health, the advancement of knowledge, or the good of society.
3. The animals selected for a procedure should be of an appropriate species and quality and the minimum number required to obtain valid results.
4. Proper use of animals, including the avoidance or minimization of discomfort, distress, and pain when consistent with sound scientific practices, is imperative.
5. Procedures with animals that may cause more than momentary or slight pain or distress should be performed with appropriate sedation, analgesia, or anesthesia. Surgical or other painful procedures should not be performed on unanaesthetized animals paralyzed by chemical agents.
6. Animals that would otherwise suffer severe or chronic pain or distress that cannot be relieved should be painlessly killed at the end of the procedure or, if appropriate, during the procedure.
7. The living conditions of animals should be appropriate for their species and contribute to their health and comfort. Normally, the housing, feeding, and care of all animals used for biomedical purposes must be directed by a veterinarian or other scientist trained and experienced in the proper care, handling, and use of the species being maintained or studied.
8. Investigators and other personnel shall be appropriately qualified and experienced for conducting procedures on living animals.
9. Where exceptions are required in relation to these Principles, the decisions should not rest with the investigators directly concerned but should be made by an appropriate review group such as an institutional animal care and use committee.

[A]For guidance throughout these Principles, the reader is referred to the Guide for the Care and Use of Laboratory Animals prepared by the Institute of Laboratory Animal Resources, National Academy of Sciences (NRC, 1996).
Data from NRC (1996).

The *National Institutes of Health* principles for animal research (**Table 12–2**) parallel the Nuremberg Code. However, because animals cannot give consent or withdraw from a study, additional monitoring and attention to relieving pain and distress are required.

In spite of the great progress made in protecting research animals, some advocacy groups still strive to abolish such animal use, as well as other uses of animals that are intended to benefit humans. Those who advocate giving *rights,* as opposed to *protections* for non-human animals are represented by animal rights organizations. In contrast, those who have concern for animal suffering and strive to protect the interests of animals belong to the *animal welfare* camp.

Air pollutants can adversely affect domestic and wild animals. Primary air standards are designed to protect human health, not animal health. Animal protection is included by the U.S. EPA in setting its secondary National Ambient Air Quality Standards (see Chapter 6).

IV. PROFESSIONAL ETHICS

Professional Associations, Societies, and Other Organizations

Professional associations and societies have several functions including:

- defining the qualifications for membership;
- helping to train and certify their members;
- publishing research, news, and other materials of interest to their members;
- holding periodic conferences for exchange of ideas;
- providing recognition of members for outstanding service;
- promoting their goals through public education and, at times, political action; and
- establishing *ethical standards* that must be followed by members in good standing.

It is this last function that will be examined further.

It must first be understood that an *essential* ethical obligation for any profession is to perform its main duties. Physicians must treat and prevent disease. Lawyers must defend their clients. Researchers must advance knowledge. Industrial hygienists must make the workplace safer and more productive. Ecologists must study and protect ecosystems. Engineers must design and maintain devices, systems, and processes that serve their employers, the public, and (in many cases) the environment. Industry and other businesses have a primary ethical obligation to provide jobs and needed goods and services. Professions and organizations also have their ethical responsibilities. The benefits institutions and professions provide to society must not be outweighed by harms that could result from unscrupulous behavior. Therefore, protecting the value of a profession benefits society by the adherence of its members to *codes of ethics.*

Sample Professional Codes of Ethics

The *Center for the Study of Ethics in the Professions* at the Illinois Institute of Technology maintains an extensive collection of about 1,000 codes of ethics (http://ethics.iit.edu). The codes come from professional and social organizations, companies, journals, and governments.

By examining a few codes of ethics (also called *codes of conduct*), one sees that they often include several basic elements:

- *Duty*—to perform one's duties competently and professionally
- *Honesty*—to report facts honestly and without bias, and not participate in deception
- *Service*—to serve individual clients, employers, society, the profession, and often the environment
- *Integrity*—to protect the dignity and prestige of the profession, and not engage in harassment or other disrespectful behavior
- *Confidentially*—to protect privacy, except in cases of illegality or where safety is at stake

Institution of Civil Engineers (http://www.ice.org.uk)

Civil engineers design and build bridges, roads, railways, stadiums, buildings, and other large structures. Their contributions are essential for supporting day-to-day life, and promoting prosperity, environmental quality, and sustainability. The *Institution of Civil Engineers* (ICE), founded in 1818, publishes numerous technical journals, organizes conferences, has offices in five continents, and has approximately 80,000 members worldwide. Their "Rules of Professional Conduct" have the following six elements:

1. All members shall discharge their professional duties with integrity.
2. All members shall only undertake work that they are competent to do.
3. All members shall have full regard for the public interest, particularly in relation to matters of health and safety, and in relation to the well-being of future generations.
4. All members shall show due regard for the environment and for the sustainable management of natural resources.
5. All members shall develop their professional knowledge, skills, and competence on a continuing basis and shall give all reasonable assistance to further the education, training and continuing professional development of others.
6. All members shall:
   a. Notify the institution if convicted of a criminal offence;
   b. Notify the institution upon becoming bankrupt or disqualified as a Company Director;
   c. Notify the Institution of any significant breach of the Rules of Professional Conduct by another member.

Note that "the institution," under 6 a. above refers to the ICE.

The Ecological Society of America (http://www.esa.org)

Ecology is a scientific discipline that deals with relationships among organisms and their environments. Ecologists not only perform research, but they are also involved in formulating policy that relates to ecosystems. The *Ecological Society of America* (ESA), founded in 1915, has chapters in Canada, Mexico, and the United States. The ESA publishes several scientific journals (e.g., *Frontiers in Ecology* and *The Environment*), and it has over 10,000 members worldwide. Its interest in environmental issues include:

- natural resources and ecosystem management;
- biological diversity and species extinction;

- ozone depletion;
- biotechnology; and
- ecosystem restoration and sustainability.

The ESA certifies and trains qualified ecologists in order to help them incorporate ecological principles in decision making. All members of the organization are expected to observe the following principles when acting professionally:

1. Ecologists will offer professional advice and guidance only on those subjects in which they are informed and qualified through professional training or experience. They will strive to accurately represent ecological understanding and knowledge and to avoid and discourage dissemination of erroneous, biased, or exaggerated statements about ecology.
2. Ecologists will not represent themselves as spokespersons for the Society without express authorization by the President of ESA.
3. Ecologists will cooperate with other researchers whenever possible and appropriate to assure rapid interchange and dissemination of ecological knowledge.
4. Ecologists will not plagiarize in verbal or written communication, but will give full and proper credit to the works and ideas of others, and make every effort to avoid misrepresentation.
5. Ecologists will not fabricate, falsify, or suppress results, deliberately misrepresent research findings, or otherwise commit scientific fraud.
6. Ecologists will conduct their research so as to avoid or minimize adverse environmental effects of their presence and activities, and in compliance with legal requirements for protection of researchers, human subjects, or research organisms and systems.
7. Ecologists will not discriminate against others in the course of their work on the basis of gender, sexual orientation, marital status, creed, religion, race, color, national origin, age, economic status, disability, or organizational affiliation.
8. Ecologists will not practice or condone harassment in any form in any professional context.
9. In communications, ecologists should clearly differentiate facts, opinions, and hypotheses.
10. Ecologists will not seek employment, grants, or gain, nor attempt to injure the reputation or professional opportunities of another scientist by false, biased, or undocumented claims, by offers of gifts or favors, or by any other malicious action.

Industrial Hygiene Associations

Industrial hygiene is a profession (and also an art and a science) devoted to protecting the health and safety of people in their workplaces and communities. Professional industrial hygienists possess college or university degrees in either engineering, physics, chemistry, biology, or a related physical or biological science. They typically practice their profession as employees in industry, government, and academia, or as private consultants. Industrial hygiene societies are engaged in training, certification, publication, and in making recommendations (for workplace air quality, safety, and stress related to heat, cold, vibration, ergonomics, etc.). Industrial hygienists are often involved in assessing and controlling industrial emission releases into the environment. Several professional organizations, including the *American Industrial Hygiene Association,* the *American Conference of Governmental Industrial Hygienists,* the *American Board of Industrial Hygiene,* and the *Academy of Industrial Hygiene,* jointly support a shared "Code of Ethics." The following eight elements are selected from a larger code of 19 elements (http://www.acgih.org):

1. Deliver competent services with objective and independent professional judgment in decision-making.
2. Recognize the limitation of one's professional ability and provide services only when qualified.
3. Follow appropriate health and safety procedures, in the course of performing professional duties, to protect clients, employers, employees and the public from conditions where injury and damage are reasonably foreseeable.
4. Assure that a conflict of interest does not compromise legitimate interests of a client, employer, employee or the public and does not influence or interfere with professional judgment.
5. Comply with laws, regulations, policies and ethical standards governing professional practice of industrial hygiene and related activities.
6. Maintain and respect the confidentiality of sensitive information obtained in the course of professional activities unless: the information is reasonably understood to pertain to unlawful activity; a court or governmental agency lawfully directs the release of

the information; the client or the employer expressly authorized the release of specific information; or, the failure to release such information would likely result in death or serious physical harm to employees and/or the public.

7. Report apparent violations of the ethics code by certificants and candidates upon a reasonable and clear factual basis.
8. Refrain from public behavior that is clearly in violation of professional, ethical or legal standards.

A Student's Pledge

Ethical codes and pledges are not limited to professional associations. The following was adopted by engineering students graduating from the University of California at Berkeley.

The Engineering Ethics Pledge:

"We, the graduating engineering class of 2004, in recognition of the affect of technology on the quality of all life throughout the world, and in accepting a personal obligation to our profession, it's members and the communities we serve, do hereby commit ourselves to the highest ethical and professional conduct and pledge to use our education only for purposes consistent with the safety, health and welfare of the public and the environment. Throughout our careers, we will consider the ethical implications of the work we do before we take action. We make this declaration because we recognize that individual responsibility is the first step on the path to a better world."

For more on the history of Berkeley's Engineering Ethics Pledge, see http://courses.cs.vt.edu/~cs3604/lib/WorldCodes/Pledge.html.

Comments on Codes of Ethics

The foregoing codes of ethics and student's pledge illustrate the commitment of individuals and their professional groups and other associations to ethical principles. When a company, society, or other group establishes an ethical code and enforces it in its membership, they thereby gain respect. As mentioned previously, adhering to strict ethical principles constrains behavior, however, it is expected of professionals. When encountering an organization for the first time, one might seek a copy of their Code of Ethics. If such a code does not exist, one does not know what behaviors are acceptable by the members.

V. PRACTICAL ETHICS

Ethical Decision Making

A *decision* may be defined as a conclusion, or a formal judgment, that is reached after due consideration. An immediate question raised by this definition is "What must be considered before making the decision?" In an ideal world, *all* of the relevant facts and *all* of the relevant consequences to *all* of those affected by each possible decision would be considered. Even in such an ideal world, certain individuals and/or groups (also called *stakeholders*) might see the decision as unfair. Such disaffected stakeholders could disagree with the relative weighting and interpretation of the facts, with the importance of the various consequences, and even the wisdom and fairness of those who made the decision.

As an example, consider a decision by a regulatory body to significantly tighten a regulation on the permissible air concentration of $PM_{2.5}$ (fine particles). The relevant facts considered by the decision makers might have included epidemiological associations between $PM_{2.5}$ and various health-related outcomes (hospital admissions, mortality rates, etc.), results of the impact of fine particles on climate model predictions, and the costs of installing particle collection devices on cars, trucks, and factories. The decision would likely be applauded by advocacy groups that focus on protecting health and the environment. On the other hand, automobile and truck drivers and some economists might challenge the decision. They might believe that the health-related associations could have been produced by factors other than anthropogenic $PM_{2.5}$ (such as co-pollutant gases, natural particles, and meteorological factors), and that the climate models were not adequately validated (see Chapter 5). The disaffected groups could also point out the adverse effects of the tightened regulations on the cost of goods and services, and even the loss of jobs (factors that also affect public health). No matter what decision is made, some would be pleased and others would be displeased. Regardless of the decision made on the air standard, litigation is very likely to follow. This is the reality of decision making in our world.

Balancing Interests

The foregoing example clearly illustrates that ethical decisions commonly involve a balancing of interests. The interests of those who might suffer adverse health outcomes was contrasted with the interests of those who would be economically affected. In the example, the regulators weighed the potential direct health effects more heavily than the potential indirect health effects (due to the costs of goods and services and the potential loss of jobs). The regulators also accepted the current climate model's long-term predictions and gave them more weight than the immediate economic concerns. Did the regulators make an ethical decision? To the extent that they seriously considered all of the relevant data and all of the significant consequences, the answer would seem to be "yes." The decision still might not be considered to be ethical in the eyes of some stakeholders. One would also need to know if the decision makers had conflicts of interest or external pressures that biased their decision. If that were the case the answer could be, "no." Balancing interests, although an ethical requirement, is a difficult task.

Public Participation in Decision Making

Given that decisions affecting air quality are based on complex scientific data and sophisticated modeling analyses, is there a role for public participation? And if the answer is, "yes," what form should that participation take? There are several models for public involvement in decision making, including:

- *Public comment*—After a tentative decision is made by an authorative body, a public solicitation is made that invites comments from all interested parties. The received comments are then "considered" prior to issuing a final decision. Governmental bodies such as the U.S. EPA use this model.
- *Public representation*—The decision-making body (e.g., a committee) has one or more voting members that represent the public. Such representatives have access to the information being considered by the decision-making committee, and they participate in deliberation and voting. This model is used by committees that approve or disapprove research protocols involving human and nonhuman research subjects. The final decision is usually based on a majority vote.
- *Public initiation*—The public, usually represented by elected government representatives or advocacy organizations, is involved in formulating a law (which includes ballot measures such as propositions) that is put to a vote. In the case of laws, the vote is conducted by elected legislators. In the case of a ballot proposition, the final decision is made by registered voters.

Each of these models has strengths and weaknesses. The public comment model allows experts to perform an analysis and arrive at what they deem to be an appropriate tentative decision. The published tentative decision will typically have an accompanying discussion of the data and analysis that supports the decision. Thus, the public has the opportunity to analyze the work of the decision makers and to provide informed input. The weakness of this model is that the public input is not binding, and may not be seriously considered prior to issuing the final decision. For this model to work well the public input should be effectively incorporated into the final decision.

The public representation model has the advantage of including pubic representatives in both the formulation and acceptance of a tentative decision. The representatives can introduce data, analysis, and arguments before a conclusion is reached by the decision-making body. For this model to work well the public representatives must devote the time and effort necessary to have effective input. Potential weaknesses of this model are:

1. That the public representatives may be overwhelmed by the complex data and analyses to the extent that they are not adequately involved; and
2. That their votes are outnumbered by the experts, making them "token" participants.

The public initiation model probably has the greatest potential for maximizing the power of the public in decision making. Professional societies, as well as informed and motivated individuals prepare laws that are presented to legislators. Elected representatives are obligated to represent the interests of their constituents and to accept input from them. After all, they face reelections in which their legislative records will be scrutinized. In the case of ballot initiatives, the voting public makes the final decision. Weaknesses of the public participation model include not only its expense, but also the possible lack of adequate technical input from experts. The final decisions may be made as a result of political deal making (in the case of laws) or biased advertising campaigns (in the case of ballot measures). Also, ballot measures may be approved by a small margin of those who vote, which can adversely affect small groups. This problem is referred to as "the tyranny of the majority."

All of these models, along with others (e.g., the use of focus-groups or public polls initiated by decision makers), permit public input. The success or failure of a given decision with respect to its public support is measured by the level of public approval or protest, and sometimes litigation.

The Common Sense Criterion

The *common sense criterion* (also called the *laughability factor*) refers to the extent to which a decision is or is not obviously fatally-flawed. For example, each year in each large city, a large number of people require emergency medical attention for choking on food. Imagine a panel of experts called together to make recommendations that would eliminate the problem. After considerable analysis they recommend outlawing all solid food—problem solved! However, the solution does not pass the common sense test, and it is even laughable. The banning of all solid food would obviously harm, if not destroy, public health. Similarly, consider air emissions that can adversely affect public health. The problem could be solved by eliminating cars, trucks, factories, electric-power plants, agriculture, animal husbandry, pollen-and spore-producing plants, and microbial life. This solution, although logical, is also laughable, as it would lead to ending all human life.

The examples given are extreme, but they demonstrate some important principles. First, a problem should not be analyzed in isolation. The problem exists in a complex world, and isolating it may not be wise. The problem is that the human brain tends to focus on problems in isolation, which is often useful (e.g., a focus on an imminent threat). Second, even the most well-reasoned step-by-step analysis can lead to a ridiculous conclusion. The scientific method actually encourages such linear stepwise analyses in which the result has been proved to be valid. However, validity and practicality are not the same. Third, and most important is this simple truth.

When a decision is made, all of the consequences will occur, not just the intended ones.

Decisions can be, and are often, made without due regard to their full range of consequences. Wise decisions are those that lead to net outcomes that do more good than harm in the real world.

VI. SUMMARY OF MAJOR POINTS

Ethics has both theoretical and practical aspects. Laws also define accepted ethical behavior. In philosophy, the nature of morality (i.e., ethical behavior) is examined. *Metaethics* deals with the large questions about ethics, and *normative* ethics deals with considerations of what *should* be done from an ethical standpoint. Should only the consequences be considered (as in teology), or should conformity to values (e.g., honesty and compassion) and principles (as in deontology) prevail in determining moral behavior? Do both teology and deontology have roles in settling air quality standards? In practical ethics, decisions must be weighed, balancing goods and harms, weighing conflicting interests of affected groups (including animals, plants, and the environment), and considering public acceptance (at a given time and in a given culture).

Research with human and non-human subjects is an area in which ethical considerations have undergone considerable evolution. Approval or disapproval of studies are made by deliberative committees with specified diversity and responsibilities. Such committees weigh the importance of studies against pain, suffering, and other risks faced by the subjects. Also maximizing the benefits and minimizing the harms associated with a study is an essential consideration.

Professional ethics are based on the elements of duty, honesty, service, integrity, and confidentially. Specific professional groups identify and enforce their ethical standards by *Codes of Conduct.* Such codes often include protecting the interests of people, animals, and the environment. Such codes exist for professional, social, business, and other groups. A code of ethics reveals what can be expected from those who adopt it.

Decision making can be a complex process that involves many ethical considerations. Balancing interests of individuals and groups that are affected is an aspect of decision making. The make-up of the group that makes decisions, and the process used, are also ethical considerations. Public influence on decision making takes several forms including opportunities to comment, participate in deliberations, and even formulate and make final decisions. The wisdom of a decision must be measured by all of its significant consequences, including the intended and the unintended consequences.

VII. QUIZ AND PROBLEMS

Quiz Questions

(select the best answer)

1. Ethical standards:
   a. are studied in a branch of philosophy.
   b. are enforced by laws.
   c. may change in a given culture over time.
   d. All of the above are true.

2. Research projects with laboratory animals typically:
   a. need not conform to any ethical standards.
   b. are only permitted if done by veterinarian researchers.
   c. are permitted only after an ethical review.
   d. do not require any control of pain or distress.
3. The Nuremberg Tribunal:
   a. established criteria for "crimes against humanity."
   b. was focused on prosecuting physicians that performed the Tuskegee Study.
   c. prosecuted Nazi physicians after World War Two.
   d. Both a. and b. are true.
   e. Both a. and c. are true.
4. Research with human subjects:
   a. must be approved by an Institutional Review Board.
   b. is illegal.
   c. can be conducted without ethical review if the degree of pain and discomfort is not severe.
   d. does not require obtaining informed consent.
5. Which statement is *not* true in metaethics?
   a. Intuitionists maintain that people know right from wrong, so that a formal analysis is irrelevant.
   b. Metaethics involves asking large questions about ethics instead of analyzing the ethics of specific issues.
   c. Metaethics does not differ from normative ethics.
   d. None of the above are true.
6. Utilitarian ethical theory:
   a. maintains that God is the final authority.
   b. maintains that "the greatest happiness for the greatest number" is a measure of the morality of an act.
   c. maintains that an act is moral if it benefits the person performing the act.
   d. None of the above are true.
7. Professional ethics:
   a. only apply to physicians.
   b. are defined by "Codes of Conduct."
   c. are the same for all scientific disciplines.
   d. are established by federal laws for each profession.
8. Members of the "Institution of Civil Engineers" have ethical obligations that include:
   a. taking into account the interests of future generations.
   b. showing regard for the environment.
   c. notifying the Institution of Civil Engineers if they have a criminal offense conviction.
   d. All of the above are true.
9. A member of the "American Industrial Hygiene Association" is ethically obligated to:
   a. provide requested services even if they are not qualified.
   b. report all unprofessional activities of their clients or employers.
   c. protect the safety of employers, employees, and the public in the course of performing their professional duties.
   d. protect the reputations of other industrial hygienists who violate the ethics code unless ordered to testify under oath.
10. The Tuskegee Syphilis Study:
    a. is an example of a study that followed sound ethical principles.
    b. stimulated congressional action to protect human subjects in research studies conducted in the United States.
    c. was conducted by Nazi scientists.
    d. Only b. and c. are true.

Problems

1. List at least five elements that you would include in a "Code of Ethics" for engineers that design and operate air pollution control systems for industry, and discuss why each element is important.
2. Review the "Engineering Ethics Pledge" and discuss how it could be applied to students graduating from your degree program.
3. How do the ethical requirements for research studies using laboratory animals differ from those using human subjects?
4. Discuss the ethical implications of establishing and enforcing air quality standards that result in substantial economic hardship for a community.
5. List the "pros" and "cons" of each of the three models of public participation with respect to establishing a new, more stringent, air pollution regulation.

6. Examine the first five elements of the Nuremberg Code. For each element, replace the concept of "experiment" with the concept of "exposure to air pollutants." Do the principles apply to air pollutant exposures?
7. As in problem 6 above, replace the concept of "experiment" with the concept of "a governmental regulation." Do our air pollution regulations meet the spirit of the Nuremberg Code?

VIII. DISCUSSION TOPICS

1. In following the ethical principle of "the greatest happiness for the greatest number":
   a. Should the happiness of non-human animals be considered?
   b. Could following this principle also lead to actions that would be considered unethical in our culture?
2. Should the ethical principle "do no harm" become law? Why or why not?
3. Suggestions have been made to distinguish "harms" (which should be minimized) from "wrongs" (which should never be permitted). Should such a differentiation be made? Why or why not?
4. Are current standards for air quality based on firm ethical principles? If so, which principles? If not, should the ethical basis be strengthened, and how?

References and Recommended Reading

Caplan, A. L., Too hard to face, *J. Am. Acad. Psych. Law,* 33:394–400, 2005.

Carson, R. L., *Silent Spring,* Houghton Mifflin, Co., Boston, MA, 1962.

Cobb, W. M., The Tuskegee Syphilis Study, *J. Nat. Med. Assoc.,* 65:345–348, 1973.

Ehrlich, P., *The Population Bomb,* Ballantine Books, New York, 1968.

Kennedy, D., Animal activism: Out of control, *Science,* 313:1541, 2006.

Malthus, T. R., *An Essay on the Principle of the Population, As it Affects the Future Improvements of Society with Remarks on the Speculations of Mr. Godwin, M. Condorcet, and Other Writers,* Printed for J. Johnson in St. Paul's Church–Yard, London, 1798.

NRC (National Research Council), *Guide for the Care and Use of Laboratory Animals,* National Academy Press, Washington, DC, 1996, pp. 1–118.

Pitts, M., A guide to the new ARENA/OLAW IACUC Guidebook, *Lab Animal,* 31(9):40–42, 2002.

Sideris, L., McCarthy, C. and Smith, D. H., Roots of concern with nonhuman animals in biomedical ethics, *ILAR Journal,* 40:3–14, 1999.

Silverman, J., Suckow, M. A., and Murthy, S., *The IACUC Handbook, 2nd Edition,* CRC Press, Boca Raton, FL, 2006.

Thomson, J. J., Chodosh, S., Fried, C., Goodman, D. S., Wax, M. L., and Wilson, J. Q., Regulations governing research on human subjects: Academic freedom and the Institutional Review Board, *Academe,* 67: 358–370, 1981.

U.S. EPA (U.S. Environmental Protection Agency), *EPA's Formative Years, 1970–1973,* http://www.epa.gov/history/publications/print/formative.htm (accessed August, 10, 2009).

White, R. M., Unraveling the Tuskegee study of untreated syphilis, *Arch. Intern. Med.,* 160:585–598, 2000.

Zurlo, J., Rudacille, D. and Goldberg, A. M., *Animal and Alternatives in Testing: History, Science, and Ethics,* Mary Ann Liebert, Inc., Larchmont, NY, 1994.

Index

Page numbers with *f* denote figures, *t* denote tables, and *x* denote exhibits. Chapter summaries, quizzes, problems, and discussion topics are not indexed.

A

B

D

E

H

I

J

K

L

M

N

O

P

Q

R

V

W

X

Y

Z

Figure and Table Credits Page

Dedication
FM-1 © National Library of Medicine

Chapter 1
1-1 Data from exhibits at the University of California Museum of Paleontology (http:/www.ucmp.berkeley.edu). Source: The University of California Air Pollution Health Effects Laboratory, with kind permission.; **1-2** The University of California Air Pollution Health Effects Laboratory, with kind permission.; **1-3** Courtesy of CDC; **Table 1-1, 1-2** Data from Clayton, G. D. and Clayton, F. E., eds., Vol 1, General Principles, Patty's Industrial Hygiene and Toxicology, 3rd Edition, John Wiley, New York, 1978.; **1-4** © Central Press/Hulton Archive/Getty Images; **1-5** Data from Wilkins, E.T., Air pollution and the London fog of December, 1952, J. Roy. Sanitary Inst., 74: 1-21, 1954. Source: The University of California Air Pollution Health Effects Laboratory, with kind permission.; **1-6, Table 1-3** The University of California Air Pollution Health Effects Laboratory, with kind permission.

Chapter 2
2-1A, B The University of California Air Pollution Health Effects Laboratory, with kind permission.; **Table 2-5** Data, in short tons (2,000 lbs, 907kg), are from various sources including: Finlayson–Pitts, B. J. and Pitts, J. N. Jr., *Atmospheric Chemistry: Fundamentals and Experimental Techniques,* John Wiley & Sons, New York, 1986); U.S. EPA (United States Environmental Protection Agency) *National Air Pollutant Emission Trends,* EPA–454/R–00–02, U.S. Environmental Protection Agency, office of Air Quality Planning and Standards, Research Triangle Park, NC, 2000.; **2-2, 2-3, 2-4, 2-5, 2-6, 2-7, 2-8, 2-9** U.S. EPA (United States Environmental Protection Agency), Our Nation's Air: Status and Trends Through 2008, EPA–454/R–09–002, U.S. Environmental Protection Agency, Research Triangle Park, NC, 2010.; **2-10, 2-11 2-12, 2-13, 2-14, 2-15, 2-16, Table 2-8** The University of California Air Pollution Health Effects Laboratory, with kind permission.

Chapter 3
Table 3-1 U.S. EPA (2010). National Ambient Air Quality Standards (NAAQS). Available at: http://www.epa.gov/air/criteria.html. Accessed October 3, 2011.; **Table 3-2** U.S. EPA (2010). The Clean Air Act Amendments of 1990 List of Hazardous Air Pollutants. Available at: http://www.epa.gov/ttn/atw/orig189.html. Accessed October 2, 2011. ; **Table 3-3** Data from Phalen, R. F., Inhalation Studies, Foundations and Techniques, 2nd Edition, Informa Healthcare, New York, 2009.; **3-1** The University of California Air Pollution Health Effects Laboratory, with kind permission.; **Table 3-5** U.S. EPA (United States Environmental Protection Agency), Air Quality Criteria for Particulate Matter, Vol. 1, EPA/600/P-99/002aF, United States Environmental Protection Agency, Washington, DC, 2004.; **3-2, 3-3A,B, 3-4**, **3-5** The University of California Air Pollution Health Effects Laboratory, with kind permission.; **Table 3-6** Data from Phalen, R. F., Inhalation Studies, Foundations and Techniques, 2nd Edition, Informa Healthcare, New York, 2009.; **Table 3-7** The University of California Air Pollution Health Effects Laboratory, with kind permission.;

Table 3-8 Data from Fuchs, N. A., The Mechanics of Aerosols: Revised and Enlarged Edition, Dover Publications, Inc. New York, 1964.; **Table 3-9** Data from Phalen, R. F., The Particulate Air Pollution Controversy: A Case Study and Lessons Learned, Kluwer Academic Publishers, Boston, MA, 2002, pp. 39–53.; **Table 3-10** Data from Barrow, G. M., Physical Chemistry, 2nd Edition, McGraw–Hill, New York, 1966, p. 521.; **Table 3-11** Data from Phalen, R. F., Inhalation Studies, Foundations and Techniques, 2nd Edition, Informa Healthcare, New York, 2009.; **3-6** Data from Finlayson–Pitts, B. J. and Pitts, J. N. Jr., Atmospheric Chemistry: Fundamentals and Techniques, John Wiley & Sons, New York, 1986.; **Table 3-12** Data from Considine, D. M., ed., Van Nostrand's Scientific Encyclopedia, 7th Edition, Van Nostrand Reinhold, New York, 1989, p. 664.; **Table 3-13** Data from U.S. EPA Emissions Inventories. Available: http://www.epa.gov/ttn/chief/eiinformation.html. Accessed September 6, 2011; **3-7** Data from Wilson, W. E., Sulfates in the atmosphere: A progress report on project MISTT*, Atmos. Environ., 12:537–547, 1978; and Whitby, K. T., The physical characteristics of sulfur aerosols, Atmos. Environ., 12:135–159, 1978. Source: The University of California Air Pollution Health Effects Laboratory, with kind permission.; **Table 3-14** Data from U.S. EPA (United States Environmental Protection Agency), Air Quality Criteria for Particulate Matter, Vol. 1, EPA/600/P-99/002aF, United States Environmental Protection Agency, Washington, DC, 2004.

Chapter 4

4-1, 4-2, 4-3, 4-4 The University of California Air Pollution Health Effects Laboratory, with kind permission.; **Table 4-3** Data from Phalen, R. F., Inhalation Studies: Foundations and Techniques, 2nd Edition, Informa, New York, 2009.; and Lodge, J. P., Jr. and Chan, T. L., eds., Cascade Impactor: Sampling and Data Analysis, AIHA Press, Akron, OH, 1986.; **Table 4-4** The University of California Air Pollution Health Effects Laboratory, with kind permission.; **4-5A** Courtesy of MSP Corporation, Shoreview, MN, USA; **4-5B, 4-6A** The University of California Air Pollution Health Effects Laboratory, with kind permission.; **4-6B** Courtesy of MSP Corporation, Shoreview, MN, USA; **4-7** Courtesy of HI-Q Environmental Products Company, Inc. (San Diego, CA; www.HI-Q.net). ; **4-8** The University of California Air Pollution Health Effects Laboratory, with kind permission.; **4-9** Courtesy of Met One Instruments, Inc.; **4-10** The University of California Air Pollution Health Effects Laboratory, with kind permission.; **Table 4-5** Data from Phalen, R. F., Inhalation Studies: Foundations and Techniques, 2nd Edition, Informa, New York, 2009.; Lodge, J. P., Jr. and Chan, T. L., eds., Cascade Impactor: Sampling and Data Analysis, AIHA Press, Akron, OH, 1986.; and Vincent, J. H., Aerosol Science for Industrial Hygienists, Elsevier Science, Inc., Tarrytown, NY, 1995.; **4-11A, B** Courtesy of Met One Instruments, Inc.; 4-12 Reproduced by permission of Restek Corporation.; **4-13** The University of California Air Pollution Health Effects Laboratory, with kind permission.; **Table 4-7** Source: U.S. EPA (U.S. Environmental Protection Agency), EPA Compendium of Methods for the Determination of toxic Organic Compounds in Ambient Air: 2nd Edition, EPA/625/R–96/010b, Center for Environmental Information, Cincinnati, OH, January 1999.; **4-14** The University of California Air Pollution Health Effects Laboratory, with kind permission.; **Table 4-10** Data from ACGIH® (American Conference of Governmental Industrial Hygienists), Advances in Air Sampling: Industrial Hygiene Science Series, Lewis Publishers, Chelsea, MI, 1988.; Cohen, B. S. and McCammon, C. S., Jr., Technical Editors, Air Sampling Instruments: For Evaluation of Atmospheric Contaminants, 9th Edition, ACGIH, Cincinnati, OH, 2001.; DiNardi, S., The Occupational Environment: Its Evaluation, Control, and Management, 2nd Edition, AIHA Press, Fairfax, VA, 2003.; Phalen, R. F., Inhalation Studies: Foundations and Techniques, 2nd Edition, Informa, New York, 2009.; and Vincent, J. H., Aerosol Science for Industrial Hygienists, Elsevier Science, Inc., Tarrytown, NY, 1995.; **Table 4-11** Data from Phalen, R. F., Inhalation Studies: Foundations and Techniques, 2nd Edition, Informa, New York, 2009.; U.S. EPA (U.S. Environmental Protection Agency), EPA Compendium of Methods for the Determination of toxic Organic Compounds in Ambient Air: 2nd Edition, EPA/625/R–96/010b, Center for Environmental Information, Cincinnati, OH, January 1999. and NIOSH (National Institute for Occupational Safety and Health), NIOSH Manual of Analytical Methods, 2nd Edition (3rd Supplement), DHHS (NIOSH Publication 2003–154), National Institute for Occupational Safety and Health, Cincinnati, OH, 2003 (http://www.cdc.gov/niosh/nmam/, accessed August 1, 2010).; **Table 4-12**

NIOSH Pocket Guide to Chemical Hazards (http://www.cdc.gov/niosh/npg); Hazardous Substance Data Bank (http://www.toxnet.nlm.nih.gov); and NIST Chemistry WebBook (http://webbook.nist.gov/chemistry). Accessed October 2, 2011. ; **4-15, 4-16, 4-17** The University of California Air Pollution Health Effects Laboratory, with kind permission.

Chapter 5

5-1, 5-2, 5-3 The University of California Air Pollution Health Effects Laboratory, with kind permission.; **5-4** U.S. EPA (U.S. Environmental Protection Agency), Integrated Science Assessment for Particulate Matter, EPA 600/R–08/139F, U.S. Environmental Protection Agency, Research Triangle Park, NC, December 2009. and Malm, W. C., Schichtel, B. A., Ames, R. B., and Gebhart, K. A., A ten-year spatial and temporal trend in sulfate across the United States, J. Geophys. Res., 107(D22):4627, doi: 10.129/2002JD002107, 2002., with permission of the American Geophysical Union.; **5-5, 5-6** The University of California Air Pollution Health Effects Laboratory, with kind permission.

Chapter 6

6-1 The University of California Air Pollution Health Effects Laboratory, with kind permission.; **Table 6-2** Source of data for China: Niu, S., Outdoor and indoor air pollution in China, Asia- Pacific J. Public Health, 1:46–50, 1987.; **Table 6-3** Data from World Health Organization (2005). Air quality guidelines- global update 2005. Available at: http://www.who.int/en. Accessed May 29, 2010; **Exhibit 6-1** U.S. EPA (2010). National Ambient Air Quality Standards (NAAQS). Available at: http://www.epa.gov/air/criteria.html. Accessed May 29, 2010.; **6-2, 6-3, 6-4, 6-5, 6-6, 6-7, 6-8** The University of California Air Pollution Health Effects Laboratory, with kind permission.; **Table 6-6** Data from Nelson, G. O., Controlled Test Atmospheres: Principles and Techniques, Ann Arbor Science, Ann Arbor, MI, 1971.; **6-9, 6-10** The University of California Air Pollution Health Effects Laboratory, with kind permission.; **Table 6-7** Data from from NIOSH (National Institute for Occupational Safety and Health). The Industrial Environment: Its Evaluation and Control, NIOSH, Cincinnati, OH, 1973, Chapter 43.; **Table 6-8** ACGIH® (American Conference of Governmental Industrial Hygienists), *Industrial Ventilation, 14th Edition,* ACGIH®, Cincinnati, OH, 1976, Fig. 11–17.; **6-11** The University of California Air Pollution Health Effects Laboratory, with kind permission.

Chapter 7

7-1 The University of California Air Pollution Health Effects Laboratory, with kind permission.; **Table 7-1** Data from Schleien, B., Slaback, L. A. Jr., and Birky, B. K., Eds., Handbook of Health Physics and Radiological Health, 3rd Edition, Williams & Wilkins, Baltimore, MD, 1998, pp. 12-23 to 12-38., Chapter; **7-2** The University of California Air Pollution Health Effects Laboratory, with kind permission.; **Table 7-3** Data from ACGIH®, Technical Committee on Air Sampling Procedures, Particle Size-Selective Sampling in the Workplace, ACGIH®, Cincinnati, OH, 1985.; ICRP (International Commission on Radiological Protection, Task Group of Committee 2), Human Respiratory Tract Model for Radiological Protection, Publication 66, Pergamon Press, New York, 1994.; NCRP (National Council on Radiation Protection and Measurements), Deposition Retention and Dosimetry of Inhaled Radioactive Substances, NCRP SC 57–2 Report, National Council on Radiation Protection and Measurements, Bethesda, MD, 1997; TGLD (Task Group on Lung Dynamics), ICRP Committee II, Deposition and retention models for internal dosimetry of the human respiratory tract, Health Phys., 12:173–207, 1966; **7-3, 7-4, 7-5** The University of California Air Pollution Health Effects Laboratory, with kind permission.; **Table 7-4** Data from ACGIH®, 2010 TLVs® and BEIs®, ACGIH®, Cincinnati, OH, 2010.; **7-6** The University of California Air Pollution Health Effects Laboratory, with kind permission., **Table 7-6** Data From NCRP (National Council on Radiation Protection and Measurements), Deposition Retention and Dosimetry of Inhaled Radioactive Substances, NCRP SC 57–2 Report, National Council on Radiation Protection and Measurements, Bethesda, MD, 1997; **7-7** The University of California Air Pollution Health Effects Laboratory, with kind permission.

Chapter 8

Table 8-1 Data from UNICEF, The State of the World's Children 2008, United Nations Children's Fund, New York, 2007. Available at: http://www.unicef.org. Accessed August 4, 2010.; **Table 8-3** Data from Ilano, A. L. and Raffin, T. A., Management of carbon monoxide poisoning, Chest, 97:165–169, 1990; Varon, J.,

Marik, P. E., Fromm, R. E. Jr., and Gueler, A., Carbon monoxide poisoning: A review for clinicians, J. Emerg. Med., 17:87–93, 1999

Chapter 9

9-1, 9-2, 9-3 The University of California Air Pollution Health Effects Laboratory, with kind permission.; **Table 9-2** Data from Eaton, D. L. and Klaassen, C. D., Principles of toxicology, in Casarett and Doull's Essentials of Toxicology, Klaassen, C. D. and Watkins III, J. B., eds., McGraw- Hill, New York, 2003, pp. 6–20.; Hodge, H. C. and Sterner, J. H., Tabulation of toxicity classes, Am. Ind. Hyg. Assoc. Q., 10:93–96, 1949.; **Table 9-3** Sources: ACGIH , 2009 TLVs and BEIs , American Conference of Governmental Industrial Hygienists, Cincinnati, OH, 2009.; Lewis, R. J., Sax's Dangerous Properties of Industrial Materials, 10th Edition, Volume. 1 & 2, John Wiley & Sons, New York, 2000.;U.S. EPA (U.S. Environmental Protection Agency), National Ambient Air Quality Standards (NAAQS), http://www.epa.gov/air/criteria.html (accessed October 3, 2010).; **9-4, 9-5, 9-6** The University of California Air Pollution Health Effects Laboratory, with kind permission.; **Table 9-4** Data from ICRP (International Commission of Radiological Protection), Report of the Task Group on Reference Man, No. 23, Pergamon Press, New York, 1975.; Guyton, A. C., ed., Textbook of Medical Physiology, 8th Edition, W. B. Saunders, Philadelphia, PA, 1991, pp. 186.; and Lidstedt, S. L. and Schaeffer, P. J., Use of allometry in predicting anatomical and physiological parameters in mammals, Lab Anim., 36:1–19, 2002.; **9-7, 9-8** The University of California Air Pollution Health Effects Laboratory, with kind permission.; **Table 9-7** Data from Madl, A. K. and Pinkerton, K. E., Health effects of inhaled engineered and incidental nanoparticles, Crit. Rev. Toxicol., 39:629–658, 2009.; **9-9** The University of California Air Pollution Health Effects Laboratory, with kind permission.

Chapter 10

10-1, 10-2 The University of California Air Pollution Health Effects Laboratory, with kind permission.; **Table 10-1** Data from Cohen, B. L. and Lee, I. S., A catalog of risks, Health Phys., 36:707–722, 1979., Cohen, B. L., Catalog of risks extended and updated, Health Phys., 61:317–335, 1991. and CIA (Central Intelligence Agency). Available at: https://www.cia.gov/library/publications/the-world-factbook/. Accessed November 16, 2009.; **Table 10-3** Data from Friis, R. H. and Sellers, T. A., Epidemiology for Public Health Practice, 4th Edition, Jones & Bartlett Learning, Burlington, MA, 2008; and Künzli, N. and Tager, I. B., The semi-individual study in air pollution epidemiology: A valid design as compared to ecologic studies, Environ. Health Perspect., 105:1078–1083, 1997.; **10-3** The University of California Air Pollution Health Effects Laboratory, with kind permission.; **10-4** Data from Thurston, G. D. and Ito, K., Epidemiological studies of ozone exposure effects, in Air Pollution and Health, Holgate, S. T., Samet, J. M., Koren, H. S., and Maynard, R. L., eds., Academic Press, San Diego, CA, 1999, pp. 485–510.. Source: The University of California Air Pollution Health Effects Laboratory, with kind permission.; **10-5** The University of California Air Pollution Health Effects Laboratory, with kind permission.

Chapter 11

Table 11-1 Data from Doll, R. and Peto, R., The causes of cancer: Quantitative estimates of avoidable risks of cancer in the United States today, J. Natl. Cancer Inst., 66:1191–1308, 1981.; Doll, R., Epidemiological evidence of the effects of behavior and the environment on the risk of human cancer, Recent Results Cancer Res., 154:3–21, 1998.; and Doll, R., The Pierre Denis memorial lecture: Nature and nurture in the control of cancer, European J. Cancer, 35:16–23, 1999.; **11-1** The University of California Air Pollution Health Effects Laboratory, with kind permission.; **Table 11-2** Data from Wilson, R., Analyzing the daily risks of life, Technology Review, 81:40–46, 1979.; **Table 11-3** Data from Cohen, B. L. and Lee, I. S., A catalog of risks, Health Physics, 36:707–722, 1979.; Cohen, B. L., Catalog of risks extended and updated, Health Physics, 61:317–335, 1991.; **11-2** The University of California Air Pollution Health Effects Laboratory, with kind permission.; **Table 11-4** Data from Hodge, H. C. and Sterner, J. H., Tabulation of toxicity classes, American Industrial Hygiene Association Quarterly, 10:93–96, 1949.; **Table 11-5** Data from Code of Federal Regulations 40 CFR § 156.62; **11-3** The University of California Air Pollution Health Effects Laboratory, with kind permission.; **Table 11-7, 11-8** U.S. EPA (2010). Integrated Risk Information System. Available at: http://www.epa.gov/iris/. Accessed September 15,

2010; **Table 11-9** Data from Slovic, P. Perception of risk, Science, 236 (April 17, 1987):280–285.

Chapter 12

12-1 © AP Photos; **Table 12-1** Reprinted from Trials of War Criminals before the Nuremberg Military Tribunals under Control Council Law No. 10, Vol. 2, pp. 181-182. Washington, D.C.: U.S. Government Printing Office, 1949.; **Table 12-2** Data from NRC (National Research Council), Guide for the Care and Use of Laboratory Animals, National Academy Press, Washington, DC, 1996, pp. 1–118.